MODERN
GLOBAL
SEISMOLOGY

This is Volume 58 in the
INTERNATIONAL GEOPHYSICS SERIES
A Series of monographs and textbooks
Edited by RENATA DMOWSKA and JAMES R. HOLTON

A complete list of the books in this series appears at the end of the volume

MODERN GLOBAL SEISMOLOGY

THORNE LAY

Institute of Tectonics
University of California, Santa Cruz
Santa Cruz, California

TERRY C. WALLACE

Geoscience Department
University of Arizona
Tucson, Arizona

ACADEMIC PRESS
San Diego   New York   Boston   London   Sydney   Tokyo   Toronto

Photo Credit: Cover Illustration realized from Figure 1.21 on Page 30.

Academic Press
An imprint of Elsevier Science
525 B Street, Suite 1900, San Diego, California 92101-4495, USA
http://www.academicpress.com

Academic Press
Harcourt Place, 32 Jamestown Road, London NW1 7BY, UK
http://www.academicpress.com

Library of Congress Cataloging-in-publication Data

Modern Global Seismology / edited by Thorne Lay, Terry C. Wallace.
        p.   cm. – (International geophysics series : v. 58)
    Includes biblioigraphical references and index.
International Standard Book Number: 0-12-732870-X
1. Seismology.   I. Lay, Thorne,   II. Wallace, Terry C.
III. Series.
QE534.2.M62   1995
551.2'2—dc20   94-33101
                                           CIP

PRINTED IN THE UNITED STATES OF AMERICA
03   04   05   06   IBT   9   8   7   6   5

We Dedicate This Book to
Professor **Hiroo Kanamori,**
Seismologist *Extraordinaire*

CONTENTS

PREFACE

This textbook is intended for upper-division undergraduate and first-year graduate survey courses in seismology. It assumes that the student is familiar with basic calculus, complex numbers, and differential equations and has some general knowledge of geology. The focus is on the fundamental theory and physics of seismic waves and the application of this theory to extract the rich information about internal structure and dynamical processes in the Earth that is contained in seismograms, instrumental recordings of mechanical vibrations of the planet. Most of the text is developed in the context of global seismology topics, meaning large-scale Earth structure and earthquake sources. However, the principles underlying elastic-wave propagation, seismic instrumentation, and techniques for extracting Earth structure and source information from seismograms are common to applications in exploration seismology, a discipline that uses seismic waves to develop high-resolution images of crustal structure for oil and mineral resource exploration. The basic principles are in no way restricted to the Earth, and in the future they will, we hope, be applied to many other celestial bodies (preliminary work has been performed on the Moon and Mars, and a specialized field called helioseismology has revealed the internal structure of the Sun).

The material in this text is derived from class notes for introductory seismology courses taught over the past 10 years by the coauthors at the University of Michigan (T.L.), the University of California at Santa Cruz (T.L.), and the University of Arizona (T.C.W.). Those class notes, in turn, have a complex legacy, in part tracing back to lecture notes of Professor Hiroo Kanamori at the Seismological Laboratory of the California Institute of Technology, who taught an inspirational introductory seismology course to the coauthors. Other material is drawn from numerous introductory geophysics textbooks and current research publications.

This effort to distill a thorough, yet accessible, introductory survey of the discipline of global seismology has clearly involved many compromises, particularly in abbreviated treatment of such topics as transient wave solutions, synthetic seismogram calculation, sociological aspects of earthquake phenomena, and rock mechanics. Fortunately, the discipline is fully spanned by several advanced theoretical seismology texts: *Quantitative Seismology* (1980) by Aki and Richards, *Seismic Waves and Sources* (1981) by Ben Menahem and Singh, *Imaging the Earth's Interior* (1985) by Claerbout, *Seismic Wave Propagation in Stratified Media* (1983) by Kennett; elementary earthquake overviews: *Earthquakes* (1988) by Bolt, *Inside the Earth* (1982) by Bolt, *Elementary Seismology* (1958) by Richter, *Nuclear Explosions and Earthquakes* (1976) by Bolt; and fracture mechanics textbooks: *Earthquake Mechanics* (1981) by Kasahara, *The Mechanics of Earthquakes and Faulting* (1990) by Scholtz, *Principles of Earthquake Source Mechanics* (1990) by Kostrov and Das. Seismology survey texts that offer alternate presentations of some material in this text, and may further enlighten the reader are: *Introduction to Seismology* (1979) by Bath, *Seismology and Plate Tectonics* (1990) by

Gubbins, *An Introduction to the Theory of Seismology* (1985) by Bullen and Bolt, and *Introduction to Seismology, Earthquakes and Earth Structure* (1994) by Stein. Many additional texts reviewing the various fields of solid Earth geophysics provide additional resources.

The field of global seismology is in a continual, rapid state of flux, and any text can at best give an instantaneous and limited version of our knowledge of Earth structure and earthquake sources. It is, therefore, up to the reader to strive independently to stay current as new discoveries are announced, which occurs as frequently as each annual meeting of the seismological research community.

We are indebted to the students who have endured preliminary, often handwritten, versions of this material and invite them to discard their early versions and replace them with this updated and improved presentation of the course material. Norm Meader typed the text and consistently improved our grammar, both being mammoth undertakings. Susie Barber greatly assisted with the preparation of figure permissions and typed earlier drafts of several chapters. Chuck Ammon, John Ebel, Paul Richards, and Susan Schwartz provided helpful comments and corrections on various chapters. Yu-Shen Zhang generated several tomography figures for us, and Rhett Butler provided up-to-date maps of global seismic networks. We also thank the many colleagues who contributed figures or gave their permission for reproduction of published figures. Finally, we thank Susan and Michelle for putting up with us while we did this.

References

Introductory References

Båth, M. (1979). "Introduction to Seismology." Birkhäuser, Basel.

Bolt, B. A. (1976). "Nuclear Explosions and Earthquakes: The Parted Veil" Freeman, San Francisco.

Bolt, B. A. (1982). "Inside the Earth: Evidence from Earthquakes." Freeman, San Francisco.

Bolt, B. A. (1988). "Earthquakes." Freeman, San Francisco.

Bullen, K. E., and Bolt, B. A. (1985). "An Introduction to the Theory of Seismology." Cambridge Univ. Press, Cambridge, UK.

Gubbins, D. (1990). "Seismology and Plate Tectonics." Cambridge Univ. Press, Cambridge, UK.

Gutenberg, B., and Richter, C. F. (1954). "Seismicity of the Earth." Princeton Univ. Press, Princeton, NJ.

Richter, C. F. (1958). "Elementary Seismology." Freeman, San Francisco.

Simon, R. B. (1981). "Earthquake Interpretations: A Manual for Reading Seismograms." Wm. Kaufmann, Los Altos, CA.

Stein, S. (1994). "Introduction to Seismology, Earthquakes, and Earth Structure." Blackwell, Boston (in preparation).

Advanced Seismology

Aki, A., and Richards, P. G. (1980). "Quantitative Seismology," 2 vols. Freeman, San Francisco.

Anderson, D. L. (1989). "Theory of the Earth." Blackwell, Boston.

Ben-Menahem, A., and Singh, S. J. (1981). "Seismic Waves and Sources." Springer-Verlag, New York.

Claerbout, J. F. (1985). "Imaging the Earth's Interior." Blackwell, Oxford.

Ewing, W. M., Jardetzky, W. S. and Press, F. (1957). "Elastic Waves in Layered Media." McGraw-Hill, New York.

James, D. E., ed. (1989). "The Encyclopedia of Solid Earth Geophysics." Van Nostrand-Reinhold, New York.

Kasahara, K. (1981). "Earthquake Mechanics." Cambridge Univ. Press, Cambridge, UK.

Kennett, B. L. N. (1983). "Seismic Wave Propagation in Stratified Media." Cambridge Univ. Press, Cambridge, UK.

Kostrov, B., and Das, S. (1988). "Principles of Earthquake Source Mechanics." Cambridge Univ. Press, Cambridge, UK.

Scholtz, C. H. (1990). "The Mechanics of Earthquakes and Faulting." Cambridge Univ. Press, Cambridge, UK.

Mathematical References

Jeffreys, Sir H., and Jeffreys, B. S. (1950). "Methods of Mathematical Physics." Cambridge Univ. Press, Cambridge, UK.

Officer, C. B. (1974). "Introduction to Theoretical Geophysics." Springer-Verlag, New York.

1

INTRODUCTION

The Earth is composed of silicate and iron-alloy materials with the remarkable property that, over the wide range of pressure and temperature conditions existing within the planet, the materials respond nearly elastically under the application of small-magnitude transient forces but viscously under the application of long-duration forces. This time dependence of the material properties means that Earth "rings like a bell" when short-term forces, such as sudden slip of rock across a fault surface or detonation of a buried explosion, are applied, even while the fluid-like flow of global convection continually reshapes the surface and interior of the planet over geological time scales. The mechanical vibrations result from the quasi-elastic behavior, which involves excitation and propagation of elastic waves in the interior. These waves are physical motions that ground-motion recording instruments called *seismometers* can preserve for scientific analysis. This text describes the nature of these elastic waves and the analysis of their recordings. It demonstrates how the elastic properties of the Earth reveal many characteristics of the present state of the Earth as well as of the long-term processes occurring in the global dynamic system. We hope that it will also provide insight into the processes producing destructive earthquakes, such as the January 17, 1994, Northridge, California event, which caused more than $20 billion in damage to Los Angeles.

Seismology is the study of the generation, propagation, and recording of elastic waves in the Earth (and other celestial bodies) and of the sources that produce them. Both natural and human-made sources of deformational energy can produce *seismic waves*, elastic disturbances that expand spherically outward from the source as a result of transient stress imbalances in the rock. The properties of seismic waves are governed by the physics of elastic solids, which is fully described by the theory of elastodynamics. Basic elastodynamics is presented in Chapters 2, 3, 4, and 8. This body of theory, rooted in continuum mechanics, linear elasticity, and applied mathematics dating back to the early 1800s, provides a quantitative framework for analysis of elastic waves in the Earth.

Seismological procedures provide the highest resolution of internal Earth structure of any geophysical method. This is because elastic waves have the shortest wavelengths of any "geophysical wave," and the physics that governs them localizes their sensitivity spatially and temporally to the precise path traveled by the energy.

These localization properties provide far higher resolution than obtainable with electrical, gravitational, magnetic, or thermal fields, which average large regions and times.

Recordings of ground motion as a function of time, or *seismograms*, provide the basic data that seismologists use to study elastic waves as they spread throughout the planet. An example of a modern seismic recording is shown in Figure 1.1. Three orthogonal components of ground motions (up–down, north–south, and east–west) are shown, as are needed to record the total (vector) ground displacement history, at station HRV (Harvard, Massachusetts). The source that produced these motions was a distant large earthquake that struck central Chile in 1985. The ground motions at HRV commenced about 10 min after the fault rupturing began, the length of time it took for the fastest seismic waves to travel through the Earth from the Chilean source region to the station. A complex sequence of slower wave arrivals caused ground motions at the station to continue for several hours. These recorded motions are quite tiny, with ground displacements of less than 0.7 mm and ground velocities of less than 60 μm/s. Such motions were imperceptible at HRV other than by sensitive instrumentation, but the waves were much stronger near the source and caused extensive damage and building collapse in Chile. Every wiggle on the seismogram has significance and contains information about the source and the Earth structure through which the waves have traveled. Seismologists strive to extract all possible information from the seismogram by understanding each wiggle.

A tremendous range in scales is considered in seismology, for both the many types of sources and the diverse seismic waves that result. The smallest detectable microearthquake has a *seismic moment* (an

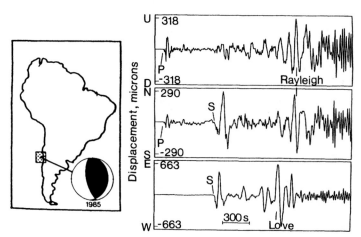

FIGURE 1.1 Recordings of the ground displacement history at station HRV (Harvard, Massachusetts) produced by seismic waves from the March 3, 1985, Chilean earthquake, which had the location shown in the inset. The three seismic traces correspond to vertical (U–D), north–south (N–S), and east–west (E–W) displacements. The direction to the source is almost due south, so all horizontal displacements transverse to the raypath appear on the east–west component. The first arrival is a *P* wave that produces ground motion along the direction of wave propagation. The *S* motion is large on the horizontal components. The Love wave occurs only on the transverse motions of the E–W component, and the Rayleigh wave occurs only on the vertical and north–south components. These motions are consistent with the predictions of Figure 1.2. (Modified from Steim, 1986.)

important physical quantity equal to the product of the fault surface area, the rigidity of the rock, and the average displacement on the fault) on the order of 10^5 N m, and great earthquakes have moments as large as 10^{23} N m. The amplitudes of seismic-wave motions are directly proportional to the seismic moments; thus seismic-wave displacements span an enormous range. Seismic waves commonly used in exploration seismology have frequencies as high as 200 Hz, while the longest-period standing waves excited by great earthquakes have frequencies around 3×10^{-4} Hz and solid Earth tide frequencies are around 2.0×10^{-5} Hz. Thus, transient ground motions spanning a frequency range of 10^7 Hz are of interest. In fact, the study of seismic sources further extends the range of interest to zero frequency, or static deformations, near faults and explosions, even while new, very high resolution shallow-imaging techniques are utilizing kilohertz frequencies. A local crustal survey may use waves that are traveling only tens of meters, while analysis of global structure may involve waves such as R_7, which travel more than 10^8 m along the Earth's surface.

One of the major challenges posed by the huge frequency range (bandwidth) and amplitude range (dynamic range) of interest for seismic observations has been to build seismometers capable of registering all useful signals against a background of ambient noise. No single instrument can record the full spectrum of motions with a linear response, so a suite of different seismic instruments that record limited portions of the seismic spectrum has been developed. However, great advances have been made in the last 10 years in developing seismic recording systems that provide remarkable bandwidth and dynamic range for the applications of global seismology to be emphasized in this text. The recording in Figure 1.1 was produced by such a system, and Chapter 5 describes the remarkable instrument technology involved in the field of *seismometry*, or recording of ground motion.

The global distribution of earthquake sources, along with the requirement of extensive surface coverage with seismometers for the unraveling and interpretation of complex seismic signals, has made global seismology a truly international discipline, with unprecedented international collaboration, seismometer development, and data exchange over its 119+-year instrumental history. Over 3000 seismological observatories are in operation around the world today, with nearly every nation participating in the effort to record seismic waves continuously. The most recent efforts to upgrade the global network instrumentation by incorporating technological advances have involved countries such as Australia, Canada, China, England, France, Germany, Holland, Italy, Japan, Norway, Russia, Switzerland, and the United States, in keeping with the historic tradition of broad international collaboration. Chapter 5 provides an overview of these efforts.

The fault that generated the 1985 Chile earthquake ruptured for about 100 km, with sliding motions on the fault lasting for only about 50 s. Thus, much of the prolonged nature of the vibrations in Figure 1.1 is due to wave interactions with the transmitting medium, which are manifested as a sequence of impulsive arrivals and longer-period oscillatory motions, including waves that repeatedly circle the globe. Most of these ground motions can now be interpreted quantitatively in the light of current knowledge of Earth structure, as shown in Chapter 6. It is the fundamental simplicity of elastic waves, which transmit disturbances over great distances through the Earth with little, or mostly predictable distortion, that allows useful information to be gleaned from the seismograms, despite the overall complexity arising from structural interactions.

Seismology is an observation-based discipline that addresses internal Earth structure and characteristics of seismic-wave sources by applying elastodynamic theory to interpret seismograms. Because of the physical constraint of being able to record seismic-wave motions only at, or very near, the surface of the Earth, seismology draws heavily upon mathematical methodologies for solving systems of equations that are collectively described as *geophysical inverse theory* (Chapter 6). Many seismological applications and results of inverse theory are described in this text. The essence of all seismic inverse problems is that inferences about the wave source or the transmitting medium are made by applying mathematical operations derived from elastodynamic theory to the observed surface ground motions. The recorded motions can be viewed as the output response of a sequence of linear filters with properties we wish to determine. We can treat instrumental, propagation, and source effects as separate filters, and we have structured this text to concentrate sequentially on each factor that shapes the observed seismogram. Inversions for filter characteristics contain many nonuniqueness problems, and strong trade-offs exist between source and propagation effects that are difficult to resolve. The history of seismological advances is one of alternating progress in describing source properties or in improving models of Earth structure, and clever strategies have been advanced to overcome the intrinsic trade-offs in the signal analysis. Remarkable resolution of deep Earth structure is now being achieved using modern inversion methods. Seismic inversion and Earth structure determination are described in Chapter 7.

In parallel with the rapid advances in our knowledge of Earth structure has come a comparable expansion of our understanding of earthquake faulting and its role in global plate tectonics. From the basic foundation of quantitative representations of shear faulting sources (described in Chapter 8) we have developed an under-standing of most faulting phenomena. Chapter 9 describes the kinematic and dynamic characteristics of shear faulting sources, their scales of variation, and measures of energy release such as seismic magnitudes and earthquake moment.

With independently derived knowledge of Earth structure, it has become possible to construct predicted ground motions to compare with observations. This serves as a basis for seismic inversion for faulting parameters. This capability has led to an appreciation that faulting is a very heterogeneous process with nonuniform stress release over the fault surface. Chapter 10 reviews the contemporary source analysis procedures used in earthquake seismology.

Seismology intrinsically provides information about active, present-day processes in the Earth. Quantification of earthquake faulting characteristics such as fault orientation, sense of slip, and cumulative displacement has played a major role in the evolution of the theory of plate tectonics. *Seismotectonics*, the study of active faulting and its relationship to plate motions and lithospheric properties, is described in Chapter 11. Seismology is the solid Earth geophysical discipline with the highest societal impact, both in assessing and reducing the danger from natural hazards and in revealing present Earth structure and buried resources. Yet the relative sluggishness of mantle convective flow, or thermal inertia of the system, ensures that knowledge of the present-day internal structure reflects processes that have been occurring in the Earth over the past several hundred million years and, to a certain extent, over the entire evolution of the planet.

1.1 Historical Development of Global Seismology

Seismology is a relatively young science, having awaited both the evolution of the theory of elasticity and the development of an instrumental data base. Although the Chinese had the first operational seismic-wave detector around 132 AD, the theoretical side of the science was considerably ahead of the observational side until the late 1800s. From the introduction in 1660 of Hooke's law, indicating a proportionality between stress and strain, to the development of equations for elasticity theory by Navier and Cauchy in 1821–1822, our understanding of the behavior of solid materials evolved rapidly. In the early 1800s the laws of conservation of energy and mass were combined to develop the equations of motion for solids. In 1830 Poisson used the equations of motion and elastic constitutive laws to show that two (and only two!) fundamental types of waves propagate through the interior of homogeneous solids: *P* waves (compressional waves involving volumetric disturbances, and directly analogous to sound waves in fluids) and *S* waves (shear waves with only shearing deformation and no volume change, which can therefore not propagate in fluids). The sense of particle motions relative to the direction of propagation for *P*- and *S*-wave disturbances is shown in Figure 1.2. These two types of motion are called body waves, because they traverse the interior of the medium. *P* (primary) waves travel faster than *S* (secondary) waves and are thus the first motion to be detected from any source in an elastic solid.

In 1887 Lord Rayleigh demonstrated the existence of additional solutions of elastic equations of motion for bodies with free surfaces. These are Rayleigh waves, involving wave motions confined to and propagating along the surface of the body. By 1911 a second type of surface-wave motion, produced in a bounded body with layered material properties, was characterized by Love and is hence called a Love wave. Rayleigh and Love waves are surface waves and result from the interaction of *P* and *S* waves with the boundary conditions on the body (i.e., vanishing shear stresses on the surface). The sense of particle motions for these surface waves is indicated in Figure 1.2. Body and surface waves are influenced by changes in material properties with depth, such as the existence of internal boundaries in the Earth that can reflect energy. These interactions can be quantitatively analyzed by solving boundary-value problems, and they are expressed in terms of reflection and transmission coefficients.

These basic elasticity solutions for a general solid medium were partial motivation for the development of instruments capable of recording time histories of the ground motion of the Earth at a fixed location. International efforts led to the invention of the first seismometer by Filippo Cecchi in Italy in 1875. The sensitivity of early seismometers improved rapidly, and by 1889 the first accurate recording of waves from a distant earthquake was obtained by an instrument in Potsdam, 15 min after the earthquake faulting occurred in Japan. The 119+ years of quantitative ground motion observations have confirmed the existence of *P*, *S*, Rayleigh, and Love waves in the Earth, as well as other, now (mostly) understood arrivals, demonstrating that the Earth behaves as a (nearly) elastic body in the frequency band of most seismic observations. The recordings in Figure 1.1 clearly exhibit distinct arrivals of *P*, *S*, Love, and Rayleigh waves, with particle motions as predicted in Figure 1.2, along with other arrivals that are explained later.

In 1892, while working in Tokyo, John Milne developed a seismometer that was sufficiently compact that it could be installed in about 40 observatories around the world. This began the systematic collection of global seismic data. Around the turn of the century, seismometer technology increased significantly, and body-wave data sets accumulated rapidly, revealing systematic behavior of body-wave arrivals as a function of distance from the sources. This began an interval of first-order discoveries about the Earth's interior and earthquake sources. Oldham discovered

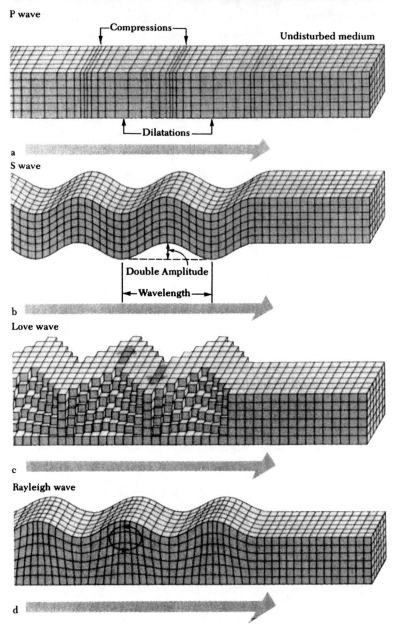

FIGURE 1.2 Schematic of the sense of particle motions during passage of the two fundamental elastic body waves, (a) *P* and (b) *S* waves, as well as the two surface waves in the Earth, (c) Love, and (d) Rayleigh waves. The waves are all propagating from left to right, with the surface of initial particle motion corresponding to the wavefront. The relative velocity of each wave type decreases from top to bottom. The passage of all four wavetypes past a single sensor is shown in Figure 1.1. (From Bolt, 1976.)

the Earth's core in 1906, and in 1913 Gutenberg determined an accurate depth to the core of about 2900 km (the current preferred value is 2889 km). In 1909 Mohorovičić discovered a sharp velocity contrast that we now refer to as the "Moho" and interpret as the base of the crust. In 1936 Inge Lehmann (an early woman seismologist) discovered the Earth's inner core. Sir Harold Jeffreys compiled the travel times of thousands of seismic arrivals and developed the first detailed cross section of the Earth from surface to center by 1939. These travel-time tables are still used routinely today to locate global earthquakes and are referred to as the Jeffreys–Bullen (J–B) tables. The J–B tables predict the arrival times of P waves to any point on the Earth's surface to within a remarkable 0.2% accuracy, limited primarily by the existence of three-dimensional variations in structure not allowed for in the tables.

In parallel with the advances in Earth structure, seismology and field observations were revealing the nature of earthquakes. In 1910, Reid enunciated the "elastic rebound theory" of earthquake faulting. The year 1928 brought the recognition of the existence of deep-focus earthquakes by Wadati. In the mid-1930s Richter developed the first quantitative measure of relative earthquake size, the local magnitude scale (M_L), referred to as the "Richter magnitude." By 1940 the global distribution of earthquakes was accurately mapped out, clearly defining major belts of activity that we now associate with boundaries between surface lithospheric plates.

Although the first half of the twentieth century revolutionized our knowledge of the Earth, seismology was still a rather obscure science, with only a small number of active seismologists. The biggest problem was that only a limited number of worldwide seismic stations existed. Furthermore, the instrument response characteristics of these stations were not stan-

dardized, making it difficult to analyze the details of the ground motion. It took the advent of underground nuclear testing for seismology to become a truly modern science.

Seismology provides a remote-sensing technique for monitoring nuclear testing, because underground explosions produce seismic waves that can be detected at great distances. In fact, a seismic station at Tucson was used by Gutenberg to determine accurately the detonation time of the first nuclear explosion (Trinity) on July 16, 1945, when timing equipment at the test site failed. (The Trinity device was suspended aboveground, but sufficient energy coupled into the ground from the blast to excite seismic waves.) The first underground nuclear explosion (designated Rainier) was detonated in 1957 by the United States, and the 1963 Limited Test Ban Treaty banned atmospheric, oceanic, and deep space testing of nuclear devices by all of its 116 signatory nations. The U.S. government recognized the need to develop a research effort to understand seismic-wave propagation in complex structures in order to monitor foreign underground tests, and so it started the VELA UNIFORM program. One of the first accomplishments of this program was the deployment of the World Wide Standard Seismograph Network in the late 1950s and early 1960s. This 120-station global network of high-quality, well-calibrated, well-timed stations caused observational seismology to leap ahead of theoretical developments, bringing about major investments in university research programs. At the same time, rapid advances in computer technology enabled sophisticated analysis of increasing volumes of seismic data. Although many first-order discoveries about the Earth had been made in the pioneering days prior to 1960, the field of global seismology truly came into its own thereafter, and we will concentrate primarily on developments of the past few decades in this text.

1.2 The Topics
of Global Seismology

Having given a brief introduction to the basic nature of seismology, we will now undertake an overview of the topics and contributions of global seismology. This text will then provide the information required for understanding how we obtain such quantitative results from seismic recordings. It is useful to state at the outset that the nature of elasticity allows us to treat mathematically the process of excitation, propagation, and recording of seismic waves as a sequence of linear filters that combine to produce observed seismograms. In other words, an observed ground displacement history, $u(t)$, can be expressed as the result of a source function, $s(t)$, operating on a propagation function, $g(t)$, combined with an instrument recording function, $i(t)$. The filter operations will later be shown to be time-domain convolutions of a transfer function $z(t)$, mapping one function, $y(t)$, into another, $x(t)$, by an integral operation:

$$x(t) = \int_{-\infty}^{\infty} y(\tau) z(t - \tau) \, d\tau \quad (1.1)$$

If we denote this integral operation as $x(t) = y(t) * z(t)$, we can express ground motion as

$$u(t) = s(t) * g(t) * i(t) \quad (1.2)$$

Modern seismology strives to describe mathematically each of the filters contributing to the observed displacements, and seismological research efforts classically bifurcate into two major categories: (1) studying the source terms and their associated phenomena, and (2) studying the propagation terms and the associated Earth structure. The instrument transfer function is always the best-known filter but involves an interesting body of theory in its own right. Much of the organization of this textbook (as well as almost every other seismology book) tends to focus sequentially on these filters. However, the convolutional nature of the preceding equation should make it clear that any analysis of ground motion must consider the combined source and propagation characteristics. Table 1.1 lists some of the many topics of classical and current interest in the two major categories. We will now survey some basic results of seismological analysis in each category before developing the theory and procedures used in global seismology.

TABLE 1.1 Major Topics of Global Seismology

Source topics	Earth structure topics
Classical objectives	
A. Source location (latitude, longitude, depth, time)	A. Basic layering (crust, mantle, core)
B. Energy release (magnitude, seismic moment)	B. Continent–ocean differences
C. Source type (earthquake, explosion, other)	C. Subduction zone geometry
D. Faulting geometry, area, displacement	D. Crustal layering, structure
E. Earthquake distribution	E. Physical state of layers (fluid, solid)
Current research objectives	
A. Slip distribution on faults	A. Lateral variations in crust, mantle, core
B. Stresses on faults and in Earth	B. Topography of internal boundaries
C. Faulting initiation/termination	C. Anelastic properties of the interior
D. Earthquake prediction	D. Compositional/thermal interpretations
E. Analysis of landslides, volcanic eruptions, etc.	E. Anisotropic properties

1.2.1 Seismic Sources

Elastic waves are generated whenever a transient stress imbalance is produced within or on the surface of an elastic medium. Almost any sudden deformation or movement of a portion of the medium results in such a source. A great variety of physical phenomena in the Earth involve rapid motions that excite detectable seismic waves. These sources can be grouped into those that are external to the solid Earth and those that are internal. Table 1.2 lists some common seismic sources, all of which involve processes of interest to Earth scientists.

Mathematical descriptions and physical theories for all of these source types have been developed, although most are kinematic descriptions rather than first-principle theories. In order to represent these complex physical phenomena mathematically, we must usually determine dynamically equivalent, idealized force systems that can be visualized as replacing the actual process. By "dynamically equivalent" we mean that the elastic motions produced by the idealized force system are the same as those of the actual process. We can then place these force representations into Newtonian equations of motion (essentially $\mathbf{F} = m\mathbf{a}$, where $\mathbf{F}$ is the force system, m is the mass of the body, and $\mathbf{a}$ is the acceleration of the body) to predict the resulting waves accurately.

External sources are usually easier to represent mathematically than internal sources. In most cases, external sources

can be treated as time-varying tractions applied to the Earth's surface (a traction is the stress vector resulting from a force applied to an element of surface area). As the traction varies with time, a stress imbalance near the source is created. This imbalance is equilibrated by motions of the medium, which in turn propagate outward as seismic waves. The mathematics of this are given in Chapter 2. Internal force systems may be relatively simple, as in the three-dipole force system needed to represent an isotropic explosion, or quite complex, as in the spatial distribution of double-couple forces needed to represent a large earthquake (to be described in Chapter 8). All sources produce body and surface waves, but the relative excitation and the frequency and amplitude characteristics of these waves depend strongly on the source type and force–time history. For example, the seismic recordings of nuclear explosions can usually be discriminated from natural earthquakes by their very strong excitation of high-frequency P waves relative to lower-frequency surface waves.

Although the sources of primary interest for this text on global seismology are shear-faulting earthquakes, many of the sources listed in Table 1.2 can produce globally observable seismic signals. Figure 1.3 shows surface ground motions produced by overhead passage of the space shuttle Columbia. As the shuttle descended for landing, it produced a sonic boom, which vibrated the ground in the Los Angeles basin. The ground motions were recorded by seismometers deployed

TABLE 1.2 Primary Sources of Seismic Waves

Internal	External	Mixed
Earthquake faulting	Wind, atmospheric pressure	Volcanic eruptions
Buried explosions	Waves and tides	Landslides
Hydrological circulation	Cultural noise (traffic, railways)	
Magma movements	Meteorite impacts	
Abrupt phase changes	Rocket launches, jet planes	
Mine bursts, rock spallation		

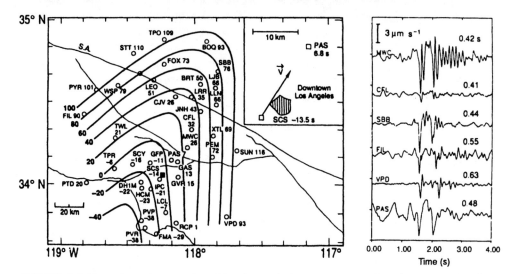

FIGURE 1.3 Ground motions produced by the sonic boom accompanying the space shuttle Columbia as it descended over the Los Angeles basin on its way to landing. The relative time of vibrations at regional seismic stations is shown on the left, with the arcuate pattern resulting from intersection of the sonic boom "mach cone" with the ground. The inset shows the trajectory of the shuttle across the basin. The actual ground-motion velocities at several stations are shown on the right. (Reprinted with permission from Kanamori *et al.* *Nature*, vol. 349, pp. 781–782; copyright©1991 Macmillan Magazines Limited.)

in the region to monitor regional earthquake activity. The time of arrival of ground vibrations at the stations allows us to determine the descending trajectory of the shuttle as the "mach cone" intersection with the ground swept across the basin (shown on the left). Actual recordings of ground-motion velocities at different stations are shown on the right and can be interpreted as the result of rapidly changing air pressure on the ground as the sonic boom front sweeps across. As exotic as this moving source may be, the resulting seismic motions behave predictably according to the theory of elastic waves. Chapter 10 describes the recovery of seismic source parameters for more conventional faulting earthquakes, and Chapter 11 discusses how we can use these parameters to learn about active tectonics.

1.2.2 Earthquake Sources Involving Shear Faulting

The development of equivalent force systems for natural earthquakes required a basic understanding of the associated process, which was not available before this century. Historically, ground breakage and surface faulting associated with Earth vibrations have often been observed, but in many instances no surface break could be associated with a tremor, confusing observers as to which was cause and which effect. It was difficult to apply any scientific method to study earthquakes because of the limited observational data base. It was not until the 1906 San Francisco earthquake that a causative theory relating the two phenomena was clearly enunciated. Reid carefully studied the well-exposed permanent ground motions that occurred at the time of the 1906 earthquake. The horizontal deformations in the vicinity of the San Andreas fault (Figure 1.4a) exhibited a simple symmetry that led him to formulate the *elastic rebound theory of earthquakes*. This partly empirical, partly intuitive theory states that crustal stresses, generally resulting from large-scale regional crustal shearing motions, cause strain to accumulate in the immediate vicinity of *faults*, which are quasiplanar breaks in the rock across which some previous displacement has occurred and which are hence relatively weak. When the strain accumulation reaches a threshold imposed by the material properties of the rock and the fault surface, abrupt frictional sliding occurs (Figure 1.4b), releasing the accumulated strain energy. Much of the strain energy is consumed in heating and fracturing of the rocks, but a portion is converted into seismic waves that propagate outward from the fault zone, communicating the disturbance to distant regions. The regional deformations continue, leading to many cycles of strain accumulation and release during the active lifetime of the fault.

The elastic rebound theory predicts permanent coseismic shear displacements (Figure 1.4) similar to the 1906 observations. We expect this particular symmetric pattern of surface displacement only for a vertical fault that slips horizontally, but shallow faults with other orientations produce easily predictable patterns of horizontal and vertical motion due to the shearing offset (the governing equations for these static deformations are derived in Chapter 8). Examples of the historical *geodetic* (measured permanent ground deformation) observations favoring this model, collected largely in Japan, where there are numerous shallow crustal faults, frequent earthquakes, and many seismologists who study faulting, are listed in

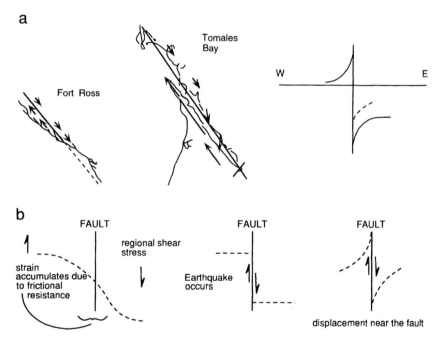

FIGURE 1.4 (a) Observed permanent ground displacements that occurred simultaneously with the 1906 San Francisco earthquake. The symmetric distribution of horizontal displacements on either side of the San Andreas fault suggests that strain energy accumulated in the vicinity of the fault and released when the fault slipped, producing the seismic vibrations that were felt as an earthquake. (b) Sketch of the process of strain accumulation in the vicinity of a fault resulting from regional shearing motions, followed by the sudden sliding of the rock on the fault surface. This is the essence of the elastic rebound theory of faulting. The coseismic distribution of actual permanent ground displacement is shown on the right. Compare this with the observations for the 1906 earthquake in Figure 1.4a.

TABLE 1.3 Classic Observations of Faulting Strain

Event	Fault length (km)	Average offset (m)	Decay distance (km)	Strain
1906 San Francisco, CA $M_S = 7.8$ (see Figure 1.4)	200	5	20	2.5×10^{-4}
1927 Tango, Japan $M_S = 7.8$	30	3	30	1.0×10^{-4}
1943 Tottori, Japan $M_S = 7.4$	40	2	15	1.3×10^{-4}
1946 Nankaido, Japan $M_S = 8.2$	80	0.7	100	1.0×10^{-5}
1971 San Fernando, CA $M_S = 6.6$	30	2	20	1.0×10^{-4}

Table 1.3. These examples indicate that the crust cannot accumulate strains much larger than about 10^{-4} without failure, where strain is calculated as slip on the fault divided by the distance perpendicular to the fault over which there are significant coseismic displacements. Most events involve strains from 10^{-5} to 10^{-4}, at least in typical continental situations, a fundamental result that we return to in Chapter 9. A large number of such observations of faulting and ground displacement have given rise to the hypothesis that most *shallow* (less than 70 km deep) earthquakes result from shear dislocations on faults,

even though most such events occur below the depth of direct observation. Systematic analysis of seismic waves from thousands of earthquakes over the past decade supports this hypothesis.

The 1906 San Francisco earthquake was also scientifically important because it was widely recorded on the early generation of seismometers available near the turn of the century. Figure 1.5 shows a horizontal component of ground motion on an Omori seismometer that was located in Tokyo. This recording shows an initial *P*-wave arrival followed by a much larger *S*-wave arrival and then a complex sequence of

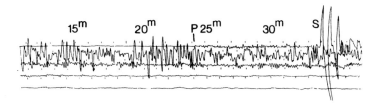

FIGURE 1.5 A classic seismic recording of the 1906 earthquake made by an Omori seismometer located in Tokyo, Japan. The ground motion is horizontal, east–west. The station is at a distance of 75.05° from the source (1° = 111 km). Time increases toward the right on the recordings, and the first arrival is a *P* wave. The *S* wave arrives about 10 min later (the tick marks indicate 60-s intervals). The record wraps around from one line to the next, as it was recorded on a rotating, translating drum.

TABLE 1.4 Characteristic Seismic Wave
Periods

Wave type	Period (s)
Body waves	0.01–50
Surface waves	10–350
Free oscillations	350–3600

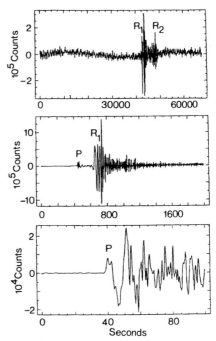

FIGURE 1.6 Broadband vertical-component recording of the 1989 Loma Prieta earthquake at station ANMO (Albuquerque, New Mexico). The top panel is 20 h in duration (the earthquake is the rider on the long-period signal); tidal effects dominate. The middle panel is for a 30-min interval, and the bottom panel is for a 100-s interval.

surface waves. The shearing nature of motions at the source is partly responsible for the greater amplitude of the S wave relative to the P wave. The combination of a conceptual model for the faulting source and the constraints on source force systems provided by observed amplitudes and polarities of P and S waves enabled the development of the *double couple* and, more recently, the *moment tensor* as general force models for shear faulting sources that are now routinely used in global seismology. This is fully described in Chapter 8.

1.2.3 Quantification of Earthquakes

In general, earthquake body waves (P and S waves) have shorter characteristic periods of vibration than surface waves (Rayleigh and Love waves), which in turn have shorter periods than *free oscillations* of the Earth (standing modes of vibration of the entire planet, which are detectable only for the largest earthquakes) (Table 1.4). Furthermore, the ground displacements for body waves generated during a large earthquake may be only 10^{-3} cm after traveling 1000 km, but long-period surface waves may have amplitudes of several meters after traveling the same distance. These differences result from source excitation and propagation interference effects that depend on the type of wave and the Earth structure.

It is important to realize that each type of seismic wave involves a spectrum of frequencies, and the ground motion from the same wave will have a different ap-

pearance depending on the filtering transfer function of the recording system. A very broadband seismometer records many frequencies of ground motion, as shown in Figure 1.6. The recordings are for the October 1989 Loma Prieta earthquake that ruptured a fault in the Santa Cruz Mountains. The top panel shows a time window of 20 h. The Loma Prieta earthquake shaking is a large rider on the long-period sinusoidal signal with a period of 12 h. The long-period signal is the solid Earth tide; the Earth rises and falls about 40 cm at station ANMO every day in response to tides caused by the gravitational attraction of the Sun and Moon. The middle panel shows a time window of 30 min containing the main signal from the Loma Prieta earthquake. The largest signal is the Rayleigh wave, which has an amplitude of about 2 cm. The bottom panel shows the first 60 s of the P arrival, which has many high-frequency oscillations. The higher-frequency energy is very complex, indicating that propagation and source effects have a strong frequency dependence. This illustrates how characterization of a seismic signal in any one frequency band may not represent the behavior in other frequency bands.

At the long-period end of the seismic spectrum, other important phenomena are observed in the seismic waves. One of the most important is caused by the spherical nature of the planet. Figure 1.7 shows long-period Rayleigh waves produced by the Loma Prieta earthquake recorded at globally distributed digital seismometers.

R_1 waves travel along the short arc of the great circle from the source to the receiver and then continue to circle the Earth, reappearing as R_3 at the same station 3 h later. R_2 travels along the long arc of the great circle and arrives at the station again as R_4 and then as R_6 etc. in 3-h shifts. These surface waves slowly decrease in amplitude as they circle the Earth because of energy losses due to *attenuation* (anelastic losses) and increasing *dispersion* (frequency dependence of velocity) of the energy. Longer-period oscillations are increasingly dominant later in the traces because both attenuation and dispersion have a strong frequency dependence. We must account for these effects when studying the source, but they reveal information about Earth structure when directly studied.

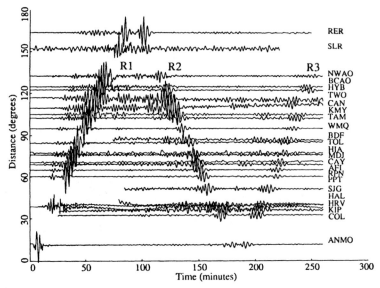

FIGURE 1.7 Long-period Rayleigh waves produced by the 1989 Loma Prieta earthquake as recorded at globally distributed digital seismometers of three global networks (GEOSCOPE, International Deployment of Accelerometers, Global Seismic Network). The vertical-component traces are filtered to include only periods longer than 125 s. The vertical axis is the angular distance along the surface from the California source, and time is from the earthquake origin time. R_1 and R_2 are Rayleigh waves traveling along the minor and major arcs of the great circle from source to station, respectively; R_3 is the next passage of the R_1 wave after circling the entire globe. (From Velasco *et al.*, 1993.)

Prior to instrumental recording, comparisons of earthquakes were based mainly on shaking damage and *seismic intensity* scales were developed based on varying damage. Intensity scales can be contoured, defining *isoseismals*, or regions of common shaking damage, typically having the highest intensities close to the fault. Although such earthquake measures are strongly influenced by proximity of the event to population centers, construction practices, and local site effects, seismic intensities are often all that we know about preinstrumental events, and they play a major role in regions such as the eastern United States, where most known large events occurred over 100 years ago. Earthquake measures based on recorded ground motions are more useful for recent events.

Until recently it has been necessary to use different seismometers, sensitive to different frequency ranges and with varying ground-motion amplification, to record the different wave types. Therefore, the various types of instruments intrinsically tend to record only those types of waves with corresponding periods, which may represent only a small part of the total ground motion. The diversity of instruments recording different wave types has led to the development of many different scales for comparing the relative size of earthquakes based on seismic waves, typically called seismic magnitudes. We use seismic waves to compare earthquake size because it can be done systematically and quantitatively and because it does not rely on damage or other macroscopic phenomena that are strongly influenced by factors other than the source (such as variable construction standards and surface topography). Almost all magnitude scales are based on the logarithmic amplitude of a particular seismicwave on a particular seismometer, with corrections for the distance to the source. Examples of the primary magnitude scales are given in Table 1.5 and compared with the period response of

TABLE 1.5 Examples of Seismic Magnitude Scales

Symbol	Name	Period of measurement (s)
M_L	Richter magnitude	0.1–1.0
m_b	Body-wave magnitude	1.0–5.0
M_S	Surface-wave magnitude	20
M_W	Moment magnitude	> 200

common seismic instruments in Figure 1.8. These show that any one earthquake can have many different seismic magnitudes, if measurements are made for different waves at different frequencies. This has often confused the news media, who (perhaps reasonably) tend to expect a given earthquake to have only a single magnitude (Richter magnitude).

A graphical presentation of the calculation of Richter magnitude is shown in Figure 1.9. The essential measurements are the peak amplitude of ground motion on a Wood–Anderson seismic recording and the difference between S and P arrival times, which is proportional to the distance to the source. Wave amplitudes decrease systematically with distance, so correction to a reference distance allows direct comparison of logarithmic amplitudes, or magnitude. Note that for a given distance, a factor of 10 difference in seismic amplitude yields a unit difference in magnitude. This relationship is empirical, with only a limited theoretical basis (described in Chapter 9), and in a strict sense this magnitude scale is restricted to events in Southern California, where it was developed, because the amplitude–distance relation varies in regions with different crustal structure. Nonetheless, seismic magnitudes have many uses in comparing earthquake properties.

Earthquakes can be quantified by determining several physical parameters, such as the fault length, rupture area, average displacement, particle velocity or accelera-

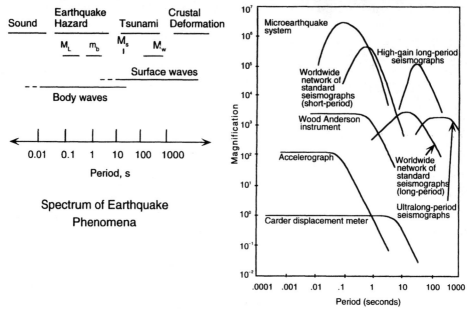

FIGURE 1.8 The range in period of seismic phenomena in the Earth is shown on the left, along with the characteristic periods of body waves, surface waves, and different seismic magnitude scales. On the right, the amplitude responses of some major seismometer systems are shown. Each magnitude scale tends to be associated with a particular instrument type; for example, the Richter magnitude, M_L, is measured on the short period Wood–Anderson instrument. (Courtesy of H. Kanamori.)

tion at the fault, duration of faulting, radiated energy, heterogeneity of slip distribution, or combinations of such quantities. Although we can determine many of these characteristics by detailed seismic-wave analysis, any given magnitude scale can only qualitatively describe the complex process at the source. We shall see that the best-defined physical quantity with which to represent the source is the seismic moment, which is controlled by static parameters of the total fault motion, with a unique value for each event. The moment magnitude scale, M_W, is based on logarithmic scaling of seismic moments to give numerical magnitudes that are roughly comparable with older magnitude scales. However, structural damage from earthquakes is often controlled by high-frequency waves, so short-period magnitudes are still very useful.

Large earthquakes have values of $M_W \geq$ 7.0, which roughly corresponds to events having more than 1 m of displacement on faults that are more than 30 km long. *Great* earthquakes have M_W values > 8.0 and involve larger faults and greater slip. The largest instrumentally recorded event is the great 1960 Chilean earthquake ($M_W = 9.5$), which involved 20 m of displacement during a few-minute-long rupture that extended over a 1000-km-long fault. The annual average number of $M_W > 7.0$ events is about 15 (Figure 1.10). One or no great events may occur each year, but more frequent smaller events can also be catastrophic in terms of loss of life and damage. An example of the awesome destructive potential of earthquakes is the 1976 ($M_W = 7.7$) Tangshan, China earthquake, which took approximately 250,000 lives (some estimates put the toll as high

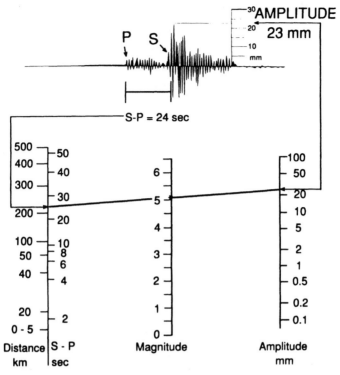

FIGURE 1.9 A graphical form of the Richter magnitude scale procedure. A recording from a local earthquake made on a Wood–Anderson seismometer must be used. The peak deflection on the record is measured, and the distance from the source is determined (it is roughly proportional to the time interval between the S and P arrivals). A line connecting these values intersects the magnitude scale at the appropriate value. The scale is logarithmic, so a factor of 10 variation in the amplitude of the seismic wave gives a unit variation in the magnitude.

as 700,000). Figure 1.10 shows that the average annual number of earthquake fatalities is about 15,000, with many areas of the world being stricken. Earthquake hazard varies dramatically with location around the world, with inferior construction practices of developing nations often accentuating earthquake damage. Circum-Pacific countries tend to have more frequent large events, resulting in greater damage potential. The 1985 Mexico City earthquake is an example of an event in a city with moderate construction standards that is located near a frequent earthquake zone. Although Mexico City was 250 km from the fault zone, at least 7000 people lost their lives, mainly due to building collapse. Soil conditions under the building foundations, construction practices in the city, and unusually long rupture duration have been blamed for the catastrophe.

1.2.4 Earthquake Distributions

One of the classical problems in global seismology has been the systematic mapping of earthquake distributions on a variety of scales. This mapping has played a key role in the evolution of the theory of *plate tectonics*, which describes the large-scale relative motions of a mosaic of lithospheric plates on the Earth's surface. It is the properties of seismic waves that allow the source location to be determined, since

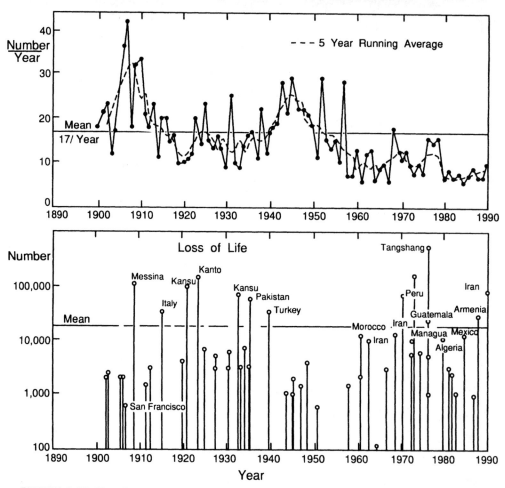

FIGURE 1.10 (Top) The annual number of large ($M_S \geq 7.0$) shallow earthquakes around the world. There are about 17 events of this size annually. (Bottom) The history of earthquake-induced fatalities in this century, with the locations of major events being indicated. Note the poor correlation with the top trace. Even small earthquakes can cause extensive loss of life in regions with poor building construction, or if secondary hazards such as fires or landslides enhance the damage. (Modified from Kanamori, 1977, 1978.)

the waves propagate through the Earth with velocities controlled by the material properties. Observations of arrival times of seismic waves and a model of the velocity structure in the Earth are needed for seismic location methods. Historically, the development of velocity models and improved source locations has evolved in a seesaw fashion, with occasional, independently known source locations and origin times providing first-order models of the structure, which could be statistically improved over time. The procedures for earthquake location are described in detail in Chapter 6.

By 1941, through work by Beno Gutenberg and Charles Richter, the global distribution of major earthquake belts was quite well determined, and the enhanced location capabilities of the modern global network now allow routine location of all events greater than magnitude 4.5 or so (Figure 1.11). The distribution of seismic events, or *seismicity*, is very nonuniform.

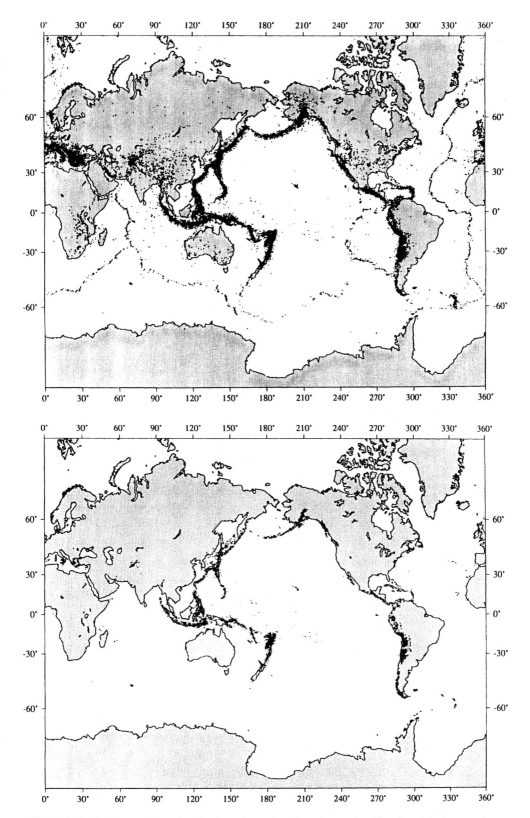

FIGURE 1.11 Maps of the distribution of earthquakes determined by the global networks for the years 1970 to 1990. At the top, the source location for events less than 100 km deep are shown; at the bottom, events with depths from 100 to 700 km are shown.

Most events occur around the Pacific margin, but midocean ridge and fracture zone structures are also quite active. Continental seismicity tends to be diffuse and is concentrated in seismic belts only along the Pacific margins. Studies by Turner and Wadati in the early 1920s revealed the occurrence of seismicity at depths greater than 70 km. The spatial distribution of such events, termed *intermediate-depth* events if they occur between 70 and 300 km depth and *deep* events if they occur between 300 and 700 km depth, is very limited. Such events are found primarily in linear belts around the Pacific, under Europe, and under Tibet. These presumably occur within downwelling portions of

oceanic plate that is sinking into the mantle. Deep events occur much less frequently and release much less energy than shallow earthquakes. The nature of their sources is also somewhat puzzling because frictional sliding supposedly cannot occur at such great depths because of the high pressures, yet the seismic radiation is similar to that for shallow shear-faulting events. These issues are discussed in Chapter 11.

The distribution of smaller-magnitude seismicity is also studied by seismologists, particularly in densely inhabited areas where the earthquake hazard is being assessed. Earthquake locations for a 10-year interval in Southern California are shown in Figure 1.12. The seismic distribution is

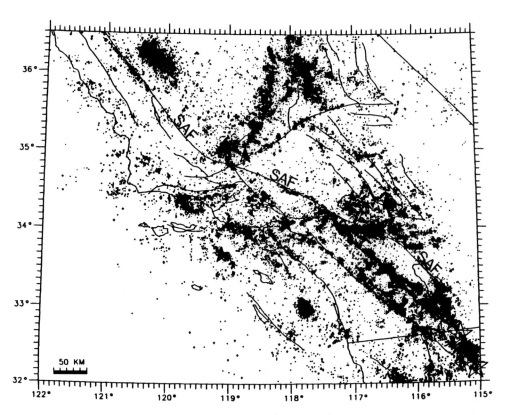

FIGURE 1.12 A map of earthquake locations in Southern California for the years 1978–1988. Most of the events are very small, and a dense network of seismometers is deployed in the region to locate all of the earthquakes accurately. The traces of known active faults observed at the surface are superimposed for comparison (as well as the borders of California), with the San Andreas fault labeled SAF. (Courtesy of Tom Heaton.)

exceedingly complex and does not strictly adhere to the mapped faults that break the surface. Dense arrays of seismometers are installed in areas of intense seismicity, or *seismogenic* zones, in order to obtain precise earthquake locations and to study the faulting motions that must be taking place in the region. Note that if we use just the short-term seismicity pattern to locate faults in this region, we may fail to identify the major fault that produces the largest earthquakes, the San Andreas fault, because it has few small events.

The need to assess large-earthquake hazard leads global seismologists to look at the historic record of large earthquakes around the world over longer periods of time. The global distribution of great earthquakes during most of this century is shown in Figure 1.13, where the seismological surface-wave magnitude scale, M_S, as well as the moment magnitude scale, M_W,

values are given for each event (when known). The availability of relatively quantitative seismic magnitudes allows us to study this historical pattern. Table 1.6 list the major events of the century. The M_S values for some events near the turn of the century have been revised downward, with new values given in parentheses. The distribution mirrors the overall seismicity pattern, with the largest events occurring around the Pacific margins, but with numerous events, many of them devastating, occurring in the Middle East as well as in China. Still, one would not identify the southern San Andreas fault as capable of producing major earthquakes, and seismologists push the historical record back to times preceding instrumental recording by using descriptive reports of historical events and by digging into near-surface faults to examine the history of motions preserved in the soil and rock disrup-

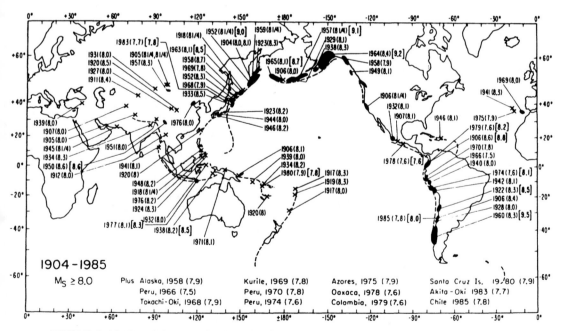

FIGURE 1.13 The global distribution of great earthquakes this century. The location and year of each event are shown, along with an M_S value in parentheses and an M_W value in brackets (if available). The filled areas are the rupture zones of the largest circum-Pacific ruptures. (From Kanamori, 1988.)

TABLE 1.6 Large Earthquakes with $M_S \geq 8.0$ for the Period 1904 to 1992

Date	Time	Region	Lat. (°N)	Long. (°E)	M_S	M_W
1904 06 25	21 00.5	Kamchatka	52	159	8.0 (7.4)	
1905 04 04	00 50.0	E. Kashmir	33	76	8.1 (7.5)	
1905 07 09	09 40.4	Mongolia	49	99	8.4 (7.6)	8.4
1905 07 23	02 46.2	Mongolia	49	98	8.4 (7.7)	8.4
1906 01 31	15 36.0	Ecuador	1	−81.5	8.7 (8.2)	8.8
1906 04 18	13 12.0	California	38	−123	8.3 (7.8)	7.9
1906 08 17	00 10.7	Aleutian Is.	51	179	8.2 (7.8)	
1906 08 17	00 40.0	Chile	−33	−72	8.4 (8.1)	8.2
1906 09 14	16 04.3	New Britain	−7	149	8.1 (7.5)	
1907 04 15	06 08.1	Mexico	17	−100	8.0 (7.7)	
1911 01 03	23 25.8	Turkestan	43.5	77.5	8.4 (7.8)	
1912 05 23	02 24.1	Burma	21	97	8.0 (7.7)	
1914 05 26	14 22.7	W. New Guinea	−2	137	8.0	
1915 05 01	05 00.0	Kurile Is.	47	155	8.0	
1917 06 26	05 49.7	Samoa Is.	−15.5	−173	8.4	
1918 08 15	12 18.2	Mindanao Is.	5.5	123	8.0	
1918 09 07	17 16.2	Kurile Is.	45.5	151.5	8.2	
1919 04 30	07 17.1	Tonga Is.	−19	−172.5	8.2	
1920 06 05	04 21.5	Taiwan	23.5	122	8.0	
1920 12 16	12 05.8	Kansu, China	36	105	8.6	
1922 11 11	04 32.6	Chile	−28.5	−70	8.3	8
1923 02 03	16 01 41	Kamchatka	54	161	8.3	8.5
1923 09 01	02 58 36	Kanto	35.25	139.5	8.2	7.9
1924 04 14	16 20 23	Mindanao	6.5	126.5	8.3	
1928 12 01	04 06 10	Chile	−35	−72	8.0	
1932 05 14	13 11 00	Molucca Passage	0.5	126	8.0	
1932 06 03	10 36 50	Mexico	19.5	−104.25	8.2	
1933 03 02	17 30 54	Sanriku	39.25	144.5	8.5	8.4
1934 01 15	08 43 18	Nepal/India	26.5	86.5	8.3	
1934 07 18	19 40 15	Santa Cruz Is.	−11.75	166.5	8.1	
1938 02 01	19 04 18	Banda Sea	−5.25	130.5	8.2	8.5
1938 11 10	20 18 43	Alaska	55.5	−158.0	8.3	8.2
1939 04 30	02 55 30	Solomon Is.	−10.5	158.5	8.0	
1941 11 25	18 03 55	N. Atlantic	37.5	−18.5	8.2	
1942 08 24	22 50 27	Peru	−15.0	−76.0	8.2	
1944 12 07	04 35 42	Tonanki	33.75	136.0	8.0	8.1
1945 11 27	21 56 50	W. Pakistan	24.5	63.0	8.0	
1946 08 04	17 51 05	Dominican Rep.	19.25	−69.0	8.0	
1946 12 20	19 19 05	Nankaido	32.5	134.5	8.2	8.1
1949 08 22	04 01 11	Queen Char. Is.	53.75	−133.25	8.1	8.1
1950 08 15	14 09 30	Assam	28.5	96.5	8.6	8.6
1951 11 18	09 35 47	Tibet	30.5	91.0	8.0	7.5
1952 03 04	01 22 43	Tokachi-Oki	42.5	143.0	8.3	8.1
1952 11 04	16 58 26	Kamchatka	52.75	159.5	8.2	9.0
1957 03 09	14 22 28	Aleutian Is.	51.3	−175.8	8.1	9.1
1957 12 04	03 37 48	Mongolia	45.2	99.2	8.0	8.1
1958 11 06	22 58 06	Kurile Is.	44.4	148.6	8.1	8.3
1960 05 22	19 11 14	Chile	−38.2	−72.6	8.5	9.5
1963 10 13	05 17 51	Kurile Is.	44.9	149.6	8.1	8.5
1964 03 28	03 36 14	Alaska	61.1	−147.5	8.4	9.2
1965 02 04	05 01 22	Aleutian Is.	51.3	178.6	8.2	8.7
1968 05 16	00 48 57	Tokachi-Oki	40.9	143.4	8.1	8.2
1977 08 19	06 08 55	Sumbawa	−11.2	118.4	8.1	8.3
1985 09 19	13 17 38	Mexico	18.2	−102.6	8.1	8.0
1989 05 23	10 54 46	Macquarie Is.	−52.3	160.6	8.2	8.2

tions. This reveals that the southern San Andreas fault has indeed had great earthquakes, the most recent in 1857, with many previous events recurring about every 130 years. Other great events revealed by historical accounts occurred in regions such as southeastern Missouri, where a sequence struck in 1811–1812, and South Carolina in 1886. In some places, such as Missouri, current small-magnitude seismicity alerts us to the local earthquake potential, whereas in others, like South Carolina, little present activity is occurring. Chapter 11 discusses the earthquake hazard issue further.

1.2.5 Global Faulting Patterns and Rupture Models

In order to understand the distribution and fundamental causes of earthquakes, we must determine the nature of the faulting motions that are involved, but only a few faults rupture the surface to give direct observation of permanent deformations that reveal the fault geometry. Again, the basic properties of seismic waves assist

us greatly. The wavefronts that expand outward from the earthquake source region retain the initial sense of deformation at the source (Figure 1.14), so that, after accounting for propagation effects, we can relate seismogram motions to near-source motions, even though the seismometer may be thousands of kilometers from the source. The near-source motions then define the source geometry, which for earthquakes involves the fault orientation and direction of slip.

Seismological analysis exploits this directional information in the wavefield to find the fault orientations for large earthquakes all over the world, even at inaccessible depths. Figure 1.15 shows the global distribution of shallow (depth less than 70 km) events with $M_{\mathrm{W}} \geq 6.5$ for the year 1989. The circular plots are stereographic projections that show fault-plane orientations and major strain axes, which reveal the sense of motion at the source (described fully in Chapter 8). Most of the large events in this year occurred around the Pacific margin and involved underthrusting of oceanic plate in convergent

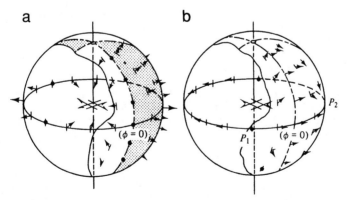

FIGURE 1.14 Elastic waves propagate away from a source with the sense of ground motions being preserved over the wavefront. The directions of P- and S-wave particle motions on the expanding wavefront are shown above, with (a) P-wave motions being perpendicular to the wavefront and reflecting initial motion either toward or away from the source. (b) S-wave motions are parallel to the wavefront, with the shearing direction being controlled by the orientation of the shearing at the source. It thus becomes possible to relate distant motions to near-source motions and determine the source geometry. (From Kasahara, 1981. Reprinted with the permission of Cambridge University Press.)

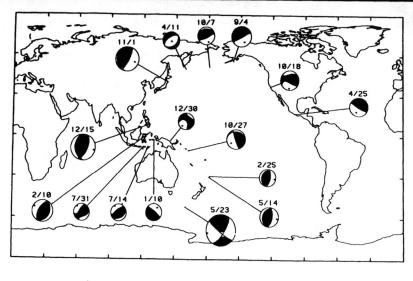

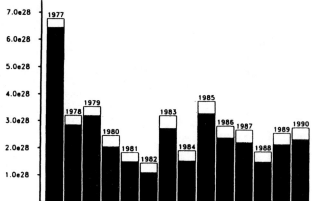

FIGURE 1.15 Top: The source mechanism for all earthquakes with $M_W \geq 6.5$ in 1989. The date and stereographic projection of the P-wave radiation pattern for each event are shown, with the size of the projections scaling with relative moment of the source. The dark areas represent compression (away from the source) motions; the white areas indicate dilatational (toward the source) motions. These source mechanisms are fully described in Chapter 8, but here they can be taken to indicate the direction of faulting associated with each source. Bottom: The annual cumulative seismic moment release from all significant seismic events, which is about 800 events per year. The darker portion of each bar indicates the contribution from just the $M_W \geq 6.5$ events, demonstrating that the small number of large events (about 20/yr) dominates. (From Dziewonski *et al.*, 1990, 1991.)

zones. The event in California is the October 18, 1989, Loma Prieta event, located in the Santa Cruz Mountains between San Jose and Santa Cruz. This event had a moderate magnitude of $M_W = 6.9$ but caused more than $7 billion in damage and killed 68 people. The relative importance of these large events is suggested by the bar graph, which shows the annual cumulative seismic moment release around the world. The darkened portions of these bars indicate the moment release contributed by the events with $M_W \geq 6.5$, which clearly dominate.

The routine determination of earthquake faulting orientation around the world by seismic-wave analysis is one of the remarkable accomplishments of global seismology. Over 10,000 earthquakes have been quantified from 1977 to 1992 by the analysis procedure used in Figure 1.15. As mentioned earlier, this is too short a time span to assess all earthquake phenomena, but it has revolutionized the fundamental data base for studying surface motions. Characterizing the average fault orientation and seismic moment of the events is only the first step. More detailed seismic analysis can be used to determine the full rupture sequence for large events, from onset to termination of faulting. Some recent results are shown in Figure 1.16 for a large earthquake in 1976 in Guatemala that resulted from the rupture of a nearly vertical fault called the Motagua fault. The rupture started at point 1 and spread in

both directions down the fault, with horizontal shearing of the two sides. Detailed analysis of very complicated P waveforms recorded around the world for this event shows that the radiation of energy was not uniform during the rupture and that the orientation of the strain release rotated slightly at different locations on the fault. The complex time history of energy release is a common attribute of large earthquake failures, as is the presence of nonuniform surface displacement along the outcrop of surface-breaking faults. Chapter 10 describes the procedures that are used in such studies.

For earthquakes more recent than 1980, the quality of global seismic data is greatly improved over earlier decades because digital recording systems became widespread, and even great earthquakes pro-

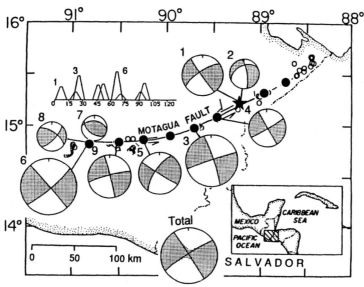

FIGURE 1.16 A map of the Motagua fault, which ruptured in the 1976 Guatemala earthquake. The seismologically determined history of energy release is shown in the upper left. Each pulse corresponds to radiation from different sections of the fault as the rupture spread away from the initiation point (star). Each subevent has a source orientation determined in the analysis, with the projections of the P-wave nodal radiation planes being shown in stereographic projections. Darkened areas represent compressional P-wave motions. The fault orientation changed during rupture, and the strength of radiation was not uniform along the fault. (From Kikuchi and Kanamori, 1991.)

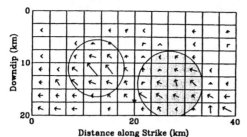

FIGURE 1.17 A model of variable slip on the fault that ruptured in the 1989 Loma Prieta, California, earthquake. This model involves some variation in the amount and direction of slip on the fault. The rupture spread outward from the initiation point at the star in the center of the fault at a depth of 18 km. Slip of the fault took place over a total of about 8 s. Two patches of primary slip are highlighted. This model was obtained by analysis of very nearby (strong motion) and distant (teleseismic) P and S waves. (From Wald *et al.*, 1991.)

duced on-scale seismograms, with greater bandwidth than previously possible. This has enabled even more detailed analysis of seismic ruptures, involving actual contouring of the variable displacement on the fault surface. An example is shown in Figure 1.17, in which seismic recordings at distant and nearby locations have been used to determine a model for the heterogeneous slip distribution on the fault causing the 1989 Loma Prieta earthquake. The data reveal two major patches of dominant slip on the fault, which ruptured in about 8 s. Studies such as these are greatly improving our understanding of earthquake rupture mechanics and are beginning to place earthquake prediction efforts on sounder physical grounds. The broad range of seismic source investigations is described further in the last four chapters of this book.

1.2.6 Radial Earth Layering

The second major branch of global seismology involves studying the structure of the Earth's interior. In order to extract the

types of information about seismic sources described above, it is critical to account for propagation effects, which requires a knowledge of the structure. In addition, most of what we know about the deep interior of the Earth regarding its composition, layering, dynamics, physical state, and temperature has been based on seismic observations of the otherwise inaccessible interior regions. Just as for seismic sources, the remarkable contributions of pioneers early in this century, such as Jeffreys, Bullen, Gutenberg, and Lehmann, solved many of the first-order Earth structure problems, such as demonstrating that the core exists and must be fluid because it does not transmit S-wave energy. But, as is the case for understanding earthquakes, resolving the second-order details is critical to understanding the dynamical processes occurring in the Earth. For example, the presence of several-hundred-degree lateral temperature differences deep in the mantle may produce only a 1% change in seismic velocity but is sufficient to drive convective flow of the interior on long time scales. Similarly, grossly different models of the chemistry of the interior differ in their elasticity parameters by a few percentages or less. Thus, there is an intense effort to determine the internal structure with very high precision, so that the composition and dynamic processes of the interior can be understood.

The key to revealing the internal structure using seismograms is the collection of large numbers of recordings at different distances from a source. A display of seismograms as a function of distance, or *seismic profile*, enables identification of coherent wave arrivals between stations. An example for a global data collection is shown in Figure 1.18. The records show good stability of the travel-time variation of a given wave type as a function of distance in the Earth. For example, at epicentral distances (measured in angular degrees along the Earth's circumference between source and receiver) of less than

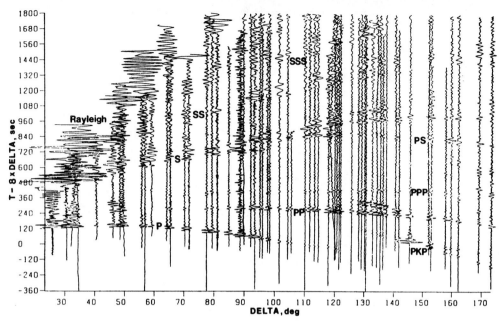

FIGURE 1.18 A collection of vertical-component seismograms for a single event that occurred near Sumatra, plotted at the angular distance to each station. The records are from the World Wide Standardized and Canadian Seismic Networks. Upward motion on each trace is toward the left. Note that coherent arrivals can be tracked from trace to trace. These define the travel-time behavior for different paths through the Earth. The start time of each trace has been reduced by a value of 8Δ s, where Δ is the angular distance. Thus, traces on the right begin much later than traces on the left. (Modified from Müller and Kind, 1976. Reprinted with permission of the Royal Astronomical Society.)

100°, a clear *P*-wave arrival occurs at the onset of ground motions. The disruption of the *P* arrival branch near 100° is due to the low-velocity core of the planet. The systematic timing as a function of distance, or *travel-time curve*, for each seismic phase can be analyzed using inverse theory to determine the internal structure of the Earth. Any radial layering will give rise to reflections and conversions of *P* and *S* waves, and fitting the travel times of later arrivals determines the depths and velocity changes of internal discontinuities.

The observed wavefield is complicated by the existence of both body and surface waves, by conversions and reflections of body waves off the core and other internal discontinuities, by the spherical geometry of the Earth and multiple reflections of body waves off the surface, as well as by

relatively small lateral variations in structure. Over the past three decades immense data bases of travel-time observations have accumulated in the routine process of locating earthquakes around the world. The United States National Earthquake Information Center (NEIC) and the International Seismological Center (ISC) in England compile earthquake bulletins with all travel-time reports from stations around the world. Simply displaying the composite travel times (Figure 1.19a) reveals a global travel-time curve. Each continuous branch of arrivals defines a particular seismic-wave path in the Earth that can be analyzed to reveal layering in the Earth. The travel-time branches are readily identified, and master travel-time curves such as those in Figure 1.19b can be determined for different source depths. Many of the complex

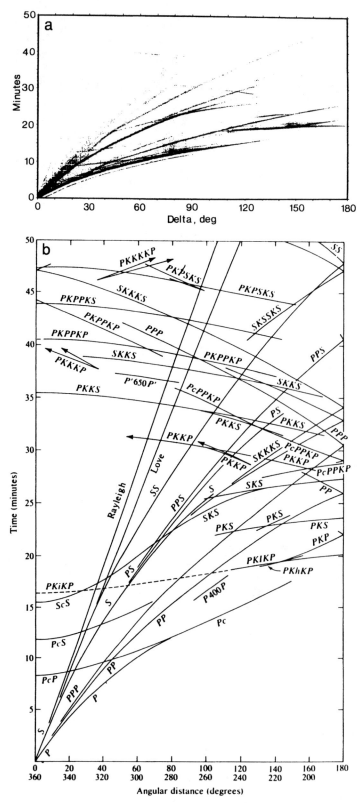

FIGURE 1.19 (a) Travel times versus angular distance from the ISC bulletin for a 5-year period. A total of 2,538,244 arrival times from phases from shallow earthquakes are shown. These define continuous arrival branches. (b) Average travel times for a surface source for various phases. The labeling indicates the general path of each phase. (From Bolt, 1982.)

interactions are low amplitude and can be observed only after stacking many observations at a given distance, which reduces background noise. This is now possible with the large data sets of digital seismograms that are accumulating (Chapter 6).

Once observed travel-time curves for seismic phases are determined, it is possible to invert for *P*- and *S*-wave velocities as a function of depth using the methods described in Chapter 7. Through analysis of body waves, long-period surface waves, and free oscillations, global seismologists have developed one-dimensional models of the elastic velocities and density of the entire Earth. One of the most frequently used models is shown in Figure 1.20. This model, different from the first generation of global models developed in the 1930s in subtle but important ways, indicates the major subdivisions of the interior: the solid inner core, the fluid outer core, the lower

mantle, and the upper mantle. The crust is a very thin veneer on the surface. Chapter 7 discusses the seismological constraints on each region. Radial models of the Earth's elastic structure are used in many applications (including earthquake rupture modeling) and are critical for efforts to determine the composition and state of the interior. However, radial models fail to express the complexity of what we know to be a dynamic, evolving system, so seismology is now striving to develop fully three-dimensional models for the interior at all scale lengths.

1.2.7 Heterogeneous Earth Models

It has long been recognized that simple layered models are a poor approximation of the Earth's crust. The obvious differ-

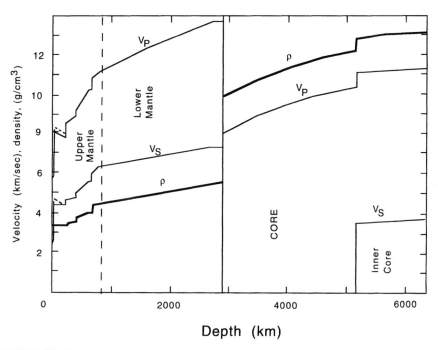

FIGURE 1.20 The Preliminary Reference Earth Model of *P* velocity (V_P), *S* velocity (V_S), and density (ρ) as a function of depth in the Earth. (After Dziewonski and Anderson, 1981.)

ence between oceanic and continental regions is one indication, but the exposed surface geology provides even clearer evidence of complexity. The geological processes that produce layering, such as sediment deposition, lava flows, and chemical precipitation, all do so on limited spatial scales, and subsequent crustal motions deform even the locally stratified rocks. Efforts to study crustal structure, driven on the one hand by resource exploration and on the other by Earth science efforts to understand how the crust evolved, have led to many attempts to develop two- and three-dimensional models for crustal regions. This requires collection of closely spaced seismic data so that coherent seismic arrivals can be detected over small horizontal ranges. An example of a dense *seismic reflection profile* (which shows energy from surface explosions reflected back from the interior) collected in the rift zone of eastern Africa is shown in Figure 1.21. A dense distribution of seismometers and very high frequency recordings are required to see the complex, laterally discontinuous arrivals reflected from deep structure, which shows tilted layering offset by faults. In many crustal locations a two-dimensional model is inadequate to inter-

pret the subsurface, particularly for complex formations that may trap oil; thus three-dimensional images are currently being developed in numerous crustal studies. Development of three-dimensional imaging has awaited and, in part, has driven the development of faster computers with massive data storage capabilities. Such high-resolution seismology efforts are still constrained by computer limitations.

During the 1950s and 1960s, the first computer-assisted analyses of long-period surface waves began to reveal systematic lateral variations in deeper Earth structure below the crust. By the 1980s many global seismologists were actively analyzing different types of seismic data to determine three-dimensional structure at depth by methods collectively identified as *seismic tomography* (based on mathematical similarities to medical imaging tomography, which is used to image internal structure of the human body without surgery), finding that every region of the interior, with the possible exception of the outer core, appears to have detectable aspherical heterogeneity. The ability to resolve this variation about the one-dimensional radial Earth models, and the recognition of its importance for internal dynamics, has

FIGURE 1.21 A seismic reflection profile from the Lake Tanganyika Rift zone. At each distance along the east–west (E–W) line, a stack of seismic traces is plotted vertically downward, with increasing time. The arrivals on the adjacent traces indicate layered structures that are cut and offset by subsurface faults, most of which are not seen at the surface. Exploration seismology develops even more detailed images of shallower structure. (Modified from Rosendahl, 1989.)

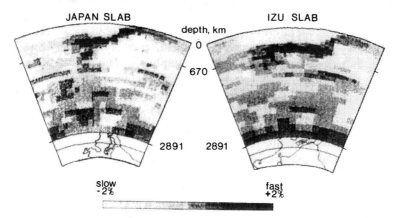

FIGURE 1.22 Vertical cross sections through a three-dimensional model of P-velocity variations in the Earth's mantle, showing seismic velocity heterogeneity in regions of downwelling oceanic lithosphere. The darker regions correspond to material that has a faster than average P velocity, resulting from low temperatures in subducting oceanic plate. Relatively low velocities are found in the wedge above subducting plates, below island arc volcanic areas. The base of the mantle in these regions is also higher velocity than average. (Modified from Fukao *et al.*, 1992.)

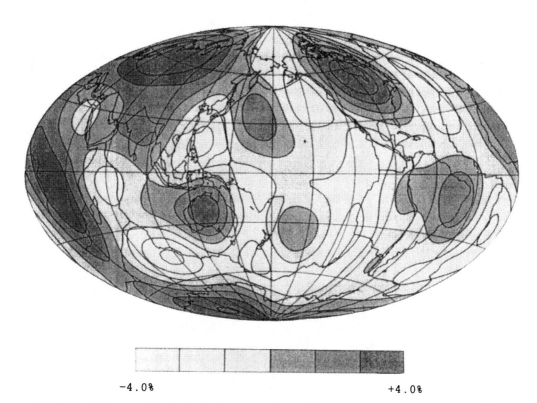

FIGURE 1.23 A model of global shear velocity variations relative to the average shear velocity at a depth of 150 km in the mantle. Darker regions correspond to higher-velocity regions. This model was obtained by analysis of body waves and surface waves, using a truncated spherical harmonic function expansion of the heterogeneity. The model can only resolve fluctuations with scale lengths of 5000 km, so small features like the slabs in Figure 1.22 are not resolved. (From Dziewonski, 1989.)

prompted a revolution in geophysical investigations of the deep interior. Three-dimensional velocity variations are now being determined for localized regions, such as the circum-Pacific downwellings where deep earthquakes occur (Figure 1.22), as well as for global models of shear-velocity structure at all depths in the interior (Figure 1.23). In addition to making deterministic maps of the large-scale structural heterogeneity, seismologists are using wave scattering theory (mostly adapted from quantum mechanics) to describe statistically small-scale heterogeneities that are detectable, but not completely resolvable, by high-frequency seismic waves.

The likelihood that the variations in velocity are at least in part due to thermal variations (higher-velocity material being colder and lower-velocity material being hotter at a given pressure), combined with the fact that any thermal variations cause density variations, suggests that the three-dimensional seismic models reveal density heterogeneity. Density heterogeneity results in long-term stresses (due to gravitational pull) that cause Earth materials to flow, with upwellings and downwellings being driven by gravity as the Earth system transports heat to the surface. Thus, remarkably, imaging the Earth with elastic waves provides a means for determining the ongoing dynamic convection of the mantle. Chapter 7 surveys these Earth structure investigations.

1.2.8 Modern Global Seismology

This introduction should make it clear that modern global seismology is a rapidly advancing, quantitative discipline that addresses a vast array of important physical phenomena in the Earth. There is beauty and elegance in the mathematical procedures used in the discipline and in the richness and complexity of seismological data. The challenge of extracting information from seismic signals continues to draw increasing numbers of researchers into the field, with both applied and basic science emphases. This text develops much of the basic theory and touches upon many of the major observations and results of modern global seismology.

References

Bolt, B. A. (1976). "Nuclear Explosions and Earthquakes: The Parted Veil." Freeman, San Francisco.

Bolt, B. A. (1982). "Inside the Earth: Evidence from Earthquakes." Freeman, San Francisco.

Dziewonski, A. M. (1989). Earth structure, global. In "The Encyclopedia of Solid Earth Geophysics" (D. E. James, ed.), pp. 331–359. Van Nostrand-Reinhold, New York.

Dziewonski, A. M., and Anderson, D. L. (1981). Preliminary reference Earth model. *Phys. Earth Planet. Inter.* **25**, 297–356.

Dziewonski, A. M., Ekström, G., Woodhouse, J. H., and Zwart, G. (1990). Centroid–moment tensor solutions for October–December 1989. *Phys. Earth Planet. Inter.* **62**, 194–207.

Dziewonski, A. M., Ekström, G., Woodhouse, J. H., and Zwart, G. (1991). Centroid–moment tensor solutions for October–December 1990. *Phys. Earth Planet. Inter.* **68**, 201–214.

Fukao, Y., Obayashi, M., Inoue, H., and Nenbai, M. (1992). Subducting slabs stagnant in the mantle transition zone. *J. Geophys. Res.* **97**, 4809–4822.

Kanamori, H. (1977). The energy release in great earthquakes. *JGR, J. Geophys. Res.* **82**, 2981–2987.

Kanamori, H. (1978). Quantification of earthquakes. *Nature (London)* **271**, 411–414.

Kanamori, H. (1988). Importance of historical seismograms for geophysical research. In "Historical Seismograms and Earthquakes of the World" (W. H. K. Lee, ed.). pp. 16–33. Academic Press. San Diego.

Kanamori, H., Mori, J., Anderson, D. L., and Heaton, T. H. (1991). Seismic excitation by the space shuttle Columbia. *Nature (London)* **349**, 781–782.

Kasahara, K. (1981). "Earthquake Mechanics." Cambridge Univ. Press, Cambridge, UK.

Kikuchi, M., and Kanamori, H. (1991). Inversion of complex body waves. III. *Bull. Seismol. Soc. Am.* **81**, 2335–2350.

Müller, G., and Kind, R. (1976). Observed and computed seismogram sections for the whole world. *Geophys. J. R. Astron. Soc.* **44**, 699–716.

Rosendahl, B. R. (1989). Continental rifts: Structural traits. *In* "The Encyclopedia of Solid Earth Geophysics" (D. E. James, ed.), pp. 104–126. Van Nostrand-Reinhold, New York.

Steim, J. M. (1986). The very broadband seismograph. Ph.D. Thesis, Harvard University, Cambridge, MA.

Velasco, A., Lay, T., and Zhang, J. (1993). Long-period surface wave inversion for source parameters of the 18 October, 1989 Loma Prieta earthquake. *Phys. Earth Planet. Inter.* **76**, 43–66.

Wald, D. J., Helmberger, D. V., and Heaton, T. H. (1991). Rupture model of the 1989 Loma Prieta earthquake from the inversion of strong-motion and broadband teleseismic data. *Bull. Seismol. Soc. Am.* **81**, 1540–1572.

2

ELASTICITY AND SEISMIC WAVES

Seismology involves analysis of ground motions produced by energy sources within the Earth, such as earthquake faulting or explosions. Except in the immediate vicinity of the source, most of the ground motion is ephemeral; the ground returns to its initial position after the transient motions have subsided. Vibrations of this type involve small elastic deformations, or *strains*, in response to internal forces in the rock, or *stresses*. The *theory of elasticity* provides mathematical relationships between the stresses and strains in the medium, and it has spawned a vast literature filled with theory and empirical documentation of elastic behavior. Here we develop only the basics of the theory of elasticity required for seismological applications, including the concepts of strain and stress, the equations of equilibrium and motion, and the fundamental nature of solutions to the equations of motion: seismic waves. Chapters 3 and 4 characterize wave interactions relevant to seismic waves in the Earth, and subsequent chapters apply these basic ideas to describe how seismologists study the Earth's interior and the sources of seismic waves.

Our development of elasticity follows that typical of texts on solid mechanics, and many more detailed discussions are available, some being listed in the References. In the study of solids, a useful, idealized concept for dealing with macroscopic phenomena is that of a *continuum*, in which matter is viewed as being continuously distributed in space. Within this continuous material we can define mathematical functions for displacement, strain, or stress fields, which have well-defined continuous spatial derivatives. We will see that applying simple laws of physics to a continuum (*continuum mechanics*) allows seismologists to explain nearly every arrival on a seismogram. We must introduce atomic-scale processes to explain some important aspects of seismology, such as the nature of anelastic-wave attenuation, but even for seismic-wave attenuation, phenomenological adaptations of continuum mechanics usually circumvent the need for detailed characterizations of microscopic phenomena. For seismology, this is critical, for we are, of course, ignorant of most of the detailed crystallographic and atomic-level structure inside the planet.

Seismology, for the most part, is concerned with very small deformations (relative length changes of $\sim 10^{-6}$) over short periods of time (< 3600 s). This greatly simplifies the mathematical framework of our elasticity theory, which is based on *infinitesimal strain theory*. In the immediate vicinity of seismic sources, or when we consider long-term, large-scale deformations of faults (as in structural geology), a more complete finite strain theory must be followed. The relationship between forces and deformations in infinitesimal strain theory is largely empirically based and given by a *constitutive law* called *Hooke's law*. The deformation is a function of material properties of the body such as density, rigidity (resistance to shear), and incompressibility (resistance to change in volume). The material properties are known as *elastic moduli*. When stress varies with time, strain varies similarly, and the balance between stress and strain results in *seismic waves*. These waves travel at velocities that depend on the elastic moduli and are governed by equations of motion. Seismic waves are loci of particle displacements, which become increasingly complex as the wave expands through the solid body. We will now proceed to show how these waves arise and how they are represented mathematically.

2.1 Strain

Because seismology is so directly associated with measurement of motions of a medium, we begin by considering how motions within a solid are described. We employ a *Lagrangian* description, in which the motion of a particular particle is followed as a function of time and space. This is a natural system for seismology, because seismograms are essentially records of particle motions at near-surface sensors as seismic waves pass by. A continuum is a continuous distribution of particles; thus a vector field, $\mathbf{u}(\mathbf{x}, t)$, is required to describe the motions of every point in the medium, where we are free to choose a convenient reference system.

A medium can undergo two fundamental types of motion: (1) whole-body translation and/or rotation, and (2) straining, or internal deformation. Translation and rotation can be described with a single

vector common to all points in the medium, and we are not concerned with such whole-body motions here. Instead, we want to describe internal deformations within the solid, which intrinsically involve spatial and temporal variations of the displacement field, $\mathbf{u}(\mathbf{x}, t)$.

Deformations within a medium are composed of components that involve length changes and angular distortions. Consider a body that is initially undeformed and unloaded with two internal points O and P (Figure 2.1a) connected by a straight line of length Δs. When forces are applied to the body, deformation moves O and P to O' and P', respectively, which are connected by a line with length $\Delta s'$. To describe the deformation of the medium, we must characterize both the change in distance between the two points and any rotation of the line $\Delta s'$ relative to the surrounding material. To do this we introduce terms for *spatial gradients of the displacement field*, or *strains*. *Normal strains* are measures of elongation, defined as

$$\varepsilon_{\text{normal}} = \lim_{\Delta s \to 0} \left(\frac{\Delta s' - \Delta s}{\Delta s} \right). \quad (2.1)$$

Normal strains involve a fractional change in distance between points. Line segment $O'P'$ might not have changed length but might have rotated with respect to the surrounding material. If we consider a perpendicular line segment OQ (Figure 2.1b) in the undistorted medium that moves to $O'Q'$, we can define the *shear strain*, a measure of internal angular distortion, as

$$\varepsilon_{\text{shear}} = \frac{1}{2} \lim_{\substack{\Delta s_1 \to 0 \\ \Delta s_2 \to 0}} \left(\frac{\pi}{2} - \theta' \right), \quad (2.2)$$

where θ' is the angle $\angle Q'O'P'$.

To be useful, the normal and shear strains must be defined with respect to a coordinate system. Since space is three-dimensional and we must describe all elongations and angular changes at every point in the medium with respect to all three dimensions, the full description of

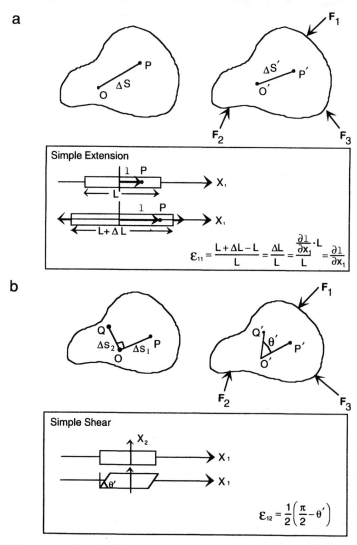

FIGURE 2.1 When a medium is deformed, we must describe both relative length changes and shearing rotations between portions of the medium. Normal strains, involving relative changes in length between points, are considered in (a). Shear strains, involving angular changes within the medium, are considered in (b).

strain involves nine terms: three normal strains, ε_{11}, ε_{22}, ε_{33}, giving relative length changes of line segments oriented in the coordinate directions, and six angular changes of each coordinate direction with respect to the other two directions, ε_{12}, ε_{13}, ε_{21}, ε_{23}, ε_{31}, ε_{32}. These nine terms have a continuous distribution throughout the medium and are functions of time. We will now define these terms for a general three-dimensional case.

2.1.1 Strain–Displacement Relationships

We seek to establish general three-dimensional relationships between nine Cartesian strain components and three Cartesian displacement components (u_1, u_2, u_3). Consider the cubic volume of material with a corner at point P in Figure 2.2, which has sides oriented perpendicular to the coordinate axes x_1, x_2, x_3.

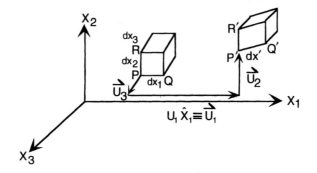

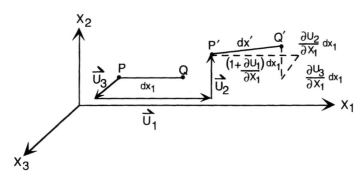

FIGURE 2.2 Displacements of a small cubic volume with a corner at point P to a new position with a corner at point P'. The displacement of P is given by (2.3), and the displacement of Q to Q' is given by (2.4). The length of $P'Q'$ is given to first order by (2.6).

The volume is infinitesimally small, so that when it is deformed, planes remain planar and lines remain straight; this is, by definition, infinitesimal strain. Point P is displaced by the displacement vector

$$\mathbf{u}(\mathbf{x}_P, t) = u_1\hat{\mathbf{x}}_1 + u_2\hat{\mathbf{x}}_2 + u_3\hat{\mathbf{x}}_3$$

$$= (u_1, u_2, u_3), \qquad (2.3)$$

moving it to point P' at time t'. Point Q is displaced to Q' by slightly different displacements, $\mathbf{u}(\mathbf{x}_Q, t)$, which we can relate to $\mathbf{u}(\mathbf{x}_P, t)$ using a first-order Taylor series expansion (omitting terms of order $\partial^2 u_1 / \partial x_1^2$ and higher):

$$\mathbf{u}(\mathbf{x}_Q, t) = \left(u_1 + \frac{\partial u_1}{\partial x_1} dx_1\right)\hat{\mathbf{x}}_1$$

$$+ \left(u_2 + \frac{\partial u_2}{\partial x_1} dx_1\right)\hat{\mathbf{x}}_2$$

$$+ \left(u_3 + \frac{\partial u_3}{\partial x_1} dx_1\right)\hat{\mathbf{x}}_3. \qquad (2.4)$$

From now on, the spatial and temporal dependence of $\mathbf{u}(\mathbf{x}, t)$ and its vector components will be implicitly assumed rather than given as arguments of the functions. If we use the definition of normal strain (2.1) for $\overline{PQ}$ and $\overline{P'Q'}$, we have

$$\varepsilon_{\text{normal}} = \frac{\overline{P'Q'} - \overline{PQ}}{\overline{PQ}} = \frac{dx' - dx_1}{dx_1}$$

$$(dx')^2 = \left[(1 + \varepsilon_{\text{normal}}) dx_1\right]^2. \qquad (2.5)$$

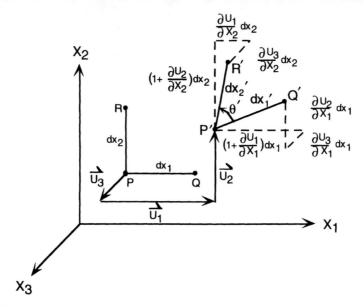

FIGURE 2.3 Angular distortion of the $x_1 x_2$ face of the cube at point P in Figure 2.2. The right angle $\angle RPQ$ is distorted to $\angle R'P'Q'$.

Since $|dx'| = |dx_1 + \mathbf{u}(\mathbf{x}_Q, t) - \mathbf{u}(\mathbf{x}_P, t)|$, from (2.4) and (2.3) we have

$$(dx')^2 = \left[\left(1 + \frac{\partial u_1}{\partial x_1}\right) dx_1\right]^2 + \left(\frac{\partial u_2}{\partial x_1} dx_1\right)^2$$

$$+ \left(\frac{\partial u_3}{\partial x_1} dx_1\right)^2. \tag{2.6}$$

Equating (2.5) and (2.6) and expanding gives

$$\left(1 + 2\varepsilon_{\text{normal}} + \varepsilon_{\text{normal}}^2\right) = 1 + 2\frac{\partial u_1}{\partial x_1}$$

$$+ \left(\frac{\partial u_1}{\partial x_1}\right)^2 + \left(\frac{\partial u_2}{\partial x_1}\right)^2 + \left(\frac{\partial u_3}{\partial x_1}\right)^2$$

$$\tag{2.7}$$

for *small* strains and *small* displacement derivatives; neglecting the squared terms leaves

$$\varepsilon_{\text{normal}} = \frac{\partial u_1}{\partial x_1}. \tag{2.8}$$

This is the same as the result we would find for the one-dimensional case in which PQ would simply change length along the x_1 direction (see Figure 2.1a). We denote this normal strain as

$$\varepsilon_{11} = \frac{\partial u_1}{\partial x_1}. \tag{2.9a}$$

The first subscript indicates the orientation of the line segment, and the second indicates the direction of length change. Similarly, two other normal strains can be defined by

$$\varepsilon_{22} = \frac{\partial u_2}{\partial x_2}, \qquad \varepsilon_{33} = \frac{\partial u_3}{\partial x_3}, \tag{2.9b}$$

corresponding to the other line segments, dx_2 and dx_3, intersecting point P in Figure 2.2. Note that these strain terms implicitly assume the spatial and temporal dependence of the displacement components.

The shear strains are slightly more complex to determine. Referring to Figure 2.3, we see that angle θ' between $\overline{P'Q'}$ and $\overline{P'R'}$

is given by a general law of cosines

$$\cos \theta' = \left[\left(1 + \frac{\partial u_1}{\partial x_1} \right) \frac{dx_1}{dx_1'} \right] \left(\frac{\partial u_1}{\partial x_2} \frac{dx_2}{dx_2'} \right)$$

$$+ \left[\left(1 + \frac{\partial u_2}{\partial x_2} \right) \frac{dx_2}{dx_2'} \right] \left(\frac{\partial u_2}{\partial x_1} \frac{dx_1}{dx_1'} \right)$$

$$+ \left[\left(\frac{\partial u_3}{\partial x_1} \frac{dx_1}{dx_1'} \right) \left(\frac{\partial u_3}{\partial x_2} \frac{dx_2}{dx_2'} \right) \right].$$

$$(2.10)$$

From (2.2), we have

$$\varepsilon_{\text{shear}} = \frac{1}{2} \left(\frac{\pi}{2} - \theta' \right) \approx \frac{1}{2} \sin \left(\frac{\pi}{2} - \theta' \right)$$

$$= \frac{1}{2} \cos \theta', \qquad (2.11)$$

where the approximation is made for small angular changes (i.e., $\theta' \approx \pi/2$). Thus

$$\varepsilon_{\text{shear}} \, dx_2' \, dx_1' = \frac{1}{2} \left[\left(1 + \frac{\partial u_1}{\partial x_1} \right) \frac{\partial u_1}{\partial x_2} \right.$$

$$+ \left(1 + \frac{\partial u_2}{\partial x_2} \right) \frac{\partial u_2}{\partial x_1}$$

$$\left. + \frac{\partial u_3}{\partial x_1} \frac{\partial u_3}{\partial x_2} \right] dx_1 \, dx_2. \quad (2.12)$$

From (2.5) and (2.9) we have

$$dx_1' = (1 + \varepsilon_{11}) \, dx_1$$
$$dx_2' = (1 + \varepsilon_{22}) \, dx_2. \qquad (2.13)$$

Thus,

$$\varepsilon_{\text{shear}} (1 + \varepsilon_{11} + \varepsilon_{22} + \varepsilon_{11}\varepsilon_{22})$$

$$= \frac{1}{2} \left(\frac{\partial u_1}{\partial x_2} + \frac{\partial u_2}{\partial x_1} + \frac{\partial u_1}{\partial x_1} \frac{\partial u_1}{\partial x_2} \right.$$

$$\left. + \frac{\partial u_2}{\partial x_2} \frac{\partial u_2}{\partial x_1} + \frac{\partial u_3}{\partial x_1} \frac{\partial u_3}{\partial x_2} \right). \quad (2.14)$$

Ignoring products of small terms

$$\varepsilon_{\text{shear}} = \frac{1}{2} \left(\frac{\partial u_1}{\partial x_2} + \frac{\partial u_2}{\partial x_1} \right). \quad (2.15)$$

We identify this angular distortion between segments in the x_1 and x_2 directions using indicial notation

$$\varepsilon_{12} = \frac{1}{2} \left(\frac{\partial u_1}{\partial x_2} + \frac{\partial u_2}{\partial x_1} \right). \quad (2.16a)$$

Similarly, we can consider distortions of other faces of the reference cube in Figure 2.2 to find

$$\varepsilon_{21} = \frac{1}{2} \left(\frac{\partial u_2}{\partial x_1} + \frac{\partial u_1}{\partial x_2} \right),$$

$$\varepsilon_{13} = \frac{1}{2} \left(\frac{\partial u_1}{\partial x_3} + \frac{\partial u_3}{\partial x_1} \right),$$

$$\varepsilon_{31} = \frac{1}{2} \left(\frac{\partial u_3}{\partial x_1} + \frac{\partial u_1}{\partial x_3} \right),$$

$$\varepsilon_{23} = \frac{1}{2} \left(\frac{\partial u_2}{\partial x_3} + \frac{\partial u_3}{\partial x_2} \right),$$

$$\varepsilon_{32} = \frac{1}{2} \left(\frac{\partial u_3}{\partial x_2} + \frac{\partial u_2}{\partial x_3} \right). \quad (2.16b)$$

Note that $\varepsilon_{ij} = \varepsilon_{ji}$.

We can represent all nine strain terms of (2.16) with compact indicial notation (see Box 2.1):

$$\varepsilon_{ij} = \frac{1}{2} (u_{i,j} + u_{j,i}) = \frac{1}{2} \left(\frac{\partial u_i}{\partial x_j} + \frac{\partial u_j}{\partial x_i} \right).$$

$$(2.17)$$

These nine terms constitute the *infinitesimal strain tensor*, a symmetric tensor with six independent quantities that can be

Box 2.1 Indicial Notation

The large number (nine) of components of the stress and strain tensors and the proliferation of terms involving their spatial derivatives make it useful to adopt a simplifying notation. We follow a conventional indicial notation. In general, the stress and strain components are prescribed with respect to some convenient reference system (e.g., the Cartesian system, x_1, x_2, x_3), and we use subscripts to indicate surfaces and directions in the reference system. A surface (e.g., the $x_2 x_3$ plane) can be indicated by the direction of the normal to the surface ($\pm x_1$). Direction, such as components of a vector, can be indicated by subscripts as follows:

$$\mathbf{u} = u_1 \hat{\mathbf{x}}_1 + u_2 \hat{\mathbf{x}}_2 + u_3 \hat{\mathbf{x}}_3 \qquad (2.1.1)$$

where $\hat{\mathbf{x}}_i$ are unit vectors in the coordinate directions. The term u_i is understood to take on values $i = 1, 2, 3$, as appropriate in a given equation. For example, the nine terms of the displacement gradient can be represented by a single indicial term:

$$\frac{\partial u_i}{\partial x_j} \begin{pmatrix} i = 1,2,3 \\ j = 1,2,3 \end{pmatrix} : \begin{cases} \dfrac{\partial u_1}{\partial x_1} & \dfrac{\partial u_1}{\partial x_2} & \dfrac{\partial u_1}{\partial x_3} \\[2mm] \dfrac{\partial u_2}{\partial x_1} & \dfrac{\partial u_2}{\partial x_2} & \dfrac{\partial u_2}{\partial x_3} \\[2mm] \dfrac{\partial u_3}{\partial x_1} & \dfrac{\partial u_3}{\partial x_2} & \dfrac{\partial u_3}{\partial x_3} \end{cases} \qquad (2.1.2)$$

where the indicial representation denotes the appropriate component for given values of i and j. This can be written even more compactly as

$$u_{i,j} \equiv \frac{\partial u_i}{\partial x_j} \qquad (2.1.3)$$

continues

ordered as

$$\varepsilon_{ij} = \begin{bmatrix} \dfrac{\partial u_1}{\partial x_1} & \dfrac{1}{2}\left(\dfrac{\partial u_2}{\partial x_1} + \dfrac{\partial u_1}{\partial x_2}\right) & \dfrac{1}{2}\left(\dfrac{\partial u_1}{\partial x_3} + \dfrac{\partial u_3}{\partial x_1}\right) \\[3mm] \dfrac{1}{2}\left(\dfrac{\partial u_2}{\partial x_1} + \dfrac{\partial u_1}{\partial x_2}\right) & \dfrac{\partial u_2}{\partial x_2} & \dfrac{1}{2}\left(\dfrac{\partial u_2}{\partial x_3} + \dfrac{\partial u_3}{\partial x_2}\right) \\[3mm] \dfrac{1}{2}\left(\dfrac{\partial u_1}{\partial x_3} + \dfrac{\partial u_3}{\partial x_1}\right) & \dfrac{1}{2}\left(\dfrac{\partial u_2}{\partial x_3} + \dfrac{\partial u_3}{\partial x_2}\right) & \dfrac{\partial u_3}{\partial x_3} \end{bmatrix}. \qquad (2.18)$$

Special functions such as the *Kronecker delta function* also benefit from indicial notation:

$$\delta_{ij} = \begin{cases} 0 & \text{for } i \neq j \\ 1 & \text{for } i = j \end{cases} \qquad i, j = 1, 2, 3 \qquad (2.1.4)$$

Throughout this text we assume the *Einstein summation notation*, in which repetition of indices within a term explicitly requires summation on that term. Thus, for a term such as

$$\Delta = \varepsilon_{11} + \varepsilon_{22} + \varepsilon_{33} = \varepsilon_{nn} \qquad (2.1.5)$$

the repeated index n implies summation. This holds for repeated indices within any single term:

$$x_i y_i = x_1 y_1 + x_2 y_2 + x_3 y_3$$

$$y_{i,i} = \frac{\partial y_1}{\partial x_1} + \frac{\partial y_2}{\partial x_2} + \frac{\partial y_3}{\partial x_3} \qquad (2.1.6)$$

When a single equation is written with indicial notation, generally a set of equations is implied, as the indices assume all of their permutations. For example, the generalized, linear, isotropic, elastic Hooke's law relating stress (σ_{ij}) and strain (ε_{ij}) terms is given by nine equations:

$$\sigma_{11} = \lambda(\varepsilon_{11} + \varepsilon_{22} + \varepsilon_{33}) + 2\mu\varepsilon_{11}$$

$$\sigma_{12} = 2\mu\varepsilon_{12}$$

$$\sigma_{13} = 2\mu\varepsilon_{13}$$

$$\sigma_{21} = 2\mu\varepsilon_{21}$$

$$\sigma_{22} = \lambda(\varepsilon_{11} + \varepsilon_{22} + \varepsilon_{33}) + 2\mu\varepsilon_{22} \qquad (2.1.7)$$

$$\sigma_{23} = 2\mu\varepsilon_{23}$$

$$\sigma_{31} = 2\mu\varepsilon_{31}$$

$$\sigma_{32} = 2\mu\varepsilon_{32}$$

$$\sigma_{33} = \lambda(\varepsilon_{11} + \varepsilon_{22} + \varepsilon_{33}) + 2\mu\varepsilon_{33},$$

which can be written as

$$\sigma_{ij} = \lambda(\varepsilon_{kk})\delta_{ij} + 2\mu\varepsilon_{ij}, \qquad (2.1.8)$$

where it is assumed that all terms $i, j = 1, 2, 3$ are explicitly considered.

Tensors are quantities that obey certain relations upon transformation of coordinate systems, as will be discussed later. Note that the strain components depend linearly on derivatives of the displacement components, a result of permitting only very small strains and small spatial derivatives in the displacement field. The strains do not depend on the absolute value of the displacements and are unitless. The normal strain terms involve volumetric changes, being compressional when negative and extensional when positive. The trace of the strain tensor is called the *cubic dilatation*, θ,

$$\theta = \varepsilon_{ii} = \frac{\partial u_1}{\partial x_1} + \frac{\partial u_2}{\partial x_2} + \frac{\partial u_3}{\partial x_3} = \nabla \cdot \mathbf{u}.$$

$$(2.19)$$

This corresponds to a fractional change in volume from $V_0 = dx_1\, dx_2\, dx_3$ to $V_1 = [(1 + \varepsilon_{11})\, dx_1 (1 + \varepsilon_{22})\, dx_2 (1 + \varepsilon_{33})\, dx_3]$, given by

$$\frac{\Delta V}{V_0} = \frac{V_1 - V_0}{V_0} \approx \varepsilon_{11} + \varepsilon_{22} + \varepsilon_{33} = \theta.$$

$$(2.20)$$

For reference we note that rigid-body rotations of the medium are expressed as

$$\frac{1}{2}\nabla \times \mathbf{u} = \frac{1}{2}\left[\left(\frac{\partial u_3}{\partial x_2} - \frac{\partial u_2}{\partial u_3}\right)\hat{\mathbf{x}}_1\right.$$

$$+ \left(\frac{\partial x_1}{\partial x_3} - \frac{\partial u_3}{\partial x_1}\right)\hat{\mathbf{x}}_2$$

$$\left. + \left(\frac{\partial u_2}{\partial x_1} - \frac{\partial u_1}{\partial x_2}\right)\hat{\mathbf{x}}_3\right], \quad (2.21)$$

which includes combinations of displacement gradients not in the strain tensor.

2.2 Stress

When a continuum is acted upon by a force, either internal or external, that force influences every point in the body. This requires a distribution of forces throughout the body. Two types of forces occur within a continuum, *body forces* and *contact forces*. Body forces are proportional to the volume of the material. The most common body force results from the acceleration due to gravity, $\mathbf{F} = m\mathbf{g}$, where the mass m depends on the volume and density of material. Contact forces are forces that depend on surface area. For example, the wind resistance a bicyclist experiences is a contact force because it depends on the cross-sectional area of the rider. Body forces have dimensions of force per unit volume; contact forces have units of force per unit area.

For a continuum that is acted on by external forces, internal contact forces must act within the medium. We visualize the medium as having an internally distributed force system, as illustrated in Figure 2.4. Imagine a plane that passes through the medium, intersecting an internal point P. If we remove one side of the medium, it is clear that maintaining the other side in equilibrium requires a distribution of forces on the plane that correspond to actual internal forces within the body. The precise force geometry depends on the direction of the fictitious plane, but some geometric consistency must exist among the various representations of all possible internal force distributions.

We subdivide the area of our fictitious plane through the medium into area elements with surface area ΔA and vector normal $\mathbf{n}$. A small force $\Delta \mathbf{F}$ acts on each element, one of which contains the point P. We define the *stress vector* or *traction vector*, $\mathbf{T(n)}$, to be

$$\mathbf{T(n)} = \lim_{\Delta A \to 0} \frac{\Delta \mathbf{F}}{\Delta A} = T_1\hat{\mathbf{x}}_1 + T_2\hat{\mathbf{x}}_2 + T_3\hat{\mathbf{x}}_3,$$

$$(2.22)$$

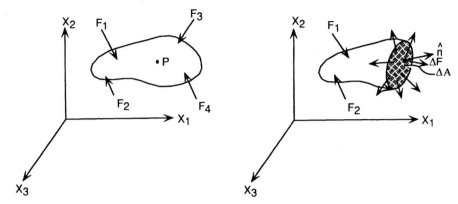

FIGURE 2.4 (Left) A continuum acted upon by external forces. (Right) Imaginary plane with normal, **n**, passing through an internal point P. A portion of the medium has been removed and replaced by a distribution of forces acting on the surface, keeping the remainder of the continuum in equilibrium. This leads to definition of internal forces and stresses on arbitrary surfaces in the medium.

which acts on the surface element at P with normal **n**. The limit is defined for the continuum model, which visualizes a continuous distribution of internal forces. Stress has physical dimensions of force per unit area and corresponds to action of part of the medium upon the other. Since our imaginary plane is arbitrary, we can choose it so that it is parallel to the $x_2 x_3$ plane for any choice of x_1. We define the stress components acting on this plane ($x_1 =$ constant), which is called the x_1 face (it has a normal in the x_1 direction), by

$$\sigma_{11} = \lim_{\Delta A_1 \to 0} \frac{\Delta F_1}{\Delta A_1}$$

$$\sigma_{12} = \lim_{\Delta A_1 \to 0} \frac{\Delta F_2}{\Delta A_1} \qquad (2.23)$$

$$\sigma_{13} = \lim_{\Delta A_1 \to 0} \frac{\Delta F_3}{\Delta A_1},$$

where

$$\Delta \mathbf{F} = \Delta F_1 \hat{\mathbf{x}}_1 + \Delta F_2 \hat{\mathbf{x}}_2 + \Delta F_3 \hat{\mathbf{x}}_3. \quad (2.24)$$

The first index of σ_{ij} in (2.23) corresponds to the direction of the normal to the plane

being acted on by the force, and the second index indicates the direction of the force. Thus σ_{11} is a stress acting normal to the plane, and σ_{12} and σ_{13} are stresses acting in the plane.

By passing two other planes through point P parallel to the $x_1 x_2$ and $x_1 x_3$ planes, we define six additional stress components

$$\sigma_{22}, \sigma_{21}, \sigma_{23} \quad \text{acting on the } x_2 \text{ face}$$

$$\sigma_{33}, \sigma_{31}, \sigma_{32} \quad \text{acting on the } x_3 \text{ face.}$$

All of these are implicitly functions of space and time. Do we need all nine terms? The answer is yes, because we want to be able to represent the complete internal force distribution at point P in sufficient generality for any possible surface that intersects P. We demonstrate this by balancing forces on a tetrahedron with three faces parallel to the coordinate planes and a fourth face with an arbitrary orientation with normal **n** (Figure 2.5). For the body to be in equilibrium, the sum of the forces on it, and the sum of the moments, must be zero. Note that we adopt a positive sign convention for stress components that are positively directed forces acting on positive

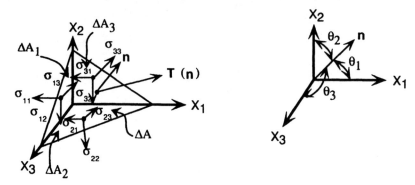

FIGURE 2.5 Balance of forces on a tetrahedron with three faces parallel to coordinate planes and an arbitrarily oriented fourth face with normal **n**. The direction of **n** is specified by direction cosines of the angles shown on the right.

faces (faces with normals in the $+x_i$ directions) and for negatively directed forces on negative faces.

In terms of the direction cosines defined in Figure 2.5, $\Delta A_i = \Delta A \cos \theta_i$, and $\mathbf{n} = n_1 \hat{\mathbf{x}}_1 + n_2 \hat{\mathbf{x}}_2 + n_3 \hat{\mathbf{x}}_3 = \cos \theta_1 \, \hat{\mathbf{x}}_1 + \cos \theta_2 \, \hat{\mathbf{x}}_2 + \cos \theta_3 \hat{\mathbf{x}}_3$. Balancing forces (stress × area) in the x_1 direction gives

$$\Sigma F_{x_1} = 0 = T_1 \Delta A - \sigma_{11} \Delta A \cos \theta_1$$

$$- \sigma_{21} \Delta A \cos \theta_2 - \sigma_{31} \Delta A \cos \theta_3 \tag{2.25}$$

$$T_1 = \sigma_{11} n_1 + \sigma_{21} n_2 + \sigma_{31} n_3. \tag{2.26a}$$

Similarly, letting $\Sigma F_{x_2} = \Sigma F_{x_3} = 0$ gives

$$T_2 = \sigma_{22} n_2 + \sigma_{12} n_1 + \sigma_{32} n_3 \tag{2.26b}$$

$$T_3 = \sigma_{33} n_3 + \sigma_{13} n_1 + \sigma_{23} n_2. \tag{2.26c}$$

or generally

$$T_i = \sigma_{ji} n_j. \tag{2.27}$$

Thus, we can linearly combine our nine components of stress defined in the coordinate planes to represent the stress on any arbitrarily oriented surface through the medium, and in general the state of stress of P depends on all nine terms.

This result leads us to define the *stress tensor* σ_{ij}

$$\sigma_{ij} = \begin{pmatrix} \sigma_{11} & \sigma_{12} & \sigma_{13} \\ \sigma_{21} & \sigma_{22} & \sigma_{23} \\ \sigma_{31} & \sigma_{32} & \sigma_{33} \end{pmatrix}. \tag{2.28}$$

The diagonal terms are called normal stresses, and the off-diagonal terms are called shear stresses. Normal stresses with positive values (directed outward from positive or negative faces as defined above) are called tensional stresses, and negative values correspond to compressional stresses. The common geophysical units for stress are *bars* (10^6 dyn/cm^2), where atmospheric pressure at sea level is approximately 1 bar. At a depth of 3–4 km in the crust the confining stress is on the order of 1 kbar. In SI units, stress is given in pascals (Pa), where 10^6 Pa = 1 MPa = 10 bars. The state of stress at depth in the Earth is nearly always compressional, and therefore all three normal stresses in (2.28) are negative. The *maximum compressive stress* is the stress with the largest absolute value, and the *minimum compressive stress* is the stress with the smallest absolute value.

Consider a cubic element in the continuum bounded by faces paralleling the co-

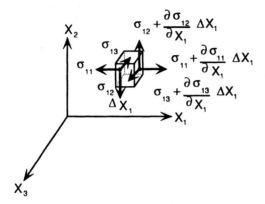

FIGURE 2.6 A cubic element in the continuum bounded by faces paralleling the coordinate planes. Balancing the stresses on each face acting in a given direction leads to the equation of equilibrium. Only the stresses acting on the $\pm x_1$ face are shown. Similar stress terms act on the other four faces.

ordinate planes (Figure 2.6). Let us assume that the cube is in static equilibrium. Then summing all of the forces that act in the x_1 direction gives

$$\sum F_{x_1} = \left[\left(\sigma_{11} + \frac{\partial \sigma_{11}}{\partial x_1} \Delta x_1 - \sigma_{11}\right) \Delta x_2 \Delta x_3\right.$$

$$+ \left(\sigma_{21} + \frac{\partial \sigma_{21}}{\partial x_2} \Delta x_2 - \sigma_{21}\right) \Delta x_1 \Delta x_3$$

$$+ \left.\left(\sigma_{31} + \frac{\partial \sigma_{31}}{\partial x_3} \Delta x_3 - \sigma_{31}\right) \Delta x_1 \Delta x_2\right] = 0$$

or

$$\left(\frac{\partial \sigma_{11}}{\partial x_1} + \frac{\partial \sigma_{21}}{\partial x_2} + \frac{\partial \sigma_{31}}{\partial x_3}\right) = 0. \quad (2.29a)$$

Similarly, letting $\sum F_{x_2} = \sum F_{x_3} = 0$ gives

$$\left(\frac{\partial \sigma_{12}}{\partial x_1} + \frac{\partial \sigma_{22}}{\partial x_2} + \frac{\partial \sigma_{32}}{\partial x_3}\right) = 0 \quad (2.29b)$$

$$\left(\frac{\partial \sigma_{13}}{\partial x_1} + \frac{\partial \sigma_{23}}{\partial x_2} + \frac{\partial \sigma_{33}}{\partial x_3}\right) = 0 \quad (2.29c)$$

or compactly,

$$\frac{\partial \sigma_{ij}}{\partial x_i} = 0 \qquad i, j = 1, 2, 3. \quad (2.30)$$

These are the *equilibrium equations*. These equations require a balance of spatial gradients of the stresses in a medium for that medium to be in stable equilibrium.

A second condition of equilibrium is that the *moments* sum to zero. Consider σ_{12} on either side of the elemental cube. The stresses are oppositely directed (no net force), thus introducing a rotational moment. Moments are given by the product of a force times the perpendicular distance from the force to a reference point. If we sum the moments about lines passing through the center of the cube in Figure 2.6 paralleling the coordinate axes, we obtain equations such as

$$\sum M_{x_3} = \left[\left(\sigma_{12} + \frac{\partial \sigma_{12}}{\partial x_1} \Delta x_1\right.\right.$$

$$\left.+ \sigma_{12}\right) \Delta x_2 \Delta x_3 \frac{\Delta x_1}{2}$$

$$- \left(\sigma_{21} + \frac{\partial \sigma_{21}}{\partial x_2} \Delta x_2\right.$$

$$\left.\left.+ \sigma_{21}\right) \frac{\Delta x_2}{2} \Delta x_3 \Delta x_1\right] = 0$$

or

$$2\sigma_{12} + \frac{\partial \sigma_{12}}{\partial x_1} \Delta x_1 - 2\sigma_{21} - \frac{\partial \sigma_{21}}{\partial x_2} \Delta x_2 = 0.$$

$$(2.31)$$

As $\Delta x_1, \Delta x_2 \to 0$, we have $\sigma_{12} = \sigma_{21}$. Similarly, letting $\sum M_{x_2} = \sum M_{x_1} = 0$ gives $\sigma_{13} = \sigma_{31}$ and $\sigma_{23} = \sigma_{32}$, or generally

$$\sigma_{ij} = \sigma_{ji} \quad (2.32)$$

This states that the stress tensor is symmetric, which reduces the number of independent components to six.

We have seen that both stress and strain are second-order tensors. A scalar is a zeroth-order tensor (magnitude, no directional property), and a vector is a first-order tensor (magnitude and directionality). Second-order tensors define interactions between vectors and directional operators, such as the orientation of the reference plane for definition of stress components. We can show that at each point in a body, three mutually perpendicular planes occur on which no shear-stress components act. This is called the principal coordinate system and is found by diagonalizing the stress tensor, as described in Box 2.2. The normals to the three planes are called *principal stress axes*. Similarly, three mutually perpendicular axes remain perpendicular for infinitesimal strains and are called the *principal axes of strain*. The trace of the stress tensor is invariant to choice of coordinate system and is related to the total stress state. The hydrostatic

---

Box 2.2 Tensor Invariants

Stress and strain are symmetric tensors (i.e., $\sigma_{ij} = \sigma_{ji}$) and thus can be diagonalized, or rotated into a principal coordinate system. Consider the stress tensor:

$$\sigma_{ij} = \begin{pmatrix} \sigma_{11} & \sigma_{12} & \sigma_{13} \\ \sigma_{21} & \sigma_{22} & \sigma_{23} \\ \sigma_{31} & \sigma_{32} & \sigma_{33} \end{pmatrix}. \tag{2.28}$$

This matrix can be diagonalized by subtracting λ from the elements of the trace, setting the determinant of the resulting matrix equal to 0, and solving for λ:

$$\begin{vmatrix} \sigma_{11} - \lambda & \sigma_{12} & \sigma_{13} \\ \sigma_{21} & \sigma_{22} - \lambda & \sigma_{23} \\ \sigma_{31} & \sigma_{32} & \sigma_{33} - \lambda \end{vmatrix} = 0, \tag{2.2.1}$$

which gives

$$\lambda^3 - \text{tr}(\sigma_{ij})\lambda^2 + \text{minor}(\sigma_{ij})\lambda - \det(\sigma_{ij}) = 0, \tag{2.2.2}$$

where $\text{tr}(\sigma_{ij}) = \sigma_{11} + \sigma_{22} + \sigma_{33}$, the trace of the original tensor matrix, $\text{minor}(\sigma_{ij})$ is the sum of the minors of the matrix $(\sigma_{11}\sigma_{22} + \sigma_{22}\sigma_{33} + \sigma_{11}\sigma_{33} - \sigma_{21}^2 - \sigma_{32}^2 - \sigma_{31}^2)$, and $\det(\sigma_{ij})$ is the determinant of the matrix $(\sigma_{11}\sigma_{22}\sigma_{33} + 2\sigma_{21}\sigma_{32}\sigma_{31} - \sigma_{11}\sigma_{32}^2 - \sigma_{22}\sigma_{31}^2 - \sigma_{33}\sigma_{21}^2)$. The parameter λ is called the *eigenvalue* and represents the values of σ_{ij} in a principal coordinate system. The symmetry of the matrix σ_{ij} ensures that the roots of (2.2.2) are real. Because the eigenvalues of a matrix are unchanged by a coordinate transformation, the coefficients of the cubic equation (2.2.2) are *invariant*. This means that the trace, minor, and determinant of the tensor are also independent of the coordinate system and, in general, have some special physical significance.

continues

Each eigenvalue has a corresponding *eigenvector*. The eigenvectors give the principal coordinate axis "directions." We can find the eigenvectors by solving the equation:

$$\begin{bmatrix} \sigma_{11} - \lambda & \sigma_{12} & \sigma_{13} \\ \sigma_{21} & \sigma_{22} - \lambda & \sigma_{23} \\ \sigma_{31} & \sigma_{32} & \sigma_{33} - \lambda \end{bmatrix} \begin{bmatrix} x_1 \\ x_2 \\ x_3 \end{bmatrix} = 0, \qquad (2.2.3)$$

where λ is one of the three roots.

stress is defined as the average of the normal stresses:

$$P = \frac{\sigma_{11} + \sigma_{22} + \sigma_{33}}{3}. \qquad (2.33)$$

The deviatoric stress is that part of the stress tensor minus the hydrostatic term

$$D_{ij} = \sigma_{ij} - P\,\delta_{ij}. \qquad (2.34)$$

A final property of the stress and strain tensors is that they obey specific rules when a coordinate system is rotated, clearly a desirable property for our generally defined terms. If we let $l_{ij} = \cos_{ij}$ be defined as the direction cosines between the new x_i' axes and the old x_j axes, then stress components obey a general transformation law given by

$$\sigma_{ij}' = l_{ip}l_{jq}\sigma_{pq}. \qquad (2.35)$$

And the strain transformation law is

$$\varepsilon_{ij}' = l_{ip}l_{jq}\varepsilon_{pq}. \qquad (2.36)$$

Physical fields that transform in this specific manner upon rotation of coordinate axes are second-order tensors.

2.3 Equation of Motion

We now consider a force balance on a cubic element in a continuum that is undergoing internal motions. Referring to Figure 2.6, the equilibrium equations (2.29) must now include inertial terms as well as any contributions from body forces. We allow the cube in Figure 2.6 to be acted on by a body force per unit volume $\mathbf{f} = f_1\hat{\mathbf{x}}_1 + f_2\hat{\mathbf{x}}_2 + f_3\hat{\mathbf{x}}_3$. The density of the material is given by ρ. Applying Newton's law to the medium gives

$$\rho\frac{\partial^2 u_i}{\partial t^2} = f_i + \frac{\partial \sigma_{ij}}{\partial x_j}. \qquad (2.37)$$

This set of three equations is called the *equation of motion* for a continuum. The inertial terms on the left relate the density-weighted accelerations to body forces and stress gradients in the medium. This is the most fundamental equation underlying the theory of seismology, as it relates forces in the medium to measurable displacements. We will see in Chapter 8 that many seismic sources can be represented by body forces that are introduced into (2.37) to fully describe resulting motions. Later in this text we will often denote derivatives with respect to time by overdots, $\partial u/\partial t = \dot{u}$, $\partial^2 u/\partial t^2 = \ddot{u}$, so the equation of motion is often found in the form

$$\rho \ddot{u}_i = f_i + \sigma_{ij,j} \qquad (2.38)$$

or in the case in which sources or body forces such as gravity are not being considered, the *homogeneous equation of motion*:

$$\rho \ddot{u}_i = \sigma_{ij,j}. \qquad (2.39)$$

In order to proceed, we need relationships between stress and displacement. There are provided by constitutive laws that relate stress to strain and hence stress to displacement gradients. In any given material, a complex relationship exists between stress and deformation, depending on parameters such as pressure, temperature, stress rate, strain history, and stress magnitude. Nearly all Earth materials flow ductilely if small, steady stresses are applied for millions of years, or they fracture or fail plastically if high stresses are applied. However, for the small-magnitude, short-duration stresses of interest in seismology, almost all Earth materials display a linear proportionality between stress and strain. This has been demonstrated empirically by applying controlled forces to rock samples and observing resulting stress–strain behavior, as shown in Figure 2.7. Note that there is a *substantial*, nearly linear interval prior to failure of the rock and that for the small strains $(10^{-5}\text{–}10^{-4})$ being considered here, this rock sample could well be represented by a linear elastic relationship (*elastic* meaning that reducing the small stress restores the medium to its original state).

The most general form of a constitutive law for linear elasticity is *Hooke's law*

$$\sigma_{ij} = C_{ijkl}\varepsilon_{kl}. \qquad (2.40)$$

The constants of proportionality, C_{ijkl}, are known as *elastic moduli* and define the material properties of the medium. In its general form, C_{ijkl} is a third-order tensor

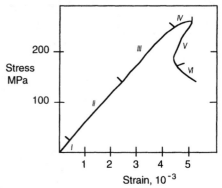

FIGURE 2.7 Stress–strain curve for a typical uniaxial compression test. Stage I involves closure of cracks; stage II is a linear elastic regime; stages III and IV involve dilatancy of the rock due to lateral expansion of the rock and microcracking; stage V involves loss of load-bearing capacity, strain localization, and development of a macroscopic shear failure; and stage VI has stress determined by residual friction on the shear zone. (Modified from Scholz, 1990.)

with 81 terms relating the nine elements of the strain tensor to the nine elements of the stress tensor by a linear sum. Note the double repeated indices in (2.40), for which we write out just the first term

$$\sigma_{11} = C_{1111}\varepsilon_{11} + C_{1112}\varepsilon_{12} + C_{1113}\varepsilon_{13}$$

$$+ C_{1121}\varepsilon_{21} + C_{1122}\varepsilon_{22}$$

$$+ C_{1123}\varepsilon_{23} + C_{1131}\varepsilon_{31}$$

$$+ C_{1132}\varepsilon_{32} + C_{1133}\varepsilon_{33}. \qquad (2.41)$$

There are nine such equations, but the symmetry of the stress and strain tensors $(\varepsilon_{ij} = \varepsilon_{ji};\ \sigma_{ij} = \sigma_{ji})$ reduces the number of independent equations to six and the number of independent coefficients to 36 $(\sigma_{ij} = \sigma_{ji} \rightarrow C_{ijkl} = C_{jikl};\ \varepsilon_{kl} = \varepsilon_{lk} \rightarrow C_{ijkl} = C_{ijlk})$. A further symmetry relation $(C_{ijkl} = C_{klij})$ follows from consideration of a strain energy density function (see Malvern, 1969), leaving 21 elastic moduli in the most general elastic material, which has *general anisotropy*, meaning the stress–strain behavior depends on the orientation of the sample.

Fortunately, the elastic properties for many materials and material composites in the Earth are independent of direction or orientation of the sample. It is possible to show (see Malvern, 1969) that an *isotropic* elastic substance has only two independent elastic moduli, called the *Lamé constants*, λ and μ. These are related to C_{ijkl} by

$$C_{ijkl} = \lambda \delta_{ij}\delta_{kl} + \mu(\delta_{ik}\delta_{jl} + \delta_{il}\delta_{jk}), \quad (2.42)$$

where the Kronecker delta function is used. For example, $C_{1111} = \lambda + 2\mu$, $C_{1122} = \lambda$, $C_{1212} = \mu$, etc. Inserting this into (2.40) gives

$$\sigma_{ij} = \left[\lambda \delta_{ij}\delta_{kl} + \mu(\delta_{ik}\delta_{jl} + \delta_{il}\delta_{jk})\right]\varepsilon_{kl}, \quad (2.43)$$

which reduces (e.g., $\delta_{kl}\varepsilon_{kl} = \varepsilon_{kk}$) to

$$\sigma_{ij} = \lambda \varepsilon_{kk}\delta_{ij} + 2\mu\varepsilon_{ij} = \lambda\theta\delta_{ij} + 2\mu\varepsilon_{ij}. \quad (2.44)$$

This form of Hooke's law for an isotropic linear elastic material was actually formulated by Navier in 1821 and Cauchy in 1823, 160 years after Hooke's work. The significance of the *shear modulus*, or *rigidity*, μ, is readily apparent as a measure of resistance to shear stress ($\sigma_{12} = 2\mu\varepsilon_{12}$; $\sigma_{13} = 2\mu\varepsilon_{13}$, etc.). For a fluid, $\mu = 0$, and for increasing values of μ, the body deforms less under stress. The second Lamé parameter, λ, is most significant in combination with other terms. Table 2.1 defines five elastic moduli that have simple physical attributes in terms of λ and μ. These include E (Young's modulus), k (bulk modulus or incompressibility), and ν (Poisson's ratio). For most seismological applications, λ or k and μ are used, with k and μ being tabulated functions for Earth parameters (see Chapter 7). For

many Earth materials, $\mu \approx \lambda$, and when they are exactly equal the material is called a *Poisson solid*, for which $\nu = 0.25$ and $k = 5/3\mu$. Table 2.2 gives algebraic relationships between the various moduli, and Table 2.3 indicates near-surface values of elastic moduli for common Earth materials. Hooke's law can be written in terms of strain components as well:

$$\varepsilon_{ij} = \frac{-\lambda\delta_{ij}}{2\mu(3\lambda + 2\mu)}\sigma_{kk} + \frac{1}{2\mu}\sigma_{ij}. \quad (2.45)$$

Introduction of Hooke's law into the equation of motion allows us to derive basic equations for displacement fields in an isotropic linear elastic material. These are extremely useful equations, but before we proceed, it is important to note that many Earth materials are in fact not isotropic, and even average upper-mantle properties require anisotropic representations. This occurs mainly because olivine, a major mineral in the upper mantle, is intrinsically very anisotropic, with elastic moduli varying by 10%, depending on orientation of the crystal. Some sedimentary rocks have fabrics that give rise to 25% anisotropy of elastic moduli. Although anisotropy can be fully analyzed, we proceed to develop our theory of seismic waves in the context of isotropic materials because it is simpler algebraically. We will return to a discussion of anisotropy later in the text, recognizing that it does give rise to observable phenomena that cannot be explained by isotropic structure.

We now combine the homogeneous equation of motion (2.39), Hooke's law (2.44), and the strain–displacement relationship (2.17) to develop an equation of motion for an isotropic linear elastic medium with no body forces. First, consider only the $i = 1$ term of (2.39):

$$\rho\frac{\partial^2 u_1}{\partial t^2} = \frac{\partial\sigma_{11}}{\partial x_1} + \frac{\partial\sigma_{12}}{\partial x_2} + \frac{\partial\sigma_{13}}{\partial x_3}. \quad (2.46)$$

TABLE 2.1 Elastic Moduli

μ Shear modulus, or rigidity. This is a measure of a material's resistance to shear.

$$\sigma_{ij} = 2\mu\varepsilon_{ij} \Rightarrow \mu = \frac{\sigma_{ij}}{2\varepsilon_{ij}}$$

Note that μ is nonnegative and has units of stress. Typical values are 2×10^{11} dyn/cm^2 or 200 kbar.

k Bulk modulus or incompressibility. k is the material resistance to a change in volume when subject to a load, and it is defined by the ratio of an applied hydrostatic pressure to the induced fractional change in volume:

$$\sigma_{ij} = -P\delta_{ij}, \quad \frac{\Delta V}{V} = \frac{-P}{k} \Rightarrow -P = k\varepsilon_{ii} \Rightarrow \frac{-P}{\varepsilon_{ii}} = \lambda + \frac{2}{3}\mu = k$$

k must be nonnegative, and as a material becomes more rigid, k increases.

λ Lamé's second constant. λ has no simple physical meaning, but it greatly simplifies Hooke's law.

E Young's modulus. E is a measure of the ratio of uniaxial stress to strain in the same direction.

$$\sigma_{11} = E\left(\frac{\Delta L}{L}\right) = E\varepsilon_{11}; \quad \text{by Hooke's Law, } E = \frac{\mu(3\lambda + 2\mu)}{(\lambda + \mu)}$$

ν Poisson's ratio. ν is the ratio of radial to axial strain when a uniaxial stress is applied ($\sigma_{11} \neq 0$, $\sigma_{22} = \sigma_{33} = 0$).

$$\varepsilon_{22} = \varepsilon_{33}, \quad \nu = \frac{-\varepsilon_{22}}{\varepsilon_{11}} = \frac{\lambda}{2(\lambda + \mu)}$$

Poisson's ratio is dimensionless and has a maximum value of 0.5. This is true for a fluid, when $\mu = 0$ (no shear resistance). The smallest value is 0—infinite shear resistance. Most Earth materials have a Poisson ratio between 0.22 and 0.35.

TABLE 2.2 Relationships between Elastic Moduli

μ	k	λ	E	ν
$\dfrac{3(k-\lambda)}{2}$	$\lambda + \dfrac{2\mu}{3}$	$k - \dfrac{2\mu}{3}$	$\dfrac{9k\mu}{3k+\mu}$	$\dfrac{\lambda}{2(\lambda+\mu)}$
$\lambda\left(\dfrac{1-2\nu}{2\nu}\right)$	$\mu\left[\dfrac{2(1+\nu)}{3(1-2\nu)}\right]$	$\dfrac{2\mu\nu}{(1-2\nu)}$	$2\mu(1+\nu)$	$\dfrac{\lambda}{(3k-\lambda)}$
$3k\left(\dfrac{1-2\nu}{2+2\nu}\right)$	$\lambda\left(\dfrac{1+\nu}{3\nu}\right)$	$3k\left(\dfrac{\nu}{1+\nu}\right)$	$\mu\left(\dfrac{3\lambda+2\mu}{\lambda+\mu}\right)$	$\dfrac{3k-2\mu}{2(3k+\mu)}$
$\dfrac{E}{2(1+\nu)}$	$\dfrac{E}{3(1-2\nu)}$	$\dfrac{E\nu}{(1+\nu)(1-2\nu)}$	$3k(1-2\nu)$	$\dfrac{3k-E}{6k}$

TABLE 2.3 Elastic Moduli for Some Common Materials

Material	k(GPa)	μ(GPa)	λ(GPa)	ν	ρ(g/cm^3)
Water	2.1	0	2.1	0.50	1.0
Sandstone	17	6	13	0.34	1.9
Olivine	129	82	74	0.24	3.2
Perovskite	266	153	164	0.26	4.1

The constitutive law and strain–displacement relations give

$$\sigma_{11} = \lambda\theta + 2\mu\varepsilon_{11}$$

$$= \lambda\left(\frac{\partial u_1}{\partial x_1} + \frac{\partial u_2}{\partial x_2} + \frac{\partial u_3}{\partial x_3}\right) + 2\mu\frac{\partial u_1}{\partial x_1}$$

$$\sigma_{12} = 2\mu\varepsilon_{12} = \mu\left(\frac{\partial u_1}{\partial x_2} + \frac{\partial u_2}{\partial x_1}\right) \qquad (2.47)$$

$$\sigma_{13} = 2\mu\varepsilon_{13} = \mu\left(\frac{\partial u_1}{\partial x_3} + \frac{\partial u_3}{\partial x_1}\right).$$

Combining these equations and assuming λ and μ are constant throughout the medium ($\partial\lambda/\partial x_i = \partial\mu/\partial x_i = 0$) gives

$$\rho\frac{\partial^2 u_1}{\partial t^2} = \lambda\frac{\partial\theta}{\partial x_1}$$

$$+ \mu\frac{\partial}{\partial x_1}\left(\frac{\partial u_1}{\partial x_1} + \frac{\partial u_2}{\partial x_2} + \frac{\partial u_3}{\partial x_3}\right)$$

$$+ \mu\left(\frac{\partial^2 u_1}{\partial x_1^2} + \frac{\partial^2 u_1}{\partial x_2^2} + \frac{\partial^2 u_1}{\partial x_3^2}\right).$$

$$(2.48)$$

Recognizing that the first term in brackets is θ and the second is the Laplacian, $\nabla^2 u_1$,

we have

$$\rho\frac{\partial^2 u_1}{\partial t^2} = (\lambda + \mu)\frac{\partial\theta}{\partial x_1} + \mu\nabla^2 u_1 \quad (2.49a)$$

and similarly from the u_2 and u_3 equations

$$\rho\frac{\partial^2 u_2}{\partial t^2} = (\lambda + \mu)\frac{\partial\theta}{\partial x_2} + \mu\nabla^2 u_2 \quad (2.49b)$$

$$\rho\frac{\partial^2 u_3}{\partial t^2} = (\lambda + \mu)\frac{\partial\theta}{\partial x_3} + \mu\nabla^2 u_3. \quad (2.49c)$$

We can write these three equations in the equivalent vector form

$$\rho\ddot{\mathbf{u}} = (\lambda + \mu)\boldsymbol{\nabla}(\boldsymbol{\nabla}\cdot\mathbf{u}) + \mu\nabla^2\mathbf{u}, \quad (2.50)$$

which is the three-dimensional homogeneous vector equation of motion for a uniform, isotropic, linear elastic medium. A common alternate form of this equation employs the vector identity (see Box 2.3)

$$\nabla^2\mathbf{u} = \boldsymbol{\nabla}(\boldsymbol{\nabla}\cdot\mathbf{u}) - (\boldsymbol{\nabla}\times\boldsymbol{\nabla}\times\mathbf{u}), \quad (2.51)$$

allowing (2.50) to be written as

$$\rho\ddot{\mathbf{u}} = (\lambda + 2\mu)\boldsymbol{\nabla}(\boldsymbol{\nabla}\cdot\mathbf{u}) - (\mu\boldsymbol{\nabla}\times\boldsymbol{\nabla}\times\mathbf{u}).$$

$$(2.52)$$

Equations (2.50) and (2.52) are complicated, three-dimensional, partial differential equations for displacements in a continuum, which we assume were initiated by an unspecified source. Although we can sometimes obtain solutions by numerical evaluation of these equations, we can proceed to gain insight into the solutions by using some standard mathematical procedures.

Box 2.3 Useful Vector Relationships

Because ground displacement has a direction and magnitude, its description is given by a vector, $\mathbf{u}(\mathbf{x}, t) = u_1(\mathbf{x}, t)\hat{\mathbf{x}}_1 + u_2(\mathbf{x}, t)\hat{\mathbf{x}}_2 + u_3(\mathbf{x}, t)\hat{\mathbf{x}}_3$, with the vector equations of motion (2.50) and (2.52) giving physically realizable displacements in a linear, elastic continuum. It is thus helpful to review a few basic vector operations that occur frequently in analysis of the vector equations of motion.

(a) The *scalar product* (*dot product* or *inner product*) of two vectors

$$\mathbf{a} = a_1\hat{\mathbf{x}}_1 + a_2\hat{\mathbf{x}}_2 + a_3\hat{\mathbf{x}}_3$$

$$\mathbf{b} = b_1\hat{\mathbf{x}}_1 + b_2\hat{\mathbf{x}}_2 + b_3\hat{\mathbf{x}}_3$$

is given by

$$\mathbf{a} \cdot \mathbf{b} = a_1 b_1 + a_2 b_2 + a_3 b_3 = a_i b_i = |\mathbf{a}|\, |\mathbf{b}| \cos\theta,$$

where θ is the angle between the two vectors. The dot product gives the length of each vector projected on the direction of the other vector. $\mathbf{a} \cdot \mathbf{b} = 0$ for perpendicular vectors ($\theta = \pi/2$); $\mathbf{a} \cdot \mathbf{b} = \mathbf{b} \cdot \mathbf{a}$.

(b) The *vector product* (*cross product* or *curl*) of $\mathbf{a}$ and $\mathbf{b}$ is

$$\mathbf{a} \times \mathbf{b} = (a_2 b_3 - a_3 b_2)\hat{\mathbf{x}}_1 + (a_3 b_1 - a_1 b_3)\hat{\mathbf{x}}_2 + (a_1 b_2 - a_2 b_1)\hat{\mathbf{x}}_3$$

$$= \begin{vmatrix} \hat{\mathbf{x}}_1 & \hat{\mathbf{x}}_2 & \hat{\mathbf{x}}_3 \\ a_1 & a_2 & a_3 \\ b_1 & b_2 & b_3 \end{vmatrix}.$$

In indicial notation we can introduce the permutation symbol

$$\varepsilon_{ijk} = \begin{cases} 0 & \text{any two indices equal} \\ 1 & i, j, k \text{ in order} \\ -1 & i, j, k \text{ not in order} \end{cases}$$

$$(\mathbf{a} \times \mathbf{b})_i = \varepsilon_{ijk} a_j b_k.$$

The cross product defines a new vector that is perpendicular to the two vectors. Properties of the dot and cross product include the following:

$$\mathbf{a} \times \mathbf{b} = -\mathbf{b} \times \mathbf{a}$$

$$\mathbf{a} \cdot (\mathbf{a} \times \mathbf{b}) = \mathbf{b} \cdot (\mathbf{a} \times \mathbf{b}) = 0$$

$$\mathbf{a} \times (\mathbf{b} + \mathbf{c}) = \mathbf{a} \times \mathbf{b} + \mathbf{a} \times \mathbf{c}$$

$$\mathbf{a} \cdot (\mathbf{b} \times \mathbf{c}) = \mathbf{b} \cdot (\mathbf{c} \times \mathbf{a}) = \mathbf{c} \cdot (\mathbf{a} \times \mathbf{b}).$$

(c) The *gradient* of a scalar field uses the "del" operator

$$\nabla = \frac{\partial}{\partial x_1}\hat{\mathbf{x}}_1 + \frac{\partial}{\partial x_2}\hat{\mathbf{x}}_2 + \frac{\partial}{\partial x_3}\hat{\mathbf{x}}_3$$

continues

applied to a scalar field $\phi(\mathbf{x})$

$$\nabla\phi = \frac{\partial\phi}{\partial x_1}\hat{\mathbf{x}}_1 + \frac{\partial\phi}{\partial x_2}\hat{\mathbf{x}}_2 + \frac{\partial\phi}{\partial x_3}\hat{\mathbf{x}}_3$$

$$(\nabla\phi)_i = \frac{\partial\phi}{\partial x_i} = \phi,i.$$

The gradient vector points in the direction of steepest slope, or rate of change, of the field ϕ.

(d) The *divergence* of a vector field $\boldsymbol{\Psi}$ is

$$\nabla\cdot\boldsymbol{\Psi} = \frac{\partial\psi_1}{\partial x_1} + \frac{\partial\psi_2}{\partial x_2} + \frac{\partial\psi_3}{\partial x_3} = \psi_{i,i}.$$

This is a scalar field that measures the flux of the vector field through a unit volume. The integral over a volume V with surface area S is

$$\int_V (\nabla\cdot\boldsymbol{\Psi})\,dV = \int_S (\mathbf{n}\cdot\boldsymbol{\Psi})\,dS,$$

where $\mathbf{n}$ is the outward-facing unit normal everywhere on S. This is *Gauss' theorem*. This states that the accumulation of the field $\boldsymbol{\Psi}$ in the volume is equal to the flux through the surface.

(e) The Laplacian of a scalar field is the divergence of the gradient:

$$\nabla^2\phi = \nabla\cdot\nabla\phi = \frac{\partial^2\phi}{\partial x_1^2} + \frac{\partial^2\phi}{\partial x_2^2} + \frac{\partial^2\phi}{\partial x_3^2} = \phi,ii,$$

which is a scalar. The Laplacian of a vector field is a vector with components that are Laplacians of the original components (if Cartesian coordinates are used). Or, for any coordinate system,

$$\nabla^2\boldsymbol{\Psi} = \nabla(\nabla\cdot\boldsymbol{\Psi}) - \nabla\times\nabla\times\boldsymbol{\Psi}.$$

(f) *Helmholtz's theorem* states that any vector field $\mathbf{u}$ can be represented in terms of a vector potential $\boldsymbol{\Psi}$ and a scalar potential ϕ by

$$\mathbf{u} = \nabla\phi + \nabla\times\boldsymbol{\Psi}$$

if

$$\nabla\times\phi = 0 \qquad (\phi \text{ is curl free})$$

$$\nabla\cdot\boldsymbol{\Psi} = 0 \qquad (\boldsymbol{\Psi} \text{ is divergence free}).$$

(g) Some useful vector identities are

$$\nabla\cdot(\nabla\times\boldsymbol{\Psi}) = 0$$

$$\nabla\times(\nabla\phi) = 0.$$

2.4 Wave Equations: *P* and *S* Waves

We can use Helmholtz's theorem (Box 2.3) to represent the displacement field as

$$\mathbf{u} = \nabla\phi + \nabla\times\mathbf{\Psi}, \qquad (2.53)$$

where ϕ is a curl-free scalar potential field ($\nabla\times\phi = 0$) and $\mathbf{\Psi}$ is a divergenceless vector potential field ($\nabla\cdot\mathbf{\Psi} = 0$). Physically, a curl-free field involves *no* shearing motion, and a divergence-free field involves no change in volume. Substituting (2.53) into (2.52) and using the vector identity ($\nabla\times\nabla\times\mathbf{\Psi} = -\nabla^2\mathbf{\Psi}$ since $\nabla\cdot\mathbf{\Psi} = 0$), we find

$$\nabla\left[(\lambda+2\mu)\nabla^2\phi - \rho\ddot{\phi}\right]$$

$$+ \nabla\times\left[\mu\nabla^2\mathbf{\Psi} - \rho\ddot{\mathbf{\Psi}}\right] = 0. \quad (2.54)$$

We can clearly satisfy this equation if each term in brackets goes to zero independently. We let

$$\alpha = \sqrt{\frac{\lambda+2\mu}{\rho}}$$

$$\qquad (2.55)$$

$$\beta = \sqrt{\frac{\mu}{\rho}}$$

and (2.54) will be solved if

$$\nabla^2\phi - \frac{1}{\alpha^2}\ddot{\phi} = 0$$

$$\qquad (2.56)$$

$$\nabla^2\mathbf{\Psi} - \frac{1}{\beta^2}\ddot{\mathbf{\Psi}} = 0,$$

where (2.56) gives a scalar wave equation for ϕ and a vector wave equation for $\mathbf{\Psi}$. α is the velocity of wave solutions, ϕ, and is called the *P*-wave velocity, and β is the *S*-wave velocity corresponding to solutions $\mathbf{\Psi}$. We will find that solving the equation of motion (2.52) in seismology generally involves solving wave equations such as

(2.56), satisfied by wave potentials from which we can determine the displacement field using (2.53). In every case the displacement field comprises two fundamental wave types, *P* and *S* waves, that propagate with distinct velocities determined by the material properties of the medium. *P* waves involve compressional motions and volumetric changes as the wave disturbance passes by, whereas *S* waves involve shearing motions without volume change. From (2.55) it is clear that $\alpha > \beta$ (for $\lambda \approx \mu$, $\alpha \approx \sqrt{3}\beta$); thus *P* waves arrive before *S* waves. The existence of solutions of the *P* and *S* wave type for motions in a solid was first recognized by Poisson in 1829. An important additional result that will not be demonstrated here is that *P* and *S* waves are in fact the *only* transient solutions for the homogeneous elastic whole space; thus together they provide a *complete* solution to the displacement equation of motion. We will now build up our insight into these wave solutions by considering one-dimensional and then three-dimensional cases.

2.4.1 One-Dimensional Wave Solutions

We can demonstrate the essence of wave behavior in a simple one-dimensional case. Let us consider longitudinal oscillations of a long, thin, elastic rod extending in the $\pm x_1$ direction (Figure 2.8). Longitudinal oscillations involve displacements only in the x_1 direction ($u_1 \neq 0$; $u_2 \approx u_3 \approx 0$). As

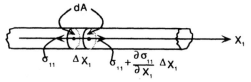

FIGURE 2.8 A very thin elastic rod extending infinitely along the x_1 axis. A stress imbalance produced by an unspecified source is assumed to exist in the rod at an instant of time.

in our general derivation, the equation of motion is derived by a balance between inertial terms and stress gradients, where we assume that an unspecified source has created a stress imbalance in the rod.

$$\Sigma F_{x_1} = m\ddot{u}_1 = \rho \, \Delta A \, \Delta x_1 \, \ddot{u}_1$$

$$= \left(\sigma_{11} + \frac{\partial \sigma_{11}}{\partial x_1} \Delta x_1 \right) \Delta A - \sigma_{11} \Delta A,$$

$$(2.57)$$

where ρ is the density of the rod. This gives

$$\rho \ddot{u}_1 = \frac{\partial \sigma_{11}}{\partial x_1}. \qquad (2.58)$$

As our constitutive law we use $\sigma_{11} = E\varepsilon_{11}$, where E is Young's modulus (Table 2.1), which gives

$$E \frac{\partial^2 u_1}{\partial x_1^2} = \rho \frac{\partial^2 u_1}{\partial t^2}. \qquad (2.59)$$

Defining $c = (E/\rho)^{1/2}$, we have a one-dimensional wave equation

$$\frac{\partial^2 u_1}{\partial x_1^2} = \frac{1}{c^2} \frac{\partial^2 u_1}{\partial t^2}. \qquad (2.60)$$

This derivation is, of course, approximate because in reality lateral strains occur in any finite rod, giving nonuniform stress across the cross section, but this is not important for wavelengths much greater than the lateral dimension of the bar. As a result of this approximation, the displacements themselves satisfy the wave equation, unlike the case of our general elastic solutions.

The general solution of (2.60) is

$$u_1(x_1, t) = f(x_1 - ct) + g(x_1 + ct),$$

$$(2.61)$$

which is called *D'Alembert's solution*. The functions f and g are *arbitrary* functions that will satisfy the initial conditions associated with a particular source that excites the initial stress imbalance, giving rise to the propagating disturbances. These disturbances propagate along the $+x_1$ (f) and $-x_1$ (g) directions with velocity $c = (E/\rho)^{1/2}$. This is made clear by considering Figure 2.9, which considers function $f(x - ct)$ at time t_0 and at some later time t' as a function of x_1, as well as a function of t for fixed $x_1 = x_0$. The arguments of f and g maintain constant functional shapes for constant values of $(x_1 \pm ct)$, with the shape translating through space with velocity c. The arguments $(x_1 \pm ct)$ are called the *phase* of the wave solution. For a given value of phase, the translating functional shape is called a *wavefront*. The velocity of the wavefront is controlled by the material properties, in this case E and ρ. A stiff rod, with a high Young's modulus, produces faster-traveling waves. Increasing density alone would tend to reduce the velocity, but in general E increases with increasing ρ, causing a compensating effect that usually gives a net increase in velocity. A seismogram would correspond to a recording of $u_1(x_0, t)$ at a fixed position $x_1 = x_0$. This will have the form $u_1(x_0, t) = f(x_0 - ct) + g(x_0 + ct)$, a function of time at x_0 (a seismogram) that records the passage of the two wave groups past position x_0.

A general procedure that we can follow to solve partial differential equations such as (2.60) is to assume the solution has a form that separates the spatial and temporal dependence. This is called the method of *separation of variables*. We assume

$$u_1(x_1, t) = X(x_1)T(t) \qquad (2.62)$$

and insert this trial solution into (2.60), giving

$$c^2 \frac{1}{X(x_1)} \frac{d^2 X(x_1)}{dx_1^2} - \frac{1}{T(t)} \frac{d^2 T(t)}{dt^2} = 0.$$

$$(2.63)$$

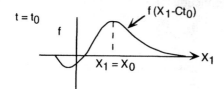
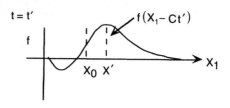

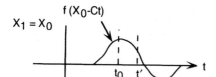

$$C = \frac{\Delta X_1}{\Delta t} = \frac{dX_1}{dt} = \text{velocity}$$

FIGURE 2.9 The one-dimensional propagating disturbance $f(x - ct)$ plotted above as a function of position for two times (t_0 and t') and below as a function of time at position $x_1 = x_0$.

Because the term on the left is a function only of x_1 and must equal the term on the right, which is a function only of t, each term must equal a constant, which we set to $-\omega^2$.

We now have two coupled ordinary differential equations

$$\frac{d^2 X(x_1)}{dx_1^2} + \frac{\omega^2}{c^2} X(x_1) = 0$$

$$\frac{d^2 T(t)}{dt^2} + \omega^2 T(t) = 0. \tag{2.64}$$

These equations can be solved by standard methods such as Fourier transforms, or in this simple case by recognizing that they have the form satisfied by simple harmonic functions. If we let

$$X(x_1) = A_1 e^{i(\omega/c)x_1} + A_2 e^{-i(\omega/c)x_1}$$

$$T(t) = B_1 e^{i\omega t} + B_2 e^{-i\omega t}, \tag{2.65}$$

we will clearly satisfy Eq. (2.64). The solution for $u_1(x_1, t)$ given by (2.62) becomes

$$u_1(x_1, t) = C_1 e^{i\omega(t + x/c)} + C_2 e^{i\omega(t - x/c)}$$

$$+ C_3 e^{-i\omega(t + x/c)} + C_4 e^{-i\omega(t - x/c)}. \tag{2.66}$$

This general solution has four arbitrary constants that will be determined by initial and boundary conditions. This solution involves general harmonic terms that have the form of D'Alembert's solution (2.61).

Harmonic wave solutions such as (2.66) are of fundamental importance in seismology. These solutions have the form

$$u(x, t) = A e^{i\omega(t \pm x/c)} = A \cos[\omega(t \pm x/c)]$$

$$+ iA \sin[\omega(t \pm x/c)], \tag{2.67}$$

comprising monochromatic harmonic sine and cosine terms. For a specified value of ω, the *angular frequency*, these harmonic terms have a *period*, $T = 2\pi/\omega$, which is the time between passage of successive peaks of the harmonic wave at a given point (Figure 2.10). If the wave is considered as a function of x alone, the *wavelength*, Λ, is the distance between peaks on the harmonic function, with $\Lambda = cT$. The term $k = (\omega/c) = (2\pi/\Lambda)$ is the

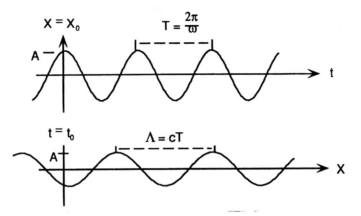

FIGURE 2.10 Definition of period, T, and wavelength, Λ, for a harmonic term $\cos[\omega(t \pm x/c)]$.

wavenumber of the harmonic wave. Table 2.4 summarizes the various variables used to describe a harmonic component. In general, seismic waves have frequencies between about 0.0003 and 100 Hz. For a typical seismic-wave velocity of 5 km/s, this involves signal wavelengths between 15,000 and 0.05 km. These waves intrinsically sample very different characteristics of the Earth.

The complex number representation of harmonic waves (2.67) does not imply the existence of "imaginary" waves. Ground displacements are real functions, and whenever actual initial and boundary conditions are applied to general solutions such as (2.66), the complex terms appear in parts of complex conjugates that eliminate the imaginary components (Box 2.4).

We conclude our discussion of one-dimensional wave solutions by considering a case in which the thin rod in Figure 2.8 does not have uniform material properties but the spatial variations in the moduli are gradual. In this case our force balance becomes

$$\rho(x_1)\ddot{u}_1 = \frac{\partial}{\partial x_1}\left(E(x_1)\frac{\partial u_1}{\partial x_1}\right)$$

$$= E(x_1)\frac{\partial^2 u_1}{\partial x_1^2} + \frac{\partial E(x_1)}{\partial x_1}\frac{\partial u_1}{\partial x_1}.$$

$$(2.68)$$

If $\partial E(x_1)/\partial x_1$, the spatial gradient of Young's modulus, is sufficiently small, the rightmost term can be ignored (the precise criteria for this approximation are discussed in the next chapter). We are left with

$$\frac{\partial^2 u_1}{\partial t^2} = c^2(x_1)\frac{\partial^2 u_1}{\partial x_1^2}, \qquad (2.69)$$

which is similar to (2.60), except that

$$c(x_1) = [E(x_1)/\rho(x_1)]^{1/2}$$

varies with position. Applying the separa-

TABLE 2.4 Relationships between Wave Variables

Period	T	$T = 1/f = 2\pi/\omega$
Frequency	f	$f = \omega/2\pi = c/\Lambda$
Wavelength	Λ	$\Lambda = cT = 2\pi/k$
Wavenumber	k	$k = 2\pi/\Lambda = \omega/c$
Velocity	c	$c = \omega/k = f\Lambda$

Box 2.4 Complex Numbers

Solutions of differential equations such as (2.64) often involve complex numbers of the form $c = a + ib$, where $i = \sqrt{-1}$. In this case, a is the real part and b is the imaginary part of the complex number c. In the imaginary plane shown in Figure 2.B4.1, a complex number, c,

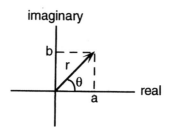

imaginary

real

FIGURE 2.B4.1 The complex plane.

is a point, and it can be represented in a *polar* coordinate form as

$$c = a + ib = re^{i\theta} = r\cos\theta + ir\sin\theta, \tag{2.4.1}$$

where the magnitude of $|c| = r = (a^2 + b^2)^{1/2}$, and the phase is angle $\theta = \tan^{-1}(b/a)$.

Addition of two complex numbers involves summation of the real and imaginary parts:

$$c + d = (a_1 + ib_1) + (a_2 + ib_2) = (a_1 + a_2) + i(b_1 + b_2). \tag{2.4.2}$$

While multiplication (note: $i \cdot i = -1$) is given by

$$c \cdot d = (a_1 + ib_1)(a_2 + ib_2) = (a_1 a_2 - b_1 b_2) + i(a_1 b_2 + b_1 a_2) \tag{2.4.3}$$

or in *polar form*

$$c \cdot d = r_1 e^{i\theta_1} r_2 e^{i\theta_2} = r_1 r_2 e^{i(\theta_1 + \theta_2)}. \tag{2.4.4}$$

The complex conjugate of a number is denoted by c^* and is given by

$$c^* = re^{-i\theta}. \tag{2.4.5}$$

The product $c \cdot c^* = re^{i\theta} re^{-i\theta} = r^2$ gives the square of the magnitude of c.

For a unit circle in the complex plane, $r = 1$, $e^{i\theta} = \cos\theta + i\sin\theta$, and $e^{-i\theta} = \cos\theta - i\sin\theta$. This representation is used in (2.67) to express a complex exponential in terms of harmonic terms. As θ assumes angles greater than π or less than $-\pi$, the value of the function repeats periodically with phase 2π, just as for a

continues

cosine or sine function. By adding and subtracting the exponentials, we obtain useful definitions:

$$\cos\theta = \frac{e^{i\theta} + e^{-i\theta}}{2}$$

$$\sin\theta = \frac{e^{i\theta} - e^{-i\theta}}{2i}. \qquad (2.4.6)$$

tion of variables (2.62) gives

$$\frac{d^2T(t)}{dt^2} + \omega^2 T = 0$$

$$\frac{d^2X(x_1)}{dx_1^2} + \frac{\omega^2}{c^2(x_1)} X(x_1) = 0. \qquad (2.70)$$

Although the temporal dependence of (2.70) is still satisfied by (2.65), we cannot simply set $X(x_1) = ce^{\pm i\alpha x_1}$ with α being a constant because we then obtain $-\alpha^2 + [\omega^2/(c(x_1))^2] = 0$, which cannot be satisfied for all x_1 using constant values of α and ω. We instead assume $X(x_1) = ce^{\pm i\alpha(x_1)}$, which leads to the equation

$$i\frac{d^2\alpha}{dx_1^2} - \left(\frac{d\alpha}{dx_1}\right)^2 + \frac{\omega^2}{c^2(x_1)} = 0. \qquad (2.71)$$

This is a nonlinear differential equation that is very difficult to solve in general. To proceed, we assume

$$\left|\frac{d^2\alpha}{dx_1^2}\right| \ll \frac{\omega^2}{c^2}, \qquad (2.72)$$

allowing us to drop the first term and solve

$$\frac{d\alpha}{dx_1} = \pm\frac{\omega}{c(x_1)}$$

$$\alpha(x_1) = \pm\omega \int_{-\infty}^{x_1} \frac{dx}{c(x)},$$

giving

$$X(x_1) = c\exp\left[\pm i\omega \int_{-\infty}^{x_1} \frac{dx}{c(x)}\right]. \qquad (2.73)$$

The condition (2.72) becomes $d^2\alpha/dx_1^2 = (\omega/c^2)(dc/dx_1) \ll \omega^2/c^2$, or $dc/dx_1 \ll \omega$. This requires that spatial derivatives of velocity be much smaller than the frequencies of interest, which must be correspondingly high, and that the velocities vary smoothly. The high-frequency approximate solution for the inhomogeneous rod is then given by

$$u(x_1, t) = A\exp\left[\pm i\omega\left(t \pm \int_{-\infty}^{x_1} \frac{dx}{c(x)}\right)\right]. \qquad (2.74)$$

Solutions of this type lead to ray theory as described in the next chapter. It is important to note that (2.74) still has a D'Alembert-type solution (2.61), where the phase function

$$t \pm \int_0^{x_1} \frac{dx}{c(x)}$$

gives the travel time of the wave through the medium from the source (at $x_1 = 0$).

2.4.2 Three-Dimensional Wave Solutions

We return our attention to (2.52), the three-dimensional equation of motion, and (2.53), the decomposition of the displacement field into P-wave and S-wave components. The displacements associated with the P wave are given in Cartesian geometry by

$$U_p = \nabla\phi = \frac{\partial\phi}{\partial x_1}\hat{x}_1 + \frac{\partial\phi}{\partial x_2}\hat{x}_2 + \frac{\partial\phi}{\partial x_3}\hat{x}_3$$

(2.75)

and ϕ satisfies (2.56)

$$\frac{\partial^2\phi}{\partial t^2} = \alpha^2\left(\frac{\partial^2\phi}{\partial x_1^2} + \frac{\partial^2\phi}{\partial x_2^2} + \frac{\partial^2\phi}{\partial x_3^2}\right). \quad (2.76)$$

Based on our experience with one-dimensional solutions of the wave equation, we seek a solution in Cartesian coordinates by separation of variables

$$\phi(x_1, x_2, x_3, t) = X(x_1)Y(x_2)Z(x_3)T(t),$$

(2.77)

which leads to a set of four coupled equations

$$\ddot{T} + \omega^2 T = 0$$

$$\ddot{X} + k_1^2 X = 0$$

$$\ddot{Y} + k_2^2 Y = 0$$

$$\ddot{Z} + k_3^2 Z = 0, \quad (2.78)$$

where $k_1^2 + k_2^2 + k_3^2 = \omega^2/\alpha^2$. Assuming harmonic solutions of (2.78) and multiplying terms together as required by (2.77) gives a general wave potential for P waves

$$\phi(x,t) = A\exp\big[\pm i(\omega t \pm k_1 x_1 \pm k_2 x_2$$

$$\pm k_3 x_3)\big], \quad (2.79)$$

which is the three-dimensional counterpart of (2.66). The solution (2.79) again assumes a D'Alembert-type functional dependence of space and time, with the exponential argument being the phase. This solution corresponds to a set of *plane waves*, free to propagate in any direction in the continuum. The requirement for a given frequency, ω, and P-wave velocity,

α, that $k_1^2 + k_2^2 + k_3^2$ is constant, defines a planar surface in Cartesian space with a normal vector $\mathbf{k}_\alpha = |\mathbf{k}_\alpha|\hat{k} = (\omega/\alpha)\hat{k}$ called the wavenumber vector. This vector defines the direction of propagation of the wave (i.e., the normal to the plane wave), and in the next chapter we use it to define seismic rays. We can write a particular choice of the solutions in (2.79) as

$$\phi(x, t) = A\exp[i(\omega t - \mathbf{k}_\alpha \cdot \mathbf{x})]. \quad (2.80)$$

Corresponding solutions to the vector wave equation in (2.56)

$$\frac{\partial^2\Psi}{\partial t^2} = \beta^2\left(\frac{\partial^2\Psi}{\partial x_1^2} + \frac{\partial^2\Psi}{\partial x_2^2} + \frac{\partial^2\Psi}{\partial x_3^2}\right) \quad (2.81)$$

are similarly given by vector solutions

$$\Psi(\mathbf{x}, t) = \mathbf{B}\exp[i(\omega t - \mathbf{k}_\beta \cdot \mathbf{x})], \quad (2.82)$$

where $|\mathbf{k}_\beta| = \omega/\beta$. Equation (2.82) gives plane-wave solutions associated with shear waves.

Let us consider a plane wave propagating with wavenumber vector $\mathbf{k}_\alpha$ contained completely in the $x_1 x_3$ plane (we can always orient our Cartesian coordinate system so that this is the case). In this case, $\partial\phi/\partial x_2 = 0$ and $k_2 = 0$. If we let the phase in (2.80) be a constant, C, we have

$$\omega t - k_1 x_1 - k_3 x_3 = C. \quad (2.83)$$

For $t = 0$, assume $C = 0$, giving $x_1 = -(k_3/k_1)x_3$, which defines a line in the $x_1 x_3$ plane along which the phase is constant (zero). This corresponds to the intersection of the plane-wave surface with the $x_1 x_3$ plane (see Figure 2.11). We have the additional requirement that $k_1^2 + k_3^2 = \omega^2/\alpha^2$. The wavenumber vector is perpendicular to the plane wave, with components k_1 and k_3 occurring along the x_1 and x_3 axes, respectively. Defining the angle between $\mathbf{k}_\alpha$ and the x_3 axis as i, we see that

$$k_1 = \frac{\omega}{\alpha} \sin i = \omega p$$

$$\hspace{6cm} (2.84)$$

$$k_3 = \frac{\omega}{\alpha} \cos i = \omega \eta_\alpha.$$

The term $(\sin i)/\alpha = p$ is called the *seismic ray parameter*, or *horizontal slowness*, and $\eta = (\cos i)/\alpha$ is called the *vertical slowness*. We will explore these parameters at great length in the next chapter.

Keeping $C = 0$ in (2.83) and increasing t in unit steps defines a sequence of parallel lines (Figure 2.11), all with the same phase, that correspond to movement of the wavefront in the $x_1 x_3$ plane in a direction defined by $\mathbf{k}_\alpha$. Similarly, if we keep $t = 0$

in (2.83), a set of parallel lines, $C = k_1 x_1 + k_3 x_3$, will exist, each line with a different phase value. Because the angle i can take on any value from 0 to 360°, our solution (2.80) actually corresponds to an infinite set of plane waves, with all possible orientations and spatial shifts filling the entire three-dimensional space.

The particular solution $A \exp[i(\omega t - k_1 x_1 - k_3 x_3)]$ corresponds to a wave propagating in the $+x_1$ and $+x_3$ directions, while $A \exp[i(\omega t - k_1 x_1 + k_3 x_3)]$ propagates in the $+x_1$ and $-x_3$ directions, $A \exp[i(\omega t + k_1 x_1 - k_3 x_3)]$ propagates in the $-x_1$ and $+x_3$ directions, and $A \exp[i(\omega t + k_1 x_1 + k_3 x_3)]$ propagates in the $-x_1$ and $-x_3$ directions. When the coefficient of ωt is negative, all of these sense of directions reverse.

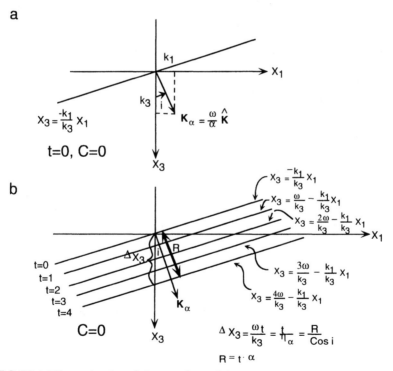

FIGURE 2.11 (a) The projection of the wavefront defined by $t = 0$, $C = 0$ in the $x_1 x_3$ plane and the associated wavenumber vector $\mathbf{k}_\alpha$. (b) Variation of the position of a wavefront of constant phase ($C = 0$) for increasing time, t. The distance that the wavefront moves after time t is $R = \alpha t$.

Box 2.5 Spherical Waves

Most of this text considers plane-wave solutions for the equations of motion, but transient wave solutions with a concentrated source location are often more readily solved using spherical waves. The three-dimensional scalar wave equation

$$\nabla^2 \Phi = \frac{1}{\alpha^2} \ddot{\Phi} \qquad (2.5.1)$$

can be solved by expressing the Laplacian operator in spherical coordinates (as defined in Figure 2.B5.1):

$$\nabla^2 \Phi = \frac{1}{r^2} \frac{\partial}{\partial r}\left(r^2 \frac{\partial \Phi}{\partial r} \right) + \frac{1}{r^2 \sin\theta} \frac{\partial}{\partial \theta}\left(\sin\theta \frac{\partial \Phi}{\partial \theta} \right) + \frac{1}{r^2 \sin^2\theta} \frac{\partial^2 \Phi}{\partial \phi^2}. \qquad (2.5.2)$$

For spherically symmetric solutions, $\Phi = \Phi(r, t)$, the homogeneous wave equation becomes

$$\frac{1}{r^2} \frac{\partial}{\partial r}\left(r^2 \frac{\partial \Phi}{\partial r} \right) = \frac{1}{\alpha^2} \frac{\partial^2 \Phi}{\partial t^2}. \qquad (2.5.3)$$

This has solutions of the form

$$\Phi(r, t) = \frac{f(t \pm r/\alpha)}{r}, \qquad (2.5.4)$$

where f is an arbitrary function, with the $(t - r/\alpha)$ phase indicating outward-propagating waves spreading spherically from the origin, and the $(t + r/\alpha)$ phase indicating inward-propagating spherical waves. The $1/r$ dependence is different from the Cartesian D'Alembert solution.

The solution for the inhomogeneous wave equation with a source at $r = 0$ [localized by the delta function defined by $\delta(r) = 0,\ r \neq 0;\ \int_V \delta(r)\, dv = 1$]

$$\nabla^2 \Phi(r, t) = \frac{1}{\alpha^2} \ddot{\Phi} - 4\pi\delta(r)f(t) \qquad (2.5.5)$$

is $\Phi(r, t) = -f(t - r/v)/r$. The displacements are given by $u_p = \nabla\Phi = (\partial\phi/\partial r)\hat{r}$. We will use this solution in Chapter 8.

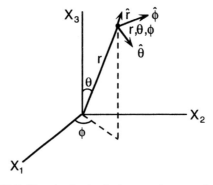

FIGURE 2.B5.1 Standard spherical-geometry coordinate system.

The potential $\phi(\mathbf{x}, t)$ is a system of waves, or a *wavefield*. If we want to determine the *P*-wave displacements, we must compute the gradient of ϕ,

$$\mathbf{U}_P = \nabla \phi = \left(\hat{\mathbf{x}}_1 \frac{\partial}{\partial x_1} + \hat{\mathbf{x}}_2 \frac{\partial}{\partial x_2} + \hat{\mathbf{x}}_3 \frac{\partial}{\partial x_3} \right)$$

$$\times A \exp\left[\pm i(\omega t \pm \mathbf{k}_\alpha \cdot \mathbf{x}) \right]. \tag{2.85}$$

For the particular choice of ϕ given by

$$\phi = A \exp\left[i(\omega t - k_1 x_1 - k_3 x_3) \right], \tag{2.86}$$

we have as a solution

$$\mathbf{U}_P(\mathbf{x}, t) = (-ik_1 A)$$

$$\times \exp\left[i(\omega t - k_1 x_1 - k_3 x_3) \right] \hat{\mathbf{x}}_1$$

$$+ 0\hat{\mathbf{x}}_2 + (-ik_3 A)$$

$$\times \exp\left[i(\omega t - k_1 x_1 - k_3 x_3) \right] \hat{\mathbf{x}}_3. \tag{2.87}$$

Thus, the *P*-wave displacements are all in the $x_1 x_3$ plane, and the *P*-wave displacement field has the same functional dependence as the *P*-wave potential field but different multiplicative constants that allow it to satisfy the equation of motion rather than the wave equation. Taking the ratio

$$\frac{U_{P_3}}{U_{P_1}} = \frac{k_3}{k_1} = \frac{\eta_\alpha}{p} \tag{2.88}$$

defines the perpendicular direction to the wavefront in Figure 2.11. This indicates that the *P-wave particle motion is perpendicular to the wavefront, and it parallels the direction in which the wave is propagating.* This characteristic of *P*-wave motion also holds for cylindrical and spherical waves. Because of the harmonic form of the mo-

tions, particles oscillate back and forth as the wave passes, alternately compressing and dilating the medium.

Let us finally consider *S*-wave particle displacements associated with vector plane-wave solutions of the form (2.82). The displacements are found from (2.53)

$$\mathbf{U}_S = \nabla \times \mathbf{\Psi} = \left(\frac{\partial \psi_3}{\partial x_2} - \frac{\partial \psi_2}{\partial x_3} \right) \hat{\mathbf{x}}_1$$

$$+ \left(\frac{\partial \psi_1}{\partial x_3} - \frac{\partial \psi_3}{\partial x_1} \right) \hat{\mathbf{x}}_2$$

$$+ \left(\frac{\partial \psi_2}{\partial x_1} - \frac{\partial \psi_1}{\partial x_2} \right) \hat{\mathbf{x}}_3. \tag{2.89}$$

We simplify the algebra by again restricting our attention to plane waves with wavenumber vectors in the $x_1 x_3$ plane, so all $\partial \psi_i / \partial x_2 \to 0$. Thus

$$\mathbf{U}_S = U_{S_1} \hat{\mathbf{x}}_1 + U_{S_2} \hat{\mathbf{x}}_2 + U_{S_3} \hat{\mathbf{x}}_3$$

$$= \left(-\frac{\partial \psi_2}{\partial x_3} \right) \hat{\mathbf{x}}_1 + \left(\frac{\partial \psi_1}{\partial x_3} - \frac{\partial \psi_3}{\partial x_1} \right) \hat{\mathbf{x}}_2$$

$$+ \left(\frac{\partial \psi_2}{\partial x_1} \right) \hat{\mathbf{x}}_3. \tag{2.90}$$

If we associate the $x_1 x_2$ plane with the Earth's surface and the x_3 axis with depth (a common convention), the U_{S_1} and U_{S_3} components comprise *S*-wave motions in the $x_1 x_3$ plane and are called the *SV* component because they entail a component of vertical (x_3) motion. The x_2 component, involving purely horizontal (x_2) motions, is called the *SH* component. Remember that for a comparable choice of coordinate system, the *P* waves had no x_2 component. The total displacement field is the sum of the P, SV, and SH waves:

$$\mathbf{U} = \mathbf{U}_P + \mathbf{U}_S = \left(\frac{\partial \phi}{\partial x_1} - \frac{\partial \psi_2}{\partial x_3} \right) \hat{\mathbf{x}}_1$$

$$+ \left(\frac{\partial \psi_1}{\partial x_3} - \frac{\partial \psi_3}{\partial x_1} \right) \hat{\mathbf{x}}_2$$

$$+ \left(\frac{\partial \phi_3}{\partial x_3} + \frac{\partial \psi_2}{\partial x_1} \right) \hat{\mathbf{x}}_3,$$

$$(2.91)$$

with both P and S waves propagating in the same medium. The *linear wave approximation* is made in assuming the P and S waves do not interfere with one another, which is valid for infinitesimal strains. This equation emphasizes the complete separation of the SH components from the P–SV components. As long as internal boundaries or free surfaces parallel the $x_1 x_2$ plane, this separation persists, as shown in later chapters.

For the SH component of motion, we let

$$V = U_{S_2} = \frac{\partial \psi_1}{\partial x_3} - \frac{\partial \psi_3}{\partial x_1}, \quad (2.92)$$

where ψ_1 and ψ_3 are both solutions of the wave equation

$$\frac{\partial^2 \psi_1}{\partial t^2} = \beta^2 \nabla^2 \psi_1$$

$$(2.93)$$

$$\frac{\partial \psi_3}{\partial t^2} = \beta^2 \nabla^2 \psi_3.$$

V itself exactly satisfies the wave equation $\partial^2 V/\partial t^2 = \beta^2 \nabla^2 V$, as is easily shown by substituting in (2.92). Thus, we have solutions for the SH wave equation of the

same form as for the P potential, ϕ:

$$V(x_1, x_3, t)$$

$$= A' \exp\left[i \left(\pm \omega t \pm k_{\beta_1} x_1 \pm k_{\beta_3} x_3 \right) \right],$$

$$(2.94)$$

where $(k_{\beta_1}^2 + k_{\beta_3}^2) = (\omega^2/\beta^2)$, $(k_{\beta_1}/\omega) = p$ $= (\sin \gamma)/\beta$, and $(k_{\beta_3}/\omega) = \eta_\beta = (\cos \gamma)/\beta$. Here γ is the angle that the wavenumber vector makes with the x_3 axis. The wavefronts move in the $x_1 x_3$ plane as discussed before, but now with velocity β. All of the SH particle displacements are in the x_2 direction, and thus they lie in the plane of the wavefront, perpendicular to the direction of propagation. For the Earth reference system the SH motions are all parallel to the surface.

For the SV displacements we use a general plane-wave solution for ψ_2:

$$\psi_2 = B' \exp\left[i \left(\pm \omega t \pm k_1 x_1 \pm k_3 x_3 \right) \right].$$

$$(2.95)$$

Thus

$$\mathbf{U}_{SV} = -\frac{\partial \psi_2}{\partial x_3} \hat{\mathbf{x}}_1 + \frac{\partial \psi_2}{\partial x_1} \hat{\mathbf{x}}_3$$

$$= \mp k_3 B' i \exp\left[i \left(\pm \omega t \pm k_1 x_1 \right. \right.$$

$$\left. \left. \pm k_3 x_3 \right) \right] \hat{\mathbf{x}}_1$$

$$\pm k_1 B' i \exp\left[i \left(\pm \omega t \pm k_1 x_1 \right. \right.$$

$$\left. \left. \pm k_3 x_3 \right) \right] \hat{\mathbf{x}}_3. \quad (2.96)$$

For a particular case, $\psi_2 = B' \exp[i(\omega t - k_1 x_1 - k_3 x_3)]$, the wavefront is given by $(\omega t - k_1 x_1 - k_3 x_3) = C$, which has a slope of $-k_1/k_3$ in the $x_1 x_3$ plane. The ratio of the corresponding SV displacement terms

Box 2.6 Seismic Waves in Anisotropic Media

The *P*- and *S*-wave behavior in isotropic homogeneous media is remarkably simple, but greater complexity arises for anisotropic media. In an anisotropic, homogeneous medium, three independent body waves are generated that have orthogonal planes of particle motions. These are usually called quasi-compressional waves (*qP*) and quasi-shear waves (*qSV* and *qSH*), with names suggestive of the isotropic counterparts. In general, the propagation direction of these waves is not perpendicular to their wavefronts, so the particle motions differ from isotropic behavior. The velocities of these waves vary with the trajectory of the wave through the medium with respect to any axes of symmetry in the structure. For a wave propagating from an isotropic medium into an anisotropic medium, one of the primary effects is the separation of the isotropic *S* wave into two quasi-shear waves, which is called shear-wave splitting.

These properties arise from the general stress–strain relationship expressed by Hooke's law, for which the most general anisotropic medium has 21 independent elastic moduli. Increasing symmetry in the structure reduces the number of moduli. If the medium has symmetry about three orthogonal planes, the medium is *orthotropic*, and only nine independent constants exist. If it has axial symmetry, yielding a *hexagonal* medium, five independent constants exist. A common case relevant to some Earth structures occurs when the symmetry axis is vertical, which is called *transverse* isotropy. If the medium exhibits direction dependence of velocity in the horizontal surface, the behavior is called *azimuthal* anisotropy.

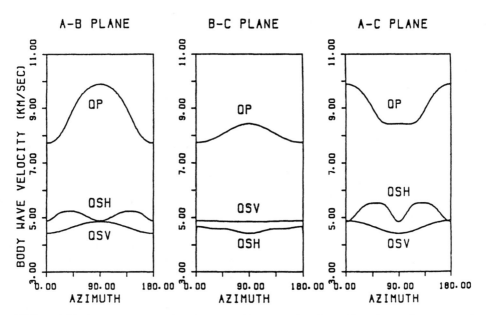

FIGURE 2.B6.1 Variations of *qP*, *qSH*, and *qSV* wave velocities within planes of symmery of single-crystal olivine. The labels A–B, B–C, and A–C denote symmetry planes that include the *a* and *b* axes, *b* and *c* axes, and *a* and *c* axes, respectively. (From Kawasaki, 1989.)

continues

One of the major components of the Earth's mantle is olivine. A single olivine crystal has orthotropic symmetry; thus the anisotropic seismic velocities have a complex behavior, as shown in Figure 2.B6.1. Since processes in the mantle may tend to partially align crystal orientations on a macroscopic scale, net seismic wave anisotropy with reduced directional dependence is often observed, as is the presence of shear-wave splitting. Figure 2.B6.1 shows the variations of α and β in a single crystal of olivine.

Anisotropic behavior may also result from structural complexities rather than intrinsic crystallographic anisotropy. The presence of networks of flattened, possibly fluid- or magma-filled cracks causes directional wave-speed dependence, with the quasi-P and -S waves being relatively slower in propagation directions perpendicular to the long axis of ellipsoidal cracks and relatively faster when propagating along the cracks' long axes, as shown in Figure 2.B6.2. Finely layered structures with alternating high- and low-velocity isotropic material can also give rise to effective anisotropic wave speeds. In later chapters, examples will be given of anisotropic body- and surface-wave observations in the Earth.

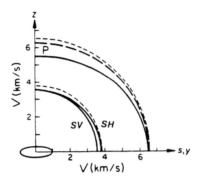

FIGURE 2.B6.2. Velocities as a function of angle and fluid properties in granite containing aligned ellipsoidal cracks (orientation shown at origin) with porosity $=0.01$ and aspect ratio $=0.05$. The short dashed lines are for the isotropic uncracked solid, the long dashes for liquid-filled cracks ($K_L = 100$ kbar), and the solid curves for gas-filled cracks ($K_L = 0.1$ kbar). (After Anderson *et al.*, 1974.)

is the same

$$\frac{U_{S_3}}{U_{S_1}} = -\frac{k_1}{k_3}. \qquad (2.97)$$

So the SV displacement is within the wavefront in the x_1x_3 plane. Our choice of discussing different components of the S vector in terms of SH and SV clearly has little significance for whole-space solutions; it merely sets the stage for subsequent discussions in the Earth coordinate system. Clearly, the total S-wave motion is

a vector displacement in the plane of the wavefront, with SH and SV components being components projected into a convenient reference system.

The overall sense of particle motions associated with the P and S waves is shown in Figure 2.12 and in the block diagram in Figure 1.2. The characteristic particle displacements associated with P and S waves result in predictable polarizations of the displacements. Most seismic stations record three components of ground motion: up–down (vertical), north–

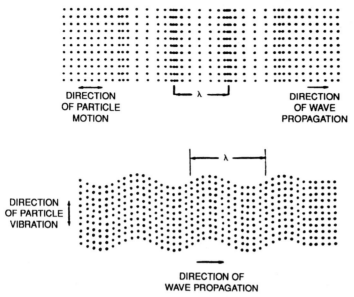

FIGURE 2.12 Sense of particle motions as a plane wave sweeps from left to right for *P* waves (top) and *S* waves (bottom). The wavelength is given by λ. (From Sheriff and Geldart, ''Exploration Seismology: Vol. 1, History, theory, and data acquisition.'' Copyright © 1982, Reprinted with the permission of Cambridge University Press.)

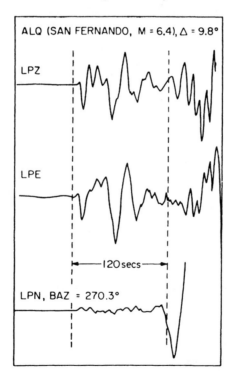

FIGURE 2.13 Three-component observation of the 1971 San Fernando earthquake recorded at ALQ (Albuquerque, New Mexico). *P* and *SV* motions are on the Z and E components, while *SH* motion is on the N component. The direction to the source is due west. (From Helmberger and Engen, 1980.)

TABLE 2.5 Compressional and Shear Velocities in Rocks

Material and source	P-wave velocity (m/s)	S-wave velocity (m/s)
Loose sand	1800	500
Clay	1100–2500	
Sandstone	1400–4300	
Anhydrite, Gulf Coast	4100	
Conglomerate, Australia	2400	
Limestone, Texas	6030	3030
Granite, Barriefield, Ontario	5640	2870
Granodiorite, Weston, Massachusetts	4780	3100
Diorite, Salem, Massachusetts	5780	3060
Basalt, Germany	6400	3200
Gabbro, Minnesota	6450	3420
Dunite, Twin Sisters, Washington	8000	4370

Source: Clark (1966).

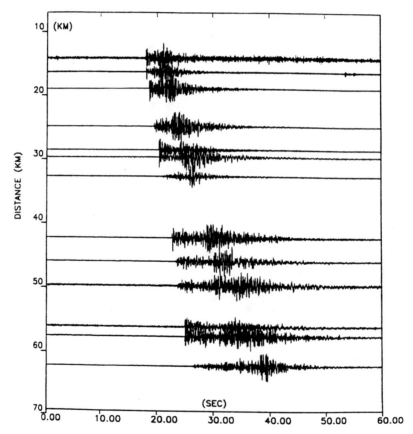

FIGURE 2.14 Vertical-component ground-motion recordings arranged from top to bottom with increasing distance from an earthquake. The first arrival on each trace is the P wave and the largest later arrival is the S wave. (Courtesy of Jim Mori.)

south, and east–west. Seismic waves arrive at the station propagating at some angle to the vertical in a direction along the great-circle path connecting the source and receiver, called the *longitudinal* or *radial* direction. A *P*-wave arrival, producing motions only in the direction of wave propagation, vibrates the ground only in the vertical and longitudinal directions, with relative strengths depending on the angle of incidence. On the other hand, the *SH* motion is entirely horizontal and perpendicular to the great-circle path direction in what is called the *transverse* or *tangential* direction. The *SV* motion is in the longitudinal and vertical plane but parallel to the wavefront. This polarization can be directly observed in seismograms when the direction to the source is either north–south or east–west from that station, as shown in Figure 2.13. These seismograms, from station ALQ (Albuquerque, New Mexico), are for the 1971 San Fernando, California, earthquake. The event was located due west of the station, so the LPN (north–south) seismogram records tangential motion only, while the LPE (east–west) component is purely longitudinal. The *P* and *SV* waves arrive on the vertical and longitudinal components, while the *SH* part of the *S* wave arrives only on the transverse component.

As mentioned earlier, *P* waves travel faster than *S* waves, and for a fluid, in which the rigidity vanishes, *S* waves cannot propagate at all. *P* waves can exist in a fluid, with acoustic waves or sound waves in the air being a form of *P* wave. Thus far, *P*- and *S*-wave velocities are independent of frequency or wavelength and depend only on the material properties of the continuum. Anelastic effects can lead to frequency dependence of velocities, as discussed in the next chapter. Table 2.5 gives examples of seismic velocities for near-surface conditions for a variety of rock types. Because $\alpha \approx 1.73\beta$, the time

separation between *P* and *S* arrivals increases with distance traveled. The ratio of travel time to distance traveled is called *moveout*. Figure 2.14 shows a sequence of seismograms at increasing distances from an earthquake. The moveout of the *S* waves is nearly twice that of the *P* waves. We will next consider how these waves have traveled through an inhomogeneous structure like the Earth.

References

Anderson, D. L., Minster, B., and Cole, D. (1974). The effect of oriented cracks on seismic velocities. *J. Geophys. Res.* **79**, 4011–4015.

Clark, S. P., Jr., ed. (1966). "Handbook of Physical Constants," rev. ed., Geol. Soc. Am. Mem., Vol. 97. Geol. Soc. Am. Boulder, CO.

Helmberger, D. V., and Engen, G. R. (1980). Modeling long-period body waves from shallow earthquakes at regional ranges. *Bull. Seismol. Soc. Am.* **70**, 1699–1714.

Kawasaki, I. (1989). Seismic anisotropy in the Earth. *In* "The Encyclopedia of Solid Earth Geophysics" (D. E. James, ed.), pp. 994–1005. Van Nostrand–Reinhold, New York.

Malvern, L. E. (1969). "Introduction to the Mechanics of a Continuous Medium." Prentice–Hall, Englewood Cliffs, NJ.

Scholz, C. H. (1990). "The Mechanics of Earthquakes and Faulting." Cambridge Univ. Press, Cambridge, UK.

Sheriff, R. E., and Geldart, L. P. (1982). "Exploration Seismology: Volume 1. History, theory and data acquisition." Cambridge Univ. Press, Cambridge, UK.

Additional Reading

Kolsky, H. (1963). "Stress Waves in Solids." Dover, New York.

Miklowitz, J., "Elastic Waves and Waveguides," North-Holland Publ., Amsterdam.

Sokolnikoff, I. S. (1956). "Mathematical Theory of Elasticity." McGraw-Hill, New York.

Sommerfeld, A. (1950). "Mechanics of Deformable Bodies," Vol. 2. Academic Press, New York.

3

BODY WAVES AND RAY THEORY

In the last chapter we derived the existence of P and S waves, the only transient solutions to a stress imbalance suddenly introduced to a homogeneous elastic space. P and S waves are known as *body waves* because they travel along paths throughout the continuum. The solutions for P and S waves, like those given in Eqs. (2.85) and (2.90), give the locations of *wavefronts*, which are loci of points that undergo the same motion at a given instant in time. *Rays* are defined as the normals to the wavefront and thus point in the direction of propagation. In the case of a plane wave, the rays are a family of parallel straight lines; in the case of a spherical wave, the rays are spokes radiating out from the seismic source. Rays provide a convenient means of tracking an expanding wavefront, and they provide an intuitive framework for extending elastic-wave solutions from homogeneous to inhomogeneous materials. If the inhomogeneities in velocity are not excessively chaotic, then the rays corresponding to P or S waves behave very much as light does in traveling through materials of varying indices of refraction. This leads to many parallels with optics: rays bend, focus, and defocus depending on the velocity distribution. Strictly speaking, we will have to approximate our displacement solutions to extract

the ray behavior, for it cannot describe all wave phenomena. These approximations are collectively known as *geometric ray theory* and are the standard basis for seismic body-wave interpretation.

In classical optics, the geometry of a wave surface is governed by Huygens' principle, which states that every point on a wavefront can be considered the source of a small secondary wavelet that travels outward in every forward direction with the velocity of the medium at that point. The wavefront at a later instant in time is found by drawing a tangent to the secondary wavelets, as shown in Figure 3.1. Thus, given the location of a wavefront at a certain instant in time, we can predict future positions of the wavefront. Portions of the wavefront which are located in relatively high-velocity material produce wavelets that travel farther in a given time interval than those produced by points in relatively low-velocity material. This causes a temporal dependence in the shape of the wavefront. Because rays are the normals to the wavefront, the rays will also change with time. Fermat's principle governs the geometry of raypaths. This usually means that the ray will follow a *minimum-time path*, which is the path that will allow the wavefront to move from point A to point B in the shortest amount of time.

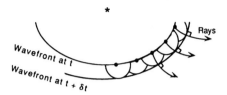

FIGURE 3.1 An expanding wavefront. Huygens' principle states that each point on the wavefront serves as a secondary source. The tangent surface of the expanding waves from the secondary sources gives the position of the wavefront at a later time.

Let us consider the approximations that must be made to the elastic-wave solutions such that ray theory is valid. Recall the equation for a plane wave:

$$\phi = Ae^{i(\pm \omega t \pm \mathbf{k} \cdot \mathbf{x})}, \qquad (3.1)$$

where $\mathbf{k}$ is a vector that points in the direction of propagation and thus, by definition, is a ray. For homogeneous material, $\mathbf{k}$ does not change as the wave propagates (it is a straight line). Now if the seismic velocity varies smoothly in space (i.e., ρ, λ, and μ have small gradients), we must solve an equation analogous to (2.69):

$$\frac{\partial^2 \phi}{\partial x_1^2} + \frac{\partial^2 \phi}{\partial x_2^2} + \frac{\partial^2 \phi}{\partial x_3^2} = \frac{1}{c^2(\mathbf{x})}\frac{\partial^2 \phi}{\partial t^2}. \quad (3.2)$$

This wave equation is an approximation of the equation of motion for heterogeneous media. As we did in the last chapter [see Eqs. (2.70) and (2.71)], we will attempt to solve this partial differential equation by assuming a functional form

$$\phi(\mathbf{x}, t) = A(\mathbf{x})e^{i\omega(W(\mathbf{x})/c_0 - t)}, \quad (3.3)$$

where $W(\mathbf{x}) \cdot \omega/c_0$, which replaces $\mathbf{k} \cdot \mathbf{x}$, is a function of position, and c_0 is a reference velocity. Substitution of (3.3) into (3.2) yields

$$\nabla^2\left[A(\mathbf{x})e^{i\omega(W(\mathbf{x})/c_0 - t)} \right]$$
$$= \frac{1}{c^2(\mathbf{x})}\frac{\partial^2}{\partial t^2}\left[A(\mathbf{x})e^{i\omega(W(\mathbf{x})/c_0 - t)} \right]. \tag{3.4}$$

The required spatial derivatives are complex; for example, $\partial^2\phi/\partial x_1^2$ is

$$\frac{\partial^2 \phi}{\partial x_1^2} = \frac{\partial}{\partial x_1}\left\{ \frac{\partial A(\mathbf{x})}{\partial x_1}e^{i\omega(W(\mathbf{x})/c_0 - t)} \right.$$
$$\left. + A(\mathbf{x})\frac{i\omega}{c_0}\frac{\partial W(\mathbf{x})}{\partial x_1}e^{i\omega(W(\mathbf{x})/c_0 - t)} \right\}$$
$$= \left[\frac{\partial^2 A(\mathbf{x})}{\partial x_1^2} - \frac{\omega^2 A(\mathbf{x})}{c_0^2}\left(\frac{\partial W(\mathbf{x})}{\partial x_1} \right)^2 \right.$$
$$+ i\left(\frac{2\omega}{c_0}\frac{\partial A(\mathbf{x})}{\partial x_1}\frac{\partial W(\mathbf{x})}{\partial x_1} \right.$$
$$\left.\left. + A(\mathbf{x})\frac{\omega}{c_0}\frac{\partial^2 W(\mathbf{x})}{\partial x_1^2} \right) \right]e^{i\omega(W(\mathbf{x})/c_0 - t)}. \tag{3.5}$$

For $\partial^2\phi/\partial x_2^2$ and $\partial^2\phi/\partial x_3^2$, we obtain similar equations with real and imaginary parts. Equating the real and imaginary parts in Eq. (3.4) gives two sets of equations:

$$\nabla^2 A(\mathbf{x}) - A(\mathbf{x})\frac{\omega^2}{c_0^2}\left[\left(\frac{\partial W(\mathbf{x})}{\partial x_1} \right)^2 + \left(\frac{\partial W(\mathbf{x})}{\partial x_2} \right)^2 + \left(\frac{\partial W(\mathbf{x})}{\partial x_3} \right)^2 \right] = \frac{-\omega^2}{c^2(\mathbf{x})}A(\mathbf{x}) \quad (3.6)$$

$$2\left(\frac{\partial W(\mathbf{x})}{\partial x_1}\frac{\partial A(\mathbf{x})}{\partial x_1} + \frac{\partial W(\mathbf{x})}{\partial x_2}\frac{\partial A(\mathbf{x})}{\partial x_2} + \frac{\partial W(\mathbf{x})}{\partial x_3}\frac{\partial A(\mathbf{x})}{\partial x_3} \right) + A(\mathbf{x})\nabla^2 W(\mathbf{x}) = 0. \quad (3.7)$$

We can rearrange the terms from Eq. (3.6) as

$$\left(\frac{\partial W(\mathbf{x})}{\partial x_1}\right)^2 + \left(\frac{\partial W(\mathbf{x})}{\partial x_2}\right)^2 + \left(\frac{\partial W(\mathbf{x})}{\partial x_3}\right)^2$$

$$-\frac{c_0^2}{c(\mathbf{x})^2} = \frac{c_0^2}{A(\mathbf{x})\omega^2}(\nabla^2 A(\mathbf{x})). \quad (3.8)$$

The right-hand side of this equation is a ratio of the spatial Laplacian of the amplitude to the amplitude divided by ω^2. For high frequencies (small wavelengths) this term is small; in fact, let it be approximately zero, and Eq. (3.6) reduces to

$$\left(\frac{\partial W(\mathbf{x})}{\partial x_1}\right)^2 + \left(\frac{\partial W(\mathbf{x})}{\partial x_2}\right)^2 + \left(\frac{\partial W(\mathbf{x})}{\partial x_3}\right)^2$$

$$= \frac{c_0^2}{c(\mathbf{x})^2}. \quad (3.9)$$

This is called the *eikonal equation*. Solutions to the eikonal equation are not exact solutions to the wave equation, but for many regions inside the real Earth, the necessary restrictions on spatial variations of the elastic parameters are satisfied, so solutions of the eikonal equation are useful.

Recall $W(\mathbf{x}) \cdot \omega/c_0$ was just $\mathbf{k} \cdot \mathbf{x}$, where $\mathbf{k}$ is a vector normal to the wavefront, or a ray. The eikonal equation is therefore a partial differential equation that relates rays to the seismic velocity distribution. The condition required for *geometric ray theory* to be a useful approximation of the wave equation is that the change in *gradient* of $A(\mathbf{x})$ over one wavelength must be much smaller than $A(\mathbf{x})$. We define a reference wavelength:

$$\lambda_0 = c_0 \frac{2\pi}{\omega}. \quad (3.10)$$

For (3.9) to hold, we required that $\lambda_0^2(\nabla^2 A(\mathbf{x})/A(\mathbf{x})) \ll \nabla W(\mathbf{x}) \cdot \nabla W(\mathbf{x})$. This

gives

$$\lambda_0^2 \frac{\nabla^2 A(\mathbf{x})}{A(\mathbf{x})} \ll \frac{c_0^2}{c(\mathbf{x})^2}. \quad (3.11)$$

For weak inhomogeneity $c_0^2/c(\mathbf{x})^2$ must be ~ 1. Therefore

$$\lambda_0^2 \frac{\nabla^2 A(\mathbf{x})}{A(\mathbf{x})} \ll 1. \quad (3.12)$$

It is possible to use a scale analysis to add physical insight into this equation. From (3.9), we see that $\nabla W(\mathbf{x}) \cdot \nabla W(\mathbf{x}) = c_0^2/c(\mathbf{x})^2$, which implies $\nabla W(\mathbf{x}) \approx c_0/c(\mathbf{x})$. From Eq. (3.7), we can write

$$\nabla^2 W(\mathbf{x}) \approx \nabla W(\mathbf{x}) \cdot \frac{\nabla A(\mathbf{x})}{A(\mathbf{x})} \quad (3.13)$$

or

$$\frac{\nabla A(\mathbf{x})}{A(\mathbf{x})} \approx \frac{\nabla^2 W(\mathbf{x})}{\nabla W(\mathbf{x})}$$

$$\approx \frac{\nabla(c_0/c(\mathbf{x}))}{c_0/c(\mathbf{x})} \approx \frac{\nabla c(\mathbf{x})}{c(\mathbf{x})}. \quad (3.14)$$

If we further compute the gradient over a wavelength and multiply by λ_0^2, we can use (3.12) to find

$$\frac{\lambda_0^2 \nabla^2 A(\mathbf{x})}{A(\mathbf{x})} \approx \frac{\lambda_0 \delta[\Delta c(\mathbf{x})]}{c(\mathbf{x})} \ll 1, \quad (3.15)$$

which states that the *eikonal equation will approximate the wave equation well if the fractional change in velocity gradient over one seismic wavelength is small compared to the velocity.*

It appears that the eikonal equations in (3.9) are complex, and they do not seem any easier to deal with than the wave equation! However, we will see that very simple equations are obtained for rays from the eikonal. The concept of rays is extremely important and is the basis of almost all body-wave interpretation. Rays allow us to track a displacement pulse from a source to a receiver, accounting for localized properties on the specific path. The conditions of validity require wavelengths smaller than a few hundred kilometers and slowly varying seismic velocities, criteria that apply to most body waves in the Earth's deep interior. An obvious question is, Are rays an adequate solution to the wave equation at boundaries between materials with different elastic moduli? Clearly condition (3.15) is violated in the presence of strong velocity gradients, but we can cast the problem as a series of discrete regions where ray representations are sufficient. The ray solutions in these regions are combined by matching boundary conditions. We discuss this in detail in later sections.

Representing a portion of a seismic wavefield as a *ray* gives rise to the concept of *seismic phases* or *arrivals*. These correspond to transient disturbances at a receiver that are P or S waves that have traveled a defined path between the seismic source and receiver. These arrivals have two primary characteristics: *travel time* and *amplitude*. The eikonal equation and its extensions can be used to quantify these two parameters. In this chapter we first develop equations for travel times and then discuss how seismic-pulse amplitude varies as it propagates.

3.1 The Eikonal Equation and Ray Geometry

Consider the three-dimensional wave surface shown in Figure 3.2. The ray, which is the normal to the wavefront, $W(\mathbf{x})$, is characterized by traveling an arc length, s, in a time, t. The direction cosines associated with the ray are given by dx_1/ds, dx_2/ds, and dx_3/ds, and must satisfy

$$\left(\frac{dx_1}{ds}\right)^2 + \left(\frac{dx_2}{ds}\right)^2 + \left(\frac{dx_3}{ds}\right)^2 = 1. \quad (3.16)$$

Now consider the physical connection between s and $W(\mathbf{x})$: $\nabla W(\mathbf{x}) \propto s$, which is just the statement that the gradient of a function (surface) is oriented normal to that function (surface). Thus we can see that dx_i/ds must be proportional to $\partial W(\mathbf{x})/\partial x_i$. This implies that we can rewrite (3.16) as

$$\left(a\frac{\partial W(\mathbf{x})}{\partial x_1}\right)^2 + \left(a\frac{\partial W(\mathbf{x})}{\partial x_2}\right)^2$$
$$+ \left(a\frac{\partial W(\mathbf{x})}{\partial x_3}\right)^2 = 1, \quad (3.17)$$

where a is a constant of proportionality. A comparison of (3.17) and (3.9) shows that (3.17) is just the eikonal equation if $a = c(\mathbf{x})/c_0$. The reciprocal, $a^{-1} = n = c_0/c(\mathbf{x})$

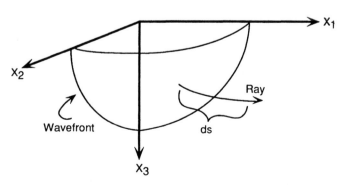

FIGURE 3.2 Three-dimensional wavefront with a ray or normal with length *ds*.

is commonly called the *index of refraction*. Equations (3.16) and (3.17) can be combined to give the *normal* equations:

$$n \frac{dx_1}{ds} = \frac{\partial W(\mathbf{x})}{\partial x_1}$$

$$n \frac{dx_2}{ds} = \frac{\partial W(\mathbf{x})}{\partial x_2}$$

$$n \frac{dx_3}{ds} = \frac{\partial W(\mathbf{x})}{\partial x_3}. \qquad (3.18)$$

Now let us consider how the normal equations change along the path of the ray. We can do this by taking the derivative of the normal equations with respect to *ds*

$$\frac{d}{ds}\left(n \frac{dx_1}{ds}\right) = \frac{d}{ds}\left(\frac{\partial W(\mathbf{x})}{\partial x_1}\right)$$

$$= \frac{\partial}{\partial x_1}\left(\frac{\partial W(\mathbf{x})}{\partial x_1} \frac{dx_1}{ds}\right.$$

$$+ \frac{\partial W(\mathbf{x})}{\partial x_2} \frac{dx_2}{ds} + \frac{\partial W(\mathbf{x})}{\partial x_3} \frac{dx_3}{ds}\right)$$

$$= \frac{\partial}{\partial x_1}\left\{n\left[\left(\frac{dx_1}{ds}\right)^2 + \left(\frac{dx_2}{ds}\right)^2\right.\right.$$

$$\left.\left. + \left(\frac{dx_3}{ds}\right)^2\right]\right\}$$

$$= \frac{\partial}{\partial x_1} n. \qquad (3.19)$$

The generalized form of this equation is called the *raypath equation*:

$$\frac{d}{ds}\left(n \frac{dx_i}{ds}\right) = \frac{\partial n}{\partial x_i}$$

$$\frac{d}{ds}\left(\frac{1}{c(\mathbf{x})} \frac{d\mathbf{x}}{ds}\right) = \nabla\left(\frac{1}{c(\mathbf{x})}\right). \qquad (3.20)$$

This is a second-order differential equation for **x**, which is just the raypath; note

that the raypath is proportional to the spatial change in the velocity distribution. Two initial conditions control the behavior of (3.20): (1) the direction in which the ray leaves some arbitrary reference point $(\partial \mathbf{x}/\partial s)|_{s_0}$ and (2) the position of the reference point s_0.

We can obtain some insight into the physics of (3.20) by considering a simple example. If we follow a ray through a material that has a change in velocity in only one direction, say depth, then $c = c(x_3)$, and thus $n = n(x_3)$. Thus $\partial n/\partial x_1 = \partial n/\partial x_2 = 0$. Then (3.20) reduces to

$$n \frac{dx_1}{ds} = c_1 = \text{constant}$$

$$n \frac{dx_2}{ds} = c_2 = \text{constant}$$

$$\frac{d}{ds}\left(n \frac{dx_3}{ds}\right) = \left(\frac{dn}{dx_3}\right). \qquad (3.21)$$

The ratio of c_1 to c_2 confines the raypath to a plane that is normal to the $x_1 x_2$ plane. (In other words, the projection of the ray into the $x_1 x_2$ plane is a straight line.) Figure 3.3 shows the geometry. For convenience, and without loss of generality, we can choose this plane to coincide with the $x_1 x_3$ plane, reducing (3.21) to

$$n \frac{dx_1}{ds} = \text{constant}$$

$$\frac{d}{ds}\left(n \frac{dx_3}{ds}\right) = \left(\frac{dn}{dx_3}\right). \qquad (3.22)$$

At a given point the direction cosine of the ray is given by

$$l_1 = \frac{dx_1}{ds} = \sin i$$

$$l_3 = \frac{dx_3}{ds} = \cos i. \qquad (3.23)$$

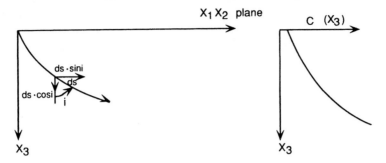

FIGURE 3.3 Raypath for a medium in which the velocity is independent of the x_2 and x_1 directions.

Thus

$$n\frac{dx_1}{ds} = \frac{c_0}{c}\sin i = \text{constant}$$

$$(3.24)$$

$$\Rightarrow \frac{\sin i}{c} = \text{constant} = p.$$

The constant p is called the *ray parameter*, or *horizontal slowness*. p varies from 0 (vertical travel path) to $1/c$ (horizontal travel path). The angle i is called the *angle*

of incidence, and it gives the *inclination* of a ray measured from the vertical (x_3 direction) at any given depth. For a prescribed reference point and takeoff angle, a ray will have a constant ray parameter, p, for the entire path. Equation (3.24) is also known as Snell's law, which can also be derived from Fermat's principle (see Box 3.1). Fermat's principle states that a raypath is a path of stationary time. Thus travel time along a raypath is a minimum (or maximum) time.

Box 3.1 Geometric Interpretation of Snell's Law

It is possible to use simple ray geometry and Fermat's principle of least time to derive *Snell's law* and the definition of seismic ray parameter. Consider a ray leaving point P in a medium of velocity α_1; what is the path the ray will take to arrive at point P' in a medium of velocity α_2? Figure 3.B1.1 shows the geometry.

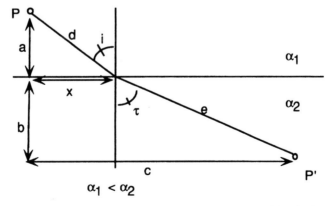

FIGURE 3.B1.1 Raypath connecting two points on either side of a boundary.

---

continues

The travel time on the path between P and P' is

$$T_{P-P'} = \frac{d}{\alpha_1} + \frac{e}{\alpha_2} = \frac{\sqrt{a^2 + x^2}}{\alpha_1} + \frac{\sqrt{b^2 + (c-x)^2}}{\alpha_2}. \tag{3.1.1}$$

The minimum-time path must satisfy $dT/dx = 0$, which implies

$$\frac{dT}{dx} = 0 = \frac{x}{\alpha_1 \sqrt{a^2 + x^2}} - \frac{c-x}{\alpha_2 \sqrt{b^2 + (c-x)^2}}; \tag{3.1.2}$$

note that $x/\sqrt{a^2 + x^2} = \sin i$, and $(c-x)/\sqrt{b^2 + (c-x)^2} = \sin \tau$. Thus

$$\frac{\sin i}{\alpha_1} = \frac{\sin \tau}{\alpha_2}. \tag{3.1.3}$$

This is the familiar expression from optics called Snell's law after Willebrod Snell (1591–1626). The generalization of Snell's law is $\sin i/v = p$, where p is called the *seismic parameter*, *ray parameter*, or *horizontal slowness*. The ray parameter is constant for the entire travel path of a ray. The consequence of a ray traversing material of changing velocity, v, is a change in inclination angle, i, with respect to a reference plane. As a ray enters material of increasing velocity, the ray is *deflected* toward the horizontal. Conversely, as a ray enters material of decreasing velocity, it is deflected toward the vertical. If the ray is traveling vertically, then $p = 0$, and the ray will experience no deflection as velocity changes.

Now let us consider the second equation in (3.22)

$$\frac{d}{ds}\left(n \frac{dx_3}{ds}\right) = \frac{d}{ds}(n \cos i) = \frac{dn}{dx_3}.$$

Rewriting this using the chain rule

$$\frac{dn}{dx_3} = -n \sin i \frac{di}{ds} + \cos i \frac{dn}{dx_3} \frac{dx_3}{ds}$$

$$= -n \sin i \frac{di}{ds} + \cos^2 i \frac{dn}{dx_3}.$$

Collecting terms,

$$\frac{dn}{dx_3}(1 - \cos^2 i) = -n \sin i \frac{di}{ds}$$

$$\Rightarrow \frac{di}{ds} = -\frac{\sin i}{n} \frac{dn}{dx_3} \tag{3.25}$$

$$\Rightarrow \frac{di}{ds} = \frac{\sin i}{c} \frac{dc}{dx_3} = p \frac{dc}{dx_3}.$$

Equation (3.25) states that the curvature of a ray is directly proportional to the

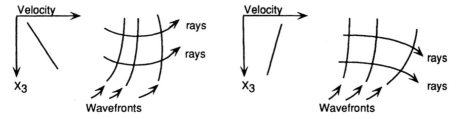

FIGURE 3.4 Ray curvature due to increasing and decreasing velocity with depth.

velocity gradient (dc/dx_3). If velocity *increases* with depth, then the ray curves *upward*. If velocity *decreases* with depth, then the ray curves *downward*. Figure 3.4 shows this by plotting the evolution of a wavefront in media with different velocity distributions.

Equation (3.22) has several interesting aspects. For each angle i, a specific ray leaves the source and follows a specific raypath. The initial angle and the velocity structure determine the distance at which the ray will emerge at the surface. For a given source–receiver geometry several possible connecting raypaths may exist, which means that a *multiplicity* of arrivals will occur, all with different initial angles and travel times. We will discuss this more fully in later sections, as it is the basis for seismic interpretation. We can use (3.22) with initial conditions to predict where and when a ray will arrive. Consider Figure 3.5. At any point along the travel path we have

$$\sin i = \frac{dx_1}{ds} = cp$$

$$\cos i = \frac{dx_3}{ds} = \sqrt{1 - \sin^2 i} = \sqrt{1 - c^2 p^2}$$

$$\text{(3.26)}$$

$$\Rightarrow dx_1 = ds \sin i = \frac{dx_3}{\cos i} cp$$

$$= \frac{cp}{\sqrt{1 - c^2 p^2}} dx_3.$$

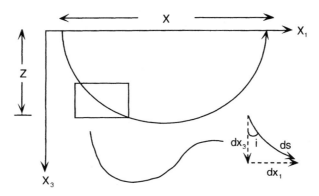

FIGURE 3.5 Geometry of the ray segment *ds*, along a path from a surface source to a surface receiver. The velocity of the medium varies only along the x_3 direction, so there will be symmetry of downgoing and upgoing legs of the raypath.

For a surface source and receiver, Eq. (3.26) can be integrated over the depth range traversed by the ray to give the distance $X(p)$, at which a ray with ray parameter p will emerge:

$$X(p) = 2\int_0^z \frac{cp}{\sqrt{1 - c^2 p^2}}\, dx_3, \quad (3.27)$$

where z is the maximum depth of penetration. The factor of 2 arises from the symmetry of the downgoing and upgoing portions of the raypath (see Figure 3.5). This is the *where* of ray equations; given the angle at which a ray leaves the source, we can calculate where it will arrive. If we generalize this to a three-dimensional case, we also require the azimuth of the raypath relative to the source. The time it takes for the ray to arrive is obtained similarly:

$$dT = \frac{ds}{c} \Rightarrow T$$

$$= 2\int_{\text{path}} \frac{ds}{c(s)}$$

$$= 2\int_0^z \frac{dx_3}{c(x_3)\cos i}$$

or

$$T = 2\int_0^z \frac{dx_3}{c^2\sqrt{1/c^2 - p^2}}, \quad (3.28)$$

where T is the travel time along the raypath to the distance defined by Eq. (3.27). We can introduce some shorthand and rewrite Eqs. (3.27) and (3.28). Let $\gamma = 1/c$; then

$$X = 2p\int_0^z \frac{dx_3}{\sqrt{\gamma^2 - p^2}} \quad (3.29)$$

and

$$T = 2\int_0^z \frac{\gamma^2}{\sqrt{\gamma^2 - p^2}}\, dx_3. \quad (3.30)$$

Note that the ray parameter, p, can be pulled outside the integral in (3.29) because it is constant along the path. Noting the similarity between X and T, we can relate the two:

$$T = 2\int_0^z \frac{\gamma^2}{\sqrt{\gamma^2 - p^2}}\, dx_3$$

$$= 2\int_0^z \left(\frac{p^2}{\sqrt{\gamma^2 - p^2}} + \sqrt{\gamma^2 - p^2} \right) dx_3$$

$$\quad (3.31)$$

$$T = pX + 2\int_0^z \sqrt{\gamma^2 - p^2}\, dx_3.$$

Equation (3.31), the travel-time equation, is a truly remarkable representation. Note that it has two terms: one depends on X and the other on z. This implies that the travel-time equation is *separable*, and the *vertical* travel time depends only on $(\gamma^2 - p^2)^{1/2}$ (usually written as η) and the horizontal travel time only on p, hence the name "horizontal slowness" for p. Similarly, η is known as the vertical slowness. Also note that $dT/dX = p$, or that the *change in travel time with distance is equal to the ray parameter*. We will use this fact extensively when we interpret the structure of the Earth.

We can also use Eq. (3.22) to give insight into the *amplitude* of a seismic arrival. Consider a spherical wave a small distance from the seismic source at the surface in a region of uniform velocity. Let the energy of the disturbance be distributed uniformly on the spherical wavefront. As the wavefront expands with time, the total energy on the surface will remain constant, but the *energy per unit surface area* will *decrease*. Define the total energy on the initially hemispherical wavefront (Figure 3.6) as K, and the energy per unit area $= K/2\pi r^2$. Now consider a bundle of rays that leave the source between the angle i_0 and $i_0 + di_0$. The fraction of en-

FIGURE 3.6 The area, which is inversely proportional to energy, for an expanding spherical wavefront.

ergy in a circular ring on the wavefront defined by these two takeoff angles is given by

$$E_0 = \frac{K}{2\pi r^2}(2\pi r \sin i_0)(di_0 r), \quad (3.32)$$

where $r \sin i_0$ is the radius of the strip, and $di_0 r$ is the width of the strip. Or

$$E_0 = K \sin i_0 \, di_0. \quad (3.33)$$

As seen in Figure 3.7, the ray bundle expands or contracts depending on the velocity profile. Upon arrival at the surface, the corresponding energy will be spread out over area $2\pi X dX \cos i_0$ (Figure 3.6). The wave energy is now spread over this larger area, so the energy density, $E(X)$, is obtained by dividing (3.33) by the new area to obtain

$$E(X) = \frac{K}{2\pi} \frac{\tan i_0}{X} \frac{di_0}{dX}. \quad (3.34)$$

This can be simplified by recalling

$$p = \frac{\sin i_0}{c_0} = \frac{dT}{dX}, \quad i_0 = \sin^{-1}\left(c_0 \frac{dT}{dX}\right). \quad (3.35)$$

Therefore

$$\frac{di_0}{dX} = \frac{c_0}{\sqrt{1 - c_0^2(dT/dX)^2}} \frac{d^2T}{dX^2}$$

$$= \frac{c_0}{\cos i_0} \frac{d^2T}{dX^2}. \quad (3.36)$$

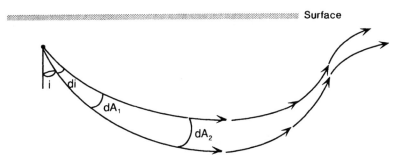

FIGURE 3.7 A bundle of rays with takeoff angles between i and $i+di$. The amplitude of the seismic signal is inversely proportional to the surface area of the wavefront subtended by the rays (dA); as the ray bundle expands or contracts due to the velocity structure, the amplitude will change. One can see how there will be a relationship between changes in takeoff angles as a function of distance and the corresponding amplitude variations with distance.

Thus Eq. (3.34) can be rewritten as

$$E(X) = \left(\frac{K}{2\pi}\right) c_0 \left(\frac{\tan i_0}{X \cos i_0}\right) \left(\frac{d^2T}{dX^2}\right).$$

$$(3.37)$$

If the source is not at the surface, then the takeoff angle at the source, i_0, will differ from the incident angle at the receiver, i. *Amplitude* is proportional to $\sqrt{E}$; thus the amplitude of a seismic arrival is proportional to the *change* in ray parameter with distance. Velocity structures for which p changes rapidly yield large amplitude variations. Conversely, constant p implies very small amplitudes.

These simple extensions of the ray equations show their utility. We will now consider some practical cases of a layered velocity structure and a continuous velocity distribution.

3.2 Travel Times in a Layered Earth

The standard method of inferring the velocity structure of the Earth is to fit the travel times of various seismic phases as a function of distance with a *layered* Earth model. The equations for travel time in a layered Earth are a discretization of Eq. (3.31). We also can derive these equations by first principles. When a ray strikes a boundary marking a change in seismic velocity (see Figure 3.8), the energy in the wave is partitioned between a *reflected* and a *refracted* ray. These two new, or *derivative*, rays will have the same ray parameter as the incident ray. The angle (i or τ) that the reflected and refracted rays make with a vertical plane is governed by Snell's law:

$$\frac{\sin i}{\alpha_1} = \frac{\sin \tau}{\alpha_2} = p. \qquad (3.38)$$

Consider the wavefront associated with the reflected ray in Figure 3.8. The wavefront will advance a distance d in a time δt; $\delta t = d/\alpha_1$. The surface intersection of the wavefront will travel along the surface at a higher velocity than the actual seismic velocity of the layer

$$\alpha_a = \frac{x}{\delta t} = \frac{d}{\sin i} \frac{1}{\delta t} = \frac{\alpha_1}{\sin i} = \frac{1}{p}, \qquad (3.39)$$

where α_a is the *apparent* velocity. From this equation it is obvious where the name *horizontal slowness* for p comes from. If

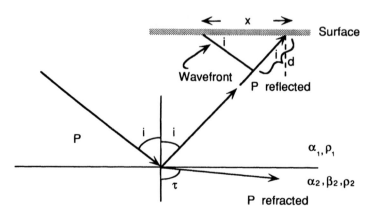

FIGURE 3.8 A *P* wave incident on a boundary between contrasting materials, in this case between a fluid layer and an underlying solid.

the ray were vertically incident on the free surface, p would be zero and the apparent velocity would be infinite.

If the velocity in layer 2 is greater than the velocity in layer 1, angle $\tau > i$. As $\tau \to 90°$, Snell's law predicts a *critical refraction*

$$\frac{\sin i_c}{\alpha_1} = \frac{\sin 90°}{\alpha_2} = \frac{1}{\alpha_2}. \quad (3.40)$$

This critical refraction is associated with a wave that is traveling horizontally (parallel to the interface between layers 1 and 2) immediately below the interface. This wave is usually referred to as a *head wave*, and it has the unique property that it transmits energy back into layer 1 continually as it travels along the interface. This energy leaves the interface with the same angle of incidence, i_c, called the *critical angle*:

$$i_c = \sin^{-1}(\alpha_1/\alpha_2). \quad (3.41)$$

Note that if $i > i_c$, no seismic energy can penetrate layer 2, and all the energy is reflected back into layer 1. If $\alpha_2 < \alpha_1$, there is no critical angle, and the refracted ray is deflected toward the vertical.

Head waves in a layered structure and their analog in a continuous velocity structure, turning rays (discussed in the next section), are extremely important in determining the velocity structure of the Earth. The travel time of these seismic waves as a

function of distance provides a direct measure of velocity at depth. Consider the three rays in the layer over a half-space structure shown in Figure 3.9. If $\alpha_2 > \alpha_1$, three *primary* travel paths exist between the source and the receiver: (1) the direct arrival, which travels in a straight line connecting source and receiver, (2) a reflected arrival, and (3) a head wave. Additional rays involving multiple reflections in the layer will also exist. The travel time for the direct arrival is given by

$$T = X/\alpha_1 = Xp, \qquad p = 1/\alpha_1. \quad (3.42)$$

The travel time for the reflected arrival is given by

$$T = \frac{2\,th}{\cos i} \frac{1}{\alpha_1}, \quad (3.43)$$

where th is the layer thickness. Finally, the travel time of the head wave is given by

$$T = \frac{r}{\alpha_2} + \frac{2th}{\cos i_c} \frac{1}{\alpha_1}. \quad (3.44)$$

These equations are all for a surface source; slight modifications are needed for sources within the layer. The second term in (3.44) is the same as (3.43) for the reflected arrival when $i = i_c$. Therefore, the refracted arrival first appears at $r = 0$, with a travel time equal to that of the reflected arrival. At closer distances only

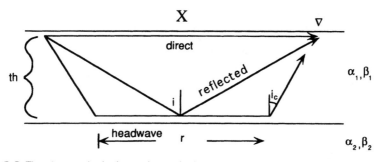

FIGURE 3.9 The three principal rays in a velocity structure that is a layer over a half-space.

the direct and reflected waves will exist. As X increases, only the r/α_2 term of (3.44) is affected; thus, the wavefront travels along the surface with apparent velocity α_2. This can be used to simplify (3.44), because $r = (X - 2th\tan i_c)$ and $\sin i_c = \alpha_1/\alpha_2$. Thus

$$T = \frac{2th}{\cos i_c}\frac{1}{\alpha_1} + \frac{1}{\alpha_2}\left(X - \frac{2th\alpha_1}{\alpha_2\cos i_c}\right)$$

$$\Rightarrow T = \frac{2th}{\cos i_c}\left(\frac{1}{\alpha_1} - \frac{\alpha_1}{\alpha_2^2}\right) + \frac{X}{\alpha_2}. \quad (3.45)$$

Recalling that $1/\alpha_2 = p$ and $\cos i_c = (1 - \sin^2 i_c)^{1/2} = (1 - \alpha_1^2 p^2)^{1/2}$, we can rewrite this as

$$T = Xp + 2th\eta_1, \quad (3.46)$$

where $\eta_1 = (1 - p^2\alpha_1^2)^{1/2}/\alpha_1$. This is the layered structure equivalent to Eq. (3.31). Equation (3.46) is an extremely useful form of the travel-time equation because it separates the travel path into a *horizontal* term and a *vertical* term. No matter how complex a raypath in a layered structure

becomes, it is possible to write the corresponding travel-time equation with a form similar to (3.46).

Equations (3.42), (3.43), and (3.46) determine a *travel-time curve*, giving the expected travel times for a given structure. Figure 3.10 shows the travel-time curve for the principal rays for the structure in Figure 3.9. At short distances only the reflected and direct arrivals exist. The direct arrival is described by a straight line, with a slope $dT/dX = p = 1/\alpha_1$. The reflected-arrival travel time is described by a hyperbola. The intercept, at $X = 0$, has a travel time of $2th/\alpha_1$. At large distances the *branch* of the travel-time curve that corresponds to the reflection becomes asymptotic to the direct arrival. The travel-time branch associated with the head wave first appears as a reflection at $X = 2th\tan i_c$. The head-wave arrival branch is a straight line with a slope $dT/dX = p = 1/\alpha_2$. Since the head wave travels with a faster apparent velocity, it eventually becomes the *first* arrival. The direct arrival is the first arrival until the *crossover distance*, X_c, after which the head wave is the first arrival. We can find this distance by realizing that at X_c the travel times of the direct arrival and head wave

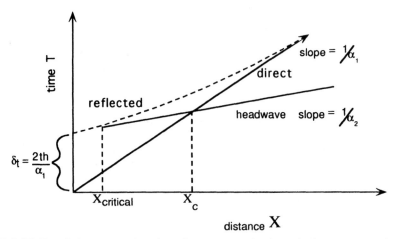

FIGURE 3.10 Travel-time curve for the primary waves in the velocity structure in Figure 3.9.

are equal:

$$T_{\text{direct}} = T_{\text{head}}$$

$$\Rightarrow \frac{X_c}{\alpha_1} = \frac{X_c}{\alpha_2} + 2th\eta_1$$

$$X_c = 2th\frac{\alpha_1\alpha_2}{\alpha_2 - \alpha_1}\eta_1$$

or

$$X_c = 2th\sqrt{\frac{\alpha_2 + \alpha_1}{\alpha_2 - \alpha_1}}. \qquad (3.47)$$

Figure 3.11 shows a seismogram from an earthquake 314 km away. Three prominent arrivals are noted: P_n, P_g, and S. Arrivals P_n and P_g correspond to the head wave and direct arrival, respectively, in Figure 3.9. In 1909 a Croatian scientist named Mohorovičić first observed these two P-wave arrivals with different apparent velocities over a several-hundred-kilometer distance. One was observed to have a velocity of 5.6 km/s, and the other a velocity of 7.9 km/s. The arrivals arise because, in a gross sense, the crust–mantle

system behaves like a layer over a half-space. The head wave, P_n, is caused by the large velocity increase at the crust–mantle boundary (known as the *Moho* discontinuity). The reflection off the Moho is known as *PmP* and is not readily identifiable in Figure 3.11.

Is the travel-time curve in Figure 3.10 complete? No, because we have only considered three rays. Clearly, S_n, S_g, and *SmS* arrivals will also occur, with travel times controlled by S velocities. Further, multiple reflections will occur in which a ray bounces between the surface and the Moho, with some arrivals having various path segments which are a mixture of P and S waves. The many possible arrivals cause the oscillations in Figure 3.11. The importance of a travel-time curve is its interpretive power. If we consider a seismic station at a given distance from a seismic source, we expect a sequence of arrivals, all with predictable travel times. Suppose we have many seismic stations that record a seismic event. If we determine the arrival times of various phases and plot them on a time–distance curve, we can infer the structure. We can deter-

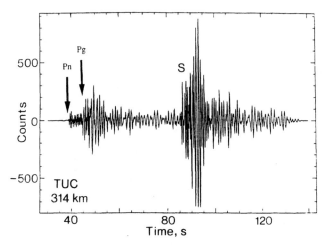

FIGURE 3.11 Vertical component of ground motion for an earthquake ($M_L = 2.7$) on the Colorado Plateau in Arizona recorded at TUC (Tucson). The head wave and direct arrival are marked P_n and P_g, respectively.

mine the layer velocity from the slope of the direct arrival and the half-space velocity from the slope of the head-wave branch. The crossover distance or the zero offset ($X = 0$) reflection time gives the layer thickness. Of course, the Earth is more complex than a layer over a half-space, so this procedure must be generalized to more than one layer. If the source is not human induced, it is also necessary to estimate the location and origin time of the source.

The travel-time equations can be generalized to the case of n layers. Consider the two-layered example shown in Figure 3.12. The crust is often approximated as a two-layer structure like this. For a surface source, the travel times for the primary waves in the top layer are just those given above for a layer over a half-space. As rays penetrate deeper into the structure, the expressions become more complicated, but they are easily built up using the single-layer equations. First, consider the portion of the head-wave travel time from point A to point B (a horizontal distance $y = X - 2\,\Delta X$):

$$t' = yp + 2\,th_2 \eta_2, \qquad (3.48)$$

where

$$p = \frac{1}{\alpha_3} = \frac{\sin i_c}{\alpha_2}. \qquad (3.49)$$

Now consider the travel time in layer 1:

$$\Delta t = \frac{D}{\alpha_1} = \frac{D(\sin^2 j + \cos^2 j)}{\alpha_1}$$

$$= D \sin j \left(\frac{\sin j}{\alpha_1} \right) + D \cos j \left(\frac{\cos j}{\alpha_1} \right)$$

$$= (\Delta X)p + th_1 \eta_1, \qquad (3.50)$$

where $\Delta X = D \sin j$, $p = (\sin j)/\alpha_1$, $th_1 = D \cos j$, and $\eta_1 = (\cos j)/\alpha_1 = (1/\alpha_1)(1 - \alpha_1^2 p^2)^{1/2}$. Thus we can write

$$T = t' + 2\,\Delta t = pX + 2 \sum_{i=1}^{2} th_i \eta_i, \qquad (3.51)$$

which can be generalized to n layers

$$T = pX + 2 \sum_{i=1}^{n} th_i \eta_i. \qquad (3.52)$$

If α in a many-layered structure increases monotonically with depth, the travel-time curve will have many branches due to head waves at each interface (see Figure 3.13). These will define a first-arrival "branch" that asymptotically corresponds to an inhomogeneous structure with a smooth velocity increase instead of layers.

We need to mention two special cases that complicate the interpretation of travel-time curves. The first of these is the case of a low-velocity zone. Consider the structure shown in Figure 3.14, where the velocity of layer 2 is *less* than that of layer

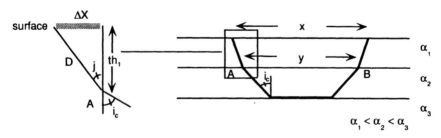

FIGURE 3.12 Head-wave raypath in a two-layered model.

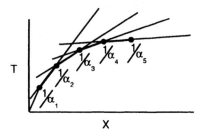

FIGURE 3.13 Travel-time curve for a finely layered Earth. The first arrival is comprised of short segments of the head-wave curves for each layer, over the limited distance range between crossover points.

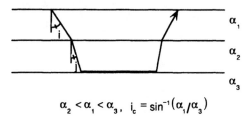

$$\alpha_2 < \alpha_1 < \alpha_3, \quad i_c = \sin^{-1}(\alpha_1/\alpha_3)$$

FIGURE 3.14 Raypath for a structure with a *low-velocity layer*. No head wave will exist on the interface between velocities α_1 and α_2.

1 and the half-space. No head wave occurs along the interface between layers 1 and 2. Therefore, we observe only a direct arrival and a head wave from the interface between layers 2 and 3 (as well as reflected arrivals from both interfaces). The corresponding travel-time equations for the direct wave and the head wave are

$$T = pX, \quad \text{where } p = 1/\alpha_1 \quad (3.53)$$

$$T = pX + 2th_1\eta_1 + 2th_2\eta_2,$$

$$\text{where } p = 1/\alpha_3. \quad (3.54)$$

Since the travel-time curve has only two branches (given no information from the reflected branches), one would interpret the curve as a *single* layer of pseudothickness $\widehat{th}$ with velocity α_1 over a half-space

of velocity α_3. The pseudothickness estimated from the crossover distance is $th_1 + th_2(\eta_2/\eta_1)$, which results in an *overestimate* of the actual depth to the half-space.

A second special case is called a *blind zone*, which arises when a layer is so thin that the head wave from it is never a first arrival. Consider the structure shown in Figure 3.15. The travel times for the two rays shown are

$$T_1 = (pX) + 2th_1\eta_1, \quad (3.55)$$

$$\text{where } p = 1/\alpha_2$$

$$T_2 = (pX) + 2th_1\eta_1 + 2th_2\eta_2, \quad (3.56)$$

$$\text{where } p = 1/\alpha_3.$$

a
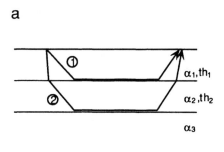

b
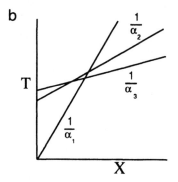

FIGURE 3.15 Travel path and corresponding travel-time curve for a blind zone. The observability of a first arrival with the slowness $1/\alpha_2$ depends on the layer thickness and velocity contrasts involved.

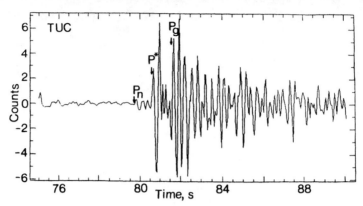

FIGURE 3.16 P* is a prominent "blind-zone" arrival generated at a midcrustal boundary. Although it travels with a faster apparent velocity than P_g, it is never a first arrival.

Note that η_1 for ray 1 *does not* equal η_1 for the second ray because p is different. For particular combinations of α_2, α_3, and th_2, the travel-time curve will look like that shown in Figure 3.15b, where the head wave with slope $1/\alpha_2$ is not observed as a first arrival. This happens if α_2/α_1 is not much larger than 1 or if th_2 is very small. Note that as th_2 *increases*, the travel-time branch associated with the half-space is *delayed*, and eventually the $1/\alpha_2$ branch will be a first arrival over a limited range. One of the most important blind zones in earthquake seismology is in the crust, due to the Conrad discontinuity. The Conrad was originally thought to represent a boundary between mafic and granitic rocks at midcrustal depths, but now it is thought to be a thermodynamically controlled interface or a rheological boundary (more on the Conrad in Chapter 7). The Conrad head wave is often denoted as P^*. Figure 3.16 shows an observed seismogram with P_n, P^*, and P_g.

3.3 Travel-Time Curves in a Continuous Medium

If we take Eq. (3.52) and let the number of layers go to infinity as each layer thick-ness goes to zero, the summation is replaced by integration, which yields Eq. (3.31). In other words, fine layering is an approximation to a continuous velocity distribution. Subtle differences occur in the character of the travel-time curves. Figure 3.17 shows the travel-time curve for a continuous, increasing velocity distribution. The slowness observed at a distance X can be found by taking the slope (dT/dX) or tangent of the travel-time curve. It is convenient to introduce the concept of *intercept*, or *delay time*,

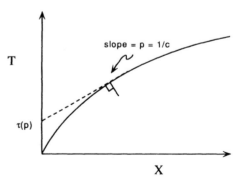

FIGURE 3.17 Travel-time curve in a continuous velocity structure. τ is defined as the intercept of the tangent to the travel-time curve at any given X, which has slope p.

Box 3.2 Travel Times for Dipping Layers

The most common complication to the travel-time equation for a plane-layered Earth is the presence of a *dipping* layer. Consider the structure shown in Figure 3.B2.1. The travel-time curve depends on whether the rays are traveling *updip* or *downdip*. The direct arrivals have the same slowness $(1/v_1)$, but the head waves have different apparent velocities. This can be seen by the area swept out by the wavefronts in a time Δt as they are incident on the surface at different angles:

$$\frac{1}{v_u} = \frac{\sin(i_c - \theta)}{v_1}$$

$$\frac{1}{v_d} = \frac{\sin(i_c + \theta)}{v_1}. \qquad (3.2.1)$$

The resulting travel-time equation can be written as

$$t_u = \frac{2h_2 \cos i_c \cos \theta}{v_1} + \frac{x \sin(i_c - \theta)}{v_1}$$

$$t_d = \frac{2h_1 \cos i_c \cos \theta}{v_1} + \frac{x \sin(i_c + \theta)}{v_1}, \qquad (3.2.2)$$

where u and d represent updip and downdip observations, respectively, Figure 3.B2.2 shows the corresponding travel-time curves. Note that the crossover distance is larger for updip travel paths. The total travel time for the source–receiver geometry must be the *same* because of *reciprocity* for interchanged source and receiver locations.

The dipping-layer problem is very common in refraction surveys, so we will give some of the equations required for their interpretation. The *true* velocity of the half-space is, of course, given by $v_1/(\sin i_c)$. We can solve for i_c and the angle of dip, θ, if we have *reversed* profiles on which we measure the apparent velocities, v_u and v_d,

$$\theta = \tfrac{1}{2}\left[\sin^{-1}(v_1/v_d) - \sin^{-1}(v_1/v_u)\right]$$

$$i_c = \tfrac{1}{2}\left[\sin^{-1}(v_1/v_d) + \sin^{-1}(v_1/v_u)\right]. \qquad (3.2.3)$$

Projecting the head-wave travel-time branches back to $x = 0$ gives intercept times, t_0, that differ for the updip and downdip directions. From these intercepts the layer thickness at each end of the profile can be determined.

$$h_1 \text{ or } h_2 = \frac{t_0 v_1 v_2}{2\cos\theta\sqrt{v_2^2 - v_1^2}}, \qquad (3.2.4)$$

continues

where t_0 is the appropriate intercept time for either updip or downdip, and v_2 is the half-space velocity.

The dip, θ, found from Eq. (3.2.3) must be interpreted carefully. This dip is the *true* dip only if the profile is perpendicular to the strike of the dipping layer. If the profile is oblique to the strike, the dip determined is actually an *apparent* dip. If the profile were parallel to the strike, the apparent dip would be zero. Without reversing the profile, one cannot be confident that layers do not dip, and incorrect structure may be inferred, so reverse profiling is a very common seismological procedure. Two-dimensional profiling can map the complete geometry.

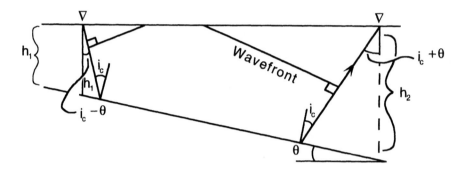

FIGURE 3.B2.1 Raypath geometry for head waves along a dipping interface.

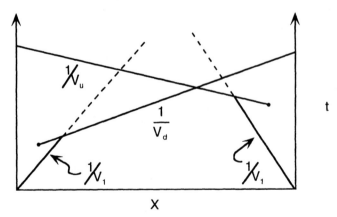

FIGURE 3.B2.2 Travel-time curves for a dipping structure. The curve with time increasing to the right is for the downdip direction. The curve with time increasing to the left is for the updip direction.

from (3.31)

$$\tau(p) = T - pX = 2\int_0^z \sqrt{\gamma^2 - p^2}\, dx_3.$$

$$(3.57)$$

where $\tau(p)$ is simply the intercept $(X = 0)$ of the tangent to the travel-time curve for a given X or p. As p increases, X decreases and τ will decrease; hence τ is a decreasing function of p:

$$\frac{d\tau}{dp} = \frac{d}{dp}\left(2\int_0^z \sqrt{\gamma^2 - p^2}\, dx_3\right)$$

$$= 2\int_0^z \frac{-p}{\sqrt{\gamma^2 - p^2}}\, dx_3$$

$$\Rightarrow \frac{d\tau}{dp} = -X. \qquad (3.58)$$

The *tau function*, $\tau(p)$, is a single-valued function of p and can simplify analysis of travel-time curves.

We will now characterize the travel-time curves for three major classes of continuous Earth structure. Figure 3.18 shows examples of three velocity models, the raypaths, the travel-time curves, p as a function of distance, and $\tau(p)$. In structure 1 (Figure 3.18a) the seismic velocity increases smoothly with depth, and the travel-time curve is a smooth, concave-downward curve. The ray parameter decreases monotonically with distance. Similarly, $\tau(p)$ is a smooth curve. In structure 2 (Figure 3.18b), the velocity gradient changes with depth; the velocity increases abruptly over a short depth interval. Seismic rays that turn above the gradient change are unaffected by it; hence the branch of the travel-time curve from A to B is identical to that for structure 1. Rays that enter the region of increased velocity gradient will be turned, or deflected, toward the horizontal. If the gradient is

strong enough, the rays will be turned such that they appear at some distance C that is *smaller* than B. Rays that bottom well below the gradient zone will have a normal concave shape. Note the similarity between the travel-time curve for structure 2 and that shown in Figure 3.10. The AB branch is analogous to the direct arrival, the CD branch is analogous to the refracted arrival, and the BC branch takes on the character of the reflected arrival. If the velocity gradient change increases to become a velocity discontinuity, the travel-time curve will approach the discrete layered case: branch AB will lengthen (point B will increase in distance X).

The distinctive "bow tie" shape of the travel-time curve shown in Figure 3.18b is called a *triplication*. The name comes from the fact that three distinct travel-time branches exist at certain distances. Seismograms at distances where the rays have passed through a structure such as that in Figure 3.18b can be quite complicated. The three different arrivals will interfere, and the character of interference will change very rapidly with distance. On the other hand, seismograms that are recorded across a triplication can be used to determine the character of the velocity change. The ray parameter is similarly multivalued in the region of the triplication, corresponding to the different branches of the travel-time curve. However, $\tau(p)$ is a single-valued function, which is one of the advantages of using it to "unfold" a triplication curve.

Structure 3 (Figure 3.18c) has a low-velocity zone beginning at a depth z_0. For rays that bottom above z_0, the travel-time curve is analogous to structure 1. As a ray penetrates below z_0, it is deflected toward the vertical, or bent down, and a *shadow* is produced at distances where no arrivals occur (distance B to D). At a depth z_1, where the velocity is equal to that at depth z_0, the shadow is terminated. Below this depth, two arrivals result from an effect

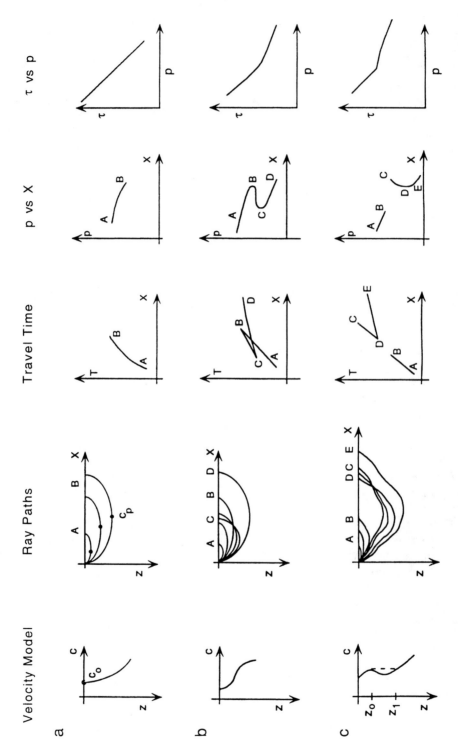

FIGURE 3.18 Three different velocity structures and corresponding ray geometries, travel-time curves, ray parameter-versus-distance variations, and τ-versus-p variations. (Modified from Officer, 1974.)

similar to the triplication. Theoretically this will result in a strong *cusp* at a distance D. The behavior of the slowness versus distance reflects the multivalued-ness associated with two arrivals. Again, $\tau(p)$ is smooth and decreases monotonically, although it will be discontinuous at the ray parameter corresponding to $1/c(z_0)$.

These three travel-time curves will be important references when we begin to interpret actual data profiles in Chapter 7. It is also important to note that the three representations of the evolving wavefield $[T(X), p(X),$ and $\tau(p)]$ are all equivalent. As we will see in Chapter 7, depending on the circumstances, we can use any of the three representations to infer structure.

3.4 Travel Times in a Spherical Earth

The travel-time equations derived in Section 3.2 are correct for a flat-layered Earth, that is, for problems in which the curvature of the Earth can be neglected. When curvature becomes important (at distances greater than about $12°$), we must modify Snell's law. Figure 3.19 shows a model of the Earth that is composed of thin, concentric shells. Across each of the shell boundaries is a discrete velocity jump. On the local scale, the surface curvature is negligible, so at position P Snell's law must be satisfied:

$$\frac{\sin \theta_1}{v_1} = \frac{\sin \theta_1'}{v_2}.$$ (3.59)

Now consider the geometry shown in Figure 3.19. Two right triangles share length d. It is clear that $d = r_1 \sin \theta_1' = r_2 \sin \theta_2$. Thus we can write

$$\frac{r_1 \sin \theta_1}{v_1} = \frac{r_2 \sin \theta_2}{v_2}.$$ (3.60)

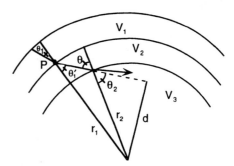

FIGURE 3.19 Ray geometry in a layered, spherical Earth.

This is a general equation along the entire raypath, since r_1 and r_2 can be any value along the raypath; thus these ratios are *constant*. We use (3.60) to define the ray parameter in a spherical Earth:

$$\frac{r \sin i}{v} = p.$$ (3.61)

Although the units of ray parameter in a spherical Earth differ from those we obtained for flat layers, the meaning is the same, with p being the slope of the travel-time curve. Consider Figure 3.20, which traces the path of two adjacent rays. The parameters of the two rays are p, Δ, and T (ray parameter, angular distance, and travel time, respectively), and $p + dp$, $\Delta + d\Delta$, and $T + dT$. From the geometry of the problem we can see that

$$\sin i_0 = \frac{v_0}{r_0}\frac{dT}{d\Delta} \Rightarrow \frac{r_0 \sin i_0}{v_0} = \frac{dT}{d\Delta} = p.$$ (3.62)

The ray parameter p is precisely the slope of the travel-time curve, as it was for the flat-Earth case except that distance is now measured in angular degrees. The ray parameter can still be identified with the inverse *apparent velocity* along the surface, or slowness. At the turning point, $p =$

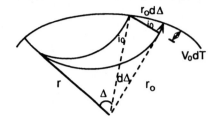

FIGURE 3.20 Raypaths for two adjacent rays in a spherical Earth.

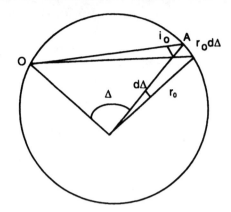

FIGURE 3.21 Travel paths in a homogeneous sphere.

$(\sin 90°)(r_t/v_t) = r_t/v_t = \xi_t$. Thus, unlike the flat-layered case, the slowness is a ratio between a velocity and a depth. This is because in a spherical Earth, every ray will return to the surface—even if the velocity decreases with depth. The units of p in a spherical Earth are s/rad or s/deg. (The "natural" units are s/rad, and care must be taken to use these units when applying inversion formulas such as the Herglotz–Wiechert technique. See Chapter 7.)

The travel-time equations in a sphere must also reflect geometric constraints. Consider a homogeneous sphere with a ray that travels from source to receiver (see Figure 3.21). The travel path is, of course, a straight line, and the travel time is given by OA/v. This can be written as

$$T(\Delta) = \frac{2r_0 \sin(\Delta/2)}{v_0}. \qquad (3.63)$$

Thus, even though the velocity is constant, the travel-time curve is not a straight line but has decreasing ray parameter with distance, $p = (r_0 \cos(\Delta/2))/v_0$.

We can derive a general equation for travel time in a sphere by considering the ray segment shown in Figure 3.22. The length of a small segment of the ray (ds) is given by

$$(ds)^2 = (dr)^2 + r^2(d\Delta)^2. \qquad (3.64)$$

Note that $\sin i = r(d\Delta/ds)$. Thus

$$p = \frac{r^2}{v} \frac{d\Delta}{ds}. \qquad (3.65)$$

Equation (3.65) can be used to eliminate ds from Eq. (3.64) to yield

$$(d\Delta)^2 = \frac{(dr)^2 p^2 v^2}{r^4 - r^2 p^2 v^2}$$

or

$$d\Delta = \frac{p}{r} \frac{dr}{\sqrt{\xi^2 - p^2}}, \qquad (3.66)$$

where $\xi = r/v$. We can integrate (3.66) to

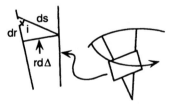

FIGURE 3.22 Geometry of ray segment ds in terms of radius r and angle $d\Delta$.

obtain

$$\Delta = 2p \int_{r_t}^{r_0} \frac{dr}{r\sqrt{\xi^2 - p^2}}, \qquad (3.67)$$

where r_0 is the radius of the Earth and r_t is the deepest point of penetration. This equation is analogous to Eq. (3.29) for a flat inhomogeneous model.

We can also eliminate $d\Delta$ from (3.64) using (3.65) to obtain

$$(ds)^2 = (dr)^2 + \frac{p^2 v^2 (ds)^2}{r^2}$$

or

$$ds = \frac{dr}{\sqrt{1 - (p^2 v^2/r^2)}} = \frac{\xi \, dr}{\sqrt{\xi^2 - p^2}}.$$

$$(3.68)$$

The travel time along any path is the path length divided by the velocity (v):

$$T = \int_{\text{path}} \frac{ds}{v} = 2\int_{r_t}^{r_0} \frac{\xi^2}{r\sqrt{\xi^2 - p^2}} \, dr. \quad (3.69)$$

Equation (3.69) is analogous to Eq. (3.30) for a flat, inhomogeneous model. Following the same logic as we used in the flat geometry, we can write (3.69) as separable travel-time equations:

$$T = 2\int_{r_t}^{r_0} \left(\frac{p^2}{r\sqrt{\xi^2 - p^2}} + \frac{\xi^2 - p^2}{r\sqrt{\xi^2 - p^2}} \right) dr$$

$$= p\Delta + 2\int_{r_t}^{r_0} \frac{\sqrt{\xi^2 - p^2}}{r} \, dr. \qquad (3.70)$$

For a given ray parameter, the first term on the right-hand side of Eq. (3.70) depends only on Δ, or *surface horizontal distance*, and the second term depends only on r, the vertical dimension. This is analogous to (3.31), with the integral corresponding to the *tau* function, $\tau(p)$, as in (3.57) for a spherical geometry.

The travel-time curves for a spherical geometry are very similar to those for a flat geometry, with the caveats that angular distance is used and ray parameter is scaled by the normalized radius. This implies that the qualitative behavior of the travel-time curves characterizing different velocity profiles in Figure 3.18 can be used to infer the gross character of velocity structure in a spherical Earth. In the real Earth prominent triplications result from velocity increases at the Moho and near 400 and 660 km depth, while the low-velocity core produces a major shadow zone (more on these in Chapter 7).

3.5 Wave Amplitude, Energy, and Geometric Spreading

Now that we have fully developed the concept of travel time for a ray, we can return to *energy* associated with an arrival. Equation (3.37) gave energy per unit surface area in a flat geometry. The variation of wave energy depends on velocity structure (d^2T/dX^2) and distance traveled (X). In general, the wave amplitude decays with distance; this is known as *geometric spreading*. We can gain some insight into geometric spreading by considering a homogeneous, spherical Earth. This requires a simple modification of (3.37); instead of a bundle of rays illuminating a ring on a flat surface, they illuminate a spherical ring. The wavefront area incident on this ring is given by

$$2\pi r_0^2 \sin \Delta |d\Delta| \cos i_0, \qquad (3.71)$$

where r_0 is the radius of the Earth (see geometry in Figure 3.21). This changes Eq. (3.37) to

$$E(\Delta) = E_0 \left(\frac{v_0}{r_0^3} \right) \left(\frac{\tan i_0}{\cos i_0} \right) \left(\frac{1}{\sin \Delta} \right) \left| \frac{d^2T}{d\Delta^2} \right|,$$

$$(3.72)$$

where $E_0 = K/2\pi$. For a homogeneous Earth, $T = [2r_0 \sin(\Delta/2)]/v_0$, which implies

$$\frac{dT}{d\Delta} = \left(\frac{r_0}{v_0}\right) \cos\left(\frac{\Delta}{2}\right) \Rightarrow \frac{d^2T}{d\Delta^2}$$

$$= \left(\frac{-r_0}{2v_0}\right) \sin\left(\frac{\Delta}{2}\right). \tag{3.73}$$

Further

$$\sin i_0 = \cos(\Delta/2)$$

$$\cos i_0 = \sin(\Delta/2)$$

$$\Rightarrow \frac{\tan i_0}{\cos i_0} = \frac{\cos(\Delta/2)}{\sin^2(\Delta/2)}. \tag{3.74}$$

Thus we can rewrite (3.72) as

$$E(\Delta) = \left(\frac{E_0}{r_0^2}\right)\left(\frac{v_0}{r_0}\right)$$

$$\times \left(\frac{1}{2\sin(\Delta/2)\cos(\Delta/2)}\right)$$

$$\times \left(\frac{\cos(\Delta/2)}{\sin^2(\Delta/2)}\right)\left(\frac{r_0}{2v_0}\right)\sin(\Delta/2)$$

$$= \frac{E_0}{4r_0^2 \sin^2(\Delta/2)}. \tag{3.75}$$

The denominator is simply the square of the length of the cord connecting the source and receiver. This implies that energy decays as $1/R^2$, where R is the distance traveled. Qualitatively, this will also hold for an inhomogeneous sphere.

Box 3.3 Caustics and the Antipode

In Figure 3.18b near the ends of the triplication, points B and C, a special amplitude behavior is predicted. In Eq. (3.37) we see that the amplitude is proportional to dp/dx, and at B and C this derivative is infinite. This represents a type of focusing called a *caustic*. The simplest way to interpret the caustic at point B is to think of rays from the AB and BC branches—the energy turning above and reflecting off the discontinuity, respectively—constructively interfering. The amplitude may be large, but it is not infinite. This is an example of how ray theory can break down; the rays are turning in a region of the Earth where the velocity gradient is rapidly changing, and our assumptions for the eikonal equation are inappropriate. Another caustic is the cusp associated with the termination of a shadow zone (see point D in Figure 3.18c).

In a spherical Earth, strong focusing occurs at the *antipode*. In the sphere, seismic waves spread in all directions. Geometric consideration shows that these spreading wavefields converge at a point exactly opposite the epicenter (see Figure 3.B3.1). If the Earth is homogeneous, then all common portions of the wavefront should arrive simultaneously at the receiver and produce strong amplification. Multiple arrivals will be observed because the wavefront has folded over on itself. Figure 3.B3.1 shows a recording of the Inangahua, New Zealand, earthquake at two stations on the Iberian Peninsula. Note how dramatic the focusing effects are for various phases within a few degrees of the antipode.

continues

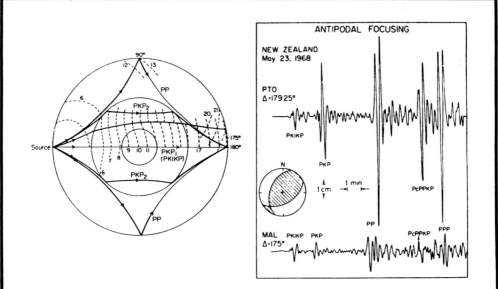

FIGURE 3.B3.1 Example of how raypaths converge at the antipode. The seismograms show the constructive interference effect on amplitudes very near the antipode ($\Delta = 180°$). (Modified from Rial and Cormier, 1980.)

How does our expression for the decay of energy relate to the ground shaking of a seismic wave? Basically, amplitude will be proportional to $\sqrt{E}$, so (3.75) tells us that the amplitudes will decrease inversely with length of the ray, $\sim 1/R$. Let us consider this in more detail. Seismic waves propagate as loci of particle motions, so a wavefront transports energy in the form of particle momentum and/or potential energy. We can calculate this energy using a simple analog, namely the restoring force of a mass suspended from a spring. This is given by $f = -kx$, where k is the spring constant. The increment of work done in moving the mass a small distance, dx, is $dW = -kx\,dx$. If the mass is initially at equilibrium, the total work is given by

$$W = \int dW = \int_0^x -kx\,dx = -\frac{1}{2}kx^2.$$

(3.76)

A similar argument can be applied to potential energy. The potential energy is the strain associated with the transient stress pulse. Thus the stored strain in a small volume is given by

$$W = \int \frac{1}{2}\sigma_{ij}\varepsilon_{ij}\,dV.$$

(3.77)

Now let us evaluate Eq. (3.77) for a particular case without loss of generality. Consider an SH plane wave propagating in the x_1 direction, with all motion in the x_2 direction:

$$u_2 = Ae^{i(\omega t - kx_1)}.$$

(3.78)

The only nonzero strains are

$$\varepsilon_{12} = \varepsilon_{21} = \frac{1}{2}\frac{\partial u_2}{\partial x_1} = -\frac{1}{2}ikAe^{i(\omega t - kx_1)}$$

(3.79)

and the stress is given by

$$\sigma_{12} = \sigma_{21} = -ik\mu Ae^{i(\omega t - kx_1)}. \quad (3.80)$$

Thus the *average* strain energy during a complete wavelength is given by

$$W = \frac{1}{\Lambda}\int_0^\Lambda \frac{1}{2}k^2A^2\mu \, dx, \quad (3.81)$$

where Λ is the wavelength. Recall that Λ equals the velocity times the period of oscillation. Further, $\mu = \rho\beta^2$ and $k = (2\pi/\Lambda)$, which can be used to obtain

$$W = \frac{1}{2}\left(\frac{2\pi}{\beta T}\right)^2 A^2\rho\beta^2 = 2\pi^2\rho\frac{A^2}{T^2}. \quad (3.82)$$

Thus the *energy* in a plane wave is proportional to the square of the pulse amplitude and inversely proportional to the square of the period. Thus, if the amplitude of two seismic signals is the same, the higher-frequency signal transports more energy.

The amplitude of a seismic signal is modified during propagation by several phenomena. We have already seen that geometric spreading decreases the amplitude. The remainder of this chapter deals with two other phenomena that affect amplitudes: reflection/refraction at a boundary and anelastic attenuation.

3.6 Partitioning of Seismic Energy at a Boundary

We have seen in the previous sections of this chapter that when a body wave encounters a boundary or discontinuity at which the seismic velocity changes, the wave will reflect or refract. As we will show, when a *P* or *SV* wave impinges on a boundary, four derivative waves result, as shown in Figure 3.23: (1) *P'*, the refracted or transmitted *P* wave (note that *P* head waves are a subset of *P'*), (2) *SV'*, the refracted *SV* (it is possible to have *P* waves generate a *SV* head wave if $\beta_2 > \alpha_1$), (3)

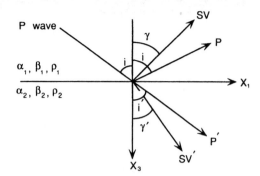

FIGURE 3.23 Ray for a *P* wave incident on a solid–solid boundary and the rays for waves generated at the interface.

P, the reflected *P* wave, and (4) *SV*, the reflected *SV* wave. The ray geometry of these *derived* waves is governed by Snell's law. By Snell's law, all of the rays must have the same ray parameter, *p*, since all the waves must move along the boundary with the same apparent velocity:

$$(\sin i)/\alpha_1 = (\sin \gamma)/\beta_1 = (\sin \gamma')/\beta_2$$

$$= (\sin i')/\alpha_2. \quad (3.83)$$

When an *SH* wave encounters a discontinuity surface parallel to the *SH* motion, only two waves are generated: (1) *SH*, reflected, and (2) *SH'*, refracted. (*SH'* can be a head wave.) The existence of multiple waves derived from a single incident wave implies that the energy of the incident wave must be partitioned. Although Snell's law and ray theory can predict the geometry of the wave interaction, we must return to a wavefield representation to determine the amplitude partitioning.

In Figure 3.23, the interface separates two materials of distinct elastic properties. Within either half-space the equations of motion for homogeneous media are valid. The physics that govern the wave propagation require that *stresses* and *displacements* be "transmitted" across the interface. Thus a stress imbalance propagating in layer 1 will result in a stress imbalance in layer 2, giving rise to a wavefield. There are several types of interface. If the interface is between two solids, all components of stress at the interface and all components of displacement are continuous. This is called a *welded interface*. If the interface is between a solid and a perfect fluid, the fluid may slip along the interface, since it has no rigidity. Thus, the tangential displacements are not continuous, and the tangential tractions must vanish. In addition, the normal traction and normal displacements at the fluid–solid contact are continuous. At a *free surface*, all the tractions must be zero, and no explicit restriction is placed on the displacements. Note that these conditions are on *tractions*, not stresses. For example, if the $x_1 x_2$ plane is a free surface, then $\sigma_{31} = \sigma_{32} = \sigma_{33} = 0$, but the other components of stress are not constrained.

These conditions on continuity of displacement and stress are the basis for predicting the partitioning of energy. Now return to Figure 3.23. Why does the P wave produce both a reflected and refracted P wave and a reflected and refracted SV wave? It makes sense that no SH wave will be produced because the particle motion of the incident P wave is confined to the $x_1 x_3$ plane, and no "refraction" of the P wave at a horizontal boundary will produce motion in the x_2 plane. Refraction of the P wave will cause particle displacements that are not parallel on opposite sides of the interface (see Figure 3.24). Thus, the P-wave displacements *alone* do not combine to give continuous displacements or tractions across the welded interface. The additional particle motion required to make the fields continuous results in SV-wave-type motion, which is also confined to the $x_1 x_3$ plane. Remember, only P- and S-wave motions exist as propagating disturbances. In a fluid, where no S waves exist, the P waves reflect and transmit purely as P waves because only *normal* displacements and *normal* tractions need to remain continuous at the boundary.

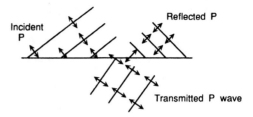

FIGURE 3.24 *P*-wave particle motions for the incident, reflected, and refracted *P* waves. Note that if this is a solid–solid boundary, the shear stress in the two layers will *not* match at the boundary, requiring the generation of *SV* motion in both media.

We can quantify the energy partitioning by using the potentials introduced in Section 2.4 for plane waves. The P-wave and SV-wave potentials for the various wave components are represented by

$$\phi^-_{(\text{layer 1})} = \phi_{\text{incident ray}} + \phi_{\text{reflected ray}}$$

$$\phi^+_{(\text{layer 2})} = \phi_{\text{refracted}}$$

$$\psi^- = \psi_{\text{reflected}}$$

$$\psi^+ = \psi_{\text{refracted}}, \qquad (3.84)$$

where ϕ and ψ are the P and SV potentials, respectively. The plane-wave potentials are of the form

$$\phi_{\text{incident}} = A_1 \exp\left[i\omega\left(px_1 + \eta_{\alpha_1} x_3 - t\right)\right].$$

$$(3.85)$$

Recall that $(kx_1)/\omega = (\sin i)/\alpha = p$. Similarly, $k_{x_3}/\omega = \eta_{\alpha_1}$. We can write similar equations for the other potentials in (3.84):

$$\phi_{\text{reflected}} = A_2 \exp\left[i\omega\left(px_1 - \eta_{\alpha_1} x_3 - t\right)\right]$$

$$\phi_{\text{refracted}} = A_3 \exp\left[i\omega\left(px_1 + \eta_{\alpha_2} x_3 - t\right)\right]$$

$$\psi_{\text{reflected}} = B_2 \exp\left[i\omega\left(px_1 - \eta_{\beta_1} x_3 - t\right)\right]$$

$$\psi_{\text{refracted}} = B_3 \exp\left[i\omega\left(px_1 + \eta_{\beta_2} x_3 - t\right)\right].$$

$$(3.86)$$

The various vertical slownesses are for the associated velocities. Note that the sign of the x_3 term changes, depending on whether the ray is refracted or reflected. This indicates the direction (down or up) in which the ray is traveling.

The ratios of the postinteraction amplitudes (A_2, A_3, B_2, B_3) over the incident amplitude (A_1) are called the *reflection* and *transmission* coefficients. These coefficients control the partitioning of amplitude among the potentials. The tractions and displacements can be calculated from the potentials by taking the derivatives with respect to x_1 and x_3, which preserves the exponential character of the potentials.

In general, the boundary conditions in a welded interface require significant algebraic manipulation (see Table 3.1), so we will consider a simplified example. A P wave incident on a fluid–fluid interface generates no S waves, so we need only consider reflected and refracted P waves. From (3.85) and (3.86) we can write down equations for the P-wave potential:

medium 1: $\phi_1 = A_1 \exp[i\omega(px_1 + \eta_1 x_3 - t)]$

                    $+ A_2 \exp[i\omega(px_1 - \eta_1 x_3 - t)]$

medium 2: $\phi_2 = A_3 \exp[i\omega(px_1 + \eta_2 x_3 - t)]$.

$$(3.87)$$

TABLE 3.1 Displacement Reflection and Transmission Coefficients

Coefficient	Formula
Solid–free surface (P–SV)	
R_{PP}	$\{-[(1/\beta^2) - 2p^2]^2 + 4p^2\eta_\alpha\eta_\beta\}/A$
R_{PS}	$\{4(\alpha/\beta)p\eta_\alpha[(1/\beta^2) - 2p^2]\}/A$
R_{SP}	$\{4(\beta/\alpha)p\eta_\beta[(1/\beta^2) - 2p^2]\}/A$
R_{SS}	$\{-[(1/\beta^2) - 2p^2]^2 + 4p^2\eta_\alpha\eta_\beta\}/A$
$R_{SS}(SH)$	1
Solid–solid (P–SV)	
R_{PP}	$[(b\eta_{\alpha_1} - c\eta_{\alpha_2})F - (a + d\eta_{\alpha_1}\eta_{\beta_2})Hp^2]/D$
R_{PS}	$-[2\eta_{\alpha_1}(ab + cd\eta_{\alpha_2}\eta_{\beta_2})p(\alpha_1/\beta_1)]/D$
T_{PP}	$[2\rho_1\eta_{\alpha_1}F(\alpha_1/\alpha_2)]/D$
T_{PS}	$[2\rho_1\eta_{\alpha_1}Hp(\alpha_1/\beta_2)]/D$
R_{SS}	$-[(b\eta_{\beta_1} - c\eta_{\beta_2})E - (a + b\eta_{\alpha_2}\eta_{\beta_1})Gp^2]/D$
R_{SP}	$-[2\eta_{\beta_1}(ab + cd\eta_{\alpha_2}\eta_{\beta_2})p(\beta_1/\alpha_1)]/D$
$R_{SS}(SH)$	$\dfrac{\mu_1\eta_{\beta_1} - \mu_2\eta_{\beta_2}}{\mu_1\eta_{\beta_1} + \mu_2\eta_{\beta_2}}$
$T_{SS}(SH)$	$\dfrac{2\mu_1\eta_{\beta_1}}{\mu_1\eta_{\beta_1} + \mu_2\eta_{\beta_2}}$

$a = \rho_2(1 - 2\beta_2^2 p^2) - \rho_1(1 - 2\beta_1^2 p^2)$ $E = b\eta_{\alpha_1} + c\eta_{\alpha_2}$

$b = \rho_2(1 - 2\beta_2^2 p^2) - 2\rho_1\beta_1^2 p^2$ $F = b\eta_{\beta_1} + c\eta_{\beta_2}$

$c = \rho_1(1 - 2\beta_1^2 p^2) + 2\rho_2\beta_2^2 p^2$ $G = a - d\eta_\alpha\eta_{\beta_2}$

$d = 2(\rho_2\beta_2^2 - \rho_1\beta_1^2)$ $H = a - d\eta_{\alpha_2}\eta_{\beta_1}$

$$D = EF + GHp^2$$

$$A = [(1/\beta^2) - 2p^2]^2 + 4p^2\eta_{\alpha_1}\eta_{\beta_1}$$

The P displacements are related to the potentials by Eq. (2.91):

$$u = \frac{\partial\phi}{\partial x_1}\mathbf{x}_1 + 0\mathbf{x}_2 + \frac{\partial\phi}{\partial x_3}\mathbf{x}_3. \quad (3.88)$$

The appropriate boundary conditions for the fluid–fluid boundary are continuity of normal stress and displacement (σ_{33} and u_3). Mathematically, the displacement condition is given by

$$\frac{\partial\phi_1}{\partial x_3} = \frac{\partial\phi_2}{\partial x_3}\bigg|_{x_3=0}. \quad (3.89)$$

Substituting (3.87) into this equation yields

$$i\omega\eta_1(A_1 - A_2)e^{i\omega(px_1-t)}$$

$$= i\omega\eta_2 A_3 e^{i\omega(px_1-t)} \quad (3.90)$$

or

$$\eta_1(A_1 - A_2) = \eta_2 A_3. \quad (3.91)$$

The condition of stress continuity is given by

$$\sigma_{33}^- = \lambda\nabla u + 2\mu\varepsilon_{33} = \sigma_{33}^+, \quad (3.92)$$

but $\mu = 0$ in a fluid. Thus

$$\lambda_1\nabla^2\phi_1 = \lambda_2\nabla^2\phi_2. \quad (3.93)$$

We can simplify (3.93) by using the fact that ϕ satisfies the wave equation:

$$\nabla^2\phi = \frac{1}{\alpha^2}\frac{\partial^2\phi}{\partial t^2} = \frac{-\omega^2}{\alpha^2}\phi. \quad (3.94)$$

Therefore, for $x_3 = 0$,

$$\frac{\lambda_1}{\alpha_1^2}(A_1 + A_2) = \frac{\lambda_2}{\alpha_2^2}A_3. \quad (3.95)$$

Now, for a fluid, $\lambda_1 = \rho_1\alpha_1^2$ and $\lambda_2 = \rho_2\alpha_2^2$, so we can rewrite (3.91) and (3.95) as a

system of equations:

$$A_1 - A_2 = \frac{\eta_2}{\eta_1}A_3$$

$$A_1 + A_2 = \frac{\rho_2}{\rho_1}A_3. \quad (3.96)$$

Thus we can solve for ratios of the amplitudes

$$\frac{A_3}{A_1} = \mathscr{T} = \frac{2\rho_1\eta_1}{\rho_1\eta_2 + \rho_2\eta_1}$$

$$\frac{A_2}{A_1} = \mathscr{R} = \frac{\rho_2\eta_1 - \rho_1\eta_2}{\rho_1\eta_2 + \rho_2\eta_1}. \quad (3.97)$$

$\mathscr{T}$ and $\mathscr{R}$ are referred to as the transmission and reflection coefficients, respectively. Note that $\mathscr{T}$ and $\mathscr{R}$ depend on η, which is $(\cos i)/\alpha$. Thus the partitioning of potential amplitudes depends on the angle at which the ray strikes the boundary. Consider the case of vertical incidence ($p = 0$, $\eta_1 = 1/\alpha_1$, $\eta_2 = 1/\alpha_2$):

$$\mathscr{R}_{i=0} = \frac{\rho_2/\alpha_1 - \rho_1/\alpha_2}{\rho_1/\alpha_2 + \rho_2/\alpha_1} = \frac{\rho_2\alpha_2 - \rho_1\alpha_1}{\rho_1\alpha_1 + \rho_2\alpha_2} \quad (3.98)$$

$$\mathscr{T}_{i=0} = \frac{2\rho_1/\alpha_1}{\rho_1/\alpha_2 + \rho_2/\alpha_1} = \frac{2\rho_1\alpha_2}{\rho_1\alpha_1 + \rho_2\alpha_2}. \quad (3.99)$$

Now at this point, the reflection and transmission coefficients are for potential, not displacement. We can obtain displacement terms by recalling $u_3 = \partial\phi/\partial x_3$:

$$\frac{u_{\text{reflected}}}{u_{\text{incident}}} = \frac{-i\omega\eta_1}{i\omega\eta_1}\frac{A_2}{A_1} = \frac{\rho_1\alpha_1 - \rho_2\alpha_2}{\rho_1\alpha_1 + \rho_2\alpha_2}$$

$$= R = -\mathscr{R}_{i=0} \quad (3.100)$$

$$\frac{u_{\text{refracted}}}{u_{\text{incident}}} = \frac{i\omega\eta_2}{i\omega\eta_1}\frac{A_3}{A_1} = \frac{1/\alpha_2}{1/\alpha_1}\frac{2\rho_1\alpha_2}{\rho_1\alpha_1 + \rho_2\alpha_2}$$

$$= T = \frac{\alpha_1}{\alpha_2}\mathscr{T}_{i=0}. \quad (3.101)$$

The R and T derived here, which are the vector displacement transmission and reflection coefficients, have extensive use in geophysics despite being derived for fluids and vertical incidence. One must be careful to keep track of the vector displacement with respect to the direction the wave is propagating in defining the sign of the motion. These reflection and transmission coefficients also hold for solid–solid interfaces at near-vertical incidence. The energy is partitioned quite simply: $T - R = 1$. The quantity $\rho\alpha$ is known as *acoustic impedance*, and depending on how acoustic impedance changes across the boundary, the reflection coefficient can have values of -1 to $+1$. Similarly, the range of the transmission coefficient is 0 to 2. A free-surface boundary will have a vertical-incidence reflection coefficient of -1 (the displacement reverses direction with respect to the direction of propgagation). The amplitude of transmitted displacement is zero.

If we return to the general form of $\mathcal{T}$ and $\mathcal{R}$ (nonvertical incidence), we can investigate the behavior of the system as the angle of incidence varies. If $\alpha_2 < \alpha_1$ and $\rho_2\alpha_2 > \rho_1\alpha_1$, then $\mathcal{R}$ will be a positive value for normal incidence. As i increases, $\mathcal{R}$ will decrease, reaching zero at an angle of incidence called the *intramission angle*:

$$\frac{\rho_2}{\rho_1} = \frac{\sqrt{(\alpha_1/\alpha_2)^2 - \sin^2 i}}{\sqrt{1 - \sin^2 i}}. \quad (3.102)$$

Beyond the intramission angle, the reflection coefficient decreases to a value of -1 at *grazing incidence* ($i = 90°$). If $\alpha_2 < \alpha_1$ and $\rho_2\alpha_2 < \rho_1\alpha_1$, the reflection coefficient is always negative and equals -1 for grazing incidence.

If $\alpha_2 > \alpha_1$, a head wave is produced at the critical angle, $i_c = \sin^{-1}(\alpha_1/\alpha_2)$. At incident angles greater than the critical angle, no P waves will *propagate* in the lower medium. This is because $p = (\sin i)/\alpha_1 = 1/c$ (where c is the apparent velocity) becomes greater than $1/\alpha_2$. Thus $\eta_2 = [(1/\alpha_2^2) - p^2]^{1/2}$ become *imaginary*.

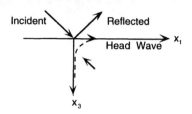

FIGURE 3.25 Exponential decay of the particle motion of a head wave propagating along the boundary.

We can write $\eta_2 = i\hat{\eta}_2 = \pm i[p^2 - (1/\alpha_2^2)]^{1/2}$, where we choose the positive sign such that the amplitude of the refracted potential (3.87) decreases exponentially away from the boundary. This keeps the wave energy bounded. Figure 3.25 illustrates the head wave with exponentially decaying displacements in the half-space. The transmission coefficient is complex, and to keep the ray parameter constant, angle i_2 becomes complex.

We can rewrite the postcritical reflection coefficient in (3.97) as

$$\mathcal{R} = \frac{\rho_2\eta_1 - \rho_1 i\hat{\eta}_2}{\rho_2\eta_1 + \rho_1 i\hat{\eta}_2}. \quad (3.103)$$

Now $\mathcal{R}$ is a complex number divided by its conjugate. This implies that the magnitude of $\mathcal{R}$ is 1, but there is a phase shift of θ

$$\mathcal{R} = e^{i\theta} \quad (3.104)$$

$$\theta = 2\tan^{-1}\left(\frac{\rho_1\hat{\eta}_2}{\rho_2\eta_1}\right). \quad (3.105)$$

Since the modulus of the reflection coefficient is 1, the postcritical reflection is referred to as *total reflection*, but it will behave differently than precritical reflections. Figure 3.26 shows a synthetic seismogram profile generated for increasing angles of incidence (increasing distance). Beyond 60 km, the reflected arrival has an angle of incidence that is greater than i_c. This is the distance at which a head wave first occurs and begins to move out from the reflected arrival. At 450 km the reflected wave is incident on the boundary at near-grazing incidence; the reflected wave-

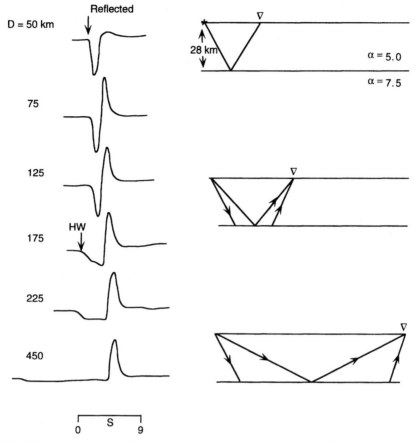

FIGURE 3.26 The change in reflected pulse shape (phase) as the incidence angle exceeds the critical angle. For the model shown, the head wave first appears at ~60 km. A comparison of seismograms at 50 and 450 km shows that the polarity has been reversed.

form is very similar to that at 50 km, except the *polarity* is completely reversed.

It is clear from Figure 3.26 that the reflected wave changes shape as the source–receiver distance increases. Although the phase shift in Eq. (3.105) explains this shape change, it is instructive to return to the equation for the reflection potential. Noting that $A_2 = A_1 \mathcal{R} = A_1 e^{i\theta}$, we can write the potential for the postcritical reflected arrival as

$$\phi = A_1 \exp[i\theta] \exp[i\omega(px_1 - \eta_1 x_3 - t)].$$

$$(3.106)$$

Now consider the behavior of θ:

$$\begin{aligned} \theta &= 0 & \text{if } i = i_c \\ \theta &< 0 & \text{for } i > i_c \\ \theta &= -\pi & i = \pi/2. \end{aligned}$$

We first rewrite (3.106) as

$$\phi = A_1 \exp[i\omega(px_1 - \eta_1 x_3 - t + (\theta/\omega))].$$

$$(3.107)$$

Now θ/ω is explicitly a *new* or additional phase term. If we apply the constant-phase argument to track the behavior of a particular wavefront, we have

$$px_1 - \eta_1 x_3 - t + (\theta/\omega) = \text{constant}.$$

$$(3.108)$$

The term $-t + (\theta/\omega) = -(t - \theta/\omega) = \hat{t}$ is an *apparent* time that now *depends on frequency*. Thus, the position of the wavefront is frequency dependent; lower frequencies (smaller ω) will have *earlier* arrival times than high frequencies (recall $\theta < 0$). As $\omega \to \infty$, $\hat{t} = t$. This implies that the wavefront is "spread out" for a post-critical reflection, each harmonic term having a separate plane wave. This behavior is called *dispersion*, a phenomenon we will become very familiar with in the next chapter. A consequence of the dispersion is that the strongest reflection coefficient occurs exactly at i_c ($R = 1$, $\theta = 0$, and wavefronts do not degrade).

Reflection and transmission at a welded interface are much more complicated than at a fluid–fluid interface. However, the *SH* system remains fairly simple because inter-action with the boundary does not produce any P or SV energy, so we will briefly consider this case. As with the fluid–fluid case, there are two boundary conditions: (1) continuity of tangential displacement ($V_2^+ = V_2^-$), and (2) continuity of shear stress ($\sigma_{23}^+ = \sigma_{23}^-$). Applying these conditions yields *SH*-displacement reflection and transmission coefficients:

$$T = \frac{2\mu_1 \eta_{\beta_1}}{\mu_1 \eta_{\beta_1} + \mu_2 \eta_{\beta_2}}$$

$$R = \frac{\mu_1 \eta_{\beta_1} - \mu_2 \eta_{\beta_2}}{\mu_1 \eta_{\beta_1} + \mu_2 \eta_{\beta_2}}. \qquad (3.109)$$

These equations are nearly identical to Eqs. (3.97), and if we consider the case of vertical incidence, then (3.109) reduces to

$$T = \frac{2\rho_1 \beta_1}{\rho_1 \beta_1 + \rho_2 \beta_2}$$

$$R = \frac{\rho_1 \beta_1 - \rho_2 \beta_2}{\rho_1 \beta_1 + \rho_2 \beta_2}. \qquad (3.110)$$

Box 3.4 Seismic Diffraction

The analogy between seismic ray theory and optics extends to the concept of diffraction. *Diffraction* is defined as the transmission of energy by nongeometric ray paths. In optics, the classic example of diffraction is light "leaking" around the edge of an opaque screen. In seismology, diffraction occurs whenever the radius of curvature of a reflecting interface is less than a few wavelengths of the propagating wave. Figure 3.B4.1a shows a plane wave incident upon an opaque (acoustic impedance is infinite) boundary. Ray theory requires that waves arriving at seismometers at points F and G have identical amplitudes; *no* energy is transmitted to the right of point G. In fact, the edge of the boundary acts like a secondary source (Huygens' principle) and radiates energy forward in all directions. These diffractions can be understood from the standpoint of *Fresnel zones*, a concept that states that waves reflect from a large region rather than just a point. Thus, the Fresnel zone causes the ray traveling to F to "see" the edge of the reflector, although the geometric raypath clearly misses the boundary. The first Fresnel zone may be thought of as a cone with the edge of the reflector as its apex. For a receiver that is a distance d beyond the reflector, the cone's radius is given by $r = d + \frac{1}{2}\lambda$, where λ is the wavelength of the seismic wave. Figure 3.B4.1b shows the amplitude variation predicted for the experiment given in 3.B4.1a.

continues

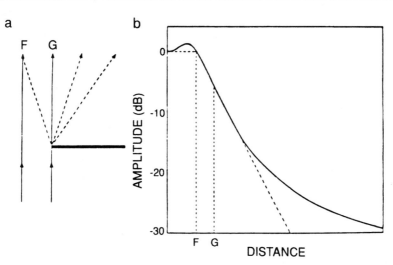

FIGURE 3.B4.1 (a) Rays incident on a grating. Energy is *diffracted* around the edge. (b) Amplitude of energy as a function of distance into the diffraction zone. (From Doornbos, 1989).

Diffraction is present at many scales within the Earth and has occasionally led to erroneous interpretations of structure. Figure 3.B4.2 shows an example from reflection seismology. Here, a high-velocity layer is sandwiched between half-spaces, and the layer is offset by a normal fault. The seismograms shown are for a source and receiver placed at each successive distance point. At $x = 2000$, the seismogram is made up of two pulses, of opposite polarity, representing reflections off the top and bottom of the layer. As x increases, later arrivals begin to appear, forming a parabola known as a "diffraction frown."

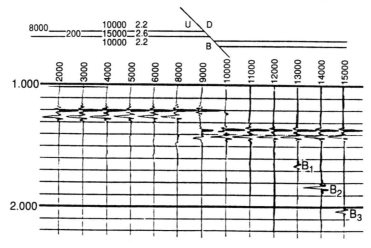

FIGURE 3.B4.2 A synthetic reflection seismic section for the structure at the top of the figure. A fault offsets a high-velocity bed. $B1$, $B2$, and $B3$ are diffracted arrivals. (From Waters, K. H. "Reflection Seismology: A Tool for Energy Resource Exploration." Copyright ©1981 John Wiley & Sons.)

The quantity $\rho\beta$ is called the *shear impedance*. The *SH* critical-angle behavior for $\beta_2 > \beta_1$ is very analogous to that described for the acoustic (fluid) case.

The *P–SV* system requires using every potential term in Eq. (3.84). In general, four derivative waves exist for each incident *P* or *SV* wave. The velocities may permit both *P* and *SV* head waves for incident *P* or *S* waves. For the welded interface, σ_{3_3}, u_1, and u_3 must be continuous (used for boundary conditions). For the case of an incident *P* wave, the displacement boundary conditions [using (3.86)], give (u_1 continuous)

$$p(A_1 + A_2) + \eta_{\beta_1} B_1 = pA_3 - \eta_{\beta_2} B_2$$

$$(3.111)$$

and (u_3 continuous)

$$\eta_{\alpha_1}(A_1 - A_2) + pB_1 = \eta_{\alpha_2} A_3 + pB_2.$$

$$(3.112)$$

The continuity of stress (σ_{33} continuous) gives

$$\lambda_1 p^2(A_1 + A_2) + \lambda_1 p\eta_{\beta_1} B_1 + (\lambda_1 + 2\mu_1)$$
$$\times \left[\eta_{\alpha_1}^2(A_1 + A_2) - \eta_{\beta_1} pB_1\right]$$
$$= \lambda_2 p^2 A_3 - p\eta_{\beta_2}\lambda_2 B_2 + (\lambda_2 + 2\mu_2)$$
$$\times \left(\eta_{\alpha_2}^2 A_3 + \eta_{\beta_2} pB_2\right) \qquad (3.113)$$

and (σ_{31} continuous)

$$\mu_1\left[2p\eta_{\alpha_1}(A_1 - A_2) + p^2 B_1 - \eta_{\beta_1}^2 B_1\right]$$
$$= \mu_2\left[2p\eta_{\alpha_2} A_3 + p^2 B_2 - \eta_{\beta_2}^2 B_2\right].$$

$$(3.114)$$

Thus we have four equations with five unknowns. It is sufficient to determine the *ratios* with respect to A_1, thus obtaining R_{PP}, R_{PS}, T_{PP}, and T_{PS}. The algebra required to obtain these coefficients is extensive, and we leave it to the reader as an exercise to obtain the final values given in

Table 3.1. Table 3.1 lists the standard reflection and transmission coefficients for solid–solid and solid–air (free-surface reflections) interfaces.

Figures 3.27 and 3.28 show the reflection and transmission coefficients for *P* waves incident from below and above a welded interface. In the first case, the wave is going from a *fast-* to a *slow-*velocity material, and there are no critical angles. The energy partitioning is dominated by R_{PP} and T_{PP} from 0° to approximately 20°. Over this range, R_{PP} and T_{PP} are nearly identical to what would be obtained from the acoustic impedance mismatch [Eqs. (3.100) and (3.101]. When the *P* wave is incident from the low-velocity medium, the critical angle is 38.5°. The *P* transmission coefficient is 0 beyond this angle. As the angle of incidence approaches 38.5°, the coefficients vary rapidly. In particular, T_{PP} gets very large before going to zero. This can be explained by a simple geometric argument, as shown in Figure 3.29. Because the amplitude of the pulse is proportional to the square root of energy *per* surface area, as surface area goes to zero, the amplitude becomes large.

The partitioning of a wave into *four* new waves at each boundary in the Earth results in seismograms that are rich in arrivals. We refer to the partitioning of *P* waves into *SV* waves or *SV* waves into *P* waves as *mode conversions*. Mode conversions provide important information about Earth structure. Figure 3.30 shows a seismogram from a deep crustal earthquake in the Mississippi embayment. A converted phase *Sp* is generated at a sediment–bedrock interface. This arrives ahead of *S* by a time proportional to the depth of the interface and the v_P/v_S ratio in the crust. Other examples of reflected and converted phases are described in Chapter 7.

3.7 Attenuation and Scattering

Thus far we have been concerned with the *elastic* properties of the Earth in our

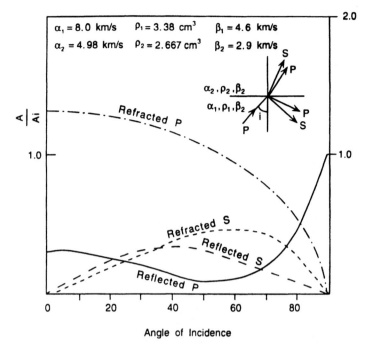

FIGURE 3.27 Reflection and refraction coefficients for a P wave incident on a boundary from a high-velocity region. For near-vertical incidence (angle $= 0°$), the reflected and refracted P-wave amplitudes approximately equal those predicted by acoustic-impedance mismatches [Eqs. (3.100) and (3.101)]. There are no critical angles in this case.

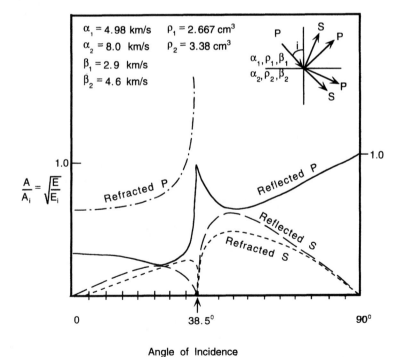

FIGURE 3.28 Reflection and refraction coefficients for a P wave incident on a boundary from a low-velocity region. i_c for the P wave occurs at 38.5°. Since the S velocity in the lower medium is lower than the upper P velocity, the refracted S wave never reaches a critical angle.

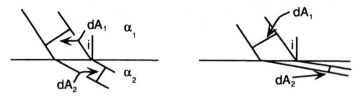

FIGURE 3.29 Schematic of ray bundles striking a boundary between low- and high-velocity material. The amplitude of the pulse is inversely proportional to the surface area dA. As i approaches the critical angle i_c, dA_2 goes to zero, and the amplitude of the refracted wave becomes very large.

discussion of wave propagation. In an idealized, purely elastic Earth, geometric spreading and the reflection and transmission of energy at boundaries control the amplitude of a seismic pulse. Once excited, these waves would persist indefinitely. The real Earth is not perfectly elastic, and propagating waves *attenuate* with time due to various energy-loss mechanisms. The successive conversion of potential energy (particle position) to kinetic energy (particle velocity) as a wave propagates is not perfectly reversible, and other work is done, such as movements along mineral dislocations or shear heating at grain boundaries, that taps the wave energy. We usually describe these processes collectively as *internal friction*, and we "model" the internal-friction effects with phenomenological descriptions because the microscopic processes are complex.

The simplest descriptions of attenuation can be developed for an oscillating mass on a spring. Consider Figure 3.31, where a

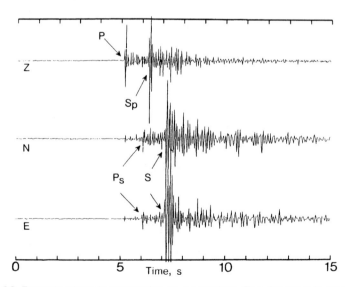

FIGURE 3.30 Example of mode conversion at a boundary. The *SV* wave is converted to a *P* wave at a sediment–bedrock interface, giving rise to the S_p precursor to *S* on the vertical (x) seismogram, while $P \rightarrow S$ conversions (P_s) are seen on the horizontals. (Courtesy of W. Mooney.)

Box 3.5 Scattering

The example seismograms in Figures 3.11 and 3.16 show a multiplicity of arrivals. Some of these arrivals can be explained in terms of reflections and mode conversions at boundaries within a simple layered model of the crust, but a one-dimensional structure cannot explain a significant amount of energy. These arrivals are produced by *scattering* caused by the wavefield's interaction with small-scale heterogeneities. Heterogeneities in material properties pervade the Earth and span many different length scales (see Chapter 7). Small-scale heterogeneity causes scattering that partitions the high-frequency wavefield into a sequence of arrivals that are often called coda waves.

Figure 3.B5.1 shows seismograms produced by the impact of a Saturn booster on the Moon's surface. These were recorded by a lunar seismometer installed during the Apollo 14 mission. The short-period three-component records ring on for more than 1 h, with waves being scattered from the highly heterogeneous region near the Moon's surface. The coda is spindle shaped, and analysis of the particle motions indicates that the energy is arriving from all directions. These differ from typical Earth recordings, for which the coda is weaker than the direct arrivals. This is because the seismic-wave attenuation on the Moon is much smaller, allowing strongly scattered waves to propagate for some time. The wave interactions with boundary irregularities and with volumetric gradients in rock properties all involve the conventional effects of refraction, conversion, reflection, and diffraction that we describe in this chapter, but the resulting overall wavefield is so complex that individual arrivals cannot be associated with a particular path through the medium given a limited number of surface recordings. Generally, seismologists attempt to characterize the statistical properties of the scattering medium in terms of the spectrum of spatial heterogeneities superimposed on any simple layered structure. Many techniques have been developed to relate the coda to the heterogeneity spectrum.

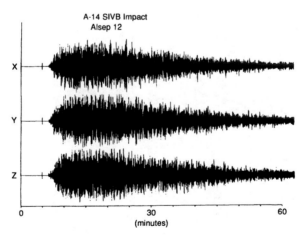

FIGURE 3.B5.1 Three-component seismograms recording the impact of an Apollo lander on the Moon. Seismograms ring for more than 1 h. (From Dainty *et al.*, 1974.)

continues

Scattering can also decrease the amplitude of a seismic phase by shifting energy from the direct arrival back into the coda. This apparent attenuation is called scattering attenuation, and is often characterized by an exponential attenuation quality factor, Q_{sc}. Unlike Q defined for anelastic processes, Q_{sc} is not a measure of energy loss per cycle but, rather, a measure of energy redistribution. Q_{sc} depends very strongly on frequency and is very path dependent, since it depends on the particular heterogeneity spectrum encountered by a wavefield propagating through the Earth. Q_{sc} is usually modeled with stochastic operators, or randomization coefficients. Figure 3.B5.2 show snapshots of a wavefield at different times as it propagates though material that has a random 10% distribution of velocity heterogeneity. Note the direct P wave remains fairly coherent, but a complex suite of later arrivals is generated by the heterogeneity. These will appear at a single station as coda scattered from all directions.

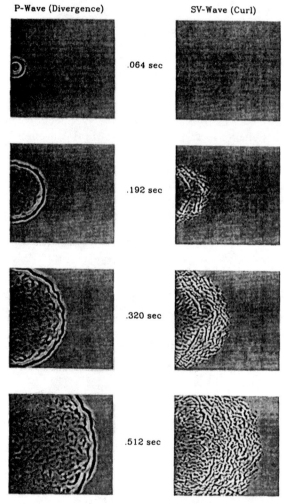

FIGURE 3.B5.2 Synthetic P waves in a heterogeneous material. After 0.512 s, the spherical wavefront is broken up and coda has been developed. (From Frankel and Clayton, 1986.)

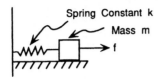

FIGURE 3.31 Phenomenological model for seismic attenuation. The spring represents elastic processes in the Earth. Force **f** represents friction opposing the motion of the mass.

mass m attached to a spring with spring constant k (k is a measure of the spring's stiffness) slides across a surface. Let us first consider a frictionless case. The equation of motion for this system relates the restoring force of the spring to the inertial force imparted by the moving mass:

$$m\ddot{x} + kx = 0. \qquad (3.115)$$

The general solution to this equation is a harmonic oscillation:

$$x = Ae^{i\omega_0 t} + Be^{-i\omega_0 t}$$

$$\omega_0 = \sqrt{k/m}. \qquad (3.116)$$

Once the motion starts, it will continue forever, oscillating at the natural frequency of the system ω_0. We can introduce *attenuation* by adding a damping force, such as friction between the moving mass and the underlying surface. In this case, there is an added force, proportional to the velocity of the mass

$$m\ddot{x} + \gamma\dot{x} + kx = 0 \qquad (3.117)$$

or rewriting,

$$\ddot{x} + \varepsilon\omega_0\dot{x} + \omega_0^2 x = 0, \qquad (3.118)$$

where $\varepsilon = (\gamma/m\omega_0)$, and $\omega_0 = (k/m)^{1/2}$. γ and ε are called *coefficients of friction*.

The solution of (3.118) is of the form

$$x(t) = A_0 e^{-\varepsilon\omega_0 t} \sin\left(\omega_0 t\sqrt{1 - \varepsilon^2}\right), \qquad (3.119)$$

where $A_0 e^{-\varepsilon\omega_0 t} = A(\varepsilon)$. This is a harmonic oscillation that decays exponentially with time. If $\varepsilon = 0$ (no attenuation), (3.119) reverts to Eq. (3.116). We can express ε in the form of a quality factor, Q:

$$\varepsilon = 1/2Q. \qquad (3.120)$$

Using (3.120), we can write the amplitude as a function of time as

$$A(t) = A_0 e^{-\omega_0 t/2Q}, \qquad (3.121)$$

where Q is defined in terms of the *fractional loss* of energy per cycle of oscillation. In other words

$$\frac{1}{Q} = \frac{-\Delta E}{2\pi E}. \qquad (3.122)$$

This is most easily understood in terms of the logarithmic decrement, δ, which is the logarithm of the ratio of amplitudes of successive cycles of oscillation

$$\delta = \ln(A_1/A_2). \qquad (3.123)$$

Since energy is proportional to the square of amplitude, then

$$2\ln A = \ln E. \qquad (3.124)$$

Combining (3.121) with (3.123), where the amplitudes are one period ($T_0 = 2\pi/\omega_0$) apart gives

$$Q = \pi/\delta. \qquad (3.125)$$

We can also write an equation for ampli-

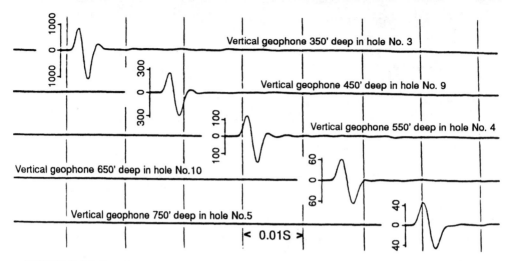

FIGURE 3.32 The effects of attenuation on a seismic pulse. Comparing the pulse width at 350 ft with that at 750 ft shows a significant pulse broadening. This is due to preferential removal of higher frequencies by attenuation. This is accompanied by a decrease in amplitude at a rate greater than expected just for geometric spreading. (After McDonal *et al.*, 1958.)

tude as a function of distance traveled:

$$A(x) = A_0 e^{-(f\pi/Qv)x}. \quad (3.126)$$

It is obvious from (3.126) that for a constant value of Q a high-frequency wave will attenuate more rapidly than a low-frequency wave. This is because for a given distance the high-frequency wave will go through more *oscillations* than a low-frequency wave will. Figure 3.32 shows the development of a wave as it travels away from its source. Notice that the pulse broadens at successive distances. The high-frequency component of the pulse has been removed through attenuation.

Energy loss through nonelastic processes is usually measured by *intrinsic attenuation* and parameterized with Q. Large values of Q imply small attenuation. As Q approaches zero, attenuation is very strong. Q for P waves in the Earth is systematically larger than Q for S waves, and we thus refer to the corresponding quantities as Q_α and Q_β, respectively. It is believed that intrinsic attenuation occurs almost entirely in shear, associated with lateral movements of lattice effects and grain

boundaries. Table 3.2 gives values of Q for several rock types. In general, Q increases with material density and velocity. For a material with all losses due to only shearing mechanisms, $Q_\alpha \approx \frac{9}{4} Q_\beta$.

Q for seismic waves is observed to be largely independent of frequency in the range from 0.001 to 1.0 Hz (Figure 3.33). At higher frequencies, Q depends on frequency and, in general, increases with frequency. To explain this frequency dependence, we must modify our phenomenological model, the oscillating spring, as shown in Figure 3.34. This model is called a *standard linear solid*. The springs

TABLE 3.2 Q for Various Rock Types

Rock type	Q_α	Q_β
Shale	30	10
Sandstone	58	31
Granite	250	70–150
Peridotite	650	280
Midmantle	360	200
Lower mantle	1200	520
Outer core	8000	0

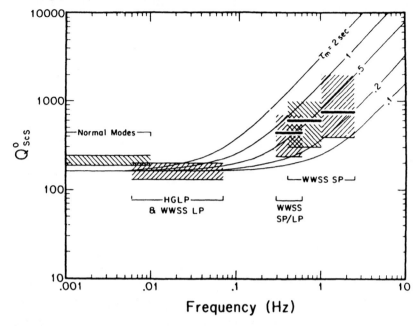

FIGURE 3.33 The frequency dependence of Q_β for observed seismic waves in the Earth. The hachured bands give the range of observations. Between 1000 s and 1 s, Q is nearly constant. (From Sipkin and Jordan, 1979.)

represent elastic behavior, and the dashpot represents nonelastic, or *viscous*, losses. Hooke's law, as written in Eq. (2.44), does not describe the constitutive relationship of a standard linear solid. Rather, the constitutive law is written

$$\sigma + \tau_\sigma \dot{\sigma} = M_r(\varepsilon + \tau_\varepsilon \dot{\varepsilon}), \quad (3.127)$$

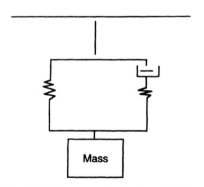

FIGURE 3.34 Phenomenological model for a standard linear solid.

where M_r is called the *relaxed elastic modulus* (appropriate for low frequencies over long times), and τ_σ and τ_ε are called the stress and strain *relaxation times*, respectively. τ_σ implies constant strain, and τ_ε implies constant stress. It is simple to understand the physics of Eq. (3.127) by returning to Figure 3.34. If you deflect the mass, it reaches a point X where it is acted on by a restoring force F. If you hold the mass at X, the force F will diminish with time as the dashpot relaxes. This reduction in restoring force is *not* recoverable. Hence the system behaves anelastically.

The dynamics of (3.127) can be investigated by looking at the ratio of stress to strain:

$$\sigma(t)/\varepsilon(t) = M^*. \quad (3.128)$$

M^* is called the *complex elastic modulus*

and is given by

$$M^* = M_r + \delta M \frac{\omega^2 \tau_\sigma^2}{1 + \omega^2 \tau_\sigma^2} + \frac{i \delta M \omega \tau_\sigma}{1 + \omega^2 \tau_\sigma^2},$$

$$(3.129)$$

where $\delta M = M_u - M_r$. $M_u = \tau_\varepsilon M_r / \tau_\sigma$ is the *unrelaxed elastic modulus* (the elastic response expected for high-frequency displacement applied over a short time—sort of like the initial deflection of the mass described above). This complex elastic modulus has several significant differences from simple elastic moduli; most important, the *behavior* of a standard linear solid depends on *frequency* (ω). This implies that waves that travel through such a solid will be *dispersed*. In other words, the different frequencies in a seismic wavelet will travel with different velocities. We can write the phase velocity as

$$v_p(\omega) = \sqrt{\frac{M_r}{\rho} \left(1 + \frac{1}{2} \frac{\delta M}{M_r} \frac{\omega^2 \tau_\sigma^2}{\left(1 + \omega^2 \tau_\sigma^2\right)} \right)}.$$

$$(3.130)$$

This equation is valid only for small δM. Note that if $\delta M = 0$, then v_p is independent of frequency and is, of course, just the velocity in the elastic case. For small δM we can also write an equation for Q:

$$\frac{1}{Q(\omega)} = \frac{\delta M}{M_r} \frac{\omega \tau_\sigma}{1 + \omega^2 \tau_\sigma^2}. \quad (3.131)$$

The foregoing expressions for phase velocity and Q can be understood by plotting them as a function of $\omega \tau_\sigma$. Figure 3.35 shows the behavior: attenuation is high when Q^{-1} is large; thus enhanced attenuation occurs over a limited range of frequencies. The peak of attenuation is called a *Debye peak*. In general, each relaxation mechanism in the Earth has a distinct Debye peak. These relaxation processes include grain boundary sliding, the forma-

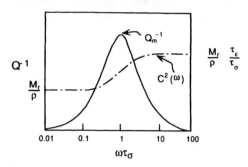

FIGURE 3.35 Q^{-1} as a function of frequency for a standard linear solid. The peak in Q^{-1} is known as a Debye peak.

tion and movement of crystal lattice defects, and thermal currents.

In the Earth we noted that measurements of seismic-wave Q indicate that Q is frequency *independent* over a large range in the seismic frequency band. How is this reconciled with the Debye peak model? Because of the great variety and scale of attenuation processes in the Earth, no single mechanism dominates. The sum or superposition of numerous Debye peaks for the various relaxation processes, each with a different frequency range, produces a broad, flattened *absorption band*. Figure 3.36 shows this superposition effect; note that Q^{-1} is basically constant for frequencies of 1.0 Hz (1.0 cycle/s) to 2.8×10^{-4} Hz (1.0 cycle/h). Phenomenologically, this corresponds to a coupled system of many standard linear solid elements.

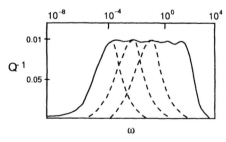

FIGURE 3.36 Superposition of numerous Debye peaks results in an absorption band—nearly constant Q over a range of frequencies.

Let us return to the single Debye peak model in Figure 3.35. For the period range shown, the change in velocity is on the order of 5%, with higher frequencies traveling faster. This result has a profound implication, namely that the velocity structure of the Earth, as determined from free oscillations or long-period surface waves, will differ from that determined by body waves. Although Q is constant in the frequency range of the absorption band, the phase velocity will still be dispersive with the following functional form:

$$c(\omega) = c_0\left[1 + \frac{1}{\pi Q_m}\ln\left(\frac{\omega}{\omega_0}\right)\right], \quad (3.132)$$

where ω_0 is some reference frequency. In general, this dispersion is minor for body waves of interest to earthquake seismology. On the other hand, it can be important for very high frequencies, and it is *very* important for seismic surface waves, which we discuss in the next chapter.

The most common way to determine Q is to compare the amplitude and frequency content of seismic rays that have traveled similar paths. This eliminates unknown source effects. An example of such a comparison is shown in Figure 3.37. For S waves that travel down to the core, reflected S (ScS) and reflected P (ScP) arrivals can be observed at the same distance. Examples are shown for short-period phases from a deep earthquake. Both ScS and ScP have about the same source radiation (S-wave energy) and similar attenuation on the path down to the core–mantle boundary. However, on the return leg through the mantle, the ScP phase is attenuated by the relatively high Q_α values in the mantle, whereas ScS is attenuated by Q_β. This causes ScS to be both lower in amplitude and depleted in high-frequency content relative to ScP, as seen at station JCT. If we account for the reflection coefficients, we can estimate Q_α

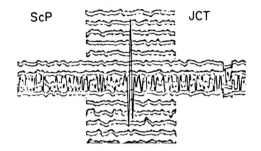

FIGURE 3.37 Short-period records from WWSSN station JCT (Junction City, Texas) for a deep South American event of March 27, 1967, showing ScP and ScS arrivals. (From Burdick, 1985. Reprinted with permission from the Royal Astronomical Society.)

and Q_β averaged over the entire mantle path by matching the amplitude and frequency content of the two signals.

Actual measurements of Q vary laterally by an order of magnitude within the Earth, much larger variation than is observed for seismic velocity (10% variations). The mechanisms of intrinsic attenuation (grain-boundary and crystal-defect sliding) are very sensitive to pressure and temperature conditions. This means that Q will vary within the Earth as a function of temperature heterogeneity. Tectonically active regions typically have relatively high heat flow and are more attenuating than "colder" regions. It has also been observed that Q variations correlate with travel-time variations. Fast travel-time paths are typically high Q, slow paths typically low Q. This is a manifestation of the thermal activation of the attenuation mechanisms. Thus, mapping Q can reveal thermal processes at depth. In Chapter 7 we will discuss the lateral variation of Q and its consequences for tectonic processes.

In body-wave studies we commonly account for the effects of attenuation by *convolving* the elastic pulse shape with an attenuation operator parameterized by the value t^*. Although we will discuss convolution later (see Chapter 10), it is instructive to introduce t^*, the travel time divided by the quality factor in a region of uniform attenuation:

$$t^* = \frac{t}{Q} = \frac{\text{travel time}}{\text{quality factor}}. \quad (3.135)$$

In the Earth, Q is a function of depth (and frequency), with the lowest Q values (highest attenuation) occurring in the upper mantle. Since $Q = Q(r)$, t^* is usually written as a path integral value

$$t^* = \int_{\text{path}} \frac{dt}{Q} = \sum_{i=1}^{N} \frac{t_i}{Q_i}, \quad (3.136)$$

where t_i and Q_i are the travel time and quality factor for the ith layer in a layered Earth. Clearly, t^* is thus the total travel time divided by the path-averaged value of Q. Observationally, we find that t^* is approximately constant for body waves with periods longer than 1 s in the distance range $30° < \Delta < 95°$. In this range, $t_\alpha^* \approx 1.0$ and $t_\beta^* \approx 4.0$. Thus, we can account for the effects of t^* by replacing t/Q in Eq. (3.125) to give

$$A = A_0 e^{-\pi f t^*}. \quad (3.137)$$

Note that t^* is much larger for S waves than for P waves; thus S waves attenuate much more rapidly with distance. Figure 3.38 shows the effects of different values of t^* on long- and short-period seismograms. Note that changing t^* by a factor of 20 changes the short-period P-wave ampli-

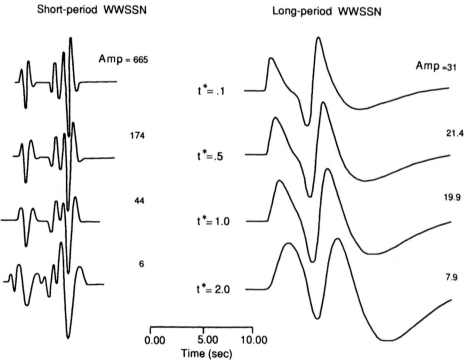

Short-period WWSSN

Long-period WWSSN

Amp = 665
$t^* = .1$
Amp = 31

174
$t^* = .5$
21.4

44
$t^* = 1.0$
19.9

6
$t^* = 2.0$
7.9

0.00   5.00   10.00
Time (sec)

FIGURE 3.38 The effect of different t_α^* on observed P waves recorded on WWSSN short-period and long-period instruments.

tude by a factor of 100, but it changes the long-period amplitude by only a factor of 6. The change in amplitude of the high-frequency energy is vastly greater, but the narrowband instrument response obscures this. Attenuation of body waves is complicated by both frequency dependence for periods of less than 1 s and strong lateral variations at all periods (see Chapter 7).

References

Burdick, L. J. (1985). Estimation of the frequency dependence of Q from ScP and ScS phases. *Geophys. J. R. Astron. Soc.* **80**, 35–55.

Dainty, A. M., Toksöz, M. N., Anderson, K. R., Pines, P. J., Nakamura, Y., and Latham, G. (1974). Seismic scattering and shallow structure of the moon in oceanus procellarum. *Moon* **9**, 11–29.

Doornbos, D. J. (1989). Seismic diffraction. In "The Encyclopedia of Solid Earth Geophysics" (D. E. James, ed.), pp. 1018–1024. Van Nostrand–Reinhold, New York.

Frankel, A., and Clayton, R. W. (1986). Finite difference simulations of seismic scattering: Implications for the propagation of short-period seismic waves in the crust and models of crustal heterogeneity. *J. Geophys. Res.* **91**, 6465-6489.

McDonal, F. J., Angona, F. A., Milles, R. L. Sengbush, R. L., Van Nostrand, R. G., and White, J. E. (1958). Attenuation of shear and compressional waves in Pierre Shale. *Geophysics* **23**, 421–439.

Officer, C. B. (1974). "Introduction to Theoretical Geophysics." Springer–Verlag, New York.

Rial, J. A., and Cormier, V. F. (1980). Seismic waves at the epicenter's antipode *J. Geophys. Res.* **85**, 2661–2668.

Sipkin, S. A., and Jordan, T. H. (1979). Frequency dependence of Q_{ScS}. *Bull. Seismol. Soc. Am.* **69**, 1055–1079.

Waters, K. H. (1981). "Reflection seismology: A tool for energy resource exploration." Second Edition, 453 pp, Wiley, New York..

Additional Reading

Aki, A., and Richards, P. G. (1980). "Quantitative Seismology," 2 vols. Freeman, San Francisco.

Bullen, K. E., and Bolt, B. A. (1985). "An Introduction to the Theory of Seismology." Cambridge Univ. Press, Cambridge, UK.

Gubbins, D. (1990). "Seismology and Plate Tectonics." Cambridge Univ. Press, Cambridge, UK.

Officer, C. B. (1974). "Introduction to Theoretical Geophysics." Springer–Verlag, New York.

Stein, S. (1993). "Introduction to Seismology, Earthquakes, and Earth Structure." Blackwell, Boston (in press).

4

SURFACE WAVES AND FREE OSCILLATIONS

The last two chapters have demonstrated the remarkably simple basic character of solutions of the equations of motion for linear-elastic, isotropic, homogeneous (or weakly inhomogeneous) unbounded media. The displacement field created by a stress imbalance is completely accounted for by propagating P and S waves, no matter what type of seismic source is involved (Chapter 8). These wavefields become increasingly complex when discontinuous material properties and localized inhomogeneities are present. Wave phenomena such as refraction, wave type conversion, frequency-dependent scattering, and diffraction take place in an inhomogeneous medium like the Earth, leading to a very complicated body wave field. The fact that the Earth's inhomogeneity is primarily one-dimensional (i.e., varies with depth) allows us to interpret most of the body-wave complexity. The Earth has two additional fundamental attributes, shared with all finite structures, that profoundly affect the seismic wavefield. These are the presence of the free surface and the finite (quasi-ellipsoidal) shape of the planet.

The free surface of an elastic medium has the special stress environment defined by the vanishing of surface tractions. For the Earth, all seismic-wave measurements are made at or near the free surface; thus it is critical to understand free-surface effects in order to interpret seismograms. At the surface both incident and reflected waves instantaneously coexist, and the total motion involves the sum of their respective amplitudes. For example, from Table 3.1 we know that a reflected SH wave has the same amplitude as the incident wave. Thus, at the free surface the amplitude of SH motion is doubled. We call this multiplicative factor the SH receiver function. Free-surface receiver functions for P and SV waves involve comparable displacement amplifications. Even more important is the interaction of incident P and SV waves with the free-surface boundary condition, which gives rise to an interference wave that effectively travels along the surface as a *Rayleigh wave*. Total reflection of SH waves at the free surface combines with internal layering of the Earth to trap SH reverberations near the surface, which interfere to produce horizontally propagating *Love waves*. Gravitationally controlled waves in water on the Earth's surface give rise to sea waves, or *tsunamis*, which often cause greater damage from earthquake faulting than any elastic waves in solid rock. We will consider the basic properties of these free-surface waves in this chapter.

The finiteness of the Earth, like its internal layering, provides scale lengths and boundary conditions on the seismic wavefield. We can view the planet as a finite elastic system with unique boundary conditions that govern the solutions of the equations of motion in the medium. This perspective leads to the definition of *normal modes* of the system, involving discrete frequencies at which the system can oscillate, in a manner analogous to the harmonic tones of an organ pipe or a vibrating guitar string. For internal sources, these normal modes are called *free oscillations*, and we will discuss these modes of whole-Earth oscillation in this chapter. Clearly, all body waves propagating in the Earth (remember, *P* and *S* waves provide *complete* solutions to the equations of motion,

although they may involve very complex wavefields) must have counterparts in both propagating surface waves or standing-wave free oscillations. Nonetheless, each representation has distinct advantages for studying Earth structure and seismic sources.

4.1 Free-Surface Interactions

Rayleigh waves involve interaction between *P* and *S* waves at a free surface; thus we must further explore the nature of body-wave reflection coefficients at the free surface. We consider the two cases shown in Figure 4.1, for incident *P* and incident *SV* plane waves impinging on a free surface. A free surface requires the tractions to vanish at $x_3 = 0$; $\sigma_{33} = \sigma_{13} = \sigma_{23} = 0$. For our choice of coordinate system, with the wavenumber vectors for the plane waves being confined to the $x_1 x_3$ plane ($u_2 = 0$,

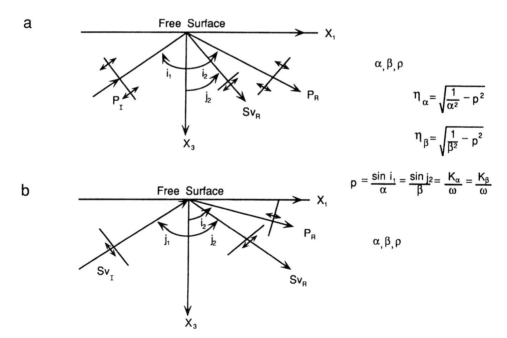

$$\alpha, \beta, \rho$$

$$\eta_\alpha = \sqrt{\frac{1}{\alpha^2} - p^2}$$

$$\eta_\beta = \sqrt{\frac{1}{\beta^2} - p^2}$$

$$p = \frac{\sin i_1}{\alpha} = \frac{\sin j_2}{\beta} = \frac{K_\alpha}{\omega} = \frac{K_\beta}{\omega}$$

FIGURE 4.1 Geometry for free-surface interactions of (a) an incident *P* wave and (b) an incident *SV* wave.

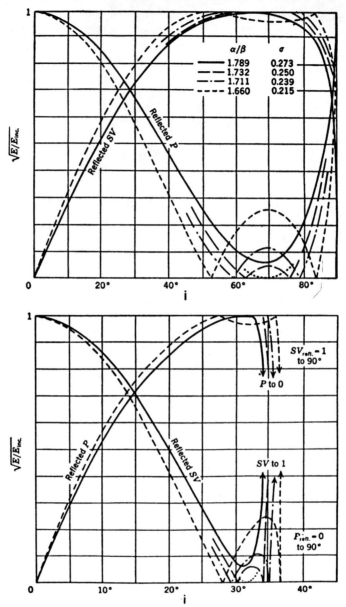

FIGURE 4.2 (Top) Square root of reflected to incident *P*-wave energy at the free surface for various half-space velocity combinations. (Bottom) Similar plot for an incident *SV* wave. (From Ewing *et al.*, 1957.)

$\partial u_i/\partial x_2 = 0$), Hooke's law (2.44) becomes

$$\sigma_{33} = \lambda \left(\frac{\partial u_1}{\partial x_1} + \frac{\partial u_3}{\partial x_3} \right) + 2\mu \frac{\partial u_3}{\partial x_3} = 0$$

$$\sigma_{13} = \mu \left(\frac{\partial u_3}{\partial x_1} + \frac{\partial u_1}{\partial x_3} \right) = 0. \qquad (4.1)$$

The displacement components are obtained from potentials by using (2.91). For the case of an incident P wave, propagating in the $-x_3$ and $+x_1$ directions (Figure 4.1a), we assume plane-wave potentials of the form

$$\phi = \phi_I + \phi_R = A \exp[i\omega(px_1 - \eta_\alpha x_3 - t)]$$

$$+ B \exp[i\omega(px_1 + \eta_\alpha x_3 - t)]$$

$$\psi = \psi_R = C \exp[i\omega(px_1 + \eta_\beta x_3 - t)].$$

$$(4.2)$$

At the free surface ($x_3 = 0$), the stress conditions [Eq. (4.1)] lead to equations relating the incident amplitude (A) to the reflected P (B) and reflected SV (C) amplitudes. Using (2.91) and (4.1), $\sigma_{33} = 0$ gives

$$(A + B)\left[(\lambda + 2\mu)\eta_\alpha^2 + p^2\lambda\right]$$

$$+ C(2\mu p\eta_\beta) = 0 \qquad (4.3)$$

and $\sigma_{13} = 0$ gives

$$(A - B)2p\eta_\alpha - C(p^2 - \eta_\beta^2) = 0. \quad (4.4)$$

Combining these equations yields the plane-wave potential reflection coefficients

These are equivalent to the expressions in Table 3.1, except that (4.5) and (4.6) are for potentials, not displacements. The value of R_{PS} vanishes when $p = 0$ (normal incidence) and when $\eta_\alpha = 0 = [(1/\alpha^2) - p^2]^{1/2} = [(1/\alpha^2) - (\sin^2 i/\alpha^2)]^{1/2}$ (i.e., at $i_1 = 90°$, grazing incidence). In general, two incident angles, i, exist at which $R_{pp} = 0$, yielding total P to SV conversion. These depend on particular values of α and β. Figure 4.2 shows calculations of the energy partitioning as a function of incidence angle for P waves for various half-space velocity parameters. The actual particle displacements at the surface consist of combined displacements due to coexisting incident and reflected P and SV motion and are obtained by computing the derivatives indicated in (2.91). This gives the incident P-wave surface response, or receiver function.

For the case of an incident SV wave (Figure 4.1b), we assume plane-wave potentials of the form

$$\phi = F \exp[i\omega(px_1 + \eta_\alpha x_3 - t)]$$

$$\psi = D \exp[i\omega(px_1 - \eta_\beta x_3 - t)]$$

$$+ E \exp[i\omega(px_1 + \eta_\beta x_3 - t)]. \quad (4.7)$$

The stress boundary conditions provide

$$R_{PP} = \frac{B}{A} = \frac{(\lambda + 2\mu)\eta_\alpha^2 + p^2\lambda + 4\mu p^2\eta_\alpha\eta_\beta/(p^2 - \eta_\beta^2)}{-\left[(\lambda + 2\mu)\eta_\alpha^2 + p^2\lambda\right] + 4\mu p^2\eta_\alpha\eta_\beta/(p^2 - \eta_\beta^2)} \qquad (4.5)$$

$$R_{PS} = \frac{C}{A} = \left(\frac{4p\eta_\alpha}{p^2 - \eta_\beta^2} \right) \left[\frac{(\lambda + 2\mu)\eta_\alpha^2 + p^2\lambda}{(\lambda + 2\mu)\eta_\alpha^2 + p^2\lambda - 4\mu p^2\eta_\alpha\eta_\beta/(p^2 - \eta_\beta^2)} \right]. \qquad (4.6)$$

potential reflection coefficients

$$R_{SS} = \frac{E}{D} = \frac{\left[(\lambda + 2\mu)\eta_\alpha^2 + p^2\lambda + 4\mu p^2 \eta_\alpha \eta_\beta / (p^2 - \eta_\beta^2)\right]}{-\left[(\lambda + 2\mu)\eta_\alpha^2 + p^2\lambda\right] + 4\mu p^2 \eta_\alpha \eta_\beta / (p^2 - \eta_\beta^2)} \qquad (4.8)$$

$$R_{SP} = \frac{F}{D} = \frac{4\mu p \eta_\beta}{\left[(\lambda + 2\mu)\eta_\alpha^2 + p^2\lambda\right] - 4\mu p^2 \eta_\alpha \eta_\beta / (p^2 - \eta_\beta^2)}. \qquad (4.9)$$

Note that $R_{PP} = R_{SS}$. The corresponding energy functions are also plotted in Figure 4.2. Since $\alpha > \beta$, an angle of incidence, $j_1 = \sin^{-1}(\beta/\alpha)$, exists such that the P-wave "reflection" travels along the free surface ($i_2 = 90°$). For angles j_1 greater than this "critical" angle, we follow the procedure introduced in Chapter 3 and allow i_2 to become complex and η_α to become purely imaginary. Thus, the ϕ potential acquires a phase shift, and the amplitude decays exponentially away from the interface, similar to the head-wave behavior discussed previously. Thus, a P wave can be "trapped" propagating along the free surface. This type of wave is known as an *evanescent wave* because it decays exponentially with depth. The postcritical SV reflection has unity magnitude and also has a phase shift, but it otherwise propagates as a plane wave. Figure 4.3a illustrates the resulting situation. The critically refracted P wave exists simultaneously with the incident SV wave, but no energy is transmitted back into the medium by the P-wave motion. This suggests that an evanescent P wave alone cannot propagate along the boundary.

The inability to trap purely P-wave energy near the boundary is demonstrated by considering the P potential

$$\phi = A \exp[i\omega(px_1 - \eta_\alpha x_3 - t)]$$

$$+ B \exp[i\omega(px_1 + \eta_\alpha x_3 - t)], \qquad (4.10)$$

for the case $p > 1/\alpha$, $\eta_\alpha = i\hat{\eta}_\alpha = i(p^2 - 1/\alpha^2)^{1/2}$, which gives

$$\phi = A \exp[i\omega(px_1 - t)] \exp[\hat{\eta}_\alpha \omega x_3]$$

$$+ B \exp[i\omega(px_1 - t)] \exp[-\omega \hat{\eta}_\alpha x_3], \qquad (4.11)$$

which diverges as $x_3 \to \infty$ unless $A = 0$. Satisfying the boundary condition $\sigma_{13} = 0$ assuming that no SV wave is present leads to $B = 0$. In other words, although a plane P wave may propagate along the boundary, an evanescent P wave alone cannot. A similar result is found for a horizontally propagating SV wave; the surface stress condition precludes the existence of purely P or SV evanescent waves on the boundary. However, we will now demonstrate that simultaneous, coupled evanescent P and SV waves do satisfy the surface boundary condition, yielding a new form of wave solution.

4.2 Rayleigh Waves

We now consider the situation in Figure 4.3b, where evanescent P and SV waves are assumed to simultaneously propagate along the free surface. Assume that the potentials have the form

$$\phi = A \exp[i\omega(px_1 + \eta_\alpha x_3 - t)]$$

$$= A \exp[-\omega \hat{\eta}_\alpha x_3] \exp[i\omega(px_1 - t)]$$

$$\psi = B \exp[i\omega(px_1 + \eta_\beta x_3 - t)]$$

$$= B \exp[-\omega \hat{\eta}_\beta x_3] \exp[i\omega(px_1 - t)], \qquad (4.12)$$

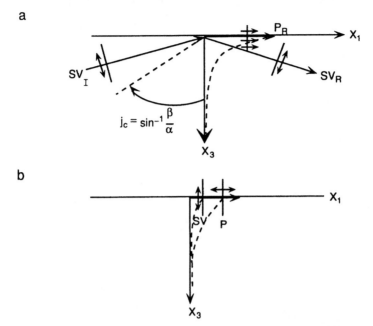

FIGURE 4.3 (a) Postcritical *SV* wave incident on a free surface gives rise to an evanescent *P* wave propagating along the boundary as well as to a phase-shifted *SV* reflection. (b) Simultaneous existence of evanescent *P*- and *SV*-wave energy traveling horizontally along a free surface produces the interference surface wave called a Rayleigh wave.

where the horizontal apparent velocity $c = (1/p) < \beta < \alpha$. This confines the energy to propagate along the surface with exponential decay of the potentials away from the $x_3 = 0$ surface:

$$\eta_\alpha = \sqrt{\frac{1}{\alpha^2} - p^2} = i\hat{\eta}_\alpha$$

$$= i\sqrt{p^2 - \frac{1}{\alpha^2}} = i\sqrt{\frac{1}{c^2} - \frac{1}{\alpha^2}}$$

$$\eta_\beta = \sqrt{\frac{1}{\beta^2} - p^2} = i\hat{\eta}_\beta$$

$$= i\sqrt{p^2 - \frac{1}{\beta^2}} = i\sqrt{\frac{1}{c^2} - \frac{1}{\beta^2}}, \quad (4.13)$$

where $1/p = c < \beta < \alpha$. If $\beta < c < \alpha$, the *SV* energy will propagate away from the free surface as a body wave, and the only way to satisfy the surface boundary condition is simultaneously to have incident *SV*

energy as shown in the last section. Lord Rayleigh (the former J. W. Stutt) explored the system in Eq. (4.12) in 1887 and found that the surface boundary condition can in fact be satisfied, leading to the existence of a coupled *P–SV* wave traveling along the surface with a velocity lower than the shear velocity and with amplitudes decaying exponentially away from the surface. These waves spread cylindrically on the surface and thus have a two-dimensional geometric decrease in amplitude with radius r from the source proportional to $1/\sqrt{r}$, compared to the three-dimensional $(1/r)$ decay for body waves. The resulting waves, *Rayleigh waves*, tend to be the largest arrivals on long-period or broadband seismograms.

Using (2.91) and (4.12), the condition $\sigma_{33}|_{x_3=0} = 0$ gives

$$A\left[(\lambda + 2\mu)\eta_\alpha^2 + \lambda p^2\right] + B(2\mu p\eta_\beta) = 0$$

$$(4.14)$$

and $\sigma_{13}|_{x_3=0} = 0$ yields

$$A(2p\eta_\alpha) + B(p^2 - \eta_\beta^2) = 0. \quad (4.15)$$

The coupled Eqs. (4.14) and (4.15) can be written in matrix form

$$\begin{bmatrix} (\lambda + 2\mu)\eta_\alpha^2 + \lambda p^2 & 2\mu p\eta_\beta \\ 2p\eta_\alpha & p^2 - \eta_\beta^2 \end{bmatrix} \begin{bmatrix} A \\ B \end{bmatrix} = \begin{bmatrix} 0 \\ 0 \end{bmatrix}$$

$$(4.16)$$

The only solutions other than the trivial solution $A = B = 0$ are given by vanishing of the determinant of the matrix

$$\left[(\lambda + 2\mu)\eta_\alpha^2 + \lambda p^2 \right] \left(p^2 - \eta_\beta^2 \right)$$

$$- 4\mu p^2 \eta_\alpha \eta_\beta = 0. \quad (4.17)$$

The term on the left in Eq. (4.17) appears in the denominators of the free-surface reflection coefficients in Eqs. (4.5) and (4.6) and again in Eqs. (4.8) and (4.9) and is hence called the *Rayleigh denominator*. If Eq. (4.17) is satisfied with a real η_α and η_β, then R_{PS} and R_{PP} will be infinite. The only possible solution to Eq. (4.17) that satisfies all conditions are imaginary η_α and η_β, which results in an evanescent wave. It is convenient to rewrite (4.17) in terms of velocities, using $\rho\alpha^2 = (\lambda + 2\mu)$, $\rho\beta^2 = \mu$ to obtain

$$\left[\alpha^2 \left(\frac{\eta_\alpha^2}{p^2} + 1 \right) - 2\beta^2 \right] \left(1 - \frac{\eta_\beta^2}{p^2} \right)$$

$$- \left(\frac{4\beta^2 \eta_\alpha \eta_\beta}{p^2} \right) = 0. \quad (4.18)$$

Since we need to satisfy (4.13), we insert corresponding expressions for η_α and η_β

into (4.18), giving

$$(c^2 - 2\beta^2)\left(2 - \frac{c^2}{\beta^2} \right)$$

$$+ 4\beta^2 \sqrt{1 - \frac{c^2}{\alpha^2}} \sqrt{1 - \frac{c^2}{\beta^2}} = 0.$$

$$(4.19)$$

This equation can be rationalized to give a final form suitable for solution:

$$\frac{c^2}{\beta^2}\left[\frac{c^6}{\beta^6} - 8\frac{c^4}{\beta^4} + c^2\left(\frac{24}{\beta^2} - \frac{16}{\alpha^2} \right) \right.$$

$$\left. - 16\left(1 - \frac{\beta^2}{\alpha^2} \right) \right] = 0. \quad (4.20)$$

For prescribed values of α and β, one solution of (4.20) for $0 < c < \beta$ can always be found. As an example, we consider the case of a Poisson solid for which $\lambda = \mu$, $\alpha^2 = 3\beta^2$. Equation (4.20) becomes

$$\left[\frac{c^6}{\beta^6} - 8\frac{c^4}{\beta^4} + \frac{56}{3}\frac{c^2}{\beta^2} - \frac{32}{3} \right] = 0, \quad (4.21)$$

which is cubic in (c^2/β^2) and has roots $(c^2/\beta^2) = 4, (2 + 2/\sqrt{3}), (2 - 2/\sqrt{3})$. Only the last root satisfies $(c/\beta) < 1$ and gives $c = 0.9194\beta$ as the velocity of a Rayleigh-wave disturbance in a Poisson solid half-space. Figure 4.4 shows solutions of (4.20) for different values of Poisson's ratio. For typical values of Poisson's ratio $(0.2 < \nu < 0.4)$, the Rayleigh-wave velocity is 0.9β to 0.95β.

We now consider the nature of the particle motions associated with a Rayleigh wave. The surface-wave motion involves a mix of P and SV motion, with relative amplitudes A and B. We can rewrite (4.14) as

$$B = \frac{-A\left[(c^2/\beta^2) - 2 \right]}{2c\eta_\beta} \quad (4.22)$$

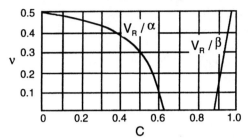

FIGURE 4.4 Half-space Rayleigh-wave velocity c as a function of Poisson's ratio, ν, where $\nu = [(\alpha^2/\beta^2)-2]/2[(\alpha^2/\beta^2)-1]$. For a fluid, $\beta = 0$ and $\nu = 0.5$, in which case $c = 0$. For a Poisson solid, $\alpha = \sqrt{3}\,\beta$, $\nu = 0.25$, and $c = 0.9194\beta$. (From Sheriff and Geldart, "Exploration Seismology," Vol. 1, History, theory, and data acquisition. Copyright©1982. Reprinted with the permission of Cambridge University Press.)

and then compute the Rayleigh-wave displacements using (4.12), (4.22), and

$$u_1 = \frac{\partial \phi}{\partial x_1} - \frac{\partial \psi}{\partial x_3}, \qquad u_3 = \frac{\partial \phi}{\partial x_3} + \frac{\partial \psi}{\partial x_1}$$

$$(4.23)$$

to find

$$u_1 = Ae^{i\omega(px_1 - t)}i\omega p$$

$$\times \left[e^{-\omega\hat{\eta}_\alpha x_3} + \frac{1}{2}\left(\frac{c^2}{\beta^2} - 2\right) e^{-\omega\hat{\eta}_\beta x_3} \right]$$

$$u_3 = -Ae^{i\omega[px_1 - t)}\omega$$

$$\times \left[\hat{\eta}_\alpha e^{-\omega\hat{\eta}_\alpha x_3} + \frac{1}{2c^2\hat{\eta}_\beta}\left(\frac{c^2}{\beta^2} - 2\right) \right.$$

$$\left. \times e^{-\omega\hat{\eta}_\beta x_3} \right]. \quad (4.24)$$

Since the Rayleigh-wave ground motion must be real, we use $\exp[i\omega(px_1 - t)] = \cos[\omega(px_1 - t)] + i\sin[\omega(px_1 - t)]$ and

retain only real terms

$$u_1 = -A\omega p \sin[\omega(px_1 - t)]$$

$$\times \left[e^{-\omega\hat{\eta}_\alpha x_3} + \frac{1}{2}\left(\frac{c^2}{\beta^2} - 2\right) e^{-\omega\hat{\eta}_\beta x_3} \right]$$

$$u_2 = -A\omega p \cos[\omega(px_1 - t)]$$

$$\times \left[c\hat{\eta}_\alpha e^{-\omega\hat{\eta}_\alpha x_3} + \frac{1}{2c\hat{\eta}_\beta}\left(\frac{c^2}{\beta^2} - 2\right) \right.$$

$$\left. \times e^{-\omega\hat{\eta}_\beta x_3} \right]. \quad (4.25)$$

For the Poisson solid, $c = 0.919\beta = 0.531\alpha$, and letting $k = \omega p = \omega/c$ be the Rayleigh wavenumber, Eq. (4.25) becomes

$$u_1 = -Ak\sin(kx_1 - \omega t)$$

$$\times (e^{-0.85kx_3} - 0.58e^{-0.39kx_3})$$

$$u_3 = -Ak\cos(kx_1 - \omega t)$$

$$\times (0.85e^{-0.85kx_3} - 1.47e^{-0.39kx_3}).$$

$$(4.26)$$

At the surface of the Poisson solid, $x_3 = 0$ and

$$u_1 = -0.42Ak\sin(kx_1 - \omega t)$$

$$u_3 = 0.62Ak\cos(kx_1 - \omega t). \quad (4.27)$$

The Rayleigh-wave displacements given by (4.26) depend harmonically on x_1 and exponentially on x_3 (depth). The displacements u_1 and u_3 are out of phase by 90° and therefore combine to give ellipsoidal particles motion, as illustrated in Figure 4.5. The surface vertical motion is larger than the horizontal motion by a factor of 1.5. At the top of the cycle (in the $-x_3$ direction) the surface horizontal motion is opposite the direction of propagation, and the elliptical motion is *retrograde*. Figure 4.6 illustrates the motion of adjacent particles on the surface and at depth as a

Rayleigh wave passes by. The horizontal distance between surface particle motions at the same point in their elliptical cycle defines the Rayleigh-wave wavelength Λ. At a depth of about $\Lambda/5$ the horizontal motion goes to zero, and at greater depth, the elliptical motion has a *prograde* sense. By a depth of $\Lambda/2$, the horizontal particle motion is about 10% of the horizontal motion at the surface, and the vertical motion is about 30% of the surface verti-

cal motion. All of the Rayleigh-wave motion is contained in the vertical plane $(x_1 x_3)$ with no tangential (u_2) component.

Since the Rayleigh-wave amplitudes have exponential dependence in the form $e^{-kx_3} = e^{(-2\pi/\Lambda)x_3}$, long-wavelength Rayleigh waves have larger displacements at greater depth than shorter-wavelength waves. In the end-member case of a homogeneous half-space, the velocity of Rayleigh waves does not depend on fre-

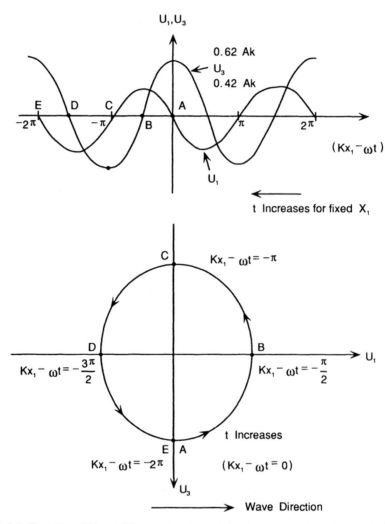

FIGURE 4.5 (Top) Plot of Eq. (4.27) as a function of the phase argument $(kx_1 - \omega t)$. (Bottom) Behavior of an individual particle as a function of time. The surface motion is retrograde elliptical.

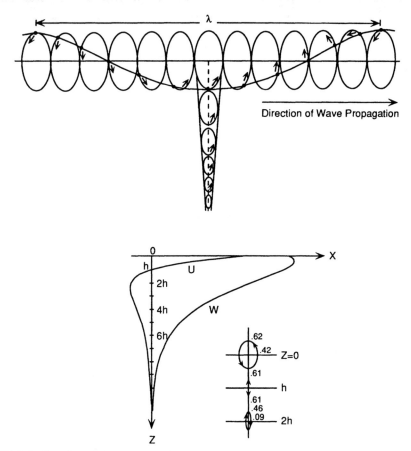

FIGURE 4.6 (Top) Rayleigh-wave particle motions over one wavelength along the surface and as a function of depth. (Bottom) Horizontal (*u*) and vertical (*w*) displacements of Rayleigh waves in a homogeneous half-space. The particle motion is retrograde elliptical above depth *h* and prograde elliptical at greater depth. (From Sheriff and Geldart, "Exploration Seismology," Vol. 1, History, theory, and acquisition. Copyright©1982. Reprinted with the permission of Cambridge University Press.)

quency, but for a layered or vertically inhomogeneous structure, the Rayleigh wave is *dispersive*. Because in general the velocity in the Earth increases with depth, the longer wavelengths tend to sample faster material, giving rise to higher Rayleigh-wave velocities for large-wavelength, low-frequency wave components, which produces dispersion. Rayleigh waves only require a free surface to be a viable solution of the equations of motion, but only a half-space produces an undispersed Rayleigh pulse (see Box 4.1). A much more

characteristic Rayleigh waveform is shown in Figure 4.7, where the Rayleigh phase labeled *LR* is spread out over more than 10 min, with lower-frequency energy arriving earlier in the waveform. We will discuss such dispersion later in this chapter. Note that the Rayleigh-wave motions are the largest of any arrivals on this seismogram, which results from the two-dimensional geometric spreading of the surface wave relative to the three-dimensional spreading that affects the body waves. Sources near the surface tend to excite

Box 4.1 Lamb's Problem

A complete theory for Rayleigh waves, even for a half-space, must include their excitation by a specific source. Chapter 8 will demonstrate how seismic sources are represented in the equations of motion and will discuss Rayleigh wave radiation from faults. At this point we show a classic result, first obtained by H. Lamb (1904), which is the transient solution to an impulsive vertical point force applied to the surface of a half-space. Part (a) of Figure 4.B1.1 shows Lamb's (1904) calculations, which are believed to be the first theoretical seismograms. The motions begin with the P arrival. The small arrival prior to the large-amplitude pulse is the S wave, and the large pulse itself is a Rayleigh-wave pulse. The Rayleigh wave shows a clear phase shift between the radial (q_0) and vertical (w_0) components and is much larger than the body-wave arrivals. The experimental result shown in part (b) is a recording of a breaking pencil lead point-force source on a piece of brass, which has a vertical motion very similar to Lamb's prediction. Recordings of natural sources approximating Lamb's solution are shown in Box 8.2, but normally Rayleigh waves in the Earth are dispersed and resemble Figure 4.7. Rayleigh-wave excitation varies substantially with source force system and depth.

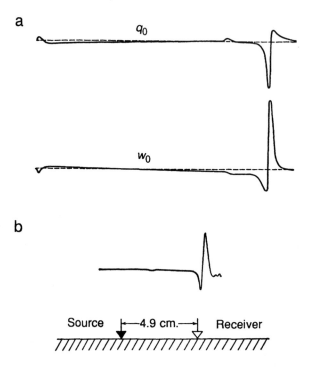

FIGURE 4.B1.1 (a) Radial (q_0) and vertical (w_0) surface ground motions calculated by Lamb (1904) for an impulsive vertical force on the surface. (b) An experimentally recorded vertical ground motion for a vertical point source. The largest motion in each case corresponds to the undispersed Rayleigh pulse. (From Ewing *et al.*, 1957).

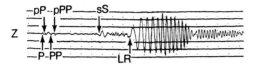

FIGURE 4.7 A characteristic vertical-component seismic recording showing body-wave arrivals (*p, pP, PP, pPP, sS*) followed by a dispersed Rayleigh wave, the onset of which is labeled *LR*. Tick marks on the record are 60 s apart, with time increasing to the right. Note that lower-frequency components of *LR* arrive earlier because of dispersion. Rayleigh-wave motions persist for over 10 min and produce the largest ground motions on the seismogram. (From Simon, "Earthquake Interpretations: A Manual for Reading Seismograms," Copyright©1981 William Kaufmann, Inc.)

strong Rayleigh waves, whereas sources deep in the Earth excite only weak Rayleigh waves. Solution of Rayleigh-wave propagation in a layered or inhomogeneous elastic medium is beyond the scope of this text (a simple case is considered in Box 4.3) but is treated fully in advanced texts by Aki and Richards (1980), Kennett (1983), and Ben-Menahem and Singh (1981). We will consider Rayleigh-wave motion in the Earth in the context of equivalent spheroidal free oscillations later in this chapter.

4.3 Love Waves

The presence of a free surface is sufficient to enable coupled *P–SV* generation of a Rayleigh-wave surface disturbance. However, the *SH* component of the *S* wave, having displacements parallel to the surface, can only have total reflections from the free surface. In order to trap any *SH* energy near the surface, the velocity structure at depth must keep turning energy toward the surface. If the *S* velocity increases with depth, a *waveguide* can be formed, in which rays are multiply reflected between the surface and deeper turning or reflection points. If the ray

strikes the reflecting horizon at *postcritical* angles, all the energy is trapped within the waveguide. The properties of an *SH* disturbance trapped in a near-surface waveguide were first explored by A. E. H. Love in 1911, and these waves are hence named *Love waves*.

We consider the nature of *SH* waves trapped in a low-velocity layer overlying a half-space, as shown in Figure 4.8. The layer has thickness H, which introduces a spatial dimension to the problem that was not present in the Rayleigh-wave solution for a half-space. This dimensionality leads to frequency dependence of velocity for the propagating interference patterns that we call a Love wave, even though the intrinsic shear-wave velocity, β_1, has no frequency dependence.

We are considering *SH*-type displacements, so we use the result found in Chapter 2, that we do not need to use potentials because the *SH* displacements satisfy the wave equation. Thus we can write plane-wave solutions of the form

$$V_1 = A \exp\left[i\omega\left(px_1 + \eta_{\beta_1}x_3 - t\right)\right]$$

$$+ B \exp\left[i\omega\left(px_1 - \eta_{\beta_1}x_3 - t\right)\right]$$

$$V_2 = C \exp\left[i\omega\left(px_1 + \eta_{\beta_2}x_3 - t\right)\right], \quad (4.28)$$

where V_1 is the *SH* displacement in the layer, composed of upward- and downward-propagating plane waves, and V_2 is the *SH* displacement in the half-space, composed of transmitted *SH* waves generated at each reflection point at the base of the layer. If $\beta_1 > \beta_2$, then $j_3 < j_1$ and transmitted energy will always propagate away from the high-velocity layer, with layer reverberations progressively diminishing. For $\beta_1 < \beta_2$, $j_3 > j_1 = j_2$ and the transmitted wave is refracted closer to the boundary but still propagates away, leaking energy out of the low-velocity layer, until $j_1 = j_c = \sin^{-1}(\beta_1/\beta_2)$, the critical angle at which the transmitted wave refracts

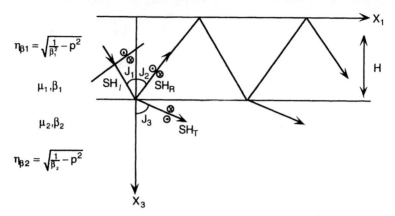

FIGURE 4.8 Geometry of SH waves that repeatedly reflect in a layer over a half-space. $x_3 = 0$ is a free surface, and the layer thickness is H. Interactions with the boundary of $x_3 = H$ involve incident (SH_I), reflected (SH_R), and transmitted (SH_T) SH waves. For $\beta_1 < \beta_2$, a critical angle $j_c = \sin^{-1}(\beta_1/\beta_2)$ will exist beyond which SH reverberations will be totally trapped in the layer ($j_1 \geq j_c$).

along the boundary as a head wave. For j_1 angles equal to, and larger than, the critical angle, the shear-wave reflection coefficient B/A has unit magnitude and acquires a phase shift, as discussed in Chapter 3. Since the SH energy is then totally reflected at both the boundary and the free surface, the postcritical SH wave in the layer will be "trapped" in the layer.

The boundary conditions for this problem are

$$\sigma_{32}|_{x_3=0} = \mu_1 \frac{\partial V_1}{\partial x_3}\bigg|_{x_3=0} = 0$$

(free surface)

$$\sigma_{32}|_{x_3=H^-} = \sigma_{32}|_{x_3=H^+}$$

(continuity of stress on boundary)

$$V_2|_{x_3=H^-} = V_2|_{x_3=H^+}$$

(continuity of displacement on boundary).

(4.29)

Applying (4.29) to (4.28) yields three resulting equations:

$$A = B \qquad (4.30)$$

$$A\mu_1\eta_{\beta_1}\left[\exp(i\omega\eta_{\beta_1}H) - \exp(-i\omega\eta_{\beta_1}H)\right]$$
$$= C\mu_2\eta_{\beta_2}\exp(i\omega\eta_{\beta_2}H) \qquad (4.31)$$

$$A\left[\exp(i\omega\eta_{\beta_1}H) + \exp(-i\omega\eta_{\beta_1}H)\right]$$
$$= C\exp(i\omega\eta_{\beta_2}H). \qquad (4.32)$$

The horizontal apparent velocity of all of the SH motions is $c = 1/p = k_1/\omega$. We can rewrite the complex exponentials in terms of trigometric functions (Box 2.4), and taking the ratio of Eqs. (4.31) and (4.32) yields

$$\tan(\omega\eta_{\beta_1}H) = \frac{\mu_2\eta_{\beta_2}}{i\mu_1\eta_{\beta_1}} = \frac{\mu_2\hat{\eta}_{\beta_2}}{\mu_1\eta_{\beta_1}}, \quad (4.33)$$

where we assume the postcritical situation for which $c = 1/p < \beta_2$, yielding $\eta_{\beta_2} = i\hat{\eta}_{\beta_2}$ with $\hat{\eta}_{\beta_2}$ being purely real. Equation (4.33) is a condition relating ω and c that must be satisfied to give a stable horizontally propagating disturbance. Because the wave velocity c explicitly depends on frequency,

ω, Eq. (4.33) is called a *dispersion equation*. Rewriting (4.33) in terms of the material parameters μ_1, μ_2, β_1, and β_2 and the variables ω and c, we have

$$\tan\left(H\omega\sqrt{1/\beta_1^2 - 1/c^2}\right)$$
$$= \frac{\mu_2\sqrt{1/c^2 - 1/\beta_2^2}}{\mu_1\sqrt{1/\beta_1^2 - 1/c^2}}. \quad (4.34)$$

Equation (4.34) indicates that for the solutions to be real numbers, $\beta_1 < c < \beta_2$.

Solutions to the Love-wave dispersion equation (4.34) are conventionally illustrated using a graphical technique. We let

$y = H[(1/\beta_1^2) - (1/c^2)]^{1/2}$, where y is defined for the interval $0 < y < H[(1/\beta_1^2) - (1/\beta_2^2)]^{1/2}$. Figure 4.9a shows a plot of $\tan \omega y$ versus the right-hand side of Eq. (4.34) over the defined interval of y. The $\tan(\omega y)$ function is periodic, resulting in discrete intersections of the two functions, corresponding to combinations of ω and c that solve Eq. (4.34). For a given value of ω, a finite number of solutions exist, which we number from left to right using n, beginning with $n = 0$. The $n = 0$ solution is called the *fundamental mode* for that frequency, and larger values of n define the *higher modes* or *overtones* of the system. The different modes have a simple physi-

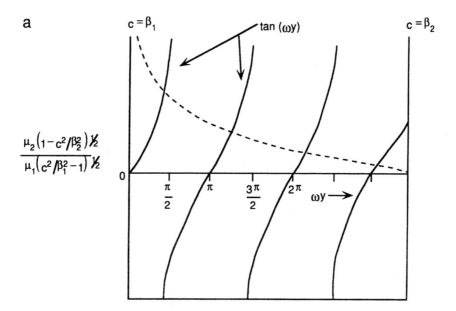

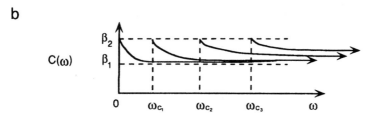

FIGURE 4.9 (a) Graphical solution of (4.34), where intersections of the dashed and solid lines yield discrete modes. (b) The phase velocity dispersion curve for fundamental and higher modes for the layer-over-a-half-space case.

cal significance that can be inferred from Figure 4.8. If we imagine harmonic displacements $V(x_2)$ distributed over the ray-paths of the reverberating SH path, the fundamental mode corresponds to one-half a harmonic cycle distributed from the surface to the intersection with the boundary. Thus, over the entire depth range of the layer along that path, the sense of motion is uniformly in the $\pm x_2$ direction. Over-

tones correspond to solutions in which the harmonic reverberations in the layer have n nodes (zero crossings) along the path from $x_3 = 0$ to $x_3 = H$, essentially dividing the layer into $n + 1$ layers oscillating in the $\pm x_2$ direction in alternating sequences separated by nodal surfaces. Box 4.2 shows that the geometry of "fitting" the oscillations into the layer gives rise to the dispersion equation.

Box 4.2 Love-Wave Optics

We can gain further insight into the Love-wave dispersion relation by explicitly considering the interference effects that underlie it. Consider (Figure 4.B2.1) a postcritical SH wavefront (PQ) at point A at time t and the wavefront at point B (P', Q'), which has just reflected from the surface. In order for the plane-wave motions of PQ and $P'Q'$ not to destructively interfere, the difference in phase must be a whole number of cycles, $2m\pi$. The difference in phase is

$$\phi_B - \phi_A = 2m\pi = \overline{AOB}(2\pi/\Lambda_0) + \phi_1 + \phi_2. \tag{4.2.1}$$

$\overline{AOB}(2\pi/\Lambda_0)$ is the differential length traveled times the wavenumber, $2\pi/\Lambda_0$, where Λ_0 is the wavelength, ϕ_1 is the phase change that the wave at $P'Q'$ underwent upon reflection at point O, and ϕ_2 is the free-surface-reflection phase change; $\phi_2 = 0$. Using the double-angle formula $\cos 2\theta = 2\cos^2 \theta - 1$, we find that $\overline{AOB} = 2H \cos j_1$. The postcritical-reflection phase change is the SH equivalent of (3.104) and (3.105)

$$\phi_1 = -2\tan^{-1}\left(\frac{\mu_2\sqrt{1/c^2 - 1/\beta_2^2}}{\mu_1\sqrt{1/\beta_1^2 - 1/c^2}}\right). \tag{4.2.2}$$

If we define the horizontal wavenumber to be $k_1 = (\omega/c) = (2\pi/\Lambda) = (2\pi/\Lambda_0)\sin j_1$, the constructive phase requirement becomes

$$2m\pi = \frac{2H \cos j_1 2\pi}{\Lambda \sin j_1} - 2\tan^{-1}\left(\frac{\mu_2\sqrt{1/c^2 - 1/\beta_2^2}}{\mu_1\sqrt{1/\beta_1^2 - 1/c^2}}\right). \tag{4.2.3}$$

Since $\sin j_1 = \beta_1/c$ and $\cos j_1 = (1 - \beta_1^2/c^2)^{1/2}$, this can be written as

$$\tan\left(H\omega\sqrt{1/\beta_1^2 - 1/c^2}\right) = \frac{\mu_2\sqrt{1/c^2 - 1/\beta_2^2}}{\mu_1\sqrt{1/\beta_1^2 - 1/c^2}}, \tag{4.2.4}$$

continues

which is the same as (4.34). A similar analysis can be done for large angles j_1, for which PQ and $P'Q'$ do not overlap, by accounting for constructive interference between spatially offset wavefronts.

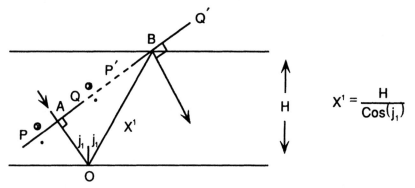

FIGURE 4.B2.1 Overlapping wavefront PQ at point A and $P'Q'$ at point B.

As ω increases in (4.34), the number of tangent functions (with zeros $\omega y = m\pi$) that fall in the defined interval of y increases. This means that the number of solutions increases (more higher modes) as ω increases. The nth overtone can only exist as a horizontally propagating wave for frequencies equal to or greater than

$$\omega_{c_n} = \frac{n\pi}{H\sqrt{(1/\beta_1^2) - (1/\beta_2^2)}}, \quad (4.35)$$

where ω_{c_n} is the cutoff frequency for the nth mode. The phase velocity of the nth overtone is $c = \beta_2$ at ω_{c_n} and approaches $c = \beta_1$ as ω increases (Figure 4.9b). This makes it clear that very high frequency waves have displacements concentrated near the surface, whereas lower-frequency components for the same mode have displacements concentrated near $x_2 = H$, giving velocities controlled by the half-space. The Love-wave displacements do extend into the half-space, but remember that their amplitudes decay exponentially below $x_3 = H$. It is not easy to visualize how this dispersion arises from plane waves

because it involves lateral interference of many upgoing and downgoing plane waves. Basically, waves with angles close to the critical angle propagate with the velocity near that of the head wave, β_2, and more horizontally propagating waves travel at velocities approaching that of the layer, β_1.

Love waves are always dispersive because they require at least a low-velocity layer over a half-space to exist. Because Love-wave particle motion is parallel to the surface, a complete separation of Love- and Rayleigh-wave surface motions occurs, with Love waves traveling faster and thus arriving on the transverse component ahead of the Rayleigh wave, which arrives on the vertical and radial components. Figure 1.2 shows block diagrams comparing the sense of motion of body and surface waves, while Figure 1.1 shows an example of a naturally rotated set of seismograms with clear body and surface-wave arrivals. The Love wave is "naturally" polarized on the transverse (E–W) component in this particular case.

The physics of Love-wave propagation in a multilayered structure like the Earth

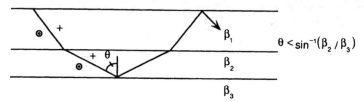

FIGURE 4.10 Love waves in a multilayered medium involve *SH*-wave reverberations trapped within the layers.

can be analyzed in much the same fashion as the simple case of a single layer over a half-space discussed earlier. Critical angles at various depths in the structure (Figure 4.10) can trap Love-wave energy in a sequence of surface waveguides. Longer-wavelength, lower-frequency waves tend to have higher velocities, because velocity usually increases with depth; however, the actual velocity gradients in the mantle cause long-period Love waves to be less dispersive than Rayleigh waves of corresponding period.

4.3.1 Surface Waves on a Spherical Earth

The particular geometry of the Earth has an important effect on surface-wave propagation—the waves spread over the spherical surface and hence converge at a point on the diametrically opposite side of the globe from the source, called the *antipode*. The waves converge from all directions at the antipode, with Rayleigh waves constructively interfering to give strong vertical amplifications, while Love waves destructively interfere to give no net Love-wave motion at the antipode. The waves "pass through" one another and diverge from the antipode, spreading over the surface again, eventually converging on the source and repeating the process. We can treat the motions of the repeated passage of Rayleigh and Love waves on the Earth's surface as *traveling waves*, as we have been discussing, or as patterns of *standing waves* or normal modes, which

are discussed later. In the perspective of a wave traveling from source to receiver, surface-wave energy obeys Fermat's principle (Chapter 3) by following the shortest travel-time path on the two-dimensional surface. If the velocity structure in question is a laterally homogeneous, flat-layered structure, the surface-wave path is a straight line on the surface from source to receiver. Lateral variations in the medium would cause the path to follow a curved trajectory, giving the least-travel-time path.

On a sphere, the surface-wave path in a laterally homogeneous, radially stratified structure is along a *great-circle path* (Box 4.4) connecting the source and receiver. Surface waves can travel in two directions along the great-circle path to the station, with the shorter path being called the *minor arc* and the longer path the *major arc*. Because waves traveling along both arcs pass the station and continue to follow the great circle, they eventually circuit the globe and pass by the station again, repeatedly. We denote long-period Rayleigh and Love waves by R and G (for Beno Gutenberg, who studied Love waves), respectively. Minor-arc arrivals are indicated with odd-number subscripts that increase with the number of passages of the station (e.g., R_1, R_3, R_5), and major-arc arrivals are indicated by even-number subscripts (R_2, R_4, R_6, etc.). Figure 4.11a shows an example of minor-arc and major-arc surface-wave arrivals on a long-period digital seismometer of the GEOSCOPE network (see Chapter 5). The horizontal ground

Box 4.3 Rayleigh Waves in a Fluid Layer over a Half-Space

The simplest case for which we can derive dispersion for Rayleigh waves is for a fluid layer ($\beta = 0$) over a half-space, which is very pertinent to the Earth. We consider the following geometry:

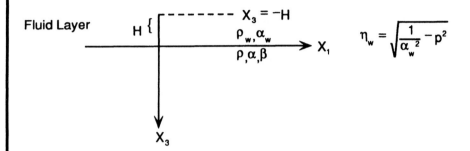

No S waves exist in the water, so we can assume that all motions are P waves traveling up and down in the water. We let the P potential in the water layer be

$$\phi_w = C_1 \exp[i\omega(px_1 - \eta_w x_3 - t)] + C_2 \exp[i\omega(px_1 + \eta_w x_3 - t)], \quad (4.3.1)$$

while in the solid we have potentials of the form (4.12)

$$\phi = A \exp[i\omega(px_1 + \eta_\alpha x_3 - t)]$$

$$\psi = B \exp[i\omega(px_1 + \eta_\beta x_3 - t)]. \quad (4.12)$$

No shear stress exists in the ideal fluid, so the boundary conditions are $\sigma_{33} = 0$ at $x_3 = -H$; σ_{33} and u_3 continuous, and $\sigma_{13} = 0$ at $x_3 = 0$.

In this example we do not require continuity of u_1 at the interface $x_3 = 0$, as this condition can be satisfied only by allowing the water to have a small finite viscosity. In a real medium, the fluid will have a finite viscosity, leading to a thin boundary layer right above the interface in which u_1 will vary rapidly, but our solution will not include this effect. The free-surface condition ($\sigma_{33} = 0$ at $x_3 = -H$) yields

$$C_1 = -C_2 \exp[-2i\omega\eta_w H]. \quad (4.3.2)$$

This result is used in the expressions derived from the interface conditions: $u_3 = \partial\phi/\partial x_3 + \partial\psi/\partial x_1$ continuity to give

$$2C_1\eta_w \exp[i\omega\eta_w H]\cos(\omega\eta_w H) = A\eta_\alpha + Bp. \quad (4.3.3)$$

Continuity of σ_{33} modifies (4.14) to give

$$-2i\rho_w \exp[i\omega\eta_w H]\sin(\omega\eta_w H) = A(\rho\alpha^2\eta_\alpha^2 + \lambda p^2) + 2B\mu p\eta_\beta, \quad (4.3.4)$$

continues

while $\sigma_{13} = 0$ is satisfied by (4.15)

$$2 A p \eta_\alpha + B\left(p^2 - \eta_\beta^2 \right) = 0. \tag{4.3.5}$$

The latter three equations relate the amplitudes of the potentials and frequency, ω, to the velocity of Rayleigh waves, $c = 1/p$. The only nontrivial solution is given by choices of ω and c that make the determinant of the coefficients of C_1, A, and B vanish. With some algebra the vanishing of the determinant gives an equation

$$\tan\left(H\omega \sqrt{\frac{1}{\alpha_w^2} - \frac{1}{c^2}} \right)$$

$$= \left[\frac{\rho \beta^4 \sqrt{c^2/\alpha_w^2 - 1}}{\rho_w c^4 \sqrt{1 - c^2/\alpha^2}} \right]$$

$$\times \left[\left(4\sqrt{1 - \frac{c^2}{\alpha^2}} \sqrt{1 - \frac{c^2}{\beta^2}} \right) - \left(2 - \frac{c^2}{\beta^2} \right)^2 \right]. \tag{4.3.6}$$

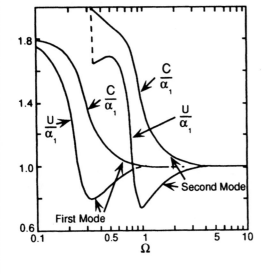

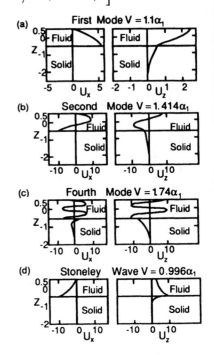

(a) First Mode $V = 1.1\alpha_1$

(b) Second Mode $V = 1.414\alpha_1$

(c) Fourth Mode $V = 174\alpha_1$

(d) Stoneley Wave $V = 0.996\alpha_1$

FIGURE 4.B3.1 (Left) Dispersion curves for the fundamental and first-overtone Rayleigh waves for a water layer over a half-space with parameters $\rho = 2.5\rho_w$, $\alpha = \sqrt{3}\,\beta$, and $\beta = 2\alpha_w$. (Right) Distribution of maximum particle motion with depth (eigenfunctions) for Rayleigh modes and the Stoneley mode. Horizontal displacements are u_x and vertical motions are u_z. (Modified from Ewing et al., 1957.)

continues

If $H = 0$, the half-space surface-wave solution (4.19) is recovered, and we have the undispersed half-space Rayleigh wave. For very large wavelengths, $\Lambda = 2\pi/k$, where $k = \omega(1/\alpha_w^2 - 1/c^2)^{1/2}$, the Rayleigh wave is insensitive to the water layer. However, for shorter wavelengths, the wave energy is partitioned between motions in the solid and motions in the water layer. As in the case of the Love-wave solution (4.34) for each frequency, discrete velocities will satisfy the dispersion equation, and higher modes with cutoff frequencies will exist as well. (Below the cutoff frequency, the higher-mode waves have velocities greater than β and leak energy into the half-space, which is called a *leaky mode*.) The horizontally propagating Rayleigh waves have velocities $\alpha_w < c < \beta < \alpha$, which ensures that they are evanescent in the half-space, with the number of nodal vertical-displacement positions ($u_3 = 0$) in the water layer corresponding to the mode number. Figure 4.B3.1 shows dispersion curves for phase velocity c and group velocity U (see Section 4.4) as a function of dimensionless frequency $\Omega = H\omega/\alpha$ for the fundamental mode and the first overtone for a specific choice of velocities. The distribution of maximum particle motions as a function of depth for three Rayleigh modes is also shown. For very large Ω, one other type of solution exists, with $c < \alpha_w$, which involves displacements that decay exponentially away from the interface in both directions. This type of interface wave is called a *Stoneley wave*. As $\Omega \to \infty$, the phase velocity of this wave approaches $0.998\,\alpha_w$.

Box 4.4 Great-Circle Paths, Azimuth, and Back Azimuth

Parameters of great-circle paths can be determined using spherical trigonometry. Consider the spherical triangle shown below. E is the source (or epicenter), S is the seismic station, and N is the north pole. A, B, and C are the three internal angles of the spherical triangle. In general, $A + B + C \neq 180°$. a, b, and c are the sides of the triangle in degrees measured between radii from an origin in the center of the sphere. If A, b, and c are given, then

$$a = \cos^{-1}(\cos b \cos c + \sin b \sin c \cos A) \tag{4.4.1}$$

$$C = \cos^{-1}\left(\frac{\cos c - \cos a \cos b}{\sin a \sin b}\right). \tag{4.4.2}$$

The angular distance a is often called Δ, the epicentral distance. For most applications, A is the difference in longitude between E and S, and b and c are the source and station colatitudes, respectively (colatitude is $90° -$ latitude). When measured clockwise from north, angle C is called the azimuth and gives the direction in which a ray must leave the source to arrive at a given station. Source

continues

radiation patterns are usually given in terms of azimuth from the source. If the station were located to the left of the epicenter in Figure 4.B4.1, the azimuth would be $360° - C$ (remember, always measure *clockwise*). The *back azimuth*, which is the angle measured from north to the direction from which energy arrives at the station, is given either by

$$B = \cos^{-1}\left(\frac{\cos b - \cos a \cos c}{\sin a \sin c}\right) \tag{4.4.3}$$

or by $360° - B$ (as in the case shown). Note that B is not simply related to C and must be calculated separately. Back azimuth is used to determine the longitudinal and transverse directions for an incoming ray at a prescribed station. The longitudinal component lies along the great circle, and the transverse component is perpendicular to the great circle.

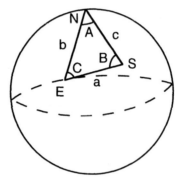

FIGURE 4.B4.1 Spherical geometry for great-circle paths.

motions are rotated to correspond to motion transverse to the great circle or along the great circle (longitudinal). Note that the G_1 and G_2 arrivals at these periods (> 100 s) are relatively impulsive, whereas the Rayleigh waves are very dispersed. The Love-wave motion is concentrated on the component transverse to the great-circle path, but some G_2 energy is visible on the longitudinal component as a result of deflection of the Love wave from the great-circle path. The Rayleigh-wave energy in R_1 and R_2 is stronger on the vertical component than on the longitudinal component by about a factor of 1.5, as found for Rayleigh waves in a Poisson half-space. The arrival labeled X_2 is a Rayleigh-wave overtone that has traveled on the major-arc path with a higher velocity than the fundamental mode. The packet of overtones on the minor arc, X_1, is weak because it mainly involves periods shorter than 100 s which have been filtered out. R_2 is more dispersed than R_1 and has lower amplitude because it has traveled farther. In general, one expects to see the amplitudes $|R_1| > |R_2| > \cdots > |R_n|$, but both propagation effects (Box 4.5) and source effects (Chapter 9) can produce anomalous amplitude behavior. The long-period Rayleigh waves travel with velocities (group velocities, as defined in Section 4.4 on dispersion) of 3.5–3.9 km/s, while long-period Love waves travel about 4.4 km/s. The

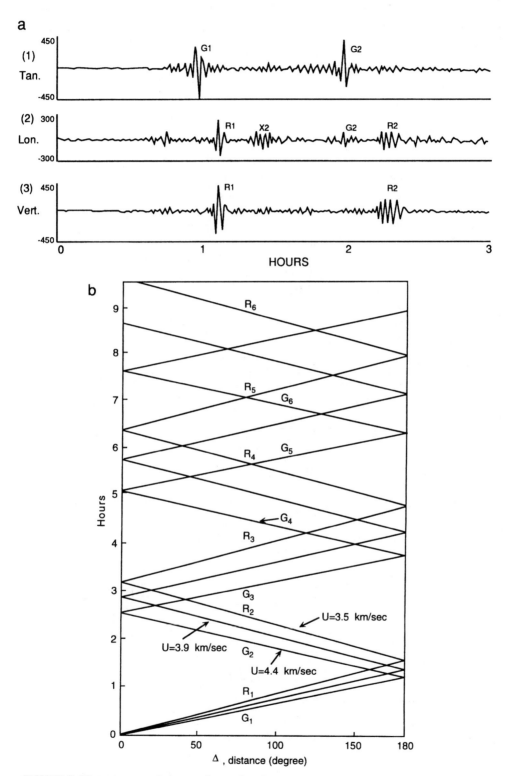

FIGURE 4.11 (a) Long-period recordings of surface waves from the May 26, 1983 Akita-Oki (Honshu) earthquake recorded by GEOSCOPE station PAF. (1) is the transverse component, (2) is the longitudinal component, and (3) is the vertical component. All traces have been filtered to remove oscillations that have periods of less than 100 s. (b) Travel time of surface waves with different group velocities for different distances. Long-period Rayleigh waves travel at a velocity of about 3.5–3.9 km/s, while Love waves travel at a velocity of about 4.4 km/s. ((b) is Courtesy of H. Kanamori.)

Box 4.5 Surface-Wave Amplitude Anomalies

Surface-wave amplitudes in a flat-layered structure decrease with increasing propagation distance because of geometric spreading, anelastic attenuation, and (generally) dispersion. On a spherical surface, surface-wave amplitudes decrease progressively with propagation distance because of anelasticity and dispersion, but geometric spreading has a more complex form. It can be shown (e.g., Aki and Richards, 1980) that away from the source or its antipode, geometric spreading is given approximately by $(\sin \Delta)^{1/2}$, where Δ is the angular distance between source and receiver. This spreading gives the lowest amplitudes near $\Delta = 90°$, i.e., when the surface wavefront is spread over the entire circumference of the planet. Curiously, R_1, R_2, and R_3, for example, all have the same geometric spreading at a given station ($\Delta = \Delta_0$). Generally, however, we expect $|R_1| > |R_2| > |R_3|$, etc. due to the dominating effects of attenuation and dispersion, as seen in Figure 4.B5.1. The seismograms in Figure 4.B5.2 show several stations with the normal behavior (HAL, RAR), but other stations (PFO, CMO, KIP) for which

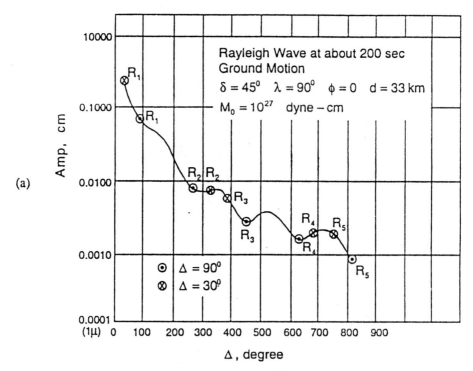

(a)

FIGURE 4.B5.1 200-s-period Rayleigh-wave amplitude on the vertical component as a function of distance. Observations at two different distances, 30° and 90°, are marked for great-circle orbits. The source is 33 km deep and has a moment of 1×10^{20} N m and a fault mechanism of strike = 0°, dip = 45°, rake = 90°. (Courtesy of H. Kanamori.)

continues

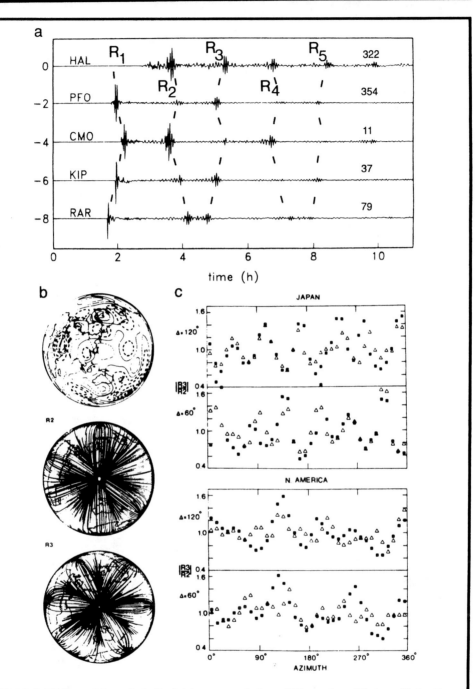

FIGURE 4.B5.2 (a) Great-circle Rayleigh-wave arrivals at IDA stations (Chapter 5) for the September 1977 Tabas, Iran earthquake. (From Masters and Ritzwoller, 1988.) (b) Projection of phase-velocity heterogeneity for 200-s-period Rayleigh waves on the hemisphere centered on Japan, along with surface-wave raypaths for R_2 arrivals at each point and R_3 arrivals at each point on the same hemisphere. (c) Calculated amplitude anomalies at different distances from two source regions for two models of surface-wave phase-velocity heterogeneity (boxes and triangles). (From Schwartz and Lay, *Geophys. Res. Lett.* **12**, 231–234, 1985; copyright by the American Geophysical Union.)

continues

strong amplitude anomalies (e.g., $|R_5| \gg |R_4|$ at PFO, KIP; $|R_4| \gg |R_3|$ at CMO) are observed. Since this earthquake did not have any source complexity that could account for these anomalies, propagation effects are probably responsible. It is now recognized that surface waves propagating on the surface of a laterally heterogeneous sphere (like the Earth) are deflected from the great-circle path, and focusing and defocusing can result. Part (b) of the figure shows raypaths on the surface of the Earth for 200-s-period surface waves traveling through a model having a laterally varying phase velocity. Instead of being straight, radial spokes, the rays bundle up, enhancing the amplitude. Part (c) shows predicted Rayleigh-wave amplitude anomalies at different distances from sources in Japan and North America plotted asfunctions of azimuth from the sources. Amplitude ratios are predicted to vary by a factor of 3, comparable with actual observations. Deflection of Love-wave energy (G_2) from the great-circle path can be observed in Figure 4.11a. While the deflections are usually minor, large-amplitude anomalies can result, and one must be cautious in assuming the surface-wave energy has propagated on the great circle.

curves in Figure 4.11b indicate approximate arrival times for sequential great-circle surface-wave groups. It takes about 2.5 h for long-period Love waves to circle the Earth and about 3 h for Rayleigh waves to do so. Additional examples of great-circle surface-wave phases are shown in profiles for the 1989 Loma Prieta earthquake in Figures 1.7 and 6.11.

4.4 Dispersion

All surface waves, except Rayleigh waves in an isotropic half-space, exhibit dispersion, with the apparent velocity along the surface depending on frequency. Almost any seismic source excites waves that comprise a continuous spectrum of frequencies, each harmonic component having a velocity, $c(\omega)$, that is called the *phase velocity*. If a monochromatic wave were somehow excited, only the phase velocity for that frequency would be needed to characterize the disturbance fully. How-

ever, when a spectrum of frequencies exists, the wave disturbances interfere, producing constructive and destructive patterns that influence the total ground motion. Constructive interference patterns behave as wave packets, which themselves propagate as disturbances along the surface with well-defined *group velocities*, $U(\omega)$. Thus, the phase velocity is directly controlled by the medium parameters (scale lengths of layering, intrinsic P and/or S velocities, rigidity, etc.) and the geometric "fit" of a particular harmonic component into the associated boundary conditions, as seen in the last section. The group velocity depends on the medium parameters through their influence on the phase velocity, but it also depends on the variation of phase velocity with frequency, which controls the interference between different harmonics.

To understand this, we begin by considering two harmonic waves with the same amplitude but slightly different frequencies (ω', ω''), wavenumbers, and phase velocities $(k' = \omega'/c', \ k'' = \omega''/c'')$. These combine to give a total displacement of

$$u = \cos(\omega't - k'x) + \cos(\omega''t - k''x).$$

$$(4.36)$$

We define ω as the average of ω'' and ω' such that $\omega' + \delta\omega = \omega = \omega'' - \delta\omega$, and $k = \omega/c$ such that $k' + \delta k = k = k'' - \delta k$, where $\delta\omega \ll \omega$, $\delta k \ll k$. By inserting these into (4.36) and using the cosine law, $2\cos x \cos y = \cos(x + y) + \cos(x - y)$, we obtain

$$u = 2\cos(\omega t - kx)\cos(\delta\omega t - \delta kx).$$

$$(4.37)$$

This is the product of two cosines, the second of which varies much more slowly than the first. Figure 4.12 shows a specific

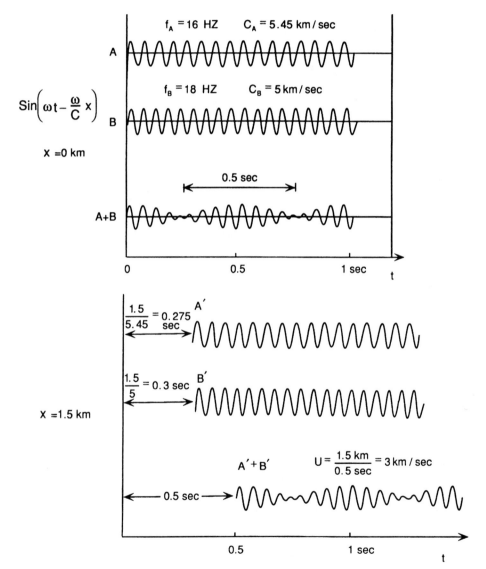

FIGURE 4.12 Example of the interference of two waves of the form (4.36) at two positions $x = 0$ and $x = 1.5$. The envelope of the interference pattern moves with group velocity $U = 3$ km/s. (Courtesy of H. Kanamori.)

example. The envelope of the modulated signal propagates with a velocity different from the phase velocity of the average harmonic term c, which is defined as the *group velocity*:

$$U = \frac{\delta\omega}{\delta k}. \qquad (4.38)$$

In the limit as $\delta\omega$ and $\delta k \to 0$

$$U = \frac{d\omega}{dk} = \frac{d(kc)}{dk} = c + k\frac{dc}{dk} = c - \Lambda\frac{dc}{d\Lambda}. \qquad (4.39)$$

From (4.39) we see that the group velocity depends on both the phase velocity and the variation of phase velocity with wavenumber. If $dc/dk = 0$, the phase and group velocities are equal. In general, in the Earth the phase velocity decreases monotonically with frequency, so $dc/dk < 0$ and $U < c$.

4.4.1 Measurement of Group and Phase Velocity

Dispersion changes the overall appearance of a surface wave as it propagates. One can visualize the surface wave as having started from the source essentially as an undispersed pulse, with each frequency component having an amplitude $A(\omega)$ and initial phase, $\phi_0(\omega)$, determined by the excitation of the source and medium. As the wave spreads outward, dispersion modifies it, spreading the energy out over a wavetrain, as shown in Figure 4.13.

The group velocity is very important in that energy propagates mainly in the constructively interfering wave packets, which move with the group velocity rather than the individual phase velocities. Box 4.6 shows that narrowband filtering of a seismogram isolates the wave packet that corresponds to the central frequency of the filter, and the group velocity for that frequency can then be determined by dividing the path length by the travel time of the wave packet. This requires knowledge of the source location and origin time. Alter-

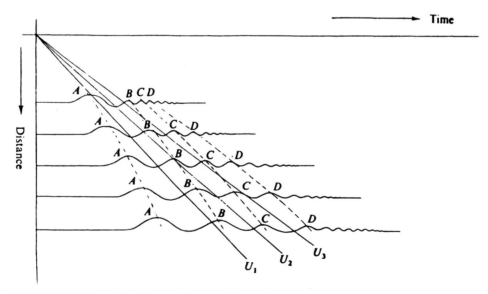

FIGURE 4.13 Example of increasing waveform dispersion with increasing distance. Solid lines indicate different group velocities that control the travel time of particular frequency motions from the origin. Dashed lines indicate phase velocities of individual harmonic components. (From Officer, 1974.)

Box 4.6 Wave Packets

An earthquake source excites surface waves with a continuum of frequencies rather than just two discrete frequencies like the example in the text. As these waves propagate away from the source, they disperse. The total surface-wave displacement involves a summation of all the propagating harmonic components. Consider the sum of a continuum of harmonic terms with uniform amplitude over a finite frequency band $\Delta\omega$ centered on average frequency ω_0 given by

$$U = \int_{\omega_0 - \Delta\omega/2}^{\omega_0 + \Delta\omega/2} \cos[\omega t - k(\omega)x] \, d\omega. \tag{4.6.1}$$

For small $\Delta\omega$, we expand $k(\omega)$ in a Taylor series:

$$k(\omega) = k(\omega_0) + \left(\frac{dk}{d\omega}\right)_{\omega_0} (\omega - \omega_0) + \cdots \tag{4.6.2}$$

and we can evaluate the integral of the first sum to order ω:

$$U = \frac{1}{t - (dk/d\omega)_{\omega_0}x} \left(\sin\left\{ \frac{\Delta\omega}{2}\left[t - \left(\frac{dk}{d\omega}\right)_{\omega_0} x \right] + \omega_0 t - k(\omega_0)x \right\} \right.$$

$$\left. - \sin\left\{ \frac{-\Delta\omega}{2}\left[t - \left(\frac{dk}{d\omega}\right)_{\omega_0} x \right] + \omega_0 t - k(\omega_0)x \right\} \right). \tag{4.6.3}$$

Using $2\sin\alpha\cos\beta = \sin(\alpha + \beta) - \sin(\beta - \alpha)$

$$U = \frac{2}{t - (dk/d\omega)_{\omega_0}x} \sin\left\{ \frac{\Delta\omega}{2}\left[t - \left(\frac{dk}{d\omega}\right)_{\omega_0} x \right] \right\} \cos(\omega_0 t - k(\omega_0)x). \tag{4.6.4}$$

If we let $Y = (\Delta\omega/2)[t - (dk/d\omega)_{\omega_0}x]$, the summation becomes

$$U = \Delta\omega \frac{\sin Y}{Y} \cos[\omega_0 t - k(\omega_0)x]. \tag{4.6.5}$$

Thus, we find a cosine harmonic term with the reference parameters modulated by a sinc function, which is peaked at $Y = 0$ and has rapidly diminishing side lobes. Thus the periodic modulations seen in Figure 4.12 are modified to a single, isolated wave packet when a continuum of frequencies is considered (Figure 4.B6.1):

$$Cos[\omega_0 t - k(\omega_0)x]$$

$$\frac{Sin\,Y}{Y}$$

Product

FIGURE 4.B6.1

continues

The envelope propagates with group velocity $U = (d\omega/dk)_{\omega_0}$. Thus, surface-wave ground motions filtered in a narrow frequency band have isolated group wave-packet arrivals, as shown in Figure 4.B6.2.

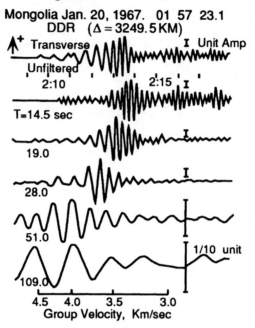

FIGURE 4.B6.2 Wave packets in narrow frequency bands obtained by filtering a Love-wave recording. The unfiltered record is shown at the top. Narrowband records with central periods shown on the left are plotted below the original seismogram, with varying amplitude scale. Note that each narrowband-filtered trace has the appearance of a wave packet. (From Kanamori and Abe, 1968.)

natively, given a single very well dispersed waveform like that in Figure 4.14, one can basically measure the arrival time of each frequency, because each oscillation corresponds to a narrow-frequency wave packet with an average period given by the period of that cycle. Knowing the origin time allows us to estimate the group-velocity dispersion curve. This procedure is not as stable as successively narrowband filtering the signal because interference over the continuous distribution of frequency components distorts each arrival.

If two stations are located on the same great-circle path, the group-velocity dispersion between the stations can be determined by measuring the difference in arrival times of filtered wave packets. This is

called the two-station method. A special application is the use of a single station to measure times between successive passes of surface waves traveling on the great circle (e.g., R_1, R_3). This yields an average group velocity over the entire great-circle path. Another way to estimate either single-station or two-station group velocities is first to determine the phase-velocity dispersion curve over the corresponding path and then use (4.39) to calculate $U(\omega)$.

Several single- and two-station methods exist for measuring phase-velocity dispersion curves. We can obtain a crude measure using well-dispersed seismograms from two nearby stations, like those in Figure 4.14b. Each harmonic term at a given point in its cycle is associated with a

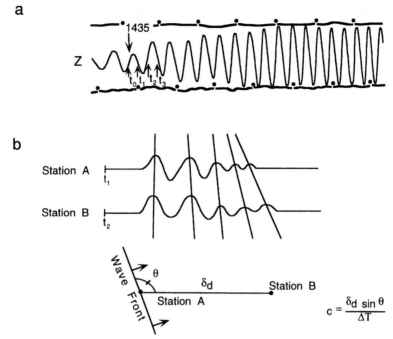

FIGURE 4.14 Examples of very well dispersed wave trains. (a) A Rayleigh wave for which individual group-arrival velocities can be made for each cycle of the waveform knowing the distance to the source and the origin time. (b) Measurement of phase velocity between two nearby stations for which common cycles of a given phase can be reliably identified and differential travel time measured.

peak or trough of a particular period of oscillation, and the differential time and propagation distance between corresponding cycles are used to estimate the phase velocity for each frequency. This procedure gives poor results unless the dispersion is so pronounced that the peaks are not actually envelopes of interfering frequencies. Typically, phase velocity is measured by taking the Fourier transform of a seismogram and obtaining the phase spectrum. A surface wave can be represented in the form

$$u(x,t) = \frac{1}{\pi} \int_0^\infty \hat{u}(\omega, x)$$
$$\times \cos\left(\omega t - \frac{\omega}{c(\omega)} x + \phi_0(\omega)\right) d\omega,$$

$$(4.40)$$

where the phase is $\phi(\omega) = \phi_0(\omega) - [\omega x/c(\omega)] + 2\pi N + \omega t$. The term $\phi_0(\omega)$ is the initial phase at the source, and the term $2\pi N$ represents the periodicity of the harmonic function. The amplitude spectrum $\hat{u}(x, \omega)$ describes the amplitude of each harmonic term that contributes to the actual time-domain waveform. If one has a single instrument-corrected seismogram that starts at time t_1 after the origin time at a distance x_1 from the source, a Fourier transform of the signal yields the phase of each frequency at the corresponding start time

$$\psi_1(\omega) = \omega t_1 + \phi_0(\omega) - \frac{\omega x_1}{c(\omega)} + 2\pi N.$$

$$(4.41)$$

If the initial phase at the source, the origin

time $(t = 0)$, and the distance traveled (x_1) are known, then $c(\omega)$ can be determined to within the uncertainty due to $2\pi N$. The value of N is usually selected by ensuring that the phase velocities for the longest-period signals converge onto globally averaged values of $c(\omega)$; long-period phase velocities vary by only a few percent, which is sufficient to constrain the choice of N. One must know the faulting mechanism and depth of the source to calculate the initial phase $\phi_0(\omega)$. In detail, additional corrections to the phase must be made because of the effects of anelasticity and polar passages (which add $\pi/2$ to the phase each time the wave passes the source location or antipode). The most accurate procedure for estimating phase velocity is to take the difference in the phase spectra at two points on a great-circle path (again, one can use a single station and look at successive great-circle orbits). In this case the initial phase cancels out, leaving

$$\psi_1(\omega) - \psi_2(\omega)$$

$$= \omega(t_1 - t_2) - \frac{\omega}{c(\omega)}(x_1 - x_2) + 2\pi M$$

$$(4.42)$$

or

$$c(\omega)$$

$$= \frac{x_1 - x_2}{(t_1 - t_2) + T[M - (1/2\pi)(\psi_1(\omega) - \psi_2(\omega))]},$$

$$(2.43)$$

where M, the difference in number of 2π cycles, is again chosen to give consistency with globally averaged values at long periods. Corrections for attenuation and polar passages between the stations are needed for precise measurements. Once the dispersion relation $f(\omega, k) = 0$ is determined, the group-velocity curve can be estimated

from the Taylor series expansion

$$f(\omega + d\omega, k + dk)$$

$$= f(\omega, k) + \left.\frac{\partial f}{\partial \omega}\right|_k d\omega + \left.\frac{\partial f}{\partial k}\right|_\omega dk$$

$$+ \cdots = 0 \qquad (4.44)$$

giving

$$U = \frac{d\omega}{dk} = -\left(\frac{\partial f}{\partial k}\right)_\omega \bigg/ \left(\frac{\partial f}{\partial \omega}\right)_k. \quad (4.45)$$

There are two main applications of dispersion-curve measurements. The most critical is the determination of velocity structure, and the second is correcting the observed phase back to the source so that the source radiation can be determined. Dispersion reflects the nature of the velocity gradients at depth, as shown in Figure 4.15a. Stronger velocity gradients produce more pronounced dispersion. Figure 4.15b shows the characteristic shape of phase- and group-velocity dispersion curves for Rayleigh waves in an elastic layer over a half-space (note the similarity to the fluid layer results in Box 4.3). Phase-velocity curves generally tend to be monotonic, whereas group-velocity curves often have a local minimum. The existence of a local minimum implies that significant energy will arrive at about the same time, producing an amplification and interference effect called an *Airy phase*. For continental paths an Airy phase with about a 20-s period often occurs, and long-period waves in the Earth have an Airy phase with approximately a 200-s period. Figure 4.16 illustrates average observed group velocities for Rayleigh waves in continental and oceanic regions. At periods longer than 80–100 s, regional near-surface differences have little effect since the waves are "seeing" deep into the upper mantle, where heterogeneity is less pronounced. The average oceanic crust is thinner than continental crust, resulting in a shift of the

a

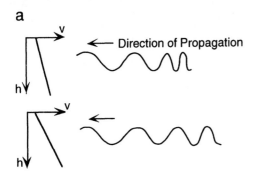

b

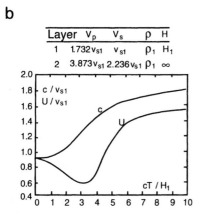

FIGURE 4.15 (a) Influence of vertical velocity gradient on dispersion of surface waves. The stronger gradient causes greater dispersion. (b) Theoretical fundamental-mode Rayleigh-wave dispersion curves for a layer over a half-space. The parameters v_P and v_S are P-wave and S-wave velocities, respectively. (Modified from Båth, 1979.)

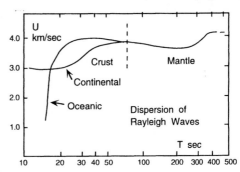

FIGURE 4.16 Observed group-velocity curves for Rayleigh waves. Averaged values for oceanic and continental paths are shown for periods less than 80 s. (Modified from Båth, 1979.)

crustal Airy phase to periods of 10–15 s. This sensitivity to crustal and upper-mantle velocity structure has led to extensive use of Rayleigh and Love waves to analyze three-dimensional Earth structure, which we will describe in Chapter 7.

Rayleigh waves in a layered structure have overtones similar to those described for Love waves in the previous section (see Box 4.3). Both Love- and Rayleigh-wave overtones have their own dispersion curves. Generally the overtone group velocities are higher than velocities for the fundamental

modes, causing overtones to arrive earlier. Figure 4.17 shows the relative contribution of the fundamental mode and the first 10 overtones to a radial-component synthetic seismogram. The overtone wave packets are identified by X_n, where odd n correspond to initial minor-arc paths and even n to initial major-arc paths. The Rayleigh-wave overtone amplitudes tend to be stronger on the horizontal component than on the vertical component, as seen in Figure 4.11. Additional overtone observations are shown in Figures 1.7 and 6.11 for the Loma Prieta earthquakes. These Rayleigh-wave overtones are useful for probing deeper structure than that sampled by fundamental modes. Love-wave overtones are not well isolated from the fundamental modes in Figure 4.11 but contribute to the long-period oscillations before the main Love-wave pulses.

4.5 Tsunamis

In our development of surface waves, we have assumed that the elastic medium has a free surface with a vacuum above it. However, 70% of the Earth's surface is covered by water of variable thickness, and all regions are overlain by the gaseous envelope of the atmosphere. Since *SH*

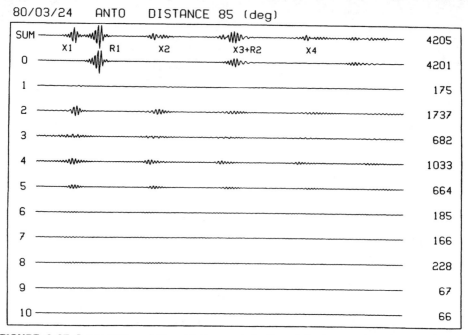

FIGURE 4.17 Computation of the relative contribution of fundamental modes ($n = 0$) and overtone branch ($n = 1$ to 10) Rayleigh waves. The total synthetic ground motion is shown at the top. This is the radial component of ground motion, on which the overtone group arrivals (X_1, X_2, X_3, X_4) have the largest amplitude relative to the fundamental-mode Rayleigh waves (R_1, R_2). (From Tanimoto, 1987. Reprinted with permission from the Royal Astronomical Society.)

waves cannot travel in the fluid media, the presence of these surface layers scarcely affects propagating Love waves, which involve only horizontal surface motions. However, the vertical surface motions caused by propagating Rayleigh-wave disturbances (as well as the vertical motions produced by incident P and SV waves) clearly must affect the water and atmospheric layers. In turn, oceanic and atmospheric disturbance, such as pressure variations, internal oscillations, winds, and tides, must produce ground motions in the solid Earth. In other words, a coupling of motion occurs across the interface despite the change in state of the medium, which leads to a number of interesting phenomena. Although some of the coupled interactions can be evaluated by treating the surface fluid layers as "elastic" layers in

which the rigidity and shear velocity go to zero (see Box 4.3 for an example), the primary restoring force for most fluid motions is gravity rather than interaction between adjacent particles. Finite-amplitude displacements of particles in ocean and atmospheric waves that are not readily described by infinitesimal strain theory can clearly occur.

Generally, we do not treat fluid motions using the Lagrangian formulation in which we have been developing elastic-wave theory for solids. Instead, the *Eulerian* formulation is used, in which we monitor the behavior of a material element according to its position at a particular time rather than keep track of particle motion. Individual particles may flux into or out of the material element. We use this formulation, developed in many texts on fluid mechanics, to describe gravitationally controlled wave behavior in media that may have large particle motions.

One of the most important gravity waves that occurs on the Earth's surface is a *tsunami*, a long-period wave in the ocean. The basic physics of these waves is like that of ordinary wind-driven ocean waves, but tsunamis are distinguished by particularly long periods (200–2000 s) and wavelengths of tens of kilometers. Tsunamis are excited by large-scale displacements of water due to submarine landslides, volcanic eruptions, or most commonly, sea-bottom displacements caused by submarine fault motions. Tsunami wave amplitudes in the deep ocean range from centimeters to 5–10 m in height, but run-up of these long-period waves on shorelines can cause enormous destruction, overwhelming the standard storm-wave coastal defenses designed for much shorter-period waves. Fortunately, truly large, damaging tsunamis are relatively rare, with about one major event occurring per decade.

Gravity waves behave differently than the elastic waves that we have been considering in that gravity is the main restoring force in the system, with gravitational energy making up more than 95% of the energy in tsunami waves (the rest is compressional energy in the slightly compressible water and compressional and shear energy in the underlying rock). A fluid-mechanics derivation provides a wave equation for tsunami wave height h as

$$\frac{\partial^2 h}{\partial t^2} = g\,\nabla \cdot (d\,\nabla h), \qquad (4.46)$$

where d is the depth of the water and g is the acceleration due to gravity. Note that gravity has not appeared in our previous elastic wave equations other than as a possible inhomogeneous body-force term.

Equation (4.46) behaves distinctively depending upon the wavelength of the tsunami wave. For wavelengths of $\Lambda_T \gg d$, long-wave (or shallow-water) theory holds, and the wave obeys

$$\frac{\partial^2 h}{\partial t^2} = c^2\,\nabla^2 h, \qquad (4.47)$$

where the velocity $c = \sqrt{gd}$ is nondispersive and depends solely on water depth. The displacements in the vertical and radial directions vary linearly with depth. The tsunami velocities for periods of 200–2000 s are on the order of 700—900 km/h in the open ocean, or about the speed of a jet airliner. At short wavelengths ($\Lambda_T \ll d$), the tsunami velocity is given by $c = (\Lambda_T g / 2\pi)^{1/2}$, giving dispersive behavior with motions that decay exponentially with depth from the surface. Theoretical tsunami group and phase-velocity curves for a homogeneous self-gravitating Earth model covered by oceans that are 2, 4, and 6 km deep are shown in Figure 4.18. The dispersive nature of tsunamis leads to calculations quite analogous to those used to produce surface-wave synthetic seismograms. Examples of synthetic tsunami waveforms at distances of 2°–20° from a 10-km-deep vertical fault with vertical shearing displacement are shown in Figure 4.19. Note the increasing spread of the tsunami wave with increasing distance due to decreasing group velocities for the shorter-period waves.

Water-pressure sensors in the deep ocean can record the passage of tsunamis larger than a few millimeters, but most tsunami records are from tide gauges in harbors. The shallowing of the water in the harbor, along with geometric effects, influences the peak amplitude of the tsunami wave. This run-up effect is due to the decrease in velocity as the depth shoals; the kinetic energy of the wave is transformed into gravitational energy by increasing the wave height. Table 4.1 lists some of the catastrophic consequences that have occurred as tsunamis came onshore.

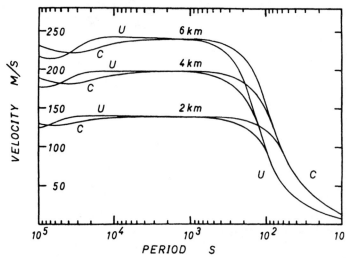

FIGURE 4.18 Tsunami dispersion curves for oceans 2, 4, and 6 km deep on a spherical planet. (From Ward, 1989.)

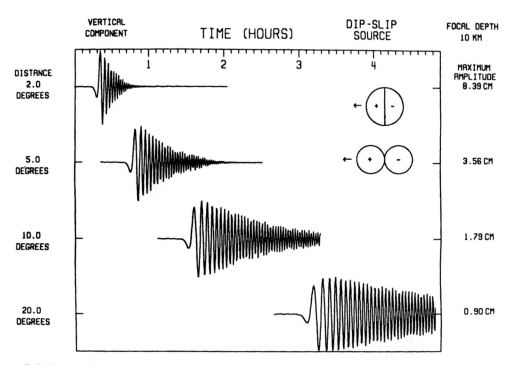

FIGURE 4.19 Synthetic vertical motions of a tsunami at distances of 2°, 5°, 10°, and 20° away from a point dip-slip source with moment $M_0 = 10^{20}$ N m and depth of 10 km. The maximum amplitude is shown on the right. The azimuthal variation in amplitude will vary with $\sin\phi$, where ϕ is the azimuth to the station measured from the strike of the fault. (From Ward, 1989.)

TABLE 4.1 Famous Tsunami Events

14th Century B.C.	Mediterranean	Thera volcanic eruption. Tsunami destroys Minoan civilization.
1755	Lisbon, Portugal	Underwater earthquake. 16-m tsunami run-up in Lisbon harbor.
1883	Indonesia	Krakatau volcanic eruption. 40-m tsunami run-up. 36,000 killed.
1896	Japan	Sanriku earthquake. 30-m tsunami run-up. 27,000 killed.
1946	Aleutian Islands	Magnitude 7.4 earthquake. 30-m tsunami locally; 16 m in Hawaii. 159 killed.
1958	Alaska Peninsula	Earthquake-triggered landslide. 550-m run-up in Lituya Bay.

An observed tide gauge recording with a tsunami signal superimposed on the 12-h tidal oscillation is shown in Figure 4.20. This recording is from the Azores Islands about 18° from an earthquake in the eastern Atlantic, located 400 km offshore from the African coast. Note that the tsunami waveform resembles the dispersed signals in Figure 4.19. About 2 h after the tsunami begins, a second arrival with a similar wave shape is seen. This corresponds to the tsunami that reflected off the coast of Africa, traveling about 800 km farther to the tide gauge. This illustrates one of the complications of tsunamis—ocean basin geometry strongly influences them.

Because the depth of ocean basins is well known, it is straightforward to determine the velocity variations that control tsunami propagation. This allows us to determine the source area that produced a tsunami (meaning the region where the seafloor was moved up or down by volcanoes, faulting, or landslides) by using the arrival times of tsunamis on tide gauges that are azimuthally distributed around the source. The seafloor motion in the source region can be estimated by correcting the observed tsunami amplitudes for any local nonlinear bathymetric effects at the tide gauge and for geometric spreading from the source region. Decay of tsunami amplitude with distance depends on source depth (see Figure 4.21), but it is approximately given by $1/\sqrt{r}$, corresponding to two-dimensional spreading. Numerical calculation of the full propagation effects allows complete modeling of tsunami waveforms to determine fault slip on submarine earthquakes (see Box 10.5). In detail, tsunami excitation depends on the geometry of faulting, the depth of faulting, and

Box 4.7 Tsunami Wavefronts

The lateral variation in ocean depth produces a laterally varying velocity structure for tsunami waves. This causes wave refraction, similar to that for surface waves (Box 4.5). Focusing and defocusing occur, resulting in nonuniform tsunami amplitudes. Modern methods account for this by either computing tsunami waves with numerical methods for a laterally varying ocean model or tracing rays along the surface through the velocity field to determine where focusing occurs. Figure 4.B7.1 illustrates the effects of actual ocean depth variations on the tsunami

continues

wavefronts relative to what they would be if the ocean depth were uniform. For a tsunami produced by an earthquake in Chile, rays converge on Japan because it is 150° away (i.e., approaching the antipode). The tsunami takes 15 h to reach Hawaii and 22 h to reach Japan. A great earthquake in Chile in 1960 produced a disastrous tsunami in Japan 1 day later.

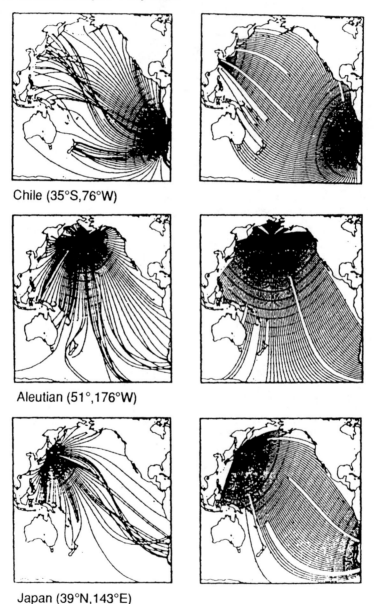

Chile (35°S,76°W)

Aleutian (51°,176°W)

Japan (39°N,143°E)

FIGURE 4.B7.1 Tsunami raypaths traced through a realistically varying ocean depth model (left) for three different source regions, compared to the simple ray patterns for a hypothetical uniform-thickness model on the right. The cross marks in the rays define the tsunami wavefront at instants in time 1 h apart. (From Satake, 1988.)

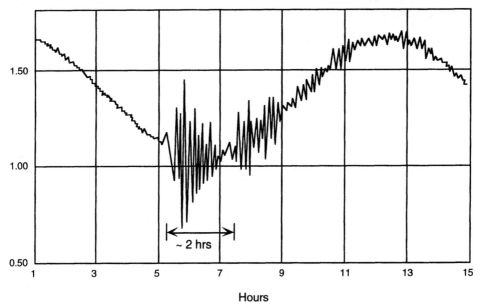

FIGURE 4.20 Record of the tsunami of the February 28, 1969 east Atlantic earthquake at the Horta (Azores Islands, Portugal) tide gauge.

the time history of faulting. Earthquakes that excite particularly strong tsunamis are called *tsunamigenic* earthquakes, and they almost always have a large seismic moment associated with shallow underwater faulting.

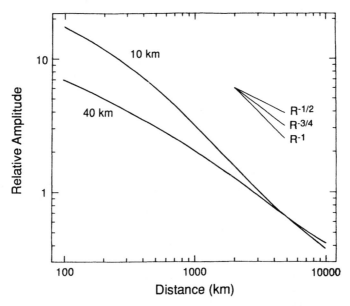

FIGURE 4.21 Computed tsunami amplitude decrease as a function of distance for earthquake sources at depths of 10 and 40 km below the ocean bottom. For distances greater than about 2000 km, the amplitude is not sensitive to source depth. (From Ward, 1989.)

4.6 Free Oscillations

We have seen how vertical scale lengths in a layered medium provide physical constraints on the types of motions than can occur for waves propagating along the surface. The finite spherical shape of the Earth intrinsically provides both radial and circumferential constraints on solutions to the equations of motion in the planet. In this perspective, only surface waves that constructively interfere after propagating around the Earth's surface will persist as long-term motions. The circumference provides a scale length into which an integral number of wavelengths can fit to produce persistent standing motions. Because only *discrete* wavelengths and frequencies fit the Earth's boundary conditions, the corresponding standing waves are called the *free oscillations* or *normal modes* of the system.

We build some insight into normal modes by considering the one-dimensional case for a string held fixed at either end (Figure 4.22). We assume that a source excites small-amplitude motions of the string that propagate as waves away from the source in the $\pm x_1$ directions, involving particle motions u in the $\pm x_3$ direction. These motions must obey the one-dimensional wave Eq. (2.60)

$$\frac{\partial^2 u}{\partial x_1^2} = \frac{1}{c^2}\frac{\partial^2 u}{\partial t^2}. \quad (4.48)$$

In Chapter 2 we derived general solutions of this equation in the form of (2.66)

$$u(x,t) = C_1 e^{i\omega(t+x/c)}$$
$$+ C_2 e^{i\omega(t-x/c)} + C_3 e^{-i\omega(t+x/c)}$$
$$+ C_4 e^{-i\omega(t-x/c)}. \quad (4.49)$$

The boundary conditions for the string are given by the fixed end points, with $u(0,t) = u(L,t) = 0$. The first gives $C_1 = -C_2$ and $C_3 = -C_4$. The condition at $x = L$

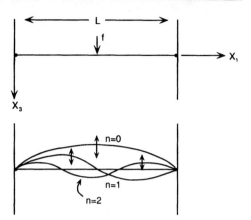

FIGURE 4.22 Geometry of a string under tension with fixed end points separated by distance L. Motions of the string excited by any source (f) comprise a weighted sum of the eigenfunctions, which are solutions that satisfy the boundary conditions with discrete eigenfrequencies. The first three eigenfunctions are shown below.

then gives

$$(C_1 e^{i\omega t} + C_3 e^{-i\omega t})2i\sin(\omega L/c) = 0. \quad (4.50)$$

The nontrivial solutions are given by zeros of the sine function, $\omega L/c = (n+1)\pi$, $n = 0, 1, 2, \ldots, \infty$. Thus, discrete frequencies of motion, $\omega_n = (n+1)\pi c/L$, called *eigenfrequencies*, exist that satisfy the boundary conditions. These eigenfrequencies have corresponding displacement patterns, $e^{i\omega_n t}\sin(\omega_n x/c)$, called *eigenfunctions* or *normal modes* of the system. The $n = 0$ mode is the fundamental mode and has no internal nodes (places where motion is zero) within the system; $n > 0$ corresponds to higher modes or overtones that each have n internal nodes. Figure 4.22 shows the first three eigenfunctions that are allowed by the boundary conditions. Oscillatory motion of each eigenfunction occurs without horizontal motion of the nodes, so that horizontal propagation of each mode alone does not appear.

Thus, these are called *standing-wave patterns*, when viewed in isolation. Any general propagating disturbance on the string can be represented by an infinite weighted sum of the eigenfunctions, because they constitute all permissible components of any solution in the medium:

$$u(x,t) = \sum_{n=0}^{\infty} \left(A_n e^{i\omega_n t} + B_n e^{-i\omega_n t} \right)$$

$$\times \sin\left(\frac{\omega_n x}{c}\right). \qquad (4.51)$$

Thus, the standing-wave representation in terms of normal modes can equivalently represent traveling waves in the system. The Fourier transform power spectrum of (4.51) will have discrete spikes at the eigenfrequencies ω_n, with relative amplitudes given by the weighting functions.

If we take a continuous-displacement recording that extends many hours or days after a large earthquake, like that in Figure 4.23a, we can view the time-domain signal as a sequential passage of surface waves traveling along great circles. When the power spectrum is computed for this signal (Figure 4.23b), we observe discrete peaks at different frequencies with variable relative amplitudes. These correspond to eigenfrequencies of the Earth system, involving standing waves that fit into the layered spherical geometry of the planet. The system is much more complex than the string, but the basic ideas are the same; the Earth can be set into global motions that make it ring like a bell. Constructive interference of the coexisting vibrations corresponds to disturbances that move along the surface as a function of time, which we view in the traveling-wave perspective as Love and Rayleigh waves. In fact, we can equivalently represent all internal body-wave motions by summing a sufficient number of normal modes, for the infinite set of modes must represent all motions in the medium.

The modes of a spherical body involve both radial and surface patterns that must fit into the geometry of the system. The viable oscillations are of two basic types: (1) *spheroidal oscillations*, analogous to the P, SV, and Rayleigh waves, which have a component of motion parallel to the radius (*radial* motion in *spherical* geometry) from the Earth's center, and (2) *toroidal* or *torsional oscillations*, involving shear motions parallel to the sphere's surface, analogous to SH- or Love-wave motions. Gravity does not influence toroidal motions at all, but long-period ($t > 500$ s) spheroidal motions do involve significant work against gravity, thereby sensing Earth's gross density structure as no other seismic wave type can.

Figure 4.24 summarizes some of the characteristics of normal-mode motions for a spherical, elastic, nonrotating medium. The easiest modes to visualize are the toroidal modes, which involve twisting motions of portions of the sphere. A nomenclature from spherical harmonics (Box 4.8) is used to identify patterns of motions. The toroidal modes are labeled $_n T_l$, where n indicates the number of zero crossing for the eigenfunction along the radius of the Earth and l indicates the number of nodal motion lines on the surface, the *angular order number* or *degree* of the spherical harmonic term. For toroidal motions the poles have no motion, counting as the $l = 1$ term. Thus $_0 T_2$ corresponds to alternating twisting of the entire upper and lower hemispheres of the body. The mode $_1 T_2$ corresponds to similar twisting of a central sphere overlain by twisting in the reverse direction of the outer hemispherical shells. The mode $_0 T_0$ is undefined, and $_0 T_1$ cannot exist because it would correspond to oscillation in the rate of rotation of the whole Earth, which violates conservation of angular momentum. Both n and l can take on integer values up to infinity, but in practice, for the Earth it is important to identify only the first few hundred values.

The nomenclature for spheroidal modes is $_n S_l$, where, in general, n and l have

Box 4.8 Spherical Harmonics

Analysis of the Earth's normal modes is most naturally performed in a spherical coordinate system. Here we consider basic mathematical solutions of the wave equation $\nabla^2 S = (1/c^2)(\partial^2 S/\partial t^2)$ in the spherical polar coordinate system (r, θ, ϕ) defined in Box 2.5. The wave equation for a homogeneous, nonrotating spherical fluid becomes (see Box 2.5)

$$\frac{1}{r^2}\frac{\partial}{\partial r}\left(r^2\frac{\partial S}{\partial r}\right) + \frac{1}{r^2 \sin\theta}\frac{\partial}{\partial\theta}\left(\sin\theta\frac{\partial S}{\partial\theta}\right) + \frac{1}{r^2\sin^2\theta}\frac{\partial^2 S}{\partial\phi^2} = \frac{1}{c^2}\frac{\partial^2 S}{\partial t^2}. \quad (4.8.1)$$

As usual, to solve this we use separation of variables, letting $S(r, \theta, \phi, t) = R(r)\Theta(\theta)\Phi(\phi)T(t)$. We find a standard solution for the time-dependent term as a harmonic function and take $T(t) = e^{-i\omega t}$, leaving

$$\frac{d^2\Phi}{d\phi^2} + m^2\Phi = 0 \quad (4.8.2)$$

$$\frac{d}{dr}\left(r^2\frac{dR}{dr}\right) + \left[\frac{\omega^2 r^2}{c^2} - l(l+1)\right]R = 0 \quad (4.8.3)$$

$$\frac{d}{d\theta}\left(\sin\theta\frac{d\Theta}{d\theta}\right) - \left[\frac{m^2}{\sin^2\theta} - l(l+1)\right](\sin\theta)\Theta = 0, \quad (4.8.4)$$

where we have introduced constants m^2 and $l(l+1)$. Equation (4.8.2) has solutions $e^{im\phi} = \cos m\phi + i \sin m\phi$, where m must be an integer for the solutions to satisfy the spherical geometry. Equation (4.8.3) for $R(r)$ involves the frequency, ω, but not m; thus in the homogeneous, nonrotating system, ω will be *independent* of m but will depend on the constant l. This is a well-studied differential equation that has solutions of the class called spherical Bessel functions. These solutions have the form

$$j_l(x) = x^l\left(\frac{-1}{x}\frac{d}{dx}\right)^l\frac{\sin x}{x}, \quad (4.8.5)$$

where here $x = \omega r/c$. For $l = 0$ and $rR(r) \propto \sin(\omega r/c)$, spherical Bessel functions have the form of decaying sinusoids, as shown in Figure 4.B8.1. Equation (4.8.4) for $\Theta(\theta)$ is also in the form of a classic equation called the associated Legendre equation, which is usually given in terms of $x = \cos\theta$ with cases $m = 0$

$$\Theta(\theta) = P_l(\cos\theta) = P_l(x), \quad (4.8.6)$$

continues

where $P_l(x)$ are Legendre polynomials. These are expressed by

$$P_l(x) = \left(\frac{1}{2^l l!} \frac{d^l}{dx^l} \right)(x^2 - 1)^l,$$

(4.8.7)

which gives $P_0(x) = 1$, $P_1(x) = x$, $P_2(x) = \frac{1}{2}(3x^2 - 1)$. Examples of Legendre polynomial functional dependence for $l = 2$ to 5 are shown in Figure 4.B8.1. For $m \neq 0$ the solutions are given by the associated Legendre functions $P_l^m(x)$, where

$$P_l^m(x) = (1 - x^2)^{m/2} \left(\frac{d^m P_l(x)}{dx^m} \right) = \left(\frac{(1 - x^2)^{m/2}}{2^l l!} \right) \left(\frac{d^{l+m}}{dx^{l+m}}(x^2 - 1)^l \right)$$

$$(-l \leq m \leq l).$$

(4.8.8)

Many mathematical texts describe the multitude of properties of these functions in detail. For $m = 0$, Φ is a constant and S will have axial symmetry.

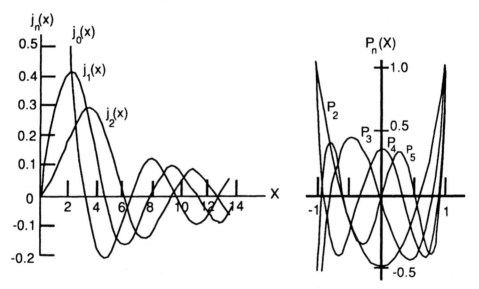

FIGURE 4.B8.1 Functional behavior of spherical Bessel functions (left) and Legendre polynomials (right).

The product $\Theta(\theta)\Phi(\phi) = P_l^m(\cos \theta)e^{im\phi}$ is called a surface spherical harmonic of degree l and order m. The most common form in seismology is the fully normalized spherical harmonic

$$Y_l^m(\theta, \phi) = (-1)^m \left[\left(\frac{2l + 1}{4\pi} \right) \frac{(l - m)!}{(l + m)!} \right]^{1/2} P_l^m(\cos \theta)e^{im\phi},$$

(4.8.9)

continues

with l and m being integers. The function $e^{im\phi}$ has zeros along $2m$ meridians of longitude (m great circles), while $P_l^m(\cos\theta)$ has zeros along $l-m$ parallels of latitude. Examples of the surface patterns produced by spherical harmonics are shown in Figure 4.B8.2. The *angular order number* l gives the total number of nodal lines on the surface with zero displacement. The parameter m gives the number of great circles through the pole with zero displacement. Thus, there are always $l-m$ nodal lines along latitude. Rotation of the coordinate system cannot change the order number l but can change m.

Application of boundary conditions on the medium prescribes the values of the eigenfrequencies in terms of the sphere geometry and constants n and l, $_n\omega_l$, where $n=0$ will correspond to the fundamental modes and $n>0$ will give the overtones of the system. Since $_n\omega_l$ does not depend on m for the case considered, all modes with angular order l (for a given n) will have the same frequency but different displacement patterns (eigenfunctions). Thus, a normal-mode power-spectrum peak for the system is actually a *multiplet*, composed of overlapping peaks of $2l+1$ singlets with different displacement patterns. This overlap of all values of m is called normal-mode *degeneracy*. Any departure of the medium from spherical symmetry, such as that produced by rotation, aspherical shape, or aspherical distribution of material properties, breaks down this degeneracy, giving each of the singlets its own frequency. This is called *splitting* of the multiplet. An actual Earth free-oscillation peak is thus composed of the overlapping peaks produced by the split multiplet, with the spread of the pulses being obscured by attenuation and limited frequency resolution, which are intrinsic in any finite time series. Mode splitting varies with path and from mode to mode.

Finally, for the elastic Earth we must use vector surface harmonics:

$$R_l^m(\theta,\phi) = Y_l^m \hat{r} \tag{4.8.10}$$

$$S_l^m(\theta,\phi) = \frac{1}{\sin\theta}\frac{\partial Y_l^m}{\partial\phi}\hat{\phi} + \frac{\partial Y_l^m}{\partial\theta}\hat{\theta} \tag{4.8.11}$$

$$T_l^m(\theta,\phi) = \frac{1}{\sin\theta}\frac{\partial Y_l^m}{\partial\phi}\hat{\theta} - \frac{\partial Y_l^m}{\partial\theta}\hat{\phi} \tag{4.8.12}$$

to represent the total ground motion. The first two terms, R_l^m and S_l^m, are needed to describe spheroidal motion, while T_l^m describes toroidal motion.

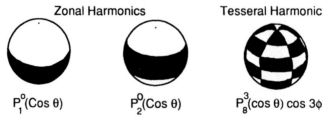

Zonal Harmonics   Tesseral Harmonic

$P_1^0(\text{Cos }\theta)$ $P_2^0(\text{Cos }\theta)$ $P_8^3(\cos\theta)\cos3\phi$

FIGURE 4.B8.2 Examples of surface spherical harmonics. $m=0$ yields *zonal* harmonics of degree l. For $l=m$, the nodal surfaces are longitudinal lines giving *sectoral* harmonics. For $0<|m|<l$, the combined latitudinal and longitudinal nodal patterns are called *tesseral* harmonics.

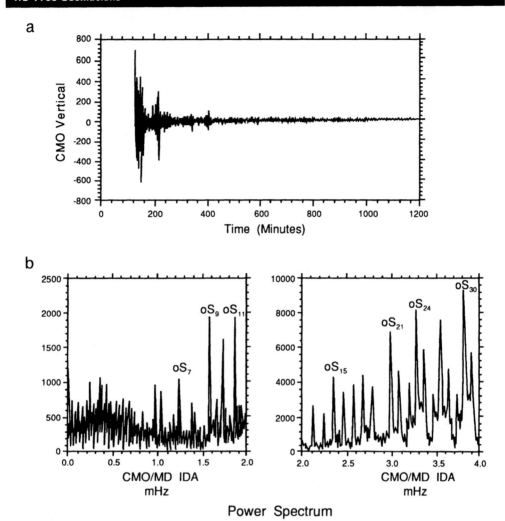

FIGURE 4.23 (a) A 20-h-long record of an IDA gravimeter, CMO (see Chapter 5), at College, Alaska recording the 1985 Mexico earthquake. Long-period Earth tides have been removed. The bursts of energy correspond to Rayleigh-wave great-circle arrivals. (b) The power spectrum for CMO showing spikes at discrete frequencies corresponding to eigenfrequencies of the Earth. (Modified from Gubbins, 1990.)

similar significance, although the poles are not positions of zero motion. Modes with $l = 0$ have no surface nodes and correspond to the subset called *radial modes*, with all motions in the radial direction. Mode $_0S_0$ involves expansion and contraction of the sphere as a whole. Mode $_1S_0$ has one internal surface of zero motion separating alternating layers moving inward or outward. For $l > 0$, nodal lines

occur on the surface along small circles parallel to the equatorial plane or along longitudinal great circles through the poles, which subdivide the surface into portions with alternating motion. Mode $_0S_1$ is undefined, as this would correspond to a horizontal shift of the center of gravity, which can happen only if the sphere is acted on by an external force. Mode $_0S_2$ is the longest-period normal mode of the

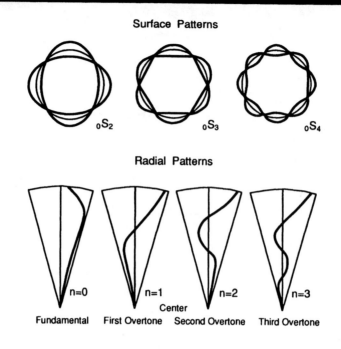

FIGURE 4.24 (Top) Surface and radial patterns of motions of spheroidal modes. (Bottom) Purely radial modes involve no nodal patterns on the surface, but overtones have nodal surfaces at depth. Toroidal modes involve purely horizontal twisting of the Earth. Toroidal overtones ($_1T_2$) have nodal surfaces at constant radii across which the sense of twisting reverses. (After Bolt, 1982.)

sphere and is sometimes called the "football" mode. It involves alternating motion from a prolate to an oblate spheroid, as shown in Figure 4.24. Mode $_0S_2$ has only two equatorial bands of zero motion, while $_0S_3$ and $_0S_4$ have three and four nodal lines, respectively.

The normal modes in the real Earth behave basically in this manner, except that complications are introduced by the variation of material properties with depth and by departures from spherical symmetry caused by rotation, aspherical shape, and material-property heterogeneity. For example, the Earth has a fluid outer core (see Chapter 7) in which the shear velocity is very small or zero. Torsional modes depend only on the shear-velocity structure and thus are confined to motions of the solid shell of the mantle. The inner core appears to be solid and may have inner-core toroidal motions, but some inner-core source must excite these and they cannot be observed at the surface. Spheroidal modes are sensitive to both P and S velocity and density structure, and the partitioning of compressional and shear energy with depth is complex. In general,

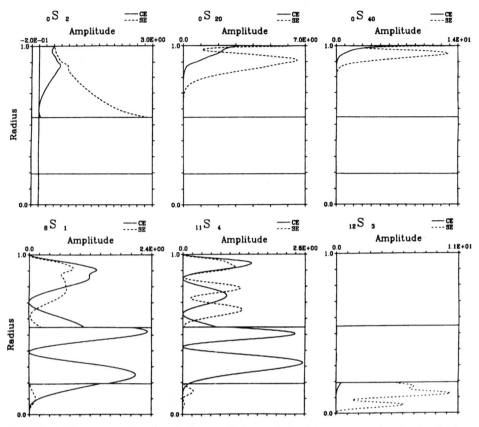

FIGURE 4.25 Compressional (solid line) and shear (dashed line) energy density for fundamental spheroidal modes (top row) and some spheroidal overtones that are sensitive to core structure. (Modified from Davis, 1989.)

fundamental modes ($n = 0$) have energy concentrated in the mantle (Figure 4.25), with the shear energy being distributed deeper into the mantle than compressional energy. Note that $_0S_2$ is sensitive to the entire mantle and hence to gravity variations over the depth extent of motions caused by the mode. Including effects of self-gravitation changes the period of this mode by almost 10 min. As l increases, the energy in both shear and compression is concentrated toward the surface. For $l > 20$, the fundamental spheroidal modes interfere to produce traveling Rayleigh-wave fundamental modes. The overtones ($n > 0$) of spheroidal motion generally involve energy sampling deeper in the Earth, includ-

ing in the inner and outer core. Examples are shown in Figure 4.25 for overtones that are sensitive to core structure. In the real Earth, radial-motion eigenfunctions of the spheroidal modes ($n, l > 0$) do not necessarily have n zero crossings along the radius, although toroidal modes do.

The normal modes of the Earth are identified primarily by computing ground-motion power spectra, as seen in Figure 4.23, and by associating the corresponding eigenfrequencies with those calculated for a model of the planet. This is clearly an iterative process, in which changes in the model can lead to reidentification of a particular mode peak. This process began in 1882 when Horace Lamb first calculated the normal modes of a homogeneous, elastic, solid sphere, finding that $_0S_2$ must have the longest period. The search for this mode of the Earth required development of very sensitive ground-motion instruments, with Hugo Benioff being a pioneer in development of ultra-long-period instrumentation. Following the great 1952 earthquake in Kamchatka, Benioff and others (1954) reported the first observation of a mode with a period of ~ 57 min, close to the ~ 1-h period expected for $_0S_2$. This observation was refined when the 1960 Chile earthquake ($M_w = 9.5$, the largest earthquake this century) occurred. About 40 normal modes were observed, and $_0S_2$ was found to have a period of 53.83 min. Because good starting Earth models existed, many modes could be confidently identified, as in Figure 4.23. Subsequently, several thousand modes have been identified and their degenerate eigenfrequencies determined. Table 4.2 lists the degenerate frequencies (see Box 4.8) of various observed modes of the Earth.

The process of identifying particular mode frequencies and finding an Earth model that is consistent with them flourished in the 1970s and continues today. Figure 4.26 shows a set of spheroidal and toroidal modes ordered by angular order number l and associated eigenfrequency; this set was obtained by data analysis by

Gilbert and Dziewonski (1975). Note that the modes sort into distinct branches for different values of n, but some branches come close together and have very similar eigenfrequencies. Groups of overtone modes along trajectories in these $\omega-l$ sets correspond to particular body-wave equivalent energy, a few of which are identified. This association is based on the modes that have appropriate phase velocities and particle motions. Given an Earth model that adequately predicts the observed eigenfrequencies, one can, of course, predict the eigenfrequencies of all modes.

This elegant procedure, of considering the entire Earth system in a boundary-value problem, is complicated by the non-spherical asymmetry of the system. The most important factor is the spinning of the Earth, which produces the Coriolis force, which is spherically asymmetric. This leads to a breakdown of the degeneracy of the eigenfrequencies for $2l + 1$ values of m for each spherical harmonic (Box 4.8). The result is called *splitting*, with the split eigenfrequencies being close together and the relative eigenvalue patterns of motions interfering with one another. The singlets are identified by the superscript m, so the multiplet $_0S_2$ is composed of singlets $_0S_2^{-2}$, $_0S_2^{-1}, _0S_2^0, _0S_2^1$, and $_0S_2^2$, each with a singlet eigenfrequency, $_n\omega_l^m$, and eigenfunction. Splitting of modes $_0S_2$ and $_0S_3$ was first observed for the 1960 Chile earthquake. Rotation splits the singlet eigenfrequencies according to the amount of angular momentum they possess about the Earth's rotation axis. The effect of rotational splitting on normal-mode peaks for stations at different latitudes is shown in Figure 4.27. The same mode has discrete multiplets at nonpolar stations (actually the $2l + 1$ multiplets are smeared together to give broadened, multiple peaks that do not resolve the individual eigenfrequencies for each l, m eigenvalue) but a single degenerate multiplet spike at high latitudes where the Coriolis force does not perturb the symmetry of the mode patterns. The strong

TABLE 4.2 Some Observed Normal-Mode Periods

Spheroidal modes	T(s)	Toroidal modes	T(s)
$_0S_0$	1227.52	$_0T_2$	2636.38
$_0S_2$	3233.25	$_0T_{10}$	618.97
$_0S_{15}$	426.15	$_0T_{20}$	360.03
$_0S_{30}$	262.09	$_0T_{30}$	257.76
$_0S_{45}$	193.91	$_0T_{40}$	200.95
$_0S_{60}$	153.24	$_0T_{50}$	164.70
$_0S_{150}$	66.90	$_0T_{60}$	139.46
$_1S_2$	1470.85	$_1T_2$	756.57
$_1S_{10}$	465.46	$_1T_{10}$	381.65
$_2S_{10}$	415.92	$_2T_{40}$	123.56

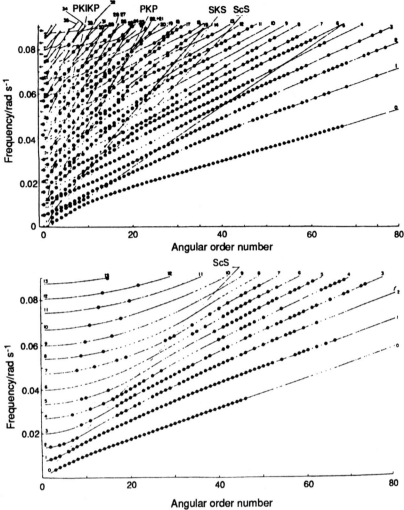

FIGURE 4.26 Spheroidal (top) and toroidal (bottom) mode eigenfrequencies as a function of angular order number, l. Note that modes align on fundamental ($n = 0$) and overtone ($n > 0$) branches. Body-wave equivalent modes, which cross branches, are indicated for a few main body-wave phases. (From Gilbert and Dziewonski, 1975.)

splitting of the modes $_{10}S_2$ and $_{11}S_4$ is greater than expected due to rotation, and these modes are sensitive to the core (see Figure 4.25); this is now attributed to anisotropy of the inner core aligned along the spin axis (Chapter 7).

In the time domain, the beating between the split singlets can strongly affect the temporal behavior of a single mode, as shown in Figure 4.28. The rotational split-ting of modes manifests itself differently for each source–receiver combination. This is true also of the effects of asphericity in the material properties, including ellipticity of the Earth. If we think of the standing-wave energy distributed over the great circle, lateral variations in velocity structure will distort the standing-wave pattern, locally perturbing the eigenfrequencies of the multiplet, as shown in

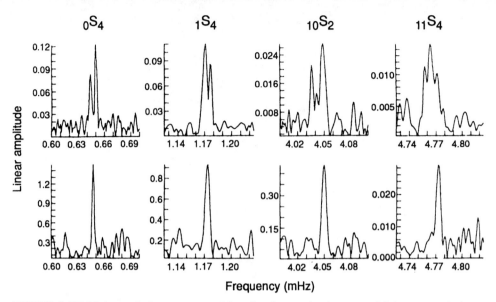

FIGURE 4.27 High-resolution spectra of four low-harmonic-degree multiplets recorded at nonpolar latitudes (top row) and polar latitudes (bottom row). The polar spectra are not obviously split, indicating effective cylindrical symmetry as produced by rotation. The low-latitude spectra are split, with rotation explaining the splitting of the $_0S_4$ and $_1S_4$ modes well but not accounting for the extent of splitting of $_{10}S_2$ and $_{11}S_4$. (From Masters and Ritzwoller, 1988.)

Figure 4.29. The mode will effectively average the great-circle velocity structure, with different average great-circle velocities leading to different multiplet frequencies for different paths. The local shift of phase at a particular distance affects the amplitude of the multiplet, leading to variations of the spectral peaks that effectively correspond to focusing and defocusing in the traveling wave–equivalent surface waves (see Box 4.5), although one must account for lateral averaging of the modes as well. The splitting and amplitude-variation properties of normal modes are used extensively in the study of Earth structure and seismic sources, as described in later chapters.

The final property of normal modes that we briefly discuss arises from the close proximity of some mode eigenfrequencies, as apparent in Figure 4.26. This can include the interactions between singlets of a given multiplet, interactions between adjacent modes on the same branch, interactions between modes on different branches, and interactions between toroidal and spheroidal modes, induced by Coriolis asymmetry. We describe such interactions as mode *coupling*. Rotation, aspherical structure, and possible anisotropy of the medium must all be included in rather complex calculations of coupling effects, but one must often do so to estimate accurately eigenfrequencies and attenuation of each mode. Figure 4.30 presents observed and synthetic seismograms, showing that coupling between spheroidal and toroidal modes can sometimes be observed (most favorably in great-circle paths traveling near the poles, with tangential motion that is very strong and spheroidal motion that is very weak on the path). The complex, ringy waveforms reflect mixing of toroidal and spheroidal energy onto this vertical (i.e., radial motion) seismogram, leading to precursory energy ahead of R_4. Such

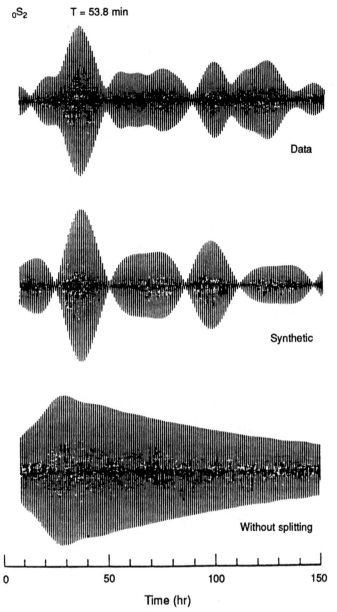

$_0S_2$ T = 53.8 min

Data

Synthetic

Without splitting

0   50   100   150

Time (hr)

FIGURE 4.28 Behavior of the mode $_0S_2$ as a function of time for an observed path, compared to synthetics with and without rotational splitting. (From Stein and Geller, 1978.)

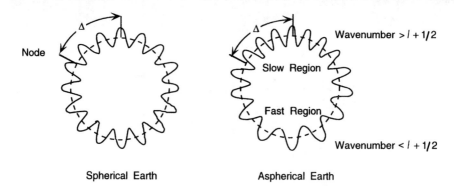

Spherical Earth   **Aspherical Earth**

FIGURE 4.29 A cartoon illustrating the distortion of the standing-wave multiplet caused by lateral heterogeneity in velocity structure. Although the number of wavelengths around any great circle remains constant, the local wavenumber, k, varies with local frequency perturbation $\delta\omega_{local}$. The spatial shift of the phase at distance Δ perturbs the observed multiplet amplitude. (Modified from Park, 1988.)

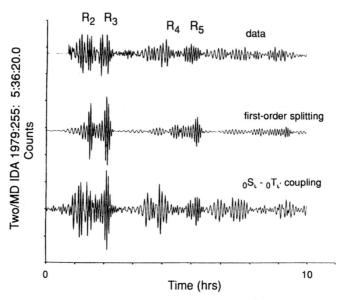

FIGURE 4.30 Data and synthetics for the September 12, 1979 New Guinea earthquake on a vertical-component recording at IDA station TWO. Coriolis coupling is high on this path, which goes within 5° of the rotation axis and leads to mixed spheroidal and toroidal motions on the seismogram. This is not included in the synthetics for first-order splitting, which account only for interactions within each multiplet, as in Figure 4.28, but is better accounted for when coupling between nearby fundamental-mode toroidal and spheroidal modes is calculated. (From Park, 1988).

complexity of free-oscillation theory is an active area of research and is revealing new aspects of Earth heterogeneity.

4.7 Attenuation of Surface Waves and Free Oscillations

Anelastic losses cause surface-wave and free-oscillation motions to attenuate with time. For body waves we characterize anelastic properties of the Earth in terms of radial and lateral variations of the P-wave attenuation quality factor Q_α and the S-wave attenuation quality factor Q_β. Since, in general, both P- and S-wave motions contribute to surface waves and standing waves, there are separate Rayleigh (Q_R), Love (Q_L), spheroidal (Q_S), and toroidal (Q_T) quality factors, all depending on frequency as well as varying from path to path. We know from Chapter 3 that the existence of anelasticity produces velocity dispersion, given by

$$c(\omega) = c_0 \left[1 + \frac{1}{\pi Q_m} \ln \frac{\omega}{\omega_0} \right], \quad (3.132)$$

where subscripts indicate a reference frequency, ω_0, and reference phase velocity, c_0, and Q_m is the wave quality factor. Since surface-wave Q values are relatively low, on the order of 100 for short-period waves and a few hundred for long-period waves, the effects of physical dispersion become important. Thus, Q is studied for long-period waves both to understand attenuation processes in the Earth and to allow models of Earth structure consistent with both body waves and surface waves or normal modes to be derived.

Measurement of surface-wave attenuation is conceptually straightforward but difficult in practice. Some of the first measurements were made for sequential great-circle passages of R_i and R_{i+2} or G_i and G_{i+2} waves. One can measure the decay coefficient γ at a given period T

$$\gamma(T) = \frac{1}{C} \ln \left(\frac{A_i}{A_{i+1}} \right), \quad (4.52)$$

where C is the circumference of the path, and the spectral amplitudes at that period are A_i (for R_i) and A_{i+2} (for R_{i+2}). This relates amplitude reduction to the attenuation factor. The corresponding inverse quality factor is given by

$$Q^{-1}(T) = \frac{\pi}{2} T U(T) \gamma(T), \quad (4.53)$$

where $U(T)$ is the group velocity on the great-circle path. This approach has been used extensively to measure surface-wave attenuation values for periods less than 500 s.

Free-oscillation attenuation measurements can be made by a variety of procedures. For an isolated split multiplet, with mean eigenfrequency ω_0, the contribution to the displacements at the surface will have the form

$$U_A(x,t) = \sum_{m=0}^{2l+1} a_m(x) \exp\left[i(\omega_0 + \delta\omega_m)t \right]$$

$$\times \exp\left[\frac{-\omega_0 t}{2Q_m} \right], \quad (4.54)$$

where $\delta\omega_m$ is the difference between the singlet eigenfrequency and the mean multiplet eigenfrequency and $a_m(x)$ is the amplitude of the singlet at the receiver. The amplitude $a_m(x)$ is a function of the source and receiver location, the Earth model, and the earthquake mechanism. The quality factor Q_m may or may not vary for each singlet.

If the multiplet is not split, then Q can be readily measured by narrowband filtering to isolate the mode and by using the temporal decay of the natural logarithm of the envelope of the time-domain trace. Figure 4.31 shows examples of this procedure. Smoothly decaying motions yield sta-

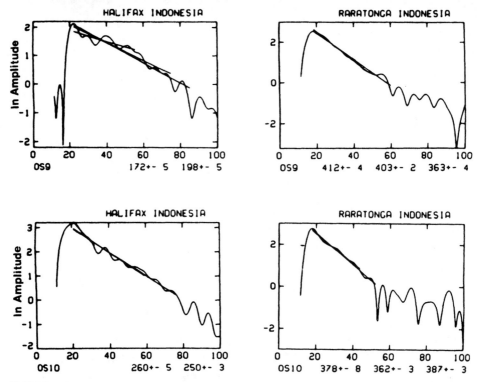

FIGURE 4.31 Q as determined by narrowband filtering of given modes ($_0S_9$ and $_0S_{10}$). Each box shows the natural logarithm of the unsmoothed envelope for a given mode as a function of time. The slope of the decay of amplitude is proportional to Q. (From Stein *et al.*, in "Anelasticity in the Earth," pp. 39–53, 1981; copyright by the American Geophysical Union.)

ble attenuation estimates. In the frequency domain, Q is estimated by the spread of the corresponding spectral peak $\Delta\omega$, with $Q = \omega_0/\Delta\omega$.

Clearly, if splitting exists, both the frequency-domain (Figure 4.27) and time-domain (Figure 4.28) signals are complex, and simple Q measurements cannot be made. The analysis used to estimate Q then depends on the relative amount of pulse broadening due to attenuation versus multiplet splitting. If one can accurately predict the individual singlet eigenfrequencies, one can estimate Q by modeling the time-domain signal or the split spectral peaks. This procedure is cur-

rently yielding attenuation values for many modes that have strong splitting, but it does have high attendant uncertainties. As high-quality digital data have increased in abundance, seismologists have even measured separate singlet attenuation values for a few strongly split modes.

The 1960 Chile earthquake commenced the analysis of free-oscillation attenuation, and it was quickly recognized that Q is higher for longer-period fundamental modes. This indicates that Q increases with depth. It is desirable to relate the particular Q value for a surface wave or normal mode to the depth-dependent values of Q_α and Q_β. For a given model, with

N layers, the toroidal-mode attenuation is given by

$$Q_T^{-1} = \sum_{l=1}^{N} \frac{\beta_l}{C_T} \left(\frac{\partial C_T}{\partial \beta_l} \right)_{k,\rho,\beta} Q_{\beta_l}^{-1} \quad (4.55)$$

and the spheroidal-mode attenuation is given by

$$Q_S^{-1} = \sum_{l=1}^{N} \left[\frac{\alpha_l}{C_S} \left(\frac{\partial C_S}{\partial \alpha_l} \right)_{k,\rho,\beta} Q_{\alpha_l}^{-1} \right.$$
$$\left. + \frac{\beta_l}{C_S} \left(\frac{\partial C_S}{\partial \beta_l} \right)_{k,\rho,\beta} Q_{\beta_l}^{-1} \right], \quad (4.56)$$

where $C_{(T,S)}$ is the mode phase velocity, k is the compressibility, ρ is the density, and

α_l, β_l are the P and S velocities in each layer. Note that these expressions give a weighted contribution of P- and S-wave attenuation in each layer corresponding to how much that layer influences the mode. These weighting factors are the *kernels* of the mode, indicating the partitioning of corresponding wave energy into each layer, where it is then attenuated according to the corresponding quality factor. The total dispersive effect on the mode due to the layered attenuation structure has a corresponding kernel. Examples of attenuation kernels for a specific Earth model for spheroidal modes are shown in Figure 4.32. The smooth shape of these kernels yields limited resolution of Q variations with depth. A model that is compatible with

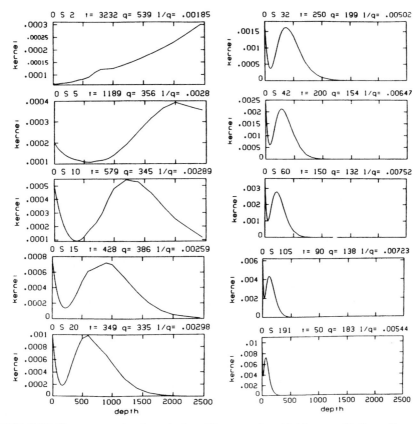

FIGURE 4.32 The attenuation kernels for different spheroidal free oscillations. For a given mode the kernel indicates the depths that are controlling the attenuation. For example, for $_0S_{191}$, all the attenuation is caused by Q in the upper 100 km of the Earth. (From Stein *et al.*, in "Anelasticity in the Earth," pp. 39–53, 1981; copyright by the American Geophysical Union.)

(although not uniquely required by) combined surface-wave, normal-mode, and body-wave attenuation measurements is shown in Figure 4.33. This model, *SL8*, shows shear attenuation, Q_μ, and bulk attenuation, Q_k (the quality factor in pure compression). These are related to Q_α and Q_β by

$$Q_\beta = Q_\mu \tag{4.57}$$

$$Q_\alpha^{-1} = L Q_\mu^{-1} + (1-L) Q_k^{-1} \tag{4.58}$$

$$Q_k = \left[(1-L) Q_\alpha Q_\beta\right] / (Q_\beta - L Q_\alpha), \tag{4.59}$$

where $L = \frac{4}{3}(\beta/\alpha)^2$. Table 8.1 also lists another attenuation model with a simpler structure that is still generally consistent with free-oscillation observations. Note the low Q in the upper mantle and the very high Q in the core. The dispersive effect on surface waves for such a Q model is illustrated in Figure 4.34, which shows the relative correction of either phase velocity (for surface waves) or period (for normal modes) as a function of period. Shorter-period waves that sample the low-Q upper mantle have strong dispersive effects. Chapter 7 will discuss ongoing efforts to map the aspherical structure of anelasticity in the Earth.

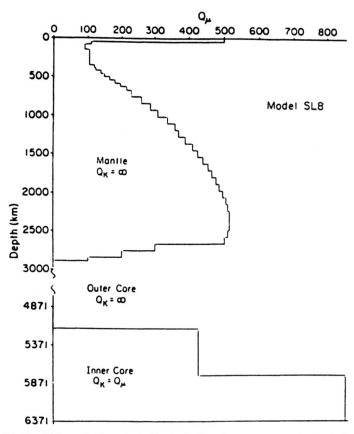

FIGURE 4.33 The SL8 model for whole-Earth Q. For the upper mantle the bulk attenuation is infinite; Q_μ is the shear Q. (From Anderson and Hart, *J. Geophys. Res.* **83**, 5869–5882, 1978; copyright by the American Geophysical Union.)

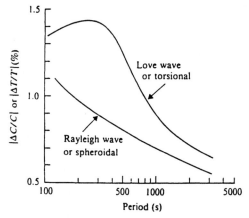

FIGURE 4.34 Fractional change in Love wave (toroidal mode) and Rayleigh wave (spheroidal mode) phase velocities (periods) as functions of period, computed for observed Q observations. (From Kanamori and Anderson, *Rev. Geophys. Space Phys.* **15**, 105–112, 1977; copyright by the American Geophysical Union.)

References

Aki, K., and Richards, P. G. (1980). "Quantitative Seismology." Freeman, San Francisco.

Anderson, D. L., and Hart, R. S. (1978). Q of the Earth. *J. Geophys. Res.* **83**, 5869–5882.

Båth, M. (1979). "Introduction to Seismology." Birkhäuser, Boston.

Benioff, H., Gutenberg, B., and Richter, C. F. (1954). Progress report, Seismological Laboratory, California Institute of Technology, 1953. *EOS, Trans. Am. Geophys. Union* **3**, 979–987.

Ben-Menahem, A., and Singh, S. J. (1981). "Seismic Waves and Sources." Springer-Verlag, New York.

Bolt, B. A. (1982). "Inside the Earth: Evidence from Earthquakes." Freeman, San Francisco.

Davis, P. (1989). Free oscillations of the Earth. *In* "The Encyclopedia of Solid Earth Geophysics" (D. E. James, ed.), pp. 431–441. Van Nostrand–Reinhold, New York.

Ewing, W. M., Jardetzky, W. S., and Press, F. (1957). "Elastic Waves in Layered Media." McGraw-Hill, New York.

Gilbert, F., and Dziewonski, A. M. (1975). An application of normal mode theory to the retrieval of structural parameters and source mechanisms from seismic spectra. *Philos. Trans. R. Soc. London, Ser. A* **278**, 187–269.

Gubbins, D. (1990). "Seismology and Plate Tectonics." Cambridge Univ. Press, Cambridge, UK.

Kanamori, H., and Abe, K. (1968). Deep structure of island arcs as revealed by surface waves. *Bull. Earthquake Res. Inst. Univ. Tokyo* **46**, 1001–1025.

Kanamori, H., and Anderson, D. L. (1977). Importance of physical dispersion in surface wave and free oscillation problems: Review. *Rev. Geophys. Space Phys.* **15**, 105–112.

Kennett, B. L. N. (1983), "Seismic Wave Propagation in Stratified Media." Cambridge Univ. Press, Cambridge, UK.

Lamb, H. (1904). On the propagation of tremors over the surface of an elastic solid. *Philos. Trans. R. Soc. London, Ser. A* **203**, 1–42.

Masters, G., and Ritzwoller, M. (1988). Low frequency seismology and three-dimensional structure—observational aspects. *In* "Mathematical Geophysics" (N. J. Vlaar, G. Nolet, M. J. R. Wortel, and S. A. P. L. Cloetingh, eds.), pp. 1–30. Reidel Publ., Dordrecht, The Netherlands.

Officer, C. B. (1974). "Introduction to Theoretical Geophysics." Springer–Verlag, New York.

Park, J. (1988). Free-oscillation coupling theory. *In* "Mathematical Geophysics" (N. J. Vlaar, G. Nolet, M. J. R. Wortel, and S. A. P. L. Cloetingh, eds.), pp. 31–52. Reidel Publ., Dordrecht, The Netherlands.

Satake, K. (1988). Effects of bathymetry on tsunami propagation: Application of ray tracing to tsunamis. *Pure Appl. Geophys.* **126**, 27–36.

Schwartz, S., and Lay, T. L. (1985). Comparison of long-period surface wave amplitude and phase anomalies for two models of global lateral heterogeneity. *Geophys. Res. Lett.* **12**, 231–234.

Sheriff, R. E., and Geldart, L. P. (1982). "Exploration Seismology, Volume 1. History, theory, and data acquisition." Cambridge Univ. Press, Cambridge.

Simon, R. B. (1981). "Earthquake Interpretations, A Manual for Reading Seismograms." William Kaufmann, Inc., Los Alto, CA, 150 pp.

Stein, S., and Geller, R. (1978). Time domain observations and synthesis of split spheroidal and torsional free oscillations of the 1960 Chilean earthquake: Preliminary results. *Bull. Seismol. Soc. Am.* **68**, 325–332.

Stein, S., Mills, J. M., Jr., and Geller, R. J. (1981). Q^{-1} models from data space inversions of fundamental spheroidal mode attenuation measurements. *In* "Anelasticity in the Earth." Geodyn. Ser. 4, (F. D. Stacey, M. S. Paterson and A. Nicholas, eds.), pp. 39–53. Am. Geophys. Union, Washington, DC.

Tanimoto, T. (1987). The three-dimensional shear wave structure in the mantle by overtone waveform inversion. I. Radial seismogram inversion. *Geophys. J. R. Astron. Soc.* **89**, 713–740.

Ward, S. (1989). Tsunamis. *In* "The Encyclopedia of Solid Earth Geophysics" (D. E. James, ed.), pp. 1279–1292. Van Nostrand–Reinhold, New York.

Additional Reading

Aki, K., and Richards, P. G. (1980). "Quantitative Seismology." Freeman, San Francisco.

Ben-Menahem, A., and Singh, S. J. (1981). "Seismic Waves and Sources." Springer-Verlag, New York.

Ewing, W. M., Jardetzky, W. S., and Press, F. (1957). "Elastic Waves in Layered Media." McGraw–Hill, New York.

Gilbert, F., and Dziewonski, A. M. (1975). An application of normal mode theory to the retrieval of structural parameters and source mechanisms from seismic spectra. *Philos. Trans. R. Soc. London, Ser. A.* **278**, 187–269.

Lapwood, E. R., and Usami, T. (1981). "Free Oscillations of the Earth." Cambridge, Univ. Press, Cambridge, UK.

Masters, G., and Ritzwoller, M. (1988). Low frequency seismology and three-dimensional structure—observational aspects. *In* "Mathematical Geophysics" (N. J. Vlaar, G. Nolet, M. J. R. Wortel, and S. A. P. L. Cloetingh, eds.), pp. 1–30. Reidel Publ., Dordrecht, The Netherlands.

5

SEISMOMETRY

The theory of elastic waves described in the previous chapters explains how the Earth vibrates as seismic waves pass through it and along its surface. Quantitative analysis of these seismic disturbances requires that the vibrations be instrumentally recorded. The instrumentation must (1) be able to detect the transient vibrations within a moving reference frame (the instrument moves with the Earth as it shakes); (2) operate continuously with a very sensitive detection capability with absolute timing so that the ground motion can be recorded as a function of time, producing a seismogram; and (3) have a fully known linear response to ground motion, or instrument calibration, which allows the seismic recording to be accurately related to the amplitude and frequency content of the causal ground motion. Such a recording system is called a *seismograph*, and the actual ground-motion sensor that converts ground motions into some form of signal is called a *seismometer*, or a *geophone* in exploration seismology. The design and development of seismic recording systems is called *seismometry*, and many successful instruments have been developed over the past 120 years, almost all based on the concept of an inertial pendulum. Different concepts are applied to study other Earth motions such as rotation, tilting, and straining.

The first known attempts to simply register the occurrence of ground motion were conducted by the Chinese as early as 132 AD. At that time, a Chinese philosopher, Chang Heng, developed the first *seismoscope*, an instrument that documents the occurrence of motion but does not produce a recording as a function of time. His instrument presumably involved a pendulum system inside a 6-ft-diameter jar, from which eight dragon heads protruded at principal compass directions. Balls were placed in the mouths of the dragons, and the internal pendulum was designed so that ground shaking would dislodge the ball from the dragon mouth in the direction of the azimuth to the source. The underlying technology for this seismoscope appears to have been lost, and significant further development of ground-motion sensors was not pursued until the 1700s.

The Italians developed numerous seismoscopes in the early eighteenth century, motivated mainly by the frequent occurrence of earthquakes in the Mediterranean. In 1751 Andrea Bina described a pendulum system with a pointer etching in sand, and increasingly sophisticated pendulum systems were incorporated in seismoscopes over the next 100 years. The first attempt to record the time of shaking was probably made in 1784, when A. Cavalli placed seismoscopes (involving bowls filled to the brim with mercury) above rotating platforms perforated with cavities, keyed to the time of day, which would collect any mercury slopped out of the bowls. In 1851 Robert Mallet applied a ground-motion sensor that used optical reflection from a basin of mercury to measure the speed of elastic waves in surface rocks, initiating the field of explosion seismology.

The first true seismograph, which recorded the relative motion of a pendulum and the Earth as a function of time, was built by Filippo Cecchi in Italy in 1875. A seismoscope was designed to start a clock and a recording device at the first onset of shaking. The oldest known seismic record produced by this system is dated February 23, 1887. A period of rapid instrument development and improvement occurred after 1875. A group of British seismologists teaching in Japan, the best known being John Milne, James Ewing, and Thomas Gray, led to the first relatively long-period systems (mainly sensitive to ground displacements for nearby events) and the first vertical-component seismographs. In these early systems, mechanical or optical systems amplified the mass motion, and friction provided the only damp-

ing of the pendulum oscillators. Europeans pursued the developments in Japan, and in 1889 the first known seismogram of a distant earthquake was made on a photographically recording, horizontal motion–sensing instrument designed by Ernst von Rebeur-Paschwitz and located in Potsdam. By 1900 the first global array of 40 photographically recording horizontal-component seismographs built by John Milne, along with other observatory instruments built in Europe and Japan, provided the initial seismogram data base for applying elastic-wave theory to begin to understand Earth vibrations.

5.1 Inertial Pendulum Systems

Almost all seismometers are based on damped inertial-pendulum systems of one form or another. Simple vertical and horizontal seismometer designs are illustrated in Figure 5.1. The frame of the seismometer is rigidly attached to the ground, and the pendulum is designed so that movement of the internal proof mass, m, is delayed relative to the ground motion by the inertia of the mass. Each pendulum

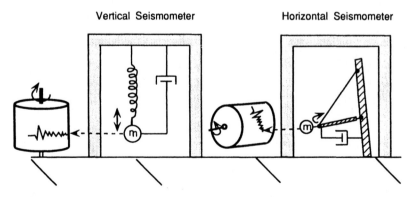

FIGURE 5.1 Schematics of inertial-pendulum vertical and horizontal seismographs. Actual ground motions displace the pendulums from their equilibrium positions, inducing relative motions of the pendulum masses. The dashpots represent a variety of possible damping mechanisms. Mechanical or optical recording systems with accurate clocks are used to produce the seismograms.

system has an equilibrium position in which the mass is at rest and to which it will return following small transitory disturbances. The orientation of the pendulum further determines which component of ground motion will induce relative pendulum motion.

Ground displacements, $U(t)$, are communicated to the proof mass via the attached springs or lever arms, with favorably oriented motions perturbing the system from its equilibrium position, leading to periodic oscillation of the mass. Friction or viscous damping, represented by the dashpots, is generally proportional to the velocity of the mass and acts to restore the system to its equilibrium position. Small-scale fluctuations in the springs and damping elements determine the intrinsic instrument noise level, below which actual ground motions cannot be detected. Although many early seismometers were designed empirically without mathematical analysis, the equation of motion for simple, damped harmonic oscillators provides insight into instrument characteristics.

The motion of the pendulum mass in an inertial reference frame is given by the sum of the ground motion plus the deviation of the mass from its equilibrium state, $y(t)$. For the vertical seismometer in Figure 5.1, the forces on the mass must act through the spring and dashpot, with recording-system friction effects included in the dashpot. The force from the spring is $-Ky(t)$, which is directly proportional to movement of the mass from its equilibrium position and which must involve stretching or contraction of the spring, which has a spring constant K. The damping force, $-D\dot{y}(t)$, is directly proportional to the velocity of the mass, with D being a damping coefficient. Newton's law ($F = ma$) is then

$$-Ky(t) - D\dot{y}(t) = m\left[\ddot{y}(t) + \ddot{U}(t)\right].$$

(5.1)

This is rearranged to give

$$\ddot{y}(t) + 2\gamma\omega_0\dot{y}(t) + \omega_0^2 y(t) = -\ddot{U}(t),$$

(5.2)

where $\omega_0 = \sqrt{K/m}$, and $\gamma = D/2\sqrt{Km}$ is the damping factor. The significance of ω_0 is shown by considering the undamped ($\gamma = 0$) system with $\ddot{U}(t) = 0$.

$$\ddot{y}(t) + \omega_0^2 y(t) = 0,$$

(5.3)

which has purely harmonic solutions of the form $\cos \omega_0 t$, $\sin \omega_0 t$, or $e^{\pm i\omega_0 t}$, where ω_0 is the natural or resonant frequency of the undamped system.

All recording systems that translate the pendulum motion into an actual seismogram, $x(t)$, involve at least a magnification coefficient, G, that gives rise to the *indicator* equation for $x(t)$:

$$\ddot{x}(t) + 2\gamma\omega_0\dot{x}(t) + \omega_0^2 x(t) = -G\ddot{U}(t).$$

(5.4)

Solutions of (5.4) for prescribed functional forms of $U(t)$ can characterize the seismograph response. This type of linear differential equation is readily solved using Laplace transforms (for transient motions) or Fourier transforms (for stationary ground oscillations). It is straightforward to consider simple harmonic forms of $U(t)$ such as

$$U(t) = e^{-i\omega t} = \cos \omega t - i \sin \omega t. \quad (5.5)$$

Of course, actual ground motion must be a real function, but it is easiest to analyze the general form of $U(t)$ and then consider the real part of $x(t)$. Inspection of (5.4) indicates that $x(t)$ will have the form $x(t) = x(\omega)e^{-i\omega t}$, giving

$$x(\omega) = \frac{-G\omega^2}{\omega^2 - \omega_0^2 + 2i\omega\gamma\omega_0}. \quad (5.6)$$

Box 5.1 Time and Frequency Domain Equivalence

In seismometry and many other aspects of seismology, it is often useful to represent transient time functions by equivalent functions in the frequency domain. This is possible using Fourier transforms, which are integral relationships that state that for an arbitrary function, $f(t)$, a set of harmonic terms exists such that

$$f(t) = \frac{1}{2\pi} \int_{-\infty}^{\infty} F(\omega) e^{i\omega t} \, d\omega, \qquad (5.1.1)$$

where

$$F(\omega) = |A(\omega)| e^{i\phi(\omega)} = \int_{-\infty}^{\infty} f(t) e^{-i\omega t} \, dt. \qquad (5.1.2)$$

These transform pairs correspond to a mapping from the time domain to the frequency domain, where ω is angular frequency, $A(\omega)$ is the amplitude of each harmonic component, and $\phi(\omega)$ is the corresponding phase shift (see Figure 5.B1.1). The integral in (5.1.1) is simply a sum, so this theorem states that an

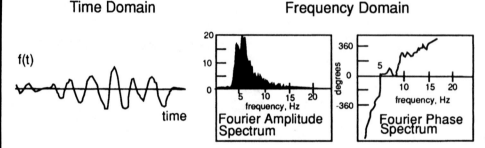

FIGURE 5.B1.1 A signal that is a function of time, as shown on the left, may be equivalently represented by its Fourier spectrum, as shown on the right. The amplitude and phase spectra are both needed to provide the complete time series.

arbitrary ground-motion time series, even an impulsive one, can be expressed as a sum of monochromatic periodic functions (Figure 5.B1.2). This is possible if the amplitude and phase alignment of the harmonic terms are chosen appropriately and the sum is over a continuous distribution of harmonic functions. Destructive and constructive interference between the harmonics is balanced so that they add up exactly to the original time series. The functions are called the signal spectrum and define the frequency-domain representation of the time-domain trace. Fourier spectra are determined using computers and digital, discretized versions of Fourier transforms. This text will often represent seismological observations by their spectra, which contain all of the information of the original seismogram, as long as both amplitude and phase are considered.

continues

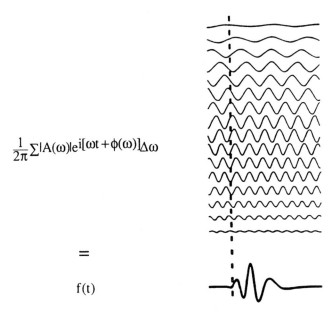

$$\frac{1}{2\pi}\sum |A(\omega)|e^{i[\omega t + \phi(\omega)]}\Delta\omega$$

$$=$$

$$f(t)$$

FIGURE 5.B1.2 A discretized version of Eq. (5.1.1), showing how a sum of harmonic terms can equal an arbitrary function. The amplitudes of each harmonic term vary, being prescribed by the amplitude spectrum. The shift of the phase of each harmonic term is given by the phase spectrum.

The complex function $x(\omega)$ can be represented in the form $x(\omega) = |x(\omega)|e^{i\phi(\omega)}$ with

$$|x(\omega)| = \frac{G\omega^2}{\sqrt{\left(\omega^2 - \omega_0^2\right)^2 + 4\omega^2\omega_0^2\gamma^2}}$$

$$\phi = -\tan^{-1}\left(\frac{2\omega\omega_0\gamma}{\omega^2 - \omega_0^2}\right) + \pi, \quad (5.7)$$

where $x(\omega)$ is called the frequency response of the instrument, and $|x(\omega)|$ is the amplitude response and $\phi(\omega)$ the phase delay. The actual physical seismogram would correspond to the real part of this solution in the time domain,

$$x(t) = \frac{1}{2\pi}\int_{-\infty}^{\infty} |x(\omega)|e^{i\phi(\omega)}e^{i\omega t}\,d\omega. \quad (5.8)$$

As $\gamma \to 0$ (undamped), the solutions have increasing amplitude as $\omega \to \omega_0$,

which is called *resonance*. Typically, the natural period of the seismometer ($T = 2\pi/\omega_0$) has the maximum amplitude response. If $\gamma \ll 1$ (underdamped), the mass responds primarily to periods near the pendulum period, and the signal tends to "ring" at that period. For $\gamma = 1$ the signal is critically damped and oscillation is minimized, with the mass quickly returning to rest as ground motion ceases. For $\gamma > 1$ (overdamped), no oscillations occur, but the mass returns to rest more slowly. Most instruments are designed to operate with near-critical damping so that the seismic record is not excessively ringy.

If the ground-motion frequency is much lower than the seismometer frequency ($\omega \ll \omega_0$), the amplitude response is proportional to ω^2/ω_0^2, and the seismogram records ground acceleration. Thus, design of accelerometers, intended to record strong acceleration at frequencies near

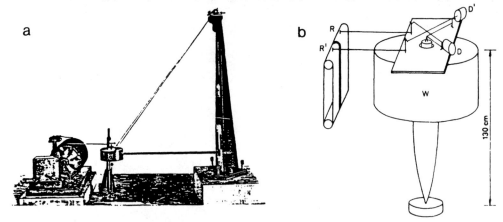

a   b

FIGURE 5.2 Early mechanical horizontal-motion seismographs: (a) The 1905 Omori 60-s horizontal-pendulum seismograph and (b) the 1904 1000-kg Wiechert inverted-pendulum seismograph. Both instruments etched a record on smoked-paper recorders. Friction on the stylus provided the only damping in the Omori system, while air pistons (D and D') damped the Wiechert instrument. Restoring springs, connected to the top of the mass, W, kept the inverted pendulum in equilibrium, with a special joint at the base of the mass permitting horizontal motion in any direction.

5–10 Hz, involves seismometers with very high resonant frequencies. If the driving frequency is much higher than the natural frequency ($\omega \gg \omega_0$), displacement on the seismogram is directly proportional to ground displacement. Much of the early developmental work in seismometry sought to reduce ω_0 to yield displacement recordings for regional-distance seismometers. Most modern seismometers are actually primarily sensitive to ground velocity because motions of the pendulum mass are converted to an output voltage signal proportional to the mass velocity. A variety of instruments with varying response characteristics will be discussed later in this chapter.

We conclude this discussion of simple harmonic oscillators by considering two classic seismic instruments developed around the turn of the century, shown in Figure 5.2. The first is the Omori horizontal pendulum seismograph, developed by a student of John Milne in Japan from 1899 to 1905. The instrument had direct re-

sponse to ground displacement for periods less than 60 s, with the long pendulum period achieved by having a nearly vertical swinging-gate pendulum (increasing the angle of the pivot arm to the vertical decreases the pendulum period). A stylus attached to the mass etched a record directly onto a rotating drum covered with smoked paper. The only damping in the system was due to the stylus friction and mechanical friction in the hinges, and, of course, the restoring force acting on the mass was simply gravity. Figure 1.5 shows a recording from an Omori instrument for the 1906 San Francisco earthquake. In 1898 E. Wiechert in Germany introduced viscous damping in a horizontal-pendulum instrument and extended this to an inverted-pendulum seismograph in 1900. A 1000-kg mass was used in the 1904 variety, shown in Figure 5.2b, with mechanical levers magnifying the signal 200 times and etching a record on smoked paper. Air-filled pistons provided the damping. Wiechert inverted-pendulum seismometers are still operated today, having provided more than 90 years of relatively uniform instrumental records.

5.2 Earth Noise

Before continuing a discussion of seismometry developments in this century we must consider an additional important aspect of ground-motion recording: the ground is never truly at rest. Because all sources of rapid deformational energy excite seismic waves and because sources such as tides, atmospheric pressure, diurnal heating of the surface, and human-induced vibrations are continuous, a continuous background noise level exists composed of small signals or *microseisms*. Any detection of transient wave arrivals must be made in the presence of this noise. Not surprisingly, the background noise level is temporally and spatially variable and is not uniform at all frequencies. This has strongly influenced the design of seismic recording systems in this century.

Wave surf and standing waves in the ocean are some of the primary sources of seismic noise, with water movements continually generating surface waves in the solid Earth. Figure 5.3 shows the variation of background noise ground acceleration for stations at varying distances from coastlines. These ground-acceleration spectra typically have noise peaks at frequencies from 0.15–0.2 Hz. The units are decibels, given by $10\log_{10}$(signal power). Because signal power is proportional to the square of the signal amplitude, 20 dB corresponds to a factor of 10 variation of the signal amplitude, in this case ground acceleration. Thus, ground-acceleration noise varies by a factor of about 10^4 over the frequency range shown, and the high-frequency noise peak will tend to swamp any seismometer with uniform sensitivity unless it has a dynamic range that can resolve very large variations in signal amplitudes. This figure also suggests that island sites (RPN is on Easter Island) will be much noisier than land sites well removed from the coast. The factor of 10 variation in noise levels in the 0.1–1.0 Hz passband also indicates that seismic-event detection will be nonuniform and measurement error will vary from station to station.

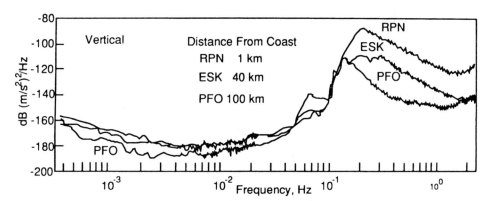

FIGURE 5.3 Power spectra of average background-noise ground acceleration recorded on vertical-motion accelerometers. Note the peak in noise near 0.2 Hz at all stations and the systematic decrease in noise with distance from the coast. Figure 5.14 shows the station locations. The units of dB (decibels) are in terms of $10\log_{10}$ (acceleration power). Thus, 20 dB corresponds to a factor of 10 variation in ground acceleration. (From Hedlin *et al.*, 1988.)

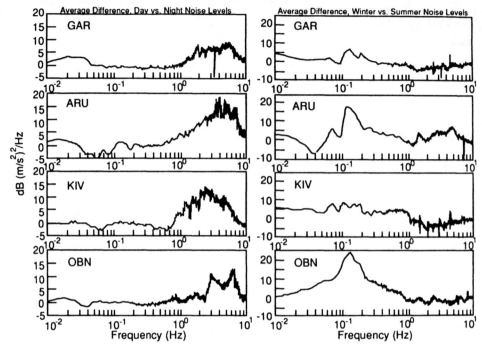

FIGURE 5.4 Differences in ground-acceleration power spectra at four stations located in the former Soviet Union between day and night (left) and winter versus summer (right). The vertical units are decibels, with 20 dB corresponding to a factor of 10 variation in ground acceleration noise level. (From Given, 1990.)

Ground noise also exhibits daily and seasonal variations at sites well removed from coastlines. Figure 5.4 shows differential noise spectra from seismic stations in the former Soviet Union, all of which display enhanced high-frequency noise during the daytime and variable seasonal noise character. Human-induced and atmospheric variations are partly responsible for the diurnal behavior. Seasonal variations can reflect ground-water freezing, changes in atmospheric patterns, and temperature variations of the recording sensors.

In the 1960s many studies of ground-motion noise characteristics were conducted to improve seismograph design. It was found that placing instruments in deep mines below the surface or in deep boreholes could significantly reduce the background noise levels, enabling better transient event detection. Figure 5.5 illustrates the factor of 10 signal-to-noise enhancement achievable by placing the sensor in a borehole. This is particularly important for noisy island sites (although horizontal tilting cannot be so easily eluded) and is driving new development of ocean-bottom borehole instrumentation.

In this text we treat seismic noise as a nuisance, limiting our ability to observe transient seismic signals, but we should note that seismologists have conducted many interesting studies of microseism sources. For example, the locations of large storm centers have been inferred from noise characteristics of sets of stations. Ground motions of microseisms vary from 10^{-8} to 10^{-3} cm, and no seismogram can ever be totally free of some background noise. Most seismological analyses must explicitly allow for noise-contaminating effects on any estimate of a given signal amplitude and phase spectrum.

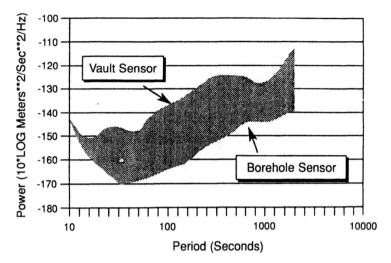

FIGURE 5.5 Illustration of the background noise reduction that can be achieved by using deep-borehole seismometers in place of vault seismometers. A value 20 dB on the vertical axis corresponds to a factor of 10 variation in ground acceleration. (From Incorporated Research Institutes for Seismology, 1991–1995.)

5.3 Electromagnetic Instruments and Early Global Networks

In 1914 a Russian, B. Galitzin, introduced an electromagnetic moving-coil transducer to convert pendulum mass motion into an electric current. Motion of a wire coil in the presence of a magnetic field generates a signal voltage that is proportional to the mass velocity, which Galitzin used to rotate a galvanometer coil. Light reflected from a mirror on the galvanometer coil was recorded on photographic paper, and a long optical lever arm was used to produce large magnifications. This type of electromagnetic system has dominated instrumentation this century, with the optical recording eliminating friction. The coupling of a seismometer pendulum, electromagnetic transducer, and galvanometer also allowed shaping the instrument response to emphasize a particular frequency passband. The electromechanical response of the galvanometer can be approximated by a solution of the form of (5.7), but with different damp-

ing and resonant frequency corresponding to the galvanometer characteristics. The product of the pendulum, transducer, and galvanometer frequency responses controls the overall instrument response, leading to responses that are peaked at the pendulum period.

Instrumental response curves for some classical mechanical and electromagnetic seismographs are shown in Figure 5.6. Note that the Galitzin responses achieve higher gains due to the optical recording, but they are more narrowband (i.e., record a narrower frequency range) than early mechanical instruments like the Wiechert, Bosch–Omori, and Milne–Shaw instruments. These instrument responses clearly show the strong falloff in response at long periods, proportional to T^{-2} (ω^2), where the response is proportional to ground acceleration. The noise spectra in Figure 5.3 show that one of the clear advantages of the Galitzin electromagnetic systems is that response at short periods, where the instruments respond directly to ground velocity (slope α T), is reduced near the large noise peaks near 5 to 6 s.

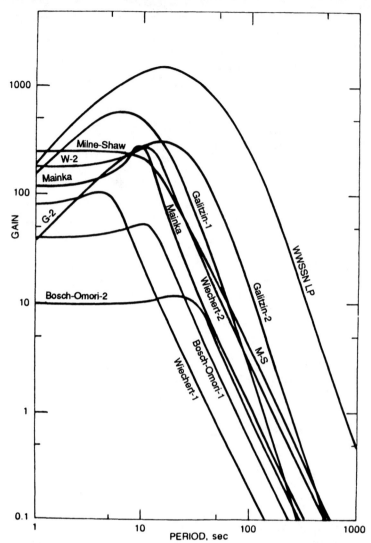

FIGURE 5.6 Instrument-response curves for a suite of classic seismometers, indicating their magnification as a function of period. For very long periods all of these instruments are sensitive to ground acceleration, with the response falling off in proportion to $1/T^2$, where T is the period. (Modified from Kanamori, 1988.)

This response tuning was critical for the 1940s development of the short-period Benioff and long-period Sprengnether electromagnetic instruments based on the Galitzin design. These were deployed in the World Wide Standardized Seismic Network (WWSSN) in the 1960s. These instruments were designed to straddle the strong Earth noise peak, with short-period instruments having 1-s pendulum periods

and 0.7-s galvanometers, while the long-period instruments had either 15- or 30-s-period pendulums with 100-s-period galvanometers.

In the early 1960s, as part of the VELA-Uniform project sponsored by the Department of Defense following the move to underground nuclear testing, a global array of these instruments was installed. Each station had three short-period and three long-period instruments to record horizontal and vertical ground motion. Initially, 30-s-period pendulums in the long-period Sprengnethers were used, but they proved to be excessively sensitive to barometric pressure variations, so more stable 15-s-period configurations were adopted by 1965. The distribution of the WWSSN stations (Figure 5.7) was extensive, reflecting the global collaboration typical of seismology, although clear gaps exist due to both political situations and ocean basins. This global network was more extensive than any preceding instrument deployment and was equipped with very accurate timing by crystal clocks and standardized instrumentation.

The instrument responses of the short-period and long-period WWSSN seismographs are shown in Figure 5.8, along with responses for other instruments that domi-nated seismic data collection from 1922, when the Wood–Anderson torsion seismograph was developed, to 1976. All of these except the Wood–Anderson instruments are electromagnetic systems with galvanometers. The torsion seismographs simply involve a copper cylinder attached to a vertical suspension wire. Shaking causes the cylinder to rotate slightly, moving a mirror that reflects a light signal to a photographic recorder. Two designs were made, one with a 0.8-s period with a magnification of 2800 and the other with a 6.0-s period and a magnification of up to 800. The short-period sensor was critical for providing regional earthquake recordings used to develop the Richter magnitude scale.

The WWSSN recordings have been very extensively utilized because the original photographic records were filmed on 35-

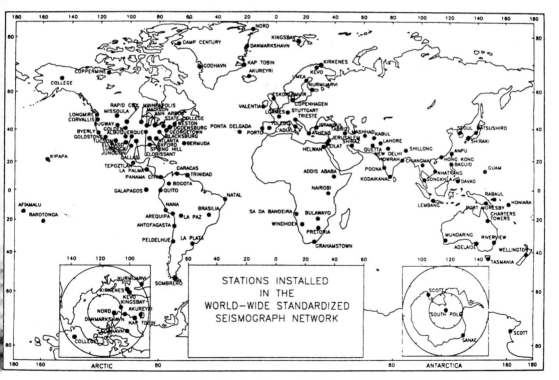

FIGURE 5.7 Global map indicating the locations of stations of the World Wide Standardized Seismograph Network (WWSSN). (Courtesy of the U.S. Geological Survey.)

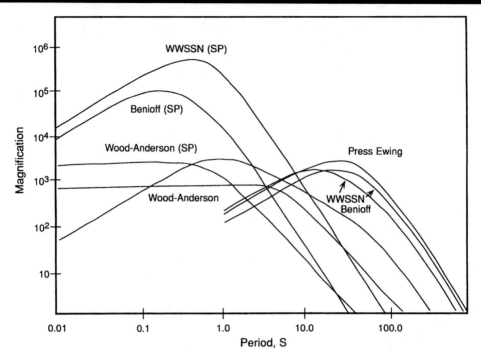

FIGURE 5.8 Instrument response curves for short- and long-period seismometers of the WWSSN, Benioff, Wood–Anderson, and Press–Ewing varieties, which dominated local and global data collection from 1950 to 1977. The instrument pairs were designed to minimize the effects of ground-noise maxima between 5 and 10 s. (Courtesy of H. Kanamori.)

or 70-mm microfiche, and copies were provided to major seismic data centers, where magnified paper copies could be made. The impact of the WWSSN was tremendous, coming at the time of the plate tectonics revolution, when accurate seismic recordings were critical for determining faulting patterns. The accurate timing and response standardization allowed many basic studies of Earth structure and earthquake sources to be conducted throughout the 1960s to 1980s.

5.4 Force-Feedback Instruments and Digital Global Networks

Beginning in the early 1970s seismic recording systems began to forgo low-dynamic-range analog recording by ink, photographic systems, or analog tape recording in favor of digital recording on magnetic tape. In essence, these systems sample the output current from the seismometer and amplification electronics, and they write the voltage at each time step to tape rather than use it to drive a mechanical or optical recording system. The first digital observatory stations were the High Gain Long Period (HGLP) stations deployed by Columbia University from 1969 to 1971 at sites in Alaska, Australia, Israel, Spain, and Thailand. The HGLP stations used sensors similar to those of the WWSSN, but they included both digital and optical recordings and had better thermal isolation. HGLP stations were the first to resolve minimum Earth noise in the 20- to 100-s-period range. Beginning around 1975, these were superceded by the Seismic Research Observatories

(SRO), with the HGLP stations being modified to become Abbreviated Seismic Research Observatories (ASRO). ASRO, SRO, and digitally upgraded WWSSN (DWWSN) made up the Global Digital Seismic Network (GDSN), with additional digital stations deployed in the Regional Seismic Test Network (RSTN).

The distribution of the GDSN is shown in Figure 5.9. Note that the total number of stations is less than that of the WWSSN. The SRO stations of the GDSN employ the KS36000 seismometer, which has three components and is deployed about 100 m deep in a borehole. The SRO and ASRO recording systems write directly to magnetic tape, and the response is still separated into short- and long-period recordings for the SRO sensors, despite the fact

that only one type of seismometer was used, unlike the separate short- and long-period sensors of the WWSSN and HGLP/ASRO. The long-period SRO response peaks at a period of 25 s, with a narrowband amplitude response. This was motivated by a desire to record 20-s-period surface waves from earthquakes and from nuclear explosions for treaty-monitoring purposes. Figure 5.10 shows examples of ground impulse responses of GDSN systems. The filtering effect of the instrument causes a spike impulse ground motion to produce a 20-s-period seismogram, which clearly limits the potential resolution of rapid ground vibration. These systems were mainly for recording global surface waves, and the convenience of digital recording prompted the first aspherical Earth model

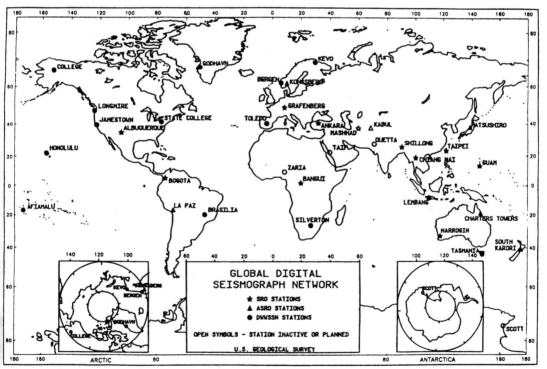

FIGURE 5.9 Global map indicating the locations of stations of the Global Digital Seismic Network (GDSN) composed of SRO, ASRO, and DWWSSN stations. These instruments dominated global data collection from 1977 to 1986. (Courtesy of the U.S. Geological Survey.)

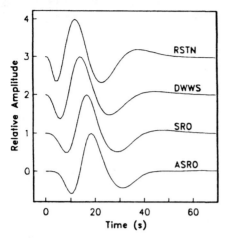

FIGURE 5.10 Examples of the impulse ground-motion response of long-period digital instruments in the GDSN. Digital seismograms are processed to account for the instrument-response distortions when analyzing seismograms. (From Shearer, 1991.)

KS54000, which intrinsically cannot accommodate large pendulum motions due to the compact size of the sensors.

Force-feedback systems of various types have actually existed at least since 1926, when de Quervain and Piccard used one in a 21-ton seismograph in Zurich. Much of the challenge in designing broadband seismometers has been in the development of stable force-generating systems that can respond accurately over the whole range of motions that a seismometer will undergo. It has also been necessary to develop recording equipment with sufficient dynamic range to exploit the capabilities of the most recent generation of sensors.

Figure 5.11 shows a schematic of the Wielandt–Streckeisen STS-1 leaf spring seismometer and a sample broadband recording system. The seismometer is a standard, remarkably compact pendulum-type design, but its capabilities are mainly due to the feedback electronics that prevent the mass from moving significantly. Digitizing the feedback-generated signal with 16- to 24-bit resolution, careful timing, filtering, and tape recording are all critical to retrieving a useful signal.

Figure 5.12 illustrates the merits of the STS-1 broadband seismograph relative to WWSSN and GDSN stations. The broadband system avoids the artificial separation of signal energy into separate short- and long-period channels as was done in the WWSSN instrumentation. The dynamic range of the system is so great that using separate channels that straddle Earth noise peaks is no longer necessary. Also, the digital filtering is far less severe than in the SRO system, allowing retrieval of much more waveform information.

The magnitude of the progress in seismograph development is dramatically illustrated in Figure 5.13, which compares the dynamic range and bandwidth of the latest generation of instrumentation with those of the WWSSN systems. The new systems being deployed by the Incorporated Research Institutions for Seismology (IRIS) jointly with the U.S. Geological

inversions (see Chapter 7). The major failing of the SRO system is that the sensors and electronics exhibit nonlinear responses for rapid accelerations such as those associated with large, impulsive body-wave arrivals. Also, the only high-frequency recordings were for triggered, short-period vertical components.

One of the critical aspects of the KS36000 and most other recent seismic sensors is that they employ force-feedback systems. This involves a negative feedback loop in which a force proportional to the inertial mass displacement is applied to the mass to cancel its relative motion. An electrical transducer converts the mass motion into an electrical signal to assess how much feedback force to apply. The amount of force required to hold the pendulum at rest corresponds to the ground acceleration. The force-feedback strategy greatly extends the bandwidth and linearity of a seismometer, because the mass cannot make large excursions that bend the springs or levers. Since 1973 all broadband seismic sensors have incorporated force feedback, particularly borehole sensors like the KS36000 or the newer

Survey (USGS) as the new Global Seismic Network (GSN) have astounding capabilities. These instruments can record both Earth tides and high-frequency body waves ranging from minimum Earth noise levels up to the strong accelerations expected for a magnitude 9.5 earthquake 3300 km away. This new instrumentation, first developed in 1986 but built on 10 years' experience with STS-1–type sensors at the Gräfenberg seismic array in Germany, is now being widely deployed around the world.

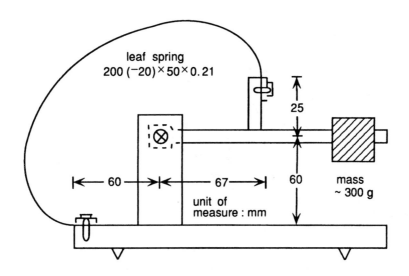

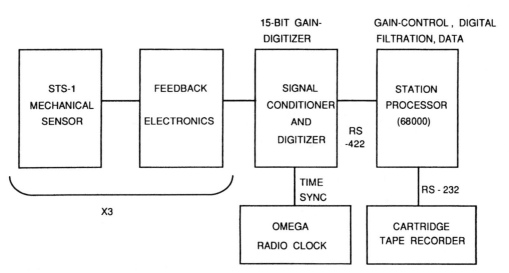

FIGURE 5.11 Schematic of the leaf-spring seismometer and system configuration involved in modern broadband digital seismographs. The Streckeisen STS-1 leaf-spring seismometers are attached to feedback electronics that adjust to minimize actual motions of the mass. The electric currents produced by the feedback are digitized, synchronized with time signals, and electronically filtered and recorded. These systems can reduce instrument noise by factors of 20–40 dB relative to GDSN-generation equipment.

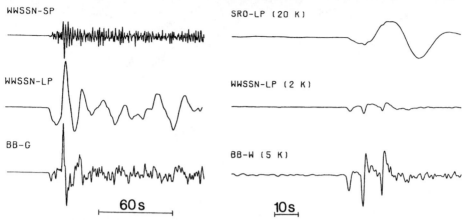

FIGURE 5.12 Comparison of seismograms with varying instrument responses for the same ground motion. The records on the left compare a teleseismic *P* wave from the March 4, 1977 Bucharest event, as it would appear on WWSSN short- and long-period seismograms, with the broadband signal (proportional to ground velocity) actually recorded at station A1 of the Gräfenberg seismic array in Germany. The broadband recording contains much more information than either WWSSN recordings alone or combined. The example on the right compares GDSN (SRO-LP), WWSSN-LP, and broadband ground-displacement recordings for a *P* wave from the April 23, 1979 Fiji earthquake that has traversed the Earth's core. The broadband recording contains much more information that can reveal details of the core structure. (Modified from Harges *et al.*, 1980.)

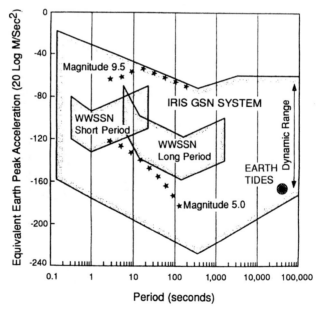

FIGURE 5.13 The range of ground acceleration (in dB) and period of ground motions spanned by the very broadband seismic system of IRIS Global Seismic Network (GSN) compared with capabilities of the WWSSN instrumentation and expected ground accelerations from magnitude 5.0 and 9.5 earthquakes at a distance of 30° (angular distance) and from Earth tide motions. GSN-type instruments have become dominant for global seismic recording since 1986. (From Incorporated Research Institutes for Seismology, 1991–1995.)

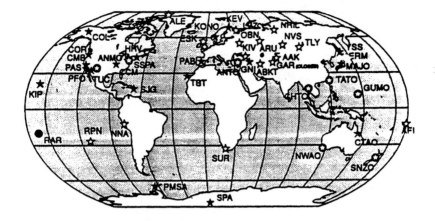

Current Sites	
★	STS-1 vault seismometer
○	KS36000i or KS54000-IRIS borehole seismometer
●	Both STS-1 and KS36000i
	+ 24-bit data logger • dial-up access (Except GNI, YSS, PMSA)
☆	STS-1 vault seismometer
	+ 16-bit dual-gain data logger or DWWSSN (AFI)

FIGURE 5.14 Global distribution of IRIS–GSN and IRIS–IDA seismic stations by the end of 1993. Broadband instrumentation with recorders of different dynamic range are differentiated in the figure. (Courtesy of R. Butler.)

Figure 5.14 shows the current distribution of the fully configured, latest-generation seismic stations operated by the United States by IRIS/USGS. The network is growing continuously, with an ultimate goal of 128 stations with relatively uniform coverage of the surface. These systems are also being deployed at another important global network (open stars in Figure 5.14) operated by the University of California at San Diego (now affiliated with IRIS), called the International Deployment of Accelerometers (IDA).

The IDA instruments were the best available from 1977 to 1987 for recording free oscillations of the Earth. The instrument used in these is a force-feedback LaCoste–Romberg vertical gravimeter, which senses vertical motion by the resulting change in gravity. The gravimeter mass is connected to the center plate of a three-plate capacitor, whose outer two plates are fixed. As the mass moves, the voltage generated between the center plate and the outer plates is proportional to the displacement. A 5-kHz alternating voltage is applied to the outer plates, so that the lower-frequency seismic signal modulates the amplitude of the 5-kHz signal. The modulated signal is fed to an amplifier, which generates a voltage that is proportional to the 5-kHz component of the signal and thus to the displacement of the mass. The signal then goes to an integrator circuit whose output is proportional to acceleration of the mass. This is the seismic systems output. This voltage is also then fed back to the outer capacitor plates

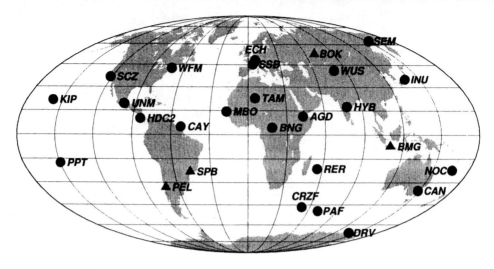

● **Stations installed**
▲ **Future sites planned**

FIGURE 5.15 Global distribution of Project GEOSCOPE stations by the end of 1990. STS-1 seismometers are located at all stations, but slightly different recording characteristics are used at different sites. (Modified from Romanowicz *et al.*, 1991.)

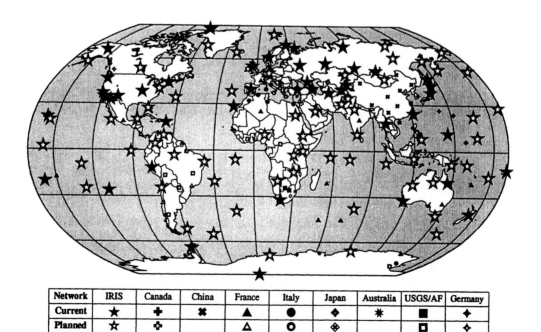

Network	IRIS	Canada	China	France	Italy	Japan	Australia	USGS/AF	Germany
Current	★	✚	✖	▲	●	✧	✳	■	✦
Planned	☆	✢		△	○	⊕		□	✧

FIGURE 5.16 Global network of very broadband seismometers planned for the end of the twentieth century, composed of various international network deployments. (Courtesy of R. Butler.)

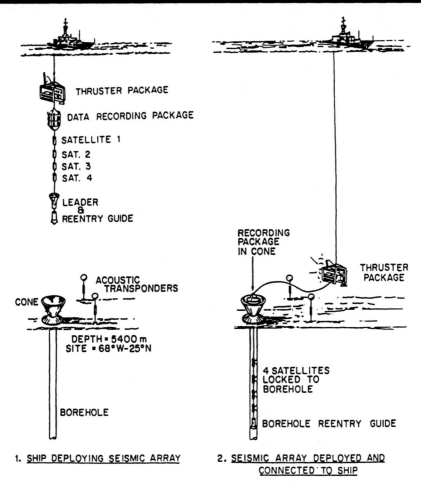

FIGURE 5.17 Schematic of ocean-bottom borehole-seismometer deployment and recording operations. Broadband instrumentation for such submarine boreholes is being designed and tested in the early 1990s. (From Stephen *et al.*, 1988.)

to stabilize the system and increase linearity.

Instrumentation comparable to that of the GSN (STS-1 seismometers with dual 16-bit digitizing system) have been deployed by France beginning in 1982 under project GEOSCOPE (Figure 5.15), and together with instruments deployed by seismologists in Europe, Australia, Canada, and Japan, a new global network (Figure 5.16) of the highest quality is evolving to finally replace the WWSSN with a complete global coverage. It will require up to

10 years to upgrade the global network fully with the new, rather expensive instrumentation. It is clear that even then, ocean basins will cause substantial gaps in coverage. To overcome this, scientists are currently developing a broadband borehole sensor for deployment in ocean basins. Both the extreme environmental conditions and the difficulty of deploying and retrieving data from the system provide major technological challenges. Figure 5.17 illustrates one concept for an ocean-bottom borehole system.

5.5 Seismic Arrays and Regional Networks

Although the first priority for seismic instrument development was global deployment of observatory instrumentation to increase knowledge of Earth's structure, the underlying principles were quickly adapted to other instrument capabilities. Small seismometers, with many hundreds of sensor channels, were developed for explosion seismology. These involve easily deployable geophones that can be laid out at regular intervals to record high-frequency seismic waves over short distances. Portable seismograph systems were designed as isolated units that could be deployed near large earthquake ruptures to record aftershocks or to study the crust locally. Ocean-bottom seismometers were designed for similar studies. Yet another seismological instrumentation development came with the VELA-Uniform project. This involved dense arrays of seismometers with either fixed locations or portable systems that were laid out in a regular pattern. In every case, these involve pendulum-based seismometers, with the most current ones having force-feedback systems to provide great bandwidth.

Major U.S.-deployed seismic arrays have included the Long Range Seismic Measurements (LRSM) program of the 1960s, the Geneva arrays of the 1960s to 1970s, the Large Aperture Seismic Array (LASA), and the Norwegian Seismic Array (NORSAR), a large array in Norway that is still operational. The LRSM involved mobile seismological observatories that used film and FM magnetic tape to record short- and long-period three-component data. Linear arrays straddling the United States were deployed primarily to record underground nuclear tests. The Geneva Arrays included five arrays around the United States installed between 1960 and 1963 that ran until 1970 or 1975. These included arrays in Oklahoma (WMO), Tennessee (CPO), Oregon (BMO), Utah (UBO), and the Tonto Forest Observatory

(TFO) in Arizona. TFO was the primary research system, but all of these were designed to study multiple-element seismic-recording procedures to assess potential advantages for seismological studies. All of them except TFO had apertures of 4 km, with 10 to 19 sensors laid out in different patterns. TFO was larger and denser and operated until 1975.

LASA was built on the experience with small arrays and involved an array of arrays with 525 seismometers over an aperture of 200 km. Twenty-one clusters, each with 25 sensors over 7-km^2 regions, were deployed, all recording vertical high-frequency (> 3 Hz) ground motion. LASA operated from the mid-1960s to 1978. This array enabled significant new analyses of high-frequency seismic waves traversing the Earth's deep interior.

NORSAR began operation in 1971 and involved the subarray cluster design developed at LASA. Twenty-two subarrays distributed over 100 km^2 were included in

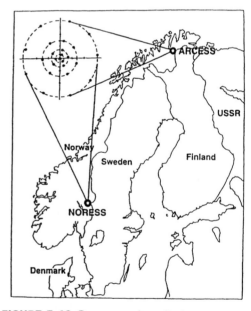

FIGURE 5.18 Geometry of small, dense arrays of high-frequency seismometers deployed at NORESS and ARCESS. The aperture of these arrays is only a few kilometers. NORESS is located in a portion of the much larger array NORSAR, which has a total aperture of about 100 km.

the original NORSAR configuration, with the array being reduced to seven subarrays with an aperture of 50 km in 1976. The primary focus of NORSAR has been monitoring underground nuclear testing in Eurasia, but many other important applications of its data have been made.

A new form of dense array is currently deployed in four locations in Europe. This involves up to 24-element arrays over a 3-km aperture with high-frequency vertical-component sensors and up to four sets of horizontal components. The first was deployed within NORSAR and is called NORESS. Figure 5.18 shows the sensor arrangement at NORESS and a similar array in northern Norway called ARCESS. A third array like this is now deployed in

Finland (FINESSA), and another is located in Germany (GERESS).

Figure 5.19 shows an example of recordings of a nearby quarry explosion at NORESS. The motions are similar across the array, which allows determination of the actual wavefront sweeping across the surface. The array signals are digitally recorded, and computers can automatically determine the direction from which the wave came, estimate the distance to the source, and identify secondary arrivals. This automation is a key advantage of small-array geometries and helps to cope with the vast number of seismic detections provided by these high-quality arrays. For deep Earth structure interpretations, arrays have been of major importance be-

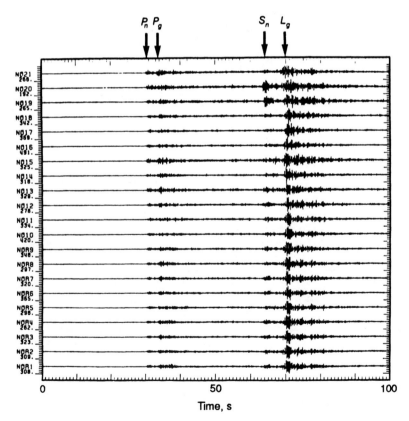

FIGURE 5.19 Example seismograms from the NORESS array for a nearby quarry blast. Individual arrivals, such as the P-wave or S-wave refracted along the top of the mantle (Pn and Sn, respectively), can be timed across the array, enabling direct measurement of apparent velocity $(dT/d\Delta)^{-1}$. The length of time shown is 100 s. (From Mykkelttveit, 1985.)

Box 5.2 Complete Ground Motion Recording

This chapter has focused on seismic instruments designed to record transient ground motions, but we must analyze other important ground motions to understand dynamic processes in the Earth. To address displacements caused by longer-term processes, specialized instruments like LaCoste accelerometers have been used to observe directly gravitational changes associated with mass redistribution, and strain and tilt meters have been developed to detect gradual displacements along faults and on or near volcanoes. Figure 5.B2.1 shows the types of ground motion and corresponding phenomena of interest that can be measured at different frequencies.

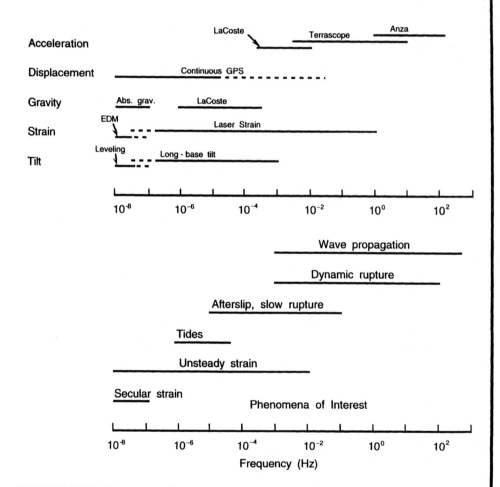

FIGURE 5.B2.1 Various ground-motion measurements and corresponding phenomena that can be studied using a variety of instrumentation. (Courtesy of D. Agnew.)

continues

Because noise processes in mechanical systems make measurement of strain and tilt difficult, strain and tilt meters have a long and interesting historical development. Agnew (1986) summarizes such systems well. The most important recent advances in permanent ground-displacement measurement have involved satellite-based systems using the Global Positioning Satellite (GPS) system or very long baseline interferometry (VLBI), which uses phase shifts between galactic radio signals to measure extremely small lateral displacements. These new instruments, which were extensively developed in the 1980s, allow us to measure directly plate tectonic motions rather than having to infer them from transient earthquake shaking. This will be discussed further in Chapter 11.

cause signals can be summed across the array with correct delay times (stacked) to enhance very small arrivals, and the slope of the travel-time curve of the individual arrivals can be measured directly.

Regional seismic networks designed to monitor small-earthquake activity across the United States began to be extensively deployed in the 1970s and continue to operate today. Figure 5.20 shows the loca-

tions of primarily short-period seismometers whose signals are digitally recorded at various research centers across the country. The networks are densest in regions of active seismicity such as California, Washington, Utah, Missouri, and New England. These also monitor areas of historically significant earthquakes with low current-day activity. The density of stations influences the lower size threshold for events

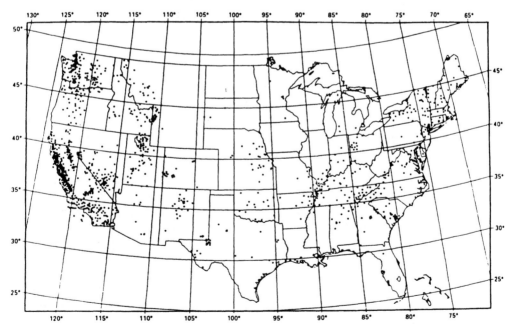

FIGURE 5.20 Regional seismic network stations in the United States, deployed to monitor local earthquake occurrence. The concentrations of stations reflect historical seismicity patterns across the country and the locations of oil fields, nuclear plants, and volcanoes. (From Heaton *et al.*, 1989.)

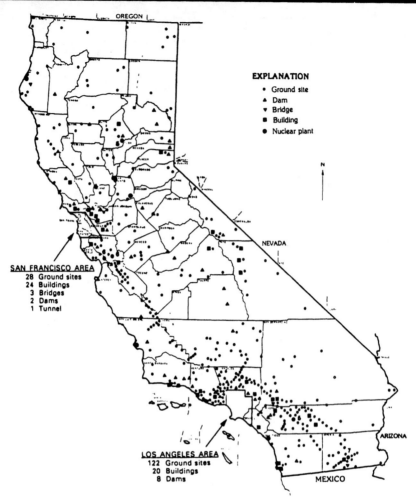

FIGURE 5.21 Locations of strong-motion accelerometers in California as of 1990, including distributions of sensors in major cities. These instruments record ground motions for nearby large events and are distributed near major seismic zones in the state. (From Heaton *et al.*, 1989.)

that can be studied. These regional seismic networks are being upgraded to increasingly sophisticated systems with automated event-location processing, broader-band and three-component recording, and accurate calibration to ground motion.

Instruments with very short natural periods, or accelerometers, are designed to record very strong ground shaking from large earthquakes, which saturates the responses of more standard seismometers like those in the regional networks. Figure 5.21 also shows the distribution of ac-

celerometer locations in California. The distribution closely follows the locations of major faults in the shallow crust, since these instruments are intended to record earthquake strong ground motions. Ground accelerations slightly exceeding 1 g have occasionally been recorded for earthquakes, and the acceleration records have played a major role in developing construction codes for buildings in regions of high earthquake risk. Accelerometers have also been deployed to study strong motions above buried explosions, some-

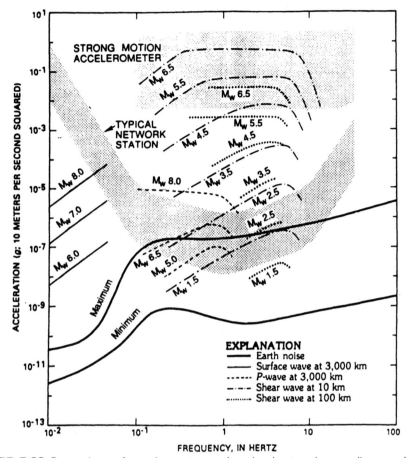

FIGURE 5.22 Comparison of accelerometer and regional-network recording capabilities relative to average noise levels and ground accelerations caused by earthquakes of various sizes at three distances. (From Heaton *et al.*, 1989.)

times recording ground accelerations of 10–30 *g*, which, of course, sends the instrument airborne!

The response characteristics of accelerometers and standard regional-network instruments are compared with expected earthquake accelerations and ground noise in Figure 5.22. Until the development of the very broadband systems currently deployed in the GSN, a spectrum of instruments was required to record the full vast range of ground motions. Several efforts are currently under way to deploy very broadband systems in conjunction with regional networks and accelerometer systems to enable on-scale recording of all local events, including magnitude 8 rup-

tures. The TERRASCOPE network, being deployed in Southern California, will have about 20 GSN-compatible stations complementing the other stations in the region. These provide the most complete recording of ground motions from local earthquakes, and several data examples from TERRASCOPE stations are shown in this text.

Broadband seismic sensors are also being deployed in a new United States National Seismograph Network (USNSN), whose planned distribution is shown in Figure 5.23, along with stations in Alaska, Hawaii, Central America, and the Caribbean. These stations include modern broadband force-feedback seismometers

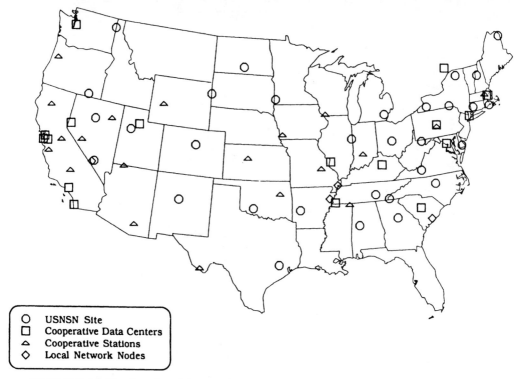

FIGURE 5.23 The broadband United States National Seismic Network (USNSN) in the process of being deployed. The network will augment regional network recording capabilities around the country, providing on-scale recordings of large, regional events. Satellite telemetry is used to transmit the signals to a central data center at the National Earthquake Information Center (NEIC) in Golden, Colorado. (Courtesy of the U.S. Geological Survey.)

with high-dynamic-range recording systems, but they also feature satellite telemetry to a central collection antenna located at the National Earthquake Information Center (NEIC) in Golden, Colorado. The network is designed to locate and analyze earthquakes larger than magnitude 2.5 anywhere in the country, with the broadband, high-dynamic-range systems providing on-scale ground motions even for the largest events.

Finally, perhaps the most flexible form of array involves portable seismographs, which are used in earthquake aftershock studies, refraction surveying, and deep-Earth investigations. From early deployable instruments which produced analog recordings (often with a stylus etching on a kerosene-smoked paper drum) a new computerized generation of portable seismographs has evolved. The IRIS organization has coordinated development of one field system, called PASSCAL instrumentation, which has superb programming flexibility and 24-bit recording capability. Together with new compact broadband sensors, these lightweight systems represent a major new tool for seismology.

References

Agnew, D. C. (1986). Strainmeters and tiltmeters. *Rev. Geophys.* **24**, 579–624.

Given, H. K. (1990). Variations in broadband seismic noise at IRIS/IDA stations in the USSR with implications for event detection. *Bull. Seismol. Soc. Am.* **80**, 2072–2088.

Harges, H.-P., Henge, M., Stork, B., Seidl, C., and Kind, R. (1980). "GRF Array Documentation," Internal Report of Seismologiscles Zentrolobservatorium, Gräfenberg.

○ USNSN Site
□ Cooperative Data Centers
△ Cooperative Stations
◇ Local Network Nodes

Heaton, T. H., Anderson, D. L., Arabasz, W. J., Buland, R., Ellsworth, W. L., Hartzell, S. H., Lay, T., and Spudich, P. (1989). National Seismic System Science Plan. *Geol. Surv. Circ. (U.S.)* **1031**.

Hedlin, M. A. H., Fels, J.-F., Berger, J., Orcutt, J. A., and Lahau, D. (1988). Seismic broadband signal and noise levels on and within the seafloor and on islands. *In* "Proceedings of a Workshop on Broad-band Downhole Seismometers in the Deep Ocean," pp. 185–192. Woods Hole Oceanographic Institution, Woods Hole, MA.

Incorporated Research Institutes for Seismology (1991). "The IRIS Proposal, 1991–1995." IRIS Consortium, Washington, DC.

Kanamori, H. (1988). The importance of historical seismograms for geophysical research. *In* "Historical Seismograms and Earthquakes of the World" (W. K. H. Lee, ed.), pp. 16–33. Academic Press, New York.

Mykkeltveit, S. (1985). A new regional array in Norway: Design work and results from analysis of data from a provisional installation. *In* "The VELA Program" (A. U. Kerr, ed.), pp. 546–553 DARPA, Washington, DC.

Romanowicz, B., Karczewski, J. F., Cara, M., Bernard, P., Borsenberger, J., Cantin, J.-M., Dole, B., Fouassier, D., Koentig, J. C., Morand, M., Pillet, R., Pyrolley, A., and Roulard, D. (1991). The GEOSCOPE Program: Present status and perspectives. *Bull. Seismol. Soc. Am.* **81**, 243–264.

Shearer, P. (1991). Constraints on upper mantle discontinuities from observations of long-period reflected and converted phases. *J. Geophys. Res.* **96**, 18,147–18,182.

Stephen, R. A., Orcutt, J. A., Berleaux, H., Koelsch, D., and Turpening, R. (1988). Low frequency acoustic seismic experiment (LFASE). *In* "Proceedings of a Workshop on Broad-band Downhole Seismometers in the Deep Ocean." pp. 216–218. Woods Hole Oceanographic Institution, Woods Hole, MA.

Additional Reading

Agnew, D. C. (1986). Strainmeters and tiltmeters. *Rev. Geophys.* **24**, 579–624.

Agnew, D. C. (1989). Seismic instrumentation. *In* "The Encyclopedia of Solid Earth Geophsyics" (D.E. James, ed.), pp. 1033–1036. Van Nostrand–Reinhold, New York.

Aki, K., and Richards, P. G. (1980). "Quantitative Seismology: Theory and Methods," Vol. 1, Chapter 10. Freeman, San Francisco.

Båth, M. (1979). "Introduction to Seismology," Chapter 2. Birkhäuser, Berlin.

Bullen, K. E., and Bolt, B. A. (1985). "An Introduction to the Theory of Seismology." Cambridge Univ. Press, Cambridge, UK.

Dewey, J., and Byerly, P. (1969). The early history of seismometry. *Bull. Seismol. Soc. Am.* **59**, 183–227.

Farrell, W. E. (1986). Sensors, systems and arrays: Seismic instrumentation under VELA Uniform. *In* "The VELA Program" (A. U. Kerr, ed.), pp. 465–505. DARPA, Executive Graphics Services, Washington, DC.

Hagiwara, T. (1958). A note on the theory of the electromagnetic seismograph. *Bull. Earthquake Res. Inst., Univ. Tokyo* **36**, 139–164.

Howell, B. F., Jr. (1989). Seismic instrumentation: History. *In* "The Encyclopedia of Solid Earth Geophysics" (D. E. James, ed.), pp. 1037–1044. Van Nostrand–Reinhold, New York.

McCowan, D. W., and La Coss, R. T. (1978). Transfer function for the seismic research observatory system. *Bull. Seismol. Soc. Am.* **68**, 501–512.

Peterson, J., Butler, H. M., Holcomb, L. T., and Hutt, C. R. (1976). The seismic research observatory. *Bull. Seismol. Soc. Am.* **66**, 2049–2068.

Steim, J. M. (1986). The very broadband seismograph, Ph.D. Thesis, Harvard University, Cambridge, MA.

Wielandt, E., and Streckeisen, G. (1982). The leaf spring seismometer: Design and performance. *Bull. Seismol. Soc. Am.* **72**, 2349–2367.

6

SEISMOGRAM INTERPRETATION

In the preceding chapters we have discussed the theory of wave propagation and how ground vibrations are recorded as seismograms. Much of the material in the remaining chapters of this book will deal with inferences extracted from seismograms. Our knowledge of the velocity structure of the Earth and of the various types of seismic sources is the result of *interpreting* seismograms. The more fully we quantify all of the ground motions in a seismogram, the more fully we understand the Earth's structure and its dynamic processes. Seismograms are a complicated mixture of source radiation effects such as the spectral content and relative amplitude of the *P*- and *S*-wave energy that is generated at the source, propagation phenomena such as multiple arrivals produced by reflection and transmission at seismic impedance boundaries or at the surface, and frequency band–limiting effects of the recording instrument. Only experience, and sound foundations in elastic-wave theory, can guide a seismologist to sort out coherent vibrations produced by reflections off deep layers from background noise or from other arrivals scattered by the Earth's three-dimensional heterogeneity. This chapter describes the essence of this procedure, with examples of how sim-

ple measurements lead to important results such as the location of the source. In modern practice many of these procedures are implemented on computers to assist with processing vast quantities of data.

Figure 6.1 shows broadband seismic recordings from a deep earthquake beneath Peru recorded at HRV (the Harvard, Massachusetts, seismic station). Vertical and horizontal ground motions are shown, with the horizontal component oriented transverse to the back azimuth to the source. The P, S, Love (L), and Rayleigh (R) waves are marked, but additional large-amplitude arrivals or phases clearly exist. The keys to identifying these arrivals involve assessing their behavior as a function of distance, measuring the type of ground motion they produce, and establishing their consistency from event to event. These additional arrivals are primarily reflections from velocity discontinuities at depth or from the free surface of the Earth. The timing of the various arrivals is a predictable function of the depth of the source and the distance between the seismic source and receiver. These signals are more complex than those in Figure 1.1 because the source is deep, which allows the surface reflections to be observed. The identification of seismic phases is by no

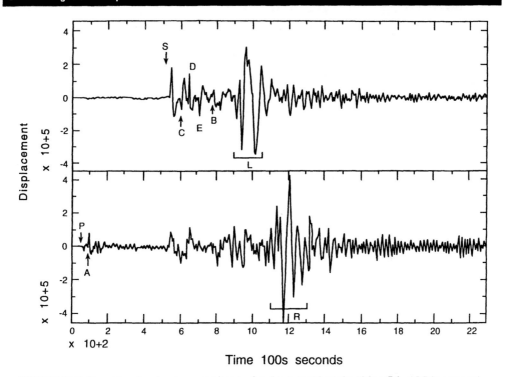

FIGURE 6.1 Broadband seismic recordings of a deep earthquake (May 24, 1991) beneath Peru recorded at HRV (Harvard, Massachusetts). The top and bottom traces are the tangential and vertical components, respectively. *P*, *S*, *R* and *L* are labeled, as are several other phases (*A*, *B*, *C*, *D*, and *E*). *A* is a *P* wave that reflected off the surface above the source (*pP*), *B* is an *S* wave that reflected off the surface halfway between the source and receiver (*SS*), *C* is an *S* wave that reflected off the surface above the source (*sS*), *D* is an *S* wave that reflected off the Earth's core (*ScS*), and *E* is an *S* wave that first reflected off the surface above the source and then off the Earth's core (*sScS*). Additional arrivals include surface and core multiple reflections and scattered surface waves.

means a trivial exercise, and in fact many modern-day seismologists have little direct experience in the routine "reading" or "picking" of seismic-phase travel times and amplitudes. Systematic cataloging of the absolute and differential travel times of all phases on seismograms provides information that we can use to determine the structure of the Earth and to generate travel-time tables that can be used to locate other earthquakes.

Nearly 3000 seismic stations distributed worldwide have been systematically reporting major seismic phase arrival times to the International Seismological Centre (ISC) since 1964. Once direct *P* arrivals at

different stations have been associated with a particular event and that event is located, one can seek to interpret the additional arrivals. The ISC data base has more than 7 million arrival times that have been attributed to more than 25 seismic phases, each with a specific structural interaction, or path, through the Earth. Figure 1.19a shows a large sample of the ISC travel-time picks as a function of epicentral distance. A smaller data set, for particularly well-located events, is shown in Figure 6.2. Clear lineaments exist that represent the travel-time branches of various phases such as direct *P* and *S*, as well as phases that have more complicated travel paths. One

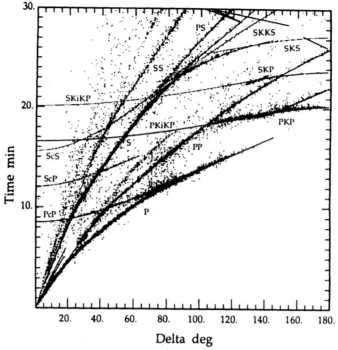

FIGURE 6.2 Six thousand travel times picked from phases of select shallow earthquakes and explosions with known or particularly well-determined locations. Superimposed on the travel times are the interpretation of the phases and the curves showing predicted arrival times based on the iasp91 Earth model. The phases are named using a convention that describes the wave's path through the Earth. For example, *PcP* is the *P* wave reflected from the Earth's core. Some of the arrivals continue to be observed beyond 180°, and they "wrap" around onto this plot. (From Kennett and Engdahl, 1991.)

can view this as the Earth's "fingerprint," uniquely characterizing the complexity imparted into seismic wavefields by the structure. A seismogram at any particular distance will record the corresponding time sequence of arrivals, although source radiation and depth differences may make seismograms at the same distance appear dissimilar. In this chapter we will develop a nomenclature for the various arrivals and some simple rules for identifying seismic phases. The fact that coherent travel-time branches are so pronounced in Figures 1.19 and 6.2 demonstrates the gross radial symmetry of the Earth's layered velocity structure. On the other hand, some of the unidentified arrivals as well as some of the scatter about the mean for any

given branch are manifestations of three-dimensional velocity heterogeneity. Assuming a radially symmetric, layered velocity structure enables us to predict the arrival times of most seismic phases to within a few percent, which provides the basis for most earthquake location procedures. Later, we will discuss several techniques for locating earthquakes, including some that can be adapted to three-dimensional structures.

For many seismic sources, the *P* and *S* waves are radiated from a concentrated volume, which can be approximated as a point source. The coordinates of an earthquake point source are known as the *hypocenter*. The hypocenter is usually given in terms of latitude, longitude, and depth below the surface. The *epicenter* is the surface projection of the hypocenter (the latitude and the longitude), and the *focal depth* is the depth below the surface. *Epicentral distance* is the distance separating the epicenter and the recording seismic station. For large earthquakes, the finiteness of the source volume is not negligible, and then these terms usually refer to the point at which the rupture initiates. Other terms such as the earthquake *centroid* will be introduced later to define the effective point of stress release of the source.

The basic character of seismograms depends strongly on the epicentral distance. At short epicentral distances the character of seismograms is dominated by the details of the highly heterogeneous crustal structure. At large distances, seismograms are dominated by the relatively simple velocity structure of the deep mantle and core. There are four general classifications of seismograms based on epicentral distance: (1) *Local distances* are defined as travel paths of less than 100 km. Seismic recordings at local distances are strongly affected by shallow crustal structure, and relatively simple direct P and S phases are followed by complex reverberations. (2) *Regional distances* are defined as $100 \leq X \leq 1400$ km ($1° \leq \Delta \leq 13°$), where X and Δ are the epicentral distance in kilometers and angular degrees, respectively. Regional-distance seismograms are dominated by seismic energy refracted along or reflected several times from the crust–mantle boundary. The corresponding waveforms tend to be complex because many phases arrive close in time. (3) *Upper-mantle distances* are defined as $13° \leq \Delta \leq 30°$, and seismograms recorded at these distances are dominated by seismic energy that turns in the depth range of 70 to 700 km below the surface. This region of the Earth has a very complex velocity distribution, with a *low-velocity zone* in the upper mantle and at least two major velocity discontinuities (400 and 660 km depths) within what is called the *transition zone*. (We will discuss

the details of these velocity structures in the next chapter.) The direct P and S phases at upper-mantle distances have complex interactions with the discontinuities. (4) *Teleseismic distances* are defined as $\Delta \geq 30°$. The direct P- and S-wave arrivals recorded at teleseismic distances out to $\Delta \approx 95°$ are relatively simple, indicating a smooth velocity distribution below the transition zone, between 700 and 2886 km depth. The simplicity of teleseismic direct phases between 30° and 95° makes them invaluable for studying earthquake sources because few closely spaced arrivals occur that would obscure the source information (Chapter 10). The overall seismogram at these distances is still complex because of the multiplicity of arrivals that traverse the mantle, mainly involving surface and core reflections (Figures 6.1, 1.19). Beyond 95°, the direct phases become complicated once again due to interactions with the Earth's core. Since the character of seismograms depends on the epicentral distance, the nomenclature for phases is also distance dependent.

6.1 Nomenclature

6.1.1 Body-Wave Nomenclature

Seismic-wave energy can travel multiple paths from a source to a receiver at a given distance. For example, as we saw in Chapter 3, energy traveling through a single, flat layer over a high-velocity half-space will result in P and S head waves, direct P and S arrivals, and many reflected arrivals. The reflected arrivals and head waves include energy that initially took off upward from the source before traversing the shallow layer to interact with the half-space. To help sort out the various phases, seismologists have developed a nomenclature to describe each phase in terms of its general raypath.

The simplest and most frequently studied body-wave phases are the direct ar-

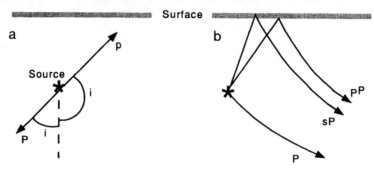

FIGURE 6.3 (a) Geometry of upgoing and downgoing rays. (b) Geometry of depth phases.

rivals. They travel the minimum-time path between source and receiver and are usually just labeled *P* or *S*. At epicentral distances greater than a few tens of kilometers in the Earth, direct arrivals usually leave the source *downward*, or away from the surface, and the increasing velocities at depth eventually refract the wave back to the surface. Figure 6.3a shows two rays leaving a seismic source. The angle, *i*, that the ray makes with a downward vertical axis through the source is known as the *takeoff angle*. If the takeoff angle of a ray is less than 90°, the phase, or that segment of the raypath, is labeled with a capital letter: *P* or *S*. If the seismic ray has a takeoff angle greater than 90°, the ray is *upgoing*, and if it reflects from the surface or is a short upgoing segment of a composite raypath, it is signified by a lowercase letter: *p* or *s*. Upgoing rays that travel from the source up to the free surface, reflect, and travel on to the receiver are known as *depth phases*.

The various portions of the path a ray takes, for example, between the source and the free surface, are known as *legs*. Each leg of a ray is designated with a letter indicating the mode of propagation as a *P* or *S* wave, and the phase is designated by stringing together the names of legs. Thus, there are four possible depth phases that have a single leg from the surface reflection point to the receiver: *pP*, *sS*, *pS*, and *sP* (see Figure 6.3b). The relative timing between the direct arrivals and the depth phases is very sensitive to the depth of the seismic source (hence the name depth phases). Figure 6.4 shows examples of the *pP* depth phase for two events. The *pP* arrivals must arrive later than direct *P* because they traverse a longer path through the Earth, but their relative amplitudes can vary due to the source radiation pattern. The *sP* phase, which always arrives after *pP*, is present but not impulsive in these examples.

At local and regional distances a special nomenclature is used to describe the travel paths. Figure 6.5a shows a very simplified crustal cross section with primary raypaths, and Figure 6.5b shows an actual regional-distance seismogram as it appears for two instrument responses. Note how different the ground motion appears for the different frequency bands. The higher-frequency signal allows ready identification of discrete arrivals, but there is a continuous flux of short-period energy, much of which is scattered in the crust. The direct arrivals at these short distances are usually referred to as P_g and S_g. Depending on the source depth, the velocity gradient within the shallow crust, and the distance between the source and the station, these arrivals may be either upgoing or downgoing phases. The *g* subscript is from early petrological models that divided the crust into two layers: an upper *granitic* layer over a basaltic layer. Arrivals

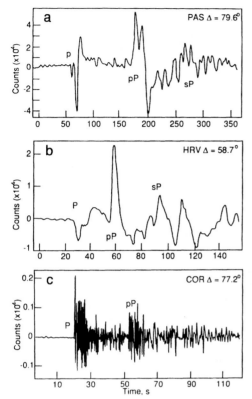

FIGURE 6.4 Examples of depth phases. (a) Broadband recording of a deep earthquake (June 23, 1991, depth =590 km) beneath Sakhalin Island recorded at PAS (Pasadena, California). (b) and (c) Recordings of the Peruvian earthquake shown in Figure 6.1. The middle trace is a broadband recording at HRV (Harvard, Massachusetts); the bottom panel is a simulated short-period recording at COR (Corvallis, Oregon).

that travel as head waves along, or just below, the Moho are known as P_n and S_n. The frequency dependence of these head waves (Chapter 3) tends to make them longer period. Moho reflections are labeled *PmP*, *PmS*, *SmP*, or *SmS*. (Note that each leg of the ray is named, and *m* denotes a reflection at the Moho.) At distances less than about 100 km, P_g is the first arrival. Beyond 100 km (depending on the crustal thickness), P_n becomes the first arrival, as in Figure 6.5. The phase labeled R_g in Figure 6.5 is a short-period Rayleigh

wave, which will be described later. In many regions of the Earth additional regional arrivals are observed that have classically been interpreted as head waves traveling along a midcrustal velocity discontinuity, usually known as the *Conrad discontinuity*. In the next chapter we discuss the Conrad discontinuity further, but here it is sufficient to state that the arrivals associated with the Conrad, called P^* and S^*, respectively, are observed only in certain regions. For example, P^* is very strong in the western United States (see Figure 3.16) but nearly absent in the eastern United States. In older literature P^* is written as *Pb* (*b* denotes the *basaltic* layer).

At distances beyond 13°, P_n amplitudes typically become too small to identify the phase, and the first arrival is a ray that has bottomed in the upper mantle. The standard nomenclature for this arrival is now just *P* or *S*, although subscripts are used to identify different triplication branches for the transition zone arrivals. Seismic phases that reflect at a boundary within the Earth are subscripted with a symbol representing the boundary. For example, *P*-wave energy that travels to the core and reflects is called *PcP*, the *c* indicating reflection at the core. In a spherical Earth it is possible for a ray to travel down through the mantle, return to the surface, reflect, and then repeat the process (Figure 6.6). Because the original ray initially traveled downward, the phase is denoted by a capital letter. The free-surface reflection is not denoted by a symbol; rather, the next leg is just written *P* or *S*. This type of phase is known as a *surface reflection*. Some common surface reflections are *PP*, *PS*, and *PPP*, where *PP* and *PS* each have one surface reflection (involving conversion for *PS*), and *PPP* has two surface reflections. Multiple reflections from both the core and surface occur as well, such as *PcPPcP*, *ScSScS* (*ScS₂*), and *ScSScSScS* (*ScS₃*) (see Figure 6.7). Both reflected phases and surface reflections can be gen-

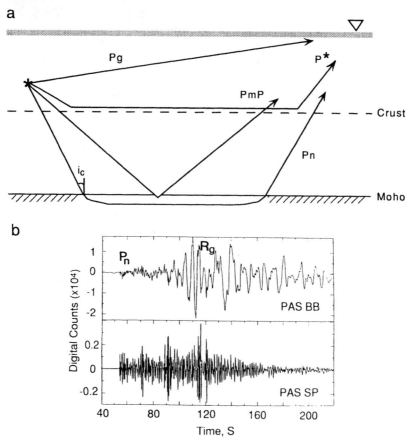

FIGURE 6.5 (a) A simplified cross section of a two-layer crust and corresponding raypaths for various phases observed at regional distances. (b) Broadband and short-period seismograms for an event located 300 km from PAS. The crustal phases are much more apparent at high frequencies. The phases are complex due to multiple travel paths within the crust.

erated by depth phases. In this case the phase notation is preceded by a lowercase *s* or *p*, for example, *pPcP* and *sPP* (Figure 6.7). All of these phases are a natural consequence of the Earth's free surface and its internal layering, combined with the behavior of elastic waves.

The amplitude of body-wave phases varies significantly with epicentral distance. This occurs both because reflection coefficients depend on the angle of incidence on a boundary and because the velocity distribution within the Earth causes focusing or defocusing of energy, depending on the behavior of geometric spreading

along different raypaths. Thus, the fact that a raypath can exist geometrically does not necessarily mean it will produce a measurable arrival. For example, the *P*-wave reflection coefficient for a vertically incident wave on the core is nearly zero (the impedance contrast is small), but at wider angles of incidence the reflection coefficient becomes larger. Thus, *PcP* can have a large amplitude in the distance range $30° < \Delta < 40°$. The surface reflections *PS* and *SP* do not appear at distances of less than 40°, but they may be the largest-amplitude body waves beyond 100°. Progressive energy losses due to at-

Box 6.1 Seismic Waves in the Ocean

In the early 1940s D. Lineham reported a class of seismic waves that were observed only on coastal and island seismic stations. These seismic waves, denoted T waves (*t*ertiary waves, compared to *p*rimary and *s*econdary waves), travel at very low phase velocities and correspond to sound waves trapped in the oceanic water layer. The normal salinity and temperature profile of the ocean conspires to decrease the compressional velocity of seawater from 1.7 km/s at the surface to about 1.5 km/s at a depth of 800–1300 m. Below this depth the velocity increases. This low-velocity channel is known as the SOFAR (sound fixing and ranging channel), and it traps sound waves very efficiently. Sound waves that enter the SOFAR channel can bounce back and forth between the top and the bottom of the channel (beyond critical angle), and since the attenuation of seawater is very low, the energy can travel very long distances, eventually coupling back into solid rock at ocean coastlines. For some shallow volcanic events the observed T waves may be larger than the P and S arrivals by a factor of 5 or more.

The multiply reflected nature of T waves results in a complex wave packet. The T phase does not have a sharp onset and may produce ringing arrivals that last longer than 2 min. They are high-frequency waves (never observed at periods larger than 2 s) and are usually monochromatic. T waves are best observed on ocean-bottom seismometers (OBS), although they are occasionally observed as converted phases at island seismic stations. These converted phases are referred as TPg, TSg, or TRg. Figure 6.B1.1 shows an example T phase. Considerable research has been done on T phases for two reasons: (*1*) submarine noise can generate T phases that have been observed up to 1000 km away, and (*2*) they are a powerful tool for discriminating between underwater nuclear explosions and natural earthquakes. In the case of nuclear explosions, the sound is injected directly into the SOFAR channel and can be 30 times larger than the P or S waves.

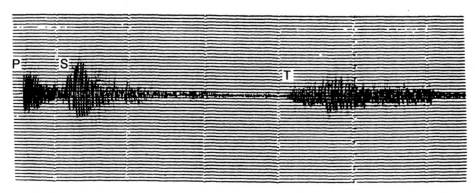

FIGURE 6.B1.1 Short-period recording showing a typical T phase recorded at an island station. (From Kulhánek, 1990).

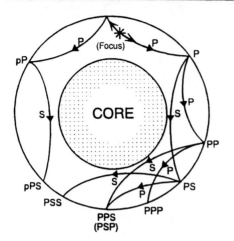

FIGURE 6.6 Raypaths for various surface reflections observed in the Earth. (Modified from Bullen and Bolt, 1985.)

tenuation cause multiple reverberations to become smaller (Figure 6.7). Amplitudes are further complicated by variability of excitation, which depends on the orientation of the seismic source. Figure 6.7 shows a three-component recording with various phases identified, showing how the polarization of ground motion also critically influences the amplitude of individual arrivals.

Direct P waves that travel beyond 95° show rapidly fluctuating, regionally variable amplitudes. Beyond 100° the amplitudes decay rapidly, and short-period energy nearly disappears beyond 103°. Short-period P waves reappear beyond 140° but with a discontinuous travel-time

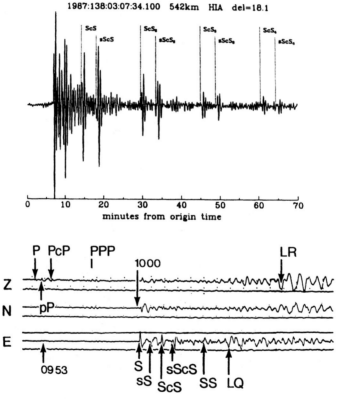

FIGURE 6.7 Examples of seismograms recorded at upper-mantle and teleseismic distances. Multiple S-wave (ScS_n) reflections off the core and free surface ($sScS_n$) are shown at the top, on a long-period transverse-component signal. Note that it takes about 15 min for an S wave to travel down to the core and back. A three-component recording is shown below, with the E–W component being naturally rotated as the transverse component. Note the different observability of phases on each component. (Lower figure from Simon, 1981).

branch (see Figure 6.2). The distance range $103° < \Delta < 140°$ is called the *core shadow zone* and is caused by a dramatic drop in seismic velocities that occurs going from the base of the mantle into the core. Body waves that pass through the core have their own nomenclature. The legs of *P* waves traversing the outer core are denoted by a *K* (from Kernwellen, the German word for core). As discussed in the next chapter, the outer core is a fluid, so only *P* waves can propagate through it. Thus a *P* wave that travels to the core, traverses it, and reemerges as a *P* wave is denoted as *PKP* (or abbreviated *P′*). Similarly, it is possible to have phases *PKS*, *SKS*, and *SKP*. The leg of a *P* wave that traverses the inner core (which is solid) is denoted with an *I* (e.g., *PKIKP*); an *S* wave that traverses the inner core is written as *J* (e.g., *PKJKP*). A reflection from the inner core–outer core boundary is denoted with an *i* (e.g., *PKiKP*). Figure 6.8 shows the raypaths for several different core phases. There is a great proliferation of phase combinations, not all of which will have significant energy.

Since the core–mantle boundary is such a strong reflector, it produces both topside (e.g., *PcP*) and bottomside (e.g., *PKKP*) reflections. *P* waves reflected once off the underside of the boundary are denoted *PKKP*, and other phases include *SKKS*, *SKKP*, and *PKKS*. Paths with multiple underside reflections are identified as *PmKP*, *SmKS*, etc., where *m* gives the number of *K* legs and *m* − 1 gives the number of underside reflections. Seismic arrays have provided observations of *P7KP* (see Figure 7.54). Figure 6.9 shows some examples of core phases. The outer core has little *P*-wave attenuation, so short-period *P* signals can be observed even for phases with long path lengths in the core. Multiple *PKP* branches can be observed at a given distance due to the spherical structure of the core and velocity gradients within it. Chapter 7 will elaborate on this. Note the decrease in amplitude of the *P*, *PcP*, and *PKiKP* phases in Figure 6.9. This results mainly from geometric spreading in the Earth and from weak reflection coefficients at different boundaries for the latter phases.

The reader should be careful not to confuse the multiplicity of seismic arrivals with complexity of the source process or with the existence of more than one initial *P* and one initial *S* spherical wavefront released from the source. First, remember that seismic rays are an artifice for tracking a three-dimensional wavefront and that wave interactions with any boundary or turning point in the Earth have frequency-dependent effects. Interactions with the Earth strongly distort the initial outgoing *P* wavefront, folding it back over on itself and begetting secondary wavefronts as energy partitions at boundaries. The body-wave nomenclature simply keeps track of the geometric complexity involved. The energy that arrives at one station as *P* may arrive at another station as *PP* with additional propagation effects. It is thus

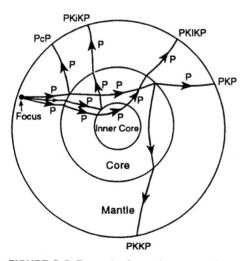

FIGURE 6.8 Raypaths for various core phases. The core–mantle boundary is at a depth of 2886 km, and the inner core–outer core boundary is at a depth of approximately 5150 km. (Modified from Bullen and Bolt, 1985.)

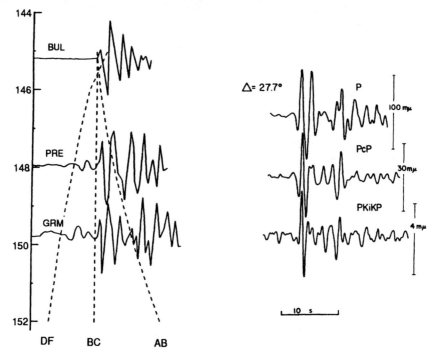

FIGURE 6.9 Examples of core phases, on short-period recordings. *PKP* can have multiple arrivals at a given station because of the geometry of the core. Note the strong amplitude variation between *P*, *PcP*, and *PKiKP*. This is caused by geometric spreading and attenuation along each path. (Left portion courtesy of X. Song; right portion from Engdahl *et al.*, 1974. Reprinted with permission of the Royal Astronomical Society.)

constructive to think of this as a wavefield that has been selectively sampled at different locations as a function of time rather than as discrete energy packets traveling from source to receiver. If we knew the Earth's structure exactly, we could reverse the propagation of the entire wavefield back to the source, successfully reconstructing the initial outgoing wavefront. Of course, sources can also have significant temporal and spatial finiteness, often visualized as subevents, each giving rise to its own full set of wave arrivals that superpose to produce very complex total ground motions. Because of our imperfect knowledge of planetary structure, as described in the next chapter, there are limits to how well we can separate source and propagation effects.

6.1.2 Surface-Wave Nomenclature

The nomenclature for surface waves is far simpler than that for body waves. This, of course, results from the fact that all surface waves travel along the surface, and the complex interference of P and S waves that yields the surface wave is treated collectively rather than as discrete arrivals. Most of the nomenclature for surface waves is related to the frequency band of the observation. At local and regional distances, short-period (< 3 s) fundamental mode Rayleigh waves are labeled R_g. R_g excitation is very dependent on the focal depth; if the source depth is greater than 3 km, R_g is usually absent. R_g propagation depends only on the seismic proper-

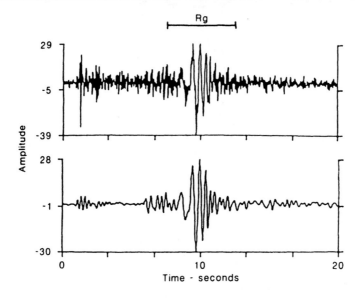

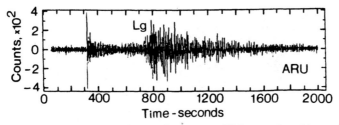

FIGURE 6.10 Examples of R_g and L_g. (a) R_g wave at 39 km produced by a shallow 2000-lb explosion in Maine. The upper trace is the raw seismogram, while the lower trace is low-pass filtered with a cutoff of 4 Hz. (b) Vertical component of a seismogram from an underground nuclear explosion at Lop Nor, China. Epicentral distance is 24°. The L_g phase is a ringy sequence of arrivals with group velocities of 3.6–3.4 km/s. (Part a is from Kafka and Ebel, 1988.)

ties of the upper crust, for which most paths have an average group velocity of about 3 km/s. In most regions, R_g is rapidly attenuated, and it is rare to identify it beyond a few hundred kilometers. High-frequency overtones, or higher-mode Rayleigh waves, as well as some high-frequency Love-wave overtone energy combine to produce a phase called L_g. L_g waves have a typical group velocity of about 3.5 km/s and can be large-amplitude arrivals on all three components of motion (vertical, radial, and transverse) out to 1000

km. L_g phases are the main high-frequency arrival at regional distances in regions of thick continental crust. Figure 6.10 shows examples of R_g and L_g.

In general, Rayleigh waves with periods of 3 to 60 s are denoted R or LR, and Love waves are denoted L or LQ (the Q is for Querwellen, a German word used to describe Love waves). Very long-period surface waves are often called *mantle waves*. The periods of mantle waves exceed 60 s, with corresponding wavelengths of several hundred to about 1200 kilome-

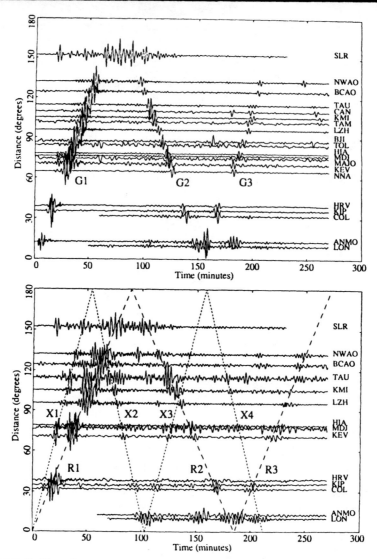

FIGURE 6.11 Profiles of transverse-component (top) and longitudinal-component (bottom) long-period seismograms for the 1989 Loma Prieta earthquake. The corresponding vertical components are shown in Figure 1.7. Great-circle arrivals of Love waves (G_i), Rayleigh waves (R_i), and Rayleigh-wave overtones (X_i) are labeled. (From Velasco *et al.*, 1994.)

ters. Mantle waves from large earthquakes can reappear at a seismic station as they make a complete circuit around the globe on a great-circle path (this was discussed in Section 4.3). Figure 1.7 and Figure 6.11 show profiles of long-period ground motions recorded globally for the 1989 Loma Prieta earthquake. Love waves are polarized such that they are seen on only the horizontal transverse component, whereas Rayleigh waves are seen on both the vertical and horizontal longitudinal components. The Rayleigh waves are labeled R_1, R_2, R_3, etc., indicating wave packets traveling along the minor arc (odd numbers) or major arc (even numbers) of the great circle. R_3 is the same packet of energy as R_1, except it has traveled an additional circuit around the Earth, and R_4 is the next passage of the R_2 wave. Long-period Love waves are labeled G_1, G_2, etc. after Gutenberg. On the radial components of motion additional arrivals between R_n arrivals correspond to higher-mode Rayleigh waves, which have group velocities that differ significantly from those of the fundamental modes. These are labeled variously as O_1, O_2 or X_1, X_2, etc. The overtone wave groups are more sensitive to deeper mantle structure than are fundamental modes of comparable period.

6.2 Travel-Time Curves

Numerous seismologists have compiled large arrival-time data sets like that shown in Figure 6.2. Average fits to the various families of arrivals are known as *travel-time curves* or *charts*. The first widely adopted empirical travel-time curves were published by Sir Harold Jeffreys and Keith Bullen in 1940; the tabular form of these travel-time curves, called *travel-time tables*, is referred to as the *J–B* tables (Jeffreys and Bullen, 1958). These represented painstaking data-collection efforts over the first four decades of the century, using a global array of diverse seismic stations. Careful statistical treatments were used to smooth the data so that meaningful average travel times are given by the tables. One can also use travel-time tables to calculate the ray parameter (the derivative of the travel-time curve) for a particular phase at a given distance and to calculate source depth. The J–B tables are remarkably accurate, and for teleseismic distances they can predict the travel times of principal seismic phases to within a few seconds. For a typical teleseismic *P*-wave travel time of 500 s, the tables are accurate to within a fraction of a percent of the total travel time. The J–B times are less useful at regional and upper-mantle dis-

tances, where strong heterogeneity affects times. Much of the inaccuracy in the travel-time tables comes from uncertainty in the origin time of the earthquake sources that generated the waves. In 1968 Eugene Herrin and colleagues attempted to improve the accuracy by using only well-located earthquakes and underground nuclear explosions. The resulting travel-time curves, known as the *1968 tables* (Herrin *et al.*, 1968), improved the J–B tables slightly at teleseismic distances and more at upper-mantle distances. Kennett and Engdahl (1991) used the complete ISC catalogue of arrival picks to construct the most accurate, radially symmetric travel-time curves yet available, known as *iasp91* (Kennett, 1991). Figure 6.12 shows the iasp91 curve for a 600-km-deep seismic source. The shape of the direct *P*-wave branch in Figure 6.12 is generally consistent with a gradual increase in velocity with depth in the mantle (see Chapter 3). On the scale of the figure, complexity of the *P*-wave branch in the distance range 15°–24° is not clear, but triplications from the transition zone are included; this complexity will be discussed in detail in the next chapter. The later branches are identified by finding paths through the Earth that are consistent with the observed times.

The details of a travel-time curve depend strongly on the depth of the source; seismic sources not at the surface have separate curves for all depth-phase branches (compare Figures 1.19b and 6.12). The depth phases are most dramatically affected, but all the travel times will change. For example, the core shadow onset is at 103° for a surface focus, but it starts at 95° for a 600-km-deep earthquake.

Travel-time curves are a primary tool for interpreting a seismogram and identifying phases. If the location of the source is not independently known, the usual procedure is first to determine an approximate epicentral distance. This usually amounts to picking the *P*-wave arrival time and the

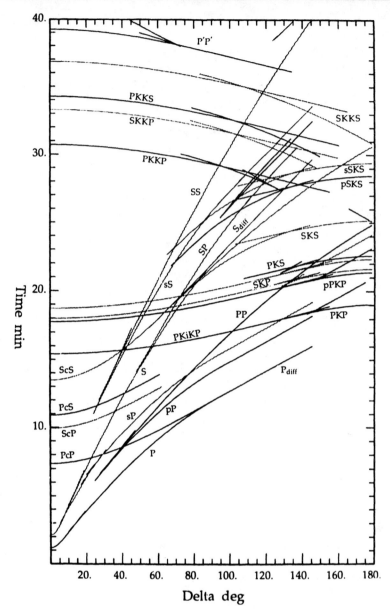

FIGURE 6.12 Travel-time curves for the empirical model iasp91 for a 600-km-deep source. This model prediction indicates the arrival times of the major depth phases. The additional depth phase travel time curves add complexity relative to the surface-focus travel time curve, one of which is shown in Figure 1.19b. Phases that extend beyond 180° have travel-time curves whose times increase to the left. (From Kennett, 1991.)

Box 6.2 Travel-Time Curves Obtained by Stacking Digital Seismograms

The availability of large data sets of digitally recorded seismograms makes it possible to construct "travel-time curves" without actually picking individual phase arrivals. If seismograms of many earthquakes are ordered in distance and plotted as a function of travel time, the corresponding figure is known as a *record section*. The moveout of the various phases in the record section produces coherent lineaments that correspond to travel-time branches. The coherence arises because the high-amplitude phases arrive in a systematic fashion, and therefore seismograms of similar epicentral distance will have a similar character. It is possible to sum together the seismograms of several events or event–station pairs over a small window of epicentral distances (e.g., $1° \pm 0.5°$), thus enhancing coherent signals and diminishing the amount of random noise. This is known as *stacking* a record section. Stacking seismograms directly has several problems; for example, the size of individual phases depends on the size of the event. This means that the stacked section will mostly depend on the largest events. Second, the polarity of various phases depends not only on propagation phenomena such as reflections but also on the orientation of the seismic source. In an attempt to correct for these factors, most stacked record sections actually sum seismograms that have been normalized to a reference phase amplitude, and only the *relative* amplitude of the signal is kept. When these corrected seismograms are stacked, coherent information gives a large-amplitude arrival. The stacked record section provides a travel-time curve that should be devoid of arrival-picking errors or systematic bias in picking procedures. Perhaps the biggest advantage of stacking is that some relative-amplitude information is preserved. Various phases will be strong at certain distances but very small at other distances, and this provides important information about the elastic properties of the Earth.

Peter Shearer (1991) developed stacking procedures for global data sets and investigated the details of the upper-mantle velocity structure. Figure 6.B2.1 shows a stacked record section of 32,376 long-period digital seismograms representing 1474 earthquakes. Comparing Figure 6.B2.1 with Figure 1.19 allows identification of the major travel-time branches (the arrival of the Rayleigh wave is marked by the strongest arrival across the section). Notice how the strength of direct P rapidly diminishes beyond $100°$. The energy that is present is called P_{diff} and represents P waves diffracted along the core surface. Another advantage of using digital data to produce stacked travel-time curves is that the data contain the frequency signature of the various arrivals. If the seismograms are high-pass filtered prior to stacking, only sharp velocity boundaries are imaged. Short-period stacks typically show strong $PKKP$ and $P'P'$ phases; long-period stacks show PPP and SSS, which lack high frequencies due to attenuation.

continues

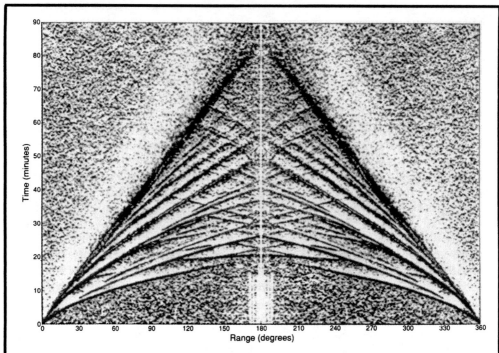

FIGURE 6.B2.1 Earth's travel time curve as defined by stacking long-period seismograms as a function of distance from the source. Shallow earthquakes were used, to avoid complexity from depth phases. (From Shearer, 1991.)

arrival time of either an S wave or Rayleigh wave, then comparing the measured differential time to the travel-time curve. Once the distance is approximated, the travel-time curve predicts times of other arrivals such as PP and ScS, and the consistency of these predicted times with the times of the observed sequence of arrivals ensures reasonably accurate identification of the reference phases. Depth phases are sought as well, with differential times such as $pP - P$ or $sS - S$ used to determine source depth. One will not observe all phases shown on a travel-time curve on any particular seismogram because of source radiation or propagation effects. Another problem results from the extreme frequency dependence of the amplitudes of some phases. A strong short-period arrival (~ 1 s) may be absent at long periods (~ 10 s), or vice versa. Further, all three components of ground motion should be used to interpret phases. For example, SKP is much stronger than PKS on the vertical component because it emerges at the surface as a P wave with a steep incidence angle, causing the ground to move mainly vertically. Even experienced seismologists can misidentify arrivals, given the many possibilities, and it is often necessary to inspect other stations that record the event to establish the travel-time moveout of the phases in question. The ISC often reidentifies phases picked by station operators who do not have accurate location estimates.

Many of the branches of the travel-time curve are related. For example, for a surface focus, the PP travel time can be equated to twice the travel time of P at half the distance: $t_{PP}(\Delta) = 2t_P(\Delta/2)$. Similarly, $t_{SP}(\Delta) = t_P(\Delta_1) + t_S(\Delta_2)$, where $\Delta = \Delta_1 + \Delta_2$, and the ray parameter of the P wave with $t_P(\Delta_1)$ equals that of the S wave with $t_S(\Delta_2)$. These types of simple relationships make it possible to predict the travel times of various phases and to determine a window in which to expect an arrival. These relationships also provide a tool for imaging the deviation of Earth structure from an ideal spherically symmetric velocity structure. Numerous investigators have mapped the differences between the observed and predicted times onto three-dimensional velocity models. These results are discussed more fully in the next chapter.

6.3 Locating Earthquakes

One of the most important tasks in observational seismology is locating seismic sources. This involves determining both the hypocentral coordinates and the source origin time. In general, determining the source location requires identification of seismic phases and measuring their arrival times, as well as knowing the velocity structure between the hypocenter and the seismic station. Given the location of a seismic source, one can calculate the travel time for any particular phase to a seismic station anywhere in an arbitrarily complex velocity model. This type of problem is known as a *forward* model; arrival times are calculated based on a parameterized model. On the other hand, finding the earthquake location is usually posed as an *inverse problem*, where we know the data (the phase arrival times) but must solve for a source location and origin time that are consistent with the data. In this section we will introduce the concept of a *generalized inverse*, perhaps the most critical modern tool for interpreting seismograms as well as for addressing other geophysical problems.

6.3.1 Single Station Locations

In general, the arrival times of various seismic phases at many seismic stations are required to determine an earthquake hypocenter and origin time accurately, but it is possible to use a single seismic station to obtain a crude estimate. Single-station methods require three-component recordings of ground motion. Since P waves are vertically and radially polarized, the vector P wave motion can be used to infer the azimuth to the epicenter. Figure 6.13 displays the nature of P-wave polarization; if the vertical motion of the P wave is upward, the radial component of the P wave is directed away from the epicenter. If the vertical component of the P wave is downward, the radial component is directed back toward the epicenter. Unless the event is at a back azimuth such that the horizontal P wave motion is naturally rotated onto a single component, both horizontal seismometers will record the radial component of the P wave. The ratio of the amplitudes on the two horizontal components can then be used to find the vector projection of the P wave along the azimuth to the seismic source.

The distance to the seismic source is obtained from the difference between the arrival time of two phases, usually P and S. If the earthquake is at local ranges, then the distance can be approximated by

$$D = \frac{t_S - t_P}{\sqrt{3} - 1} \alpha. \qquad (6.1)$$

Equation (6.1) assumes a Poisson solid. For most crustal events, the rule of thumb is $D = (t_S - t_P) \times 8.0$. At larger distances one simply uses the travel-time tables to estimate the distance. Knowing the distance, one can estimate the P travel time and thereby determine the origin time. Comparing differential times between multiple sets of phases with times from the travel-time curves can improve the dis-

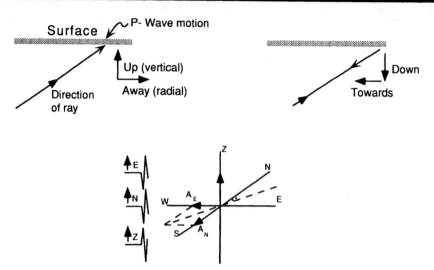

FIGURE 6.13 Procedure for determining the azimuth to the source of a recorded P wave by using the three-component vector ground motion and the fact that P-wave motions are polarized in the vertical and radial plane. (After Båth, 1979).

tance estimation. If clear depth phases are present, one may even reasonably estimate source depth from a single station. This simple procedure for estimating location is not accurate at distances greater than about 20° because the P wave arrives steeply and its horizontal component is too small to give a reliable estimate of the azimuth to the source.

6.3.2 Multiple Station Locations

When several stations are available, an accurate location can be determined by using P and/or S arrival times alone. If the event is at local distances, the two principal phases on the seismogram are P and S. The origin time of the earthquake can be determined with a very simple graphical technique called a Wadati diagram. (K. Wadati used location methods to discover the existence of deep earthquakes, in seismic bands called Wadati–Benioff zones, which were later interpreted as events in subducting slabs.) The time separation of the S and P phase

$(t_S - t_P)$ is plotted against the absolute arrival time of the P wave. Since $t_S - t_P$ goes to zero at the hypocenter, a straight-line fit on the Wadati diagram gives the approximate origin time at the intercept with the P arrival time axis. Figure 6.14 shows an example Wadati diagram. The slope of the trend is $m = (\alpha/\beta - 1)$, which

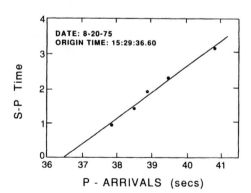

FIGURE 6.14 An example of the Wadati diagram method for determining the origin time of a local earthquake. The origin time is given by the intercept with the P arrival time axis.

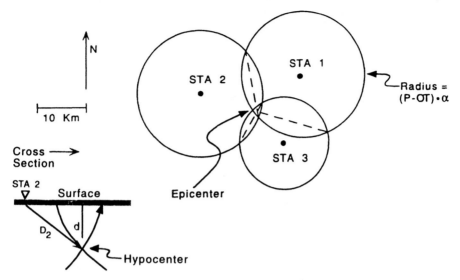

FIGURE 6.15 Circle method for triangulation of a hypocenter.

can be related to Poisson's ratio as follows:

$$\frac{\alpha}{\beta} = \sqrt{\frac{1 - \nu}{\frac{1}{2} - \nu}} \rightarrow \nu = \frac{1 - n/2}{1 - n}, \quad (6.2)$$

where $n = (m + 1)^2$.

Once the origin time (OT) has been estimated, the epicentral distance for the ith station can be estimated by taking the travel time of the P wave and multiplying it by an estimate for the average P velocity

$$D_i = \left(t_p^i - OT\right)\alpha. \quad (6.3)$$

The epicenter must lie on a hemisphere of radius D_i centered on the ith station. In map view this corresponds to a circle of radius D_i. Figure 6.15 shows an example of this method for three stations. Since a single hypocenter must account for all three P-wave arrivals, the hemispheres for all the stations must intersect at a point. The epicenter can be found by drawing the cord of intersecting sections of the circles. The intersection of the cords will give the epicenter. The focal depth, d, can be determined by taking the square root of the difference between the squares of propagation distance, D_i, and the distance along the surface to the epicenter, Δ: $d = (D^2 - \Delta^2)^{1/2}$. Including more observations will give additional intersections that theoretically should pass through the epicenter. In practice, error is always present, both in the data and in the assumptions that raypaths are straight and that the velocity is known perfectly, so scatter in the intersection usually occurs.

This method for determining the hypocenter of an earthquake is called the *method of circles*. For our example we assumed a homogeneous half-space. The method will still work for an inhomogeneous velocity structure as long as it is flat-layered. We can extend the method to a spherical Earth, but we will consider a slight variation that will help us conceptualize the inverse problem. Consider several globally distributed seismic stations that have recorded an earthquake. We need to determine four unknowns: the three coordinates of the hypocenter and the origin time. We can guess a solution for these and calculate expected P-wave

Map View

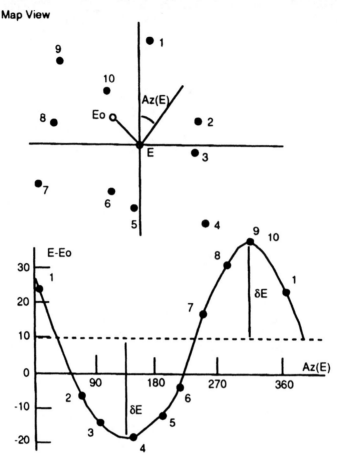

FIGURE 6.16 Adjustments of ''epicentral guesses'' are made by plotting the difference between predicted and observed travel times and epicentral distances. The sine pattern is due to a systematic distance error between the estimated location, *E*, and the actual location, E_0. (From Båth, ''Introduction to Seismology,'' Copyright ©1979. With permission from Birkhäuser Verlag.)

arrival times. If we compare these predictions to the observed times, we can determine how much our guesses are in error. We then correct our guesses and repeat the process until we have obtained acceptably small differences between calculated and observed arrival times.

In Figure 6.16, 10 seismic stations are shown distributed around a presumed epicenter (*E*). The actual epicenter (E_0) lies to the northwest. The predicted arrival times of seismic waves at stations to the northwest of the presumed epicenter will

be later than observed, and, conversely, predictions at stations to the southeast will be earlier than observed. Travel times of waves arriving at seismic stations to the northeast and southwest will not be greatly affected by this particular source mislocation. We can use these relationships to estimate a correction in the presumed epicenter, using a five-step process:

1. Determine a predicted travel time, t_i, and distance, D_i, for each station based on the presumed epicenter.

2. Determine a distance, $\hat{D}_i$, to each station by taking the difference between the observed P arrival time and the presumed origin time, $\hat{t}_i$, and converting this time difference to distance by consulting a travel-time table.

3. Plot the difference, $D_i - \hat{D}_i$, against the calculated azimuth from the presumed epicenter for each station. If the presumed epicenter is exactly right, then $D_i - \hat{D}_i$ will be zero. Otherwise, the variation of $D_i - \hat{D}_i$ with azimuth will be sinusoidal, with the maximum and minimum of the sinusoid aligned along the vector that points from the presumed to the true epicenter.

4. Shift the origin time by an amount equal to the average value of $\hat{t}_i - t_i$, and shift the location by the amplitude of the sine-curve variations, with the direction of shift being along the azimuth of the maximum of the sine curve.

5. Once a new epicenter is found, it can be used as a new starting model, and the five-step procedure is repeated. In general, a single iteration will give epicentral locations accurate to within ± 100 km.

The procedure just described is a series of forward-modeling exercises that generate data that can be compared to observations. When we find a forward model that closely approximates the observations, we declare that the model sufficiently describes the earthquake location for given model assumptions. Mathematically, we can think of this as a series of equations

$$t_{i_{\text{predicted}}} = f(\tilde{x}_i, v) = t_{i_{\text{observed}}}, \quad (6.4)$$

where $\tilde{x}_i$ is the location of the earthquake, v is the velocity structure, and f is a function which calculates the arrival time, $t_{i_{\text{predicted}}}$, given $\tilde{x}_i$ and v. If we have n stations at which we have actually measured arrival times, we can think of $t_{i_{\text{observed}}}$ as the ith component of a data vector $\mathbf{d}$ that has n components, which we write $\mathbf{d} =$

$(t_1, t_2, \ldots, t_n)$. The variable $\tilde{x}_i$ gives the model parameters, and we can also consider it a vector $\mathbf{m}$ that has m components (in general, $m = 4$, for the spatial and temporal coordinates of the earthquake). We can rewrite Eq. (6.4) as a series of equations of the form

$$\mathbf{F(m)} = \mathbf{d}. \quad (6.5)$$

$\mathbf{F}$ is defined as an operator which uses the elements of the model vector to give the data vector. If $\mathbf{F}$ is a series of linear equations (which is not the case for the location problem), then $\mathbf{F}$ is a matrix called the *data kernel*.

6.3.3 The Inverse Problem

If we could rearrange the terms in (6.5) such that we could divide $\mathbf{d}$ by some operator $\mathbf{F}^{-1}$ to give $\mathbf{m}$ directly, we would be solving an inverse problem. To develop how an inverse problem is done, it is instructive to follow the example of earthquake location in a homogeneous material with velocity v. The homogeneous medium gives us simple straight-line raypaths to simplify the algebra. The Cartesian coordinates of the true hypocenter and the ith seismic station are (x, y, z) and (x_i, y_i, z_i), respectively. Let t and t_i be the origin time of the earthquake and the arrival time at the ith station, respectively. Then

$$t_i = t + \frac{\sqrt{(x_i - x)^2 + (y_i - y)^2 + (z_i - z)^2}}{v}.$$

$$(6.6)$$

It is obvious that t_i is an element of the data vector $\mathbf{d}$, and x, y, z, and t are the elements of the model vector $\mathbf{m}$ that we wish to determine. Ideally, only one unique combination of hypocentral parameters fits the observed times. Individual data elements, d_i, are related to the model vector by the right-hand side of (6.6), which we write as

$$F(x, y, z, t) = \mathbf{d}. \qquad (6.7)$$

How can we find $\mathbf{m}$? The equation for $\mathbf{d}$ is nonlinear, which precludes us from obtaining a linear least-squares solution of the equation. The standard procedure is to linearize the problem and iteratively improve guesses of $\mathbf{m}$. The first step is to guess a solution $\mathbf{m}^0$ for which the predicted times, $\mathbf{d}^0$, can be calculated and investigate the behavior of $\mathbf{d}^0$ in the neighborhood of $\mathbf{m}^0$, just as in the last section. We approximate changes in $\mathbf{m}^0$ with a Taylor series approximation

$$m_j^1 = m_j^0 + \delta m_j^0, \qquad (6.8)$$

where δm_j^0 is an incremental variation of the jth model parameter that moves the model toward a better fit to the data. For our example problem this amounts to guessing a solution (x_0, y_0, z_0, t_0) and then determining incremental changes in that guess: $\delta x_0 = (x_1 - x_0)$, $\delta y_0 = (y_1 - y_0)$, $\delta z_0 = (z_1 - z_0)$, and $\delta t_0 = (t_1 - t_0)$. Subscripts correspond to the iteration number in the procedure. The corresponding change in the predicted data vector can be found by expanding (6.8) in a Taylor series about $\mathbf{m}^0 + \delta \mathbf{m}^0$:

$$\left(\frac{\partial F}{\partial x_0}\right)\delta x_0 + \left(\frac{\partial F}{\partial y_0}\right)\delta y_0 + \left(\frac{\partial F}{\partial z_0}\right)\delta z_0 + \left(\frac{\partial F}{\partial t_0}\right)\delta t_0$$

$$= d_i - F_i^0(x_0, y_0, z_0, t_0). \qquad (6.9)$$

Examination of (6.9) shows that the difference in the observed and predicted travel times [the right-hand side of (6.9)] is now linearly related to *changes* we require in the hypocentral coordinates to make the model better predict the data. Using only the first term of a truncated Taylor series provides the linearization, but this also precludes the perturbations from immediately converging to the true $\mathbf{m}$. The derivatives are evaluated at the guessed solution,

m_j^0. Substituting (6.7) into (6.9) and rewriting gives

$$\delta d_i = \frac{\partial d_i}{\partial m_j}\delta m_j \qquad (6.10)$$

or defining $\partial d_i / \partial m_j$ as a partial derivative matrix

$$G_{ij} = \frac{\partial d_i}{\partial m_j}, \qquad (6.11)$$

we can write a system of equations that maps changes in model parameters onto improvements in the fit to the data:

$$\Delta \mathbf{d} = \mathbf{G}\,\Delta \mathbf{m}. \qquad (6.12)$$

It is standard notation to drop the Δ notation and write (6.12) as $\mathbf{d} = \mathbf{Gm}$. Be cautious, as this form also holds for a purely linear problem, not just our linearized version. From this point forward in this discussion, the data and model vectors are understood to be vectors of *changes* in model and data space.

Returning to Eq. (6.9), we have a system of equations with four unknowns that are multiplied by constant coefficients. ($\partial F_i / \partial m_j$ are functions evaluated at $\mathbf{m}^0$ and, assuming they exist, are constants.) If there are four observed arrival times, we have four equations and can solve the system by Gaussian elimination, giving either no solution or an exact result for δm_j^0. Any errors in the data will lead to an incorrect solution, or inconsistent equations. Once δx, δy, δz, and δt are calculated, we can "correct" the source parameter guesses:

$$x_1 = x_0 + \delta x_0, \qquad y_1 = y_0 + \delta y_0,$$

$$z_1 = z_0 + \delta z_0, \qquad t_1 = t_0 + \delta t_0. \quad (6.13)$$

This new guess (x_1, y_1, z_1, t_1) is now used to repeat the entire process [Eqs. (6.8) to (6.13)] to estimate a refined model (x_2, y_2, z_2, t_2). This iterative process is continued until the $\Delta \mathbf{d}$ becomes acceptably small.

This procedure is sometimes called Geiger's method. Unfortunately, the rate at which it converges depends strongly on the accuracy of the initial guess (called the *starting model*). Further, this process does not guarantee convergence.

6.4 Generalized Inverse

In the last section we developed a system of simultaneous equations that related the changes of model parameters to improvements in the fit to the data. Equation (6.12) is valid for any problem in seismology for which we have a set of measurements that depends on a set of model parameters. A whole branch of mathematics has been developed to study the solution of such systems, known as *inverse theory*. The details of inversion are beyond the scope of this text, but we will develop some basic formulations because they are important for later chapters. For more detailed geophysical inversion theory, see Menke (1989) and Tarantolla (1987).

Equation (6.12) relates a data vector of dimension n (number of observations) to a model vector of dimension m (number of model parameters). In general, most earthquake location problems are overdetermined: there are more observations than unknowns ($n > m$). For Earth structure inversions (discussed in Chapter 7), the continuous functions of material properties are approximated by a finite model simplified to ensure $n > m$. For ($n > m$) the matrix G is not square (i.e., the matrix has more rows than columns), which we will consider later. If G is square, which implies we have a system of $n = m$ equations and $n = m$ unknowns, we could simply multiply both sides of the equation by G^{-1}, the *inverse* of G, assuming it exists. By definition, $G^{-1}G = I$, where I is the *identity matrix* (see Box 6.3), so a new system of equations is formed:

$$G^{-1}d = m. \qquad (6.14)$$

Thus we could solve for m directly. This is a simple equation and is easy to solve. Unfortunately, we never have this case in seismology. We are dealing with data that have errors, such as those associated with picking the arrival times. Similarly, Eq. (6.14) assumes that we can *perfectly* predict the data. In the case of travel times, this means that we must know the velocity structure between the source and receiver extremely well. In fact, we usually do not; thus, we are dealing with *inconsistent equations*, making it impossible to use (6.14). Despite these problems, all is not lost, because if we measure many data, we can find an *overdetermined* solution, which is the *best* model fit to an "average" of the data.

Before we discuss overdetermined solutions, let us return to Eq. (6.12), where G is a square ($n = m$) $n \times n$ matrix, and develop some definitions. In this equation we can think of G as an operator which maps the model parameters into predicted data vectors. In other words, G *transforms* an *n-dimensional* vector into another *n-dimensional* vector. This is analogous to a coordinate transformation with position vectors. We can extend the analogy to introduce the concept of *eigenvalue* problems.

The eigenvalue problem can be defined by transforming the model vector onto a parallel vector:

$$Gd = \lambda d. \qquad (6.15)$$

Physically, this means that we want to find the set of data vectors that, when operated on by G, returns vectors that point in the same direction, with a length scaled by the constant λ. We can use this to define the homogeneous equation

$$[G - \lambda]d = O_n, \qquad (6.16)$$

where O_n is an $n \times 1$ vector of zeros. This system of homogeneous equations has a nontrivial solution if, and only if, the

Box 6.3 Linear Algebra and Matrix Operations

The manipulation of matrices is extremely important in seismology, especially in inverse problems. Operations on matrices, such as addition, multiplication, and division, are known as *linear algebra*. A review of the basics of linear algebra is in order to understand Section 6.4 on the generalized inverse. In general, a matrix is described by the number of rows (M) and columns (N), or an $M \times N$ matrix. Consider such a matrix **A**:

$$
\mathbf{A} = \begin{bmatrix} a_{11} & a_{12} & \cdots & a_{1N} \\ a_{21} & a_{22} & \cdots & a_{2N} \\ \vdots & \vdots & \ddots & \vdots \\ a_{M1} & a_{M2} & \cdots & a_{MN} \end{bmatrix}. \tag{6.4.1}
$$

A second matrix, **B**, can be added to **A** if it has *exactly* the same dimensions:

$$
\mathbf{A} + \mathbf{B} = \begin{bmatrix} a_{11} + b_{11} & a_{12} + b_{12} & \cdots & a_{1N} + b_{1N} \\ a_{21} + b_{21} & a_{22} + b_{22} & \cdots & a_{2N} + b_{2N} \\ \vdots & \vdots & \ddots & \vdots \\ a_{M1} + b_{M1} & a_{M2} + b_{M2} & \cdots & a_{MN} + b_{MN} \end{bmatrix}. \tag{6.4.2}
$$

Matrix multiplication is defined by the following:

$$
\mathbf{AB} = \begin{bmatrix} a_{11} & a_{12} & \cdots & a_{1N} \\ a_{21} & a_{22} & \cdots & a_{2N} \\ \vdots & \vdots & \ddots & \vdots \\ a_{M1} & a_{M2} & \cdots & a_{MN} \end{bmatrix} \begin{bmatrix} b_{11} & b_{12} & \cdots & b_{1P} \\ b_{21} & b_{22} & \cdots & b_{2P} \\ \vdots & \vdots & \ddots & \vdots \\ b_{N1} & b_{N2} & \cdots & b_{NP} \end{bmatrix}
$$

$$
= \begin{bmatrix} c_{11} & c_{12} & \cdots & c_{1P} \\ c_{21} & c_{22} & \cdots & c_{2P} \\ \vdots & \vdots & \ddots & \vdots \\ c_{M1} & c_{M2} & \cdots & c_{MP} \end{bmatrix}, \tag{6.4.3}
$$

where $c_{ij} = a_{ik} b_{kj}$. Note that when we write out the dimensions of **A** and **B**, we see $M \times N$ and $N \times P$, respectively. Multiplication is possible *only* when the number of rows of matrix **B** matches the number of columns of matrix **A**. The resulting matrix **C** has M rows and P columns. In general, $\mathbf{AB} \neq \mathbf{BA}$. In fact, if **AB** exists, there is no reason to believe that **BA** exists.

continues

Another very important matrix operation is called the *transpose*. The transpose of **A** is formed by taking the columns and turning them into rows. In other words,

$$\mathbf{A}^{\mathrm{T}} = \begin{bmatrix} a_{11} & a_{21} & \cdots & a_{M1} \\ a_{12} & a_{22} & \cdots & a_{M2} \\ \vdots & \vdots & \ddots & \vdots \\ a_{1N} & a_{2N} & \cdots & a_{MN} \end{bmatrix}. \tag{6.4.4}$$

Thus, $\mathbf{A}^{\mathrm{T}}$ is an $N \times M$ matrix. The product $\mathbf{A}\mathbf{A}^{\mathrm{T}}$ always exists and is a square $M \times M$ matrix. If the matrix is *symmetric* ($a_{ij} = a_{ji}$), then $\mathbf{A}^{\mathrm{T}} = \mathbf{A}$. This is, of course, the case for both stress and strain.

The *determinant* of a matrix is a measure of its "size." It is written as $|\mathbf{A}|$ and is defined for *square* matrices as a complex product of entries from each column. For example, if **A** is a 3×3 matrix, then

$$|\mathbf{A}| = \begin{vmatrix} a_{11} & a_{12} & a_{13} \\ a_{21} & a_{22} & a_{23} \\ a_{31} & a_{32} & a_{33} \end{vmatrix} = a_{11}a_{22}a_{33} + a_{12}a_{23}a_{31} + a_{13}a_{21}a_{32}$$

$$- a_{13}a_{22}a_{31} - a_{12}a_{21}a_{33} - a_{11}a_{23}a_{32}. \tag{6.4.5}$$

In general, the determinant can be written

$$|\mathbf{A}| = \sum (\pm) a_{1j_1} a_{2j_2} \cdots a_{nj_n}, \tag{6.4.6}$$

summed over all permutations of $j_1, j_2, \ldots, j_n$. If the order of j_i is an *even* permutation, then the $+$ sign is used; if the order of j_i is an *odd* permutation, then the $-$ sign is used. Note that this requires $N!$ terms. The determinant is a very important concept for solving systems of equations, but it obviously can be very tedious to calculate. Fortunately, many properties of determinants greatly reduce the difficulty of computation. Detailed discussion of these procedures is beyond the scope of this seismology text, so we refer the reader to a linear algebra text such as *Elementary Linear Algebra* by Kolman.

Some of the useful facts and properties of determinants are (1) if two rows or columns in **A** are identical, $|\mathbf{A}| = 0$, and (2) if two rows or columns of **A** are interchanged, $|\mathbf{A}|$ remains the same absolute value, but its sign changes. Most important, $|\mathbf{A}| = |\mathbf{A}^{\mathrm{T}}|$.

The *inverse* of **A** is defined such that

$$\mathbf{A}\mathbf{A}^{-1} = \mathbf{A}^{-1}\mathbf{A} = \mathbf{I}, \tag{6.4.7}$$

where **I** is called the identity matrix, which is a square matrix whose elements are all zero except those along the diagonal, where the values are 1. The inverse matrix can be found by the equation

$$\mathbf{A}^{-1} = (1/|\mathbf{A}|)(\mathrm{adj}\, A), \tag{6.4.8}$$

continues

> where adj **A** is called the adjoint of **A**. The adjoint is determined from the cofactors of **A**, which are written
>
> $$\text{adj } \mathbf{A} = \mathbf{B} \qquad b_{ij} = (-)^{i+j} |\mathbf{M}_{ij}| \tag{6.4.9}$$
>
> where $\mathbf{M}_{ij}$ are matrices derived from **A** with row i and column j *removed*.

determinant of the system equals 0:

$$|\mathbf{G} - \lambda \mathbf{I}|$$

$$= \begin{vmatrix} g_{11} - \lambda & g_{12} & g_{13} & \cdots & g_{1n} \\ g_{21} & g_{22} - \lambda & g_{23} & \cdots & g_{2n} \\ \vdots & \vdots & & & \vdots \\ g_{n1} & g_{n2} & g_{n3} & \cdots & g_{nn} - \lambda \end{vmatrix} = 0. \tag{6.17}$$

The determinant is given by the polynomial of order n in λ, called the *characteristic polynomial*

$$\lambda^n + a_{n-1}\lambda^{n-1} + a_{n-2}\lambda^{n-2} + \cdots + a_0 = 0. \tag{6.18}$$

The roots of this equation are called *eigenvalues* of **G**. Thus there are n eigenvalues $(\lambda_1, \lambda_2, \ldots, \lambda_n)$, each of which can be used to find a solution of the system of homogeneous equations. The solution can be tabulated as follows:

$$\lambda = \lambda_1 \qquad \mathbf{u}_1 = \left[u_1^1, u_2^1 \cdots u_n^1 \right]^{\mathrm{T}}$$

$$= \lambda_2 \qquad \mathbf{u}_2 = \left[u_1^2, u_2^2 \cdots u_n^2 \right]^{\mathrm{T}}$$

$$\vdots \qquad \qquad \vdots \qquad \qquad \vdots$$

$$= \lambda_n \qquad \mathbf{u}_n = \left[u_1^n, u_2^n \cdots u_n^n \right]^{\mathrm{T}}. \tag{6.19}$$

The solutions **u** represent n distinct vectors called the *eigenvectors* of **G**.

The eigenvalues and eigenvectors are used to define two matrices:

$$\mathbf{\Lambda} = \begin{bmatrix} \lambda_1 & 0 & 0 & \cdots & 0 \\ 0 & \lambda_2 & 0 & \cdots & 0 \\ \vdots & \vdots & \ddots & & \vdots \\ 0 & 0 & 0 & \cdots & \lambda_n \end{bmatrix} \tag{6.20}$$

$$\mathbf{U} = \begin{bmatrix} u_1^1 & u_1^2 & \cdots & u_1^n \\ u_2^1 & u_2^2 & \cdots & u_2^n \\ \vdots & \vdots & & \vdots \\ u_n^1 & u_n^2 & \cdots & u_n^n \end{bmatrix}. \tag{6.21}$$

Thus, $\mathbf{\Lambda}$ is a diagonal matrix with the eigenvalues of **G**, and **U** is an $n \times n$ matrix with the eigenvectors of **G** in the columns. The eigenvectors define a new coordinate system, equally valid for describing the solution to our original system of equations. The advantage of the new coordinate system is mainly in the computational simplicity of determining the generalized inverse, but some physical insight can be gleaned by considering the role of **G**. As we stated before, **G** transforms, or links, the model parameters (or *model space*) to predicted data (or *data space*). A change in one model parameter will affect certain elements of data space. For example, in the earthquake location problem, a change in depth will affect the travel times at all observing stations. In the eigenvector coordinate system, the original model parameters are mapped to a new model space; likewise for data space. In this coordinate system, a change in *one* model parameter will affect only *one* element of the data vector. This one-to-one linkage is extremely valuable when one tries to understand which combination of model parameters is best resolved.

In Eq. (6.18) we assume that the characteristic polynomial has n roots. In general, these roots may be complex, but for the types of problems we are interested in solving in seismology, all the roots are real. However, we have no guarantee that the roots will not be zero or repeated, problems which must be addressed. If an eigenvalue is zero, this implies that one "coordinate axis" in the transform space does not exist, and only $n-1$ model parameters are resolvable. If an eigenvalue is repeated, we refer to the eigenvalues as *degenerate*. In this case, the eigenvalue does not correspond to a single eigenvector but rather a plane defined by the two eigenvectors. In this case, the *uniqueness* of the eigenvector vanishes; any two vectors in the plane will describe the new model space.

One application of eigenvectors is to find the *principal coordinate system*. This amounts to the diagonalization of G in our example. From (6.15) we can write

$$GU = U\Lambda. \tag{6.22}$$

Multiplication by the inverse of U yields

$$U^{-1}GU = \Lambda, \tag{6.23}$$

with Λ being the diagonalized matrix (6.20). As an example, consider the case where G is an arbitrary stress tensor. Then the eigenvalues are the magnitude of the principle stresses, and the eigenvectors give the orientation of coordinate axes for the principal stress system.

We can use eigenvalue analysis to find a formulation of G^{-1}. We start by performing a similar eigenvalue analysis on G^T (the transpose of G) to define a matrix of eigenvectors V. The eigenvalues of G^T are identical to those of G (the value of the determinant does not change if the columns and rows are interchanged). Thus

$$G^T V = V\Lambda. \tag{6.24}$$

We can manipulate Eq. (6.24) by taking the transpose of both sides:

$$V^T G = \Lambda V^T, \tag{6.25}$$

multiplying both sides by U:

$$V^T G U = \Lambda V^T U. \tag{6.26}$$

Now we can use Eq. (6.22) to write (6.26) as

$$V^T U \Lambda = \Lambda V^T U. \tag{6.27}$$

This is an important equation and is the basis of a technique called *singular value decomposition*. The only way in which (6.27) can be true is for $V^T U = I$. This implies that V^T and U are orthogonal. Using our coordinate transformation analogy, this orthogonality means the coordinate axes (eigenvectors) are orthogonal. Recall that we defined the inverse matrix as $U^{-1}U = I$, which leads to several relations between the eigenvectors of G and G^T:

$$V^T = U^{-1}$$

$$U = (V^T)^{-1}$$

$$V = (U^T)^{-1}$$

$$U^T = V^{-1}$$

$$U^T U = V^T V = I. \tag{6.28}$$

If we substitute $V^T = U^{-1}$ in Eq. (6.25) and multiply by U, we obtain what is known as the singular value decomposition of G:

$$G = U\Lambda V^T = U\Lambda U^{-1} = U\Lambda U^T. \tag{6.29}$$

A similar equation can be written for $\mathbf{G}^{-1}$:

$$\mathbf{G}^{-1} = \mathbf{U}\boldsymbol{\Lambda}^{-1}\mathbf{V}^{\mathrm{T}}, \qquad (6.30)$$

where $\boldsymbol{\Lambda}^{-1}$ is given by

$$\boldsymbol{\Lambda}^{-1} = \begin{bmatrix} 1/\lambda_1 & 0 & 0 & \cdots & 0 \\ 0 & 1/\lambda_2 & 0 & \cdots & 0 \\ 0 & 0 & \ddots & & \vdots \\ \vdots & \vdots & & & 0 \\ 0 & 0 & \cdots & 0 & 1/\lambda_n \end{bmatrix}. \qquad (6.31)$$

These manipulations are a rather complex approach to inverting a square matrix, although insight can always be gained from considering the eigenvector and eigenvalue structure of a problem. However, for most problems in geophysics, we are involved with situations with many more data than model parameters. If the seismic data and our models for predicting observations were perfect, then an overdetermined problem would be *redundant*. For a redundant set of equations, some eigenvalues will be zero. In this case $\mathbf{G}^{-1}$ does not exist, but you could eliminate the redundancy and still find a complete solution. If the data and model contain noise and errors, perfect redundancy will not exist, and the inverse problem can be formulated as an overdetermined problem. In this case, $\mathbf{G}$ is not a square matrix and we cannot use (6.14) or (6.31) directly, so we must further manipulate our basic formulation of the problem (6.12).

An example of a overdetermined problem is an unknown earthquake location with more than four arrival times. No location will *perfectly* predict all the arrival times, so we seek a location which will provide the *best* prediction. The best fit.is usually defined as the model with the smallest *residual*, or difference between observed and predicted data. From Eq. (6.12), we can write an equation which measures the *misfit* of the model:

$$\mathbf{E} = [\mathbf{d} - \mathbf{Gm}]. \qquad (6.32)$$

If the model exactly fit all the data, then $\mathbf{E}$ would be a vector of dimension n with all elements equal to zero. Since this will not usually be the case, the inverse problem is designed to find a model that minimizes $\mathbf{E}$. One of the most common ways to do this is to write an equation for the *squared error*

$$E^2 = \sum_{i=1}^{n} \left(d_i - \sum_{j=1}^{m} G_{ij}m_j \right)^2 \qquad (6.33)$$

and force E^2 to be a minimum. We do this by taking the derivative of E^2 with respect to the model parameters and setting it equal to zero:

$$\frac{\partial E^2}{\partial m_k} = 2E\frac{\partial E}{\partial m_k}$$

$$= -2\sum_{i=1}^{n} \left(d_i - \sum_{j=1}^{m} G_{ij}m_j \right)G_{ik} = 0 \qquad (6.34)$$

or collecting terms,

$$\sum_{i=1}^{n} d_i G_{ik} = \sum_{i=1}^{n} \left(\sum_{j=1}^{m} G_{ij}m_j \right)G_{ik}, \qquad (6.35)$$

which can be rewritten in matrix notation as

$$\mathbf{G}^{\mathrm{T}}\mathbf{d} = \mathbf{G}^{\mathrm{T}}\mathbf{Gm}, \qquad (6.36)$$

a very useful form called the *normal equations*. $\mathbf{G}^{\mathrm{T}}\mathbf{G}$ is now a square matrix, so it has an inverse (as long as it is not singular!). Further, $\mathbf{G}^{\mathrm{T}}\mathbf{G}$ is symmetric, which means that its eigenvalues are real and nonnegative. Therefore we can write an equation of the form

$$\mathbf{m} = [\mathbf{G}^{\mathrm{T}}\mathbf{G}]^{-1}\mathbf{G}^{\mathrm{T}}\mathbf{d}, \qquad (6.37)$$

where $[\mathbf{G}^{\mathrm{T}}\mathbf{G}]^{-1}\mathbf{G}^{\mathrm{T}} = \mathbf{G}^{-\mathrm{g}}$ is called the *generalized inverse* of $\mathbf{G}$. (Strictly speaking, this is called the *least-squares inverse*; if G^TG is nonsingular, then it is the generalized inverse.) Equation (6.37) provides the best solution to **m** in a least-squares sense (the squared error is minimized). Equation (6.37) (and its modifications) is one of the most important equations in geophysics for both linear and nonlinear problems. In general, most problems that are posed are not linear. Thus, **m** will only serve as a correction to a starting model, and we must repeat the inverse process, with **G** updated for the new model.

We can solve (6.37) by the method of singular value decomposition. The non-square matrix **G** has dimensions $n \times m$, so G^TG is of dimension $m \times m$, and GG^T is of dimension $n \times n$. Since these matrices are square, we can use (6.29) to solve for the inverse of G^TG, in terms of eigenvector matrices **V** and **U** for G^TG and $(G^TG)^T = GG^T$, respectively. From (6.29) we can write

$$G^TG = \left[V\Lambda_{m(2)}V^T \right], \qquad (6.38)$$

where **V** is $m \times m$, and the eigenvalue matrix $\Lambda_{m(2)}$ is $m \times m$. The eigenvalues are actually just the squared values of the eigenvalues of **G** itself, so we denote them with the subscript (2). This corresponds to the singular-value decomposition of G^TG. One singular value exists for each model parameter, although these are not guaranteed to be nonzero. The matrix **V** contains the eigenvectors associated with these singular values, and we say that it "spans the model space." We can write a similar decomposition for GG^T:

$$GG^T = U\Lambda_{n(2)}U^T, \qquad (6.39)$$

where **U** is now an $n \times n$ matrix of eigenvectors "spanning the data space." Note that GG^T has dimensions of $n \times n$, with $n > m$, but only up to m eigenvalues are nonzero, and they are the same as those in (6.38). The extra $n - m$ rows and columns of Λ_n are just zeros, although the corresponding eigenvectors need not be zero.

In Eq. (6.37) we need $[G^TG]^{-1}$. Our eigenvalue formulation allows us to carry out the necessary inverse directly:

$$[G^TG]^{-1} = [V^T]^{-1}[\Lambda_{m(2)}]^{-1}V^{-1}. \quad (6.40)$$

The inverse of V^T is just **V**, and similarly, V^{-1} is V^T. Thus, we can rewrite Eq. (6.40) as

$$[G^TG]^{-1} = V[\Lambda_{m,(2)}]^{-1}V^T. \quad (6.41)$$

Now we can use (6.41) to write the generalized inverse, $[G^TG]^{-1}G^T = G^{-g}$. First, consider G^T. While this is a nonsquare matrix of dimension $m \times n$, we can decompose it using (6.29), if we consider only the up to m eigenvectors in the matrix Λ_m. This allows us to use (6.29) as

$$G = U\Lambda_m V^T \qquad (6.42)$$

$$G^T = V\Lambda_m U^T. \qquad (6.43)$$

Combining (6.41) and (6.43), recognizing that the eigenvector matrices are the same for G^TG and G^T, we have

$$G^{-g} = [G^TG]^{-1}G^T$$

$$= \left\{ V[\Lambda_{m,(2)}]^{-1}V^T \right\}\left\{ V\Lambda_m U^T \right\} \quad (6.44)$$

$$= V[\Lambda_{m,(2)}]^{-1}\Lambda_m U^T. \quad (6.45)$$

Recalling that the eigenvalues in $\Lambda_{m(2)}$ are simply the square of those in Λ_m, we find

$$G^{-g} = [G^TG]^{-1}G^T = V\Lambda_m^{-1}U^T, \quad (6.46)$$

where the eigenvector matrix is of the form of (6.31). The squares of the eigenvalues in (6.46) are called the singular values (most algorithms will just compute the inverse of G^TG) and these are arranged such that $\lambda_1 \geq \lambda_2 \geq \cdots \lambda_m \geq 0$.

6.4.1 Errors, Redundant Data, and Resolution

In the previous discussion we assumed that all the data were independent and that each model parameter affected all the data. For the earthquake location problem this assumption is valid, but for many other geophysical inverse problems this will not be the case. In the latter situation, $G^T G$ will be singular, with at least one eigenvalue equal to zero. In this case the form of Eq. (6.46) is slightly modified. The matrices V, Λ, and U are redefined as systems of p nonzero eigenvalues:

$$\Lambda_p = \begin{bmatrix} \lambda_1 & 0 & 0 & \cdots & 0 \\ 0 & \lambda_2 & 0 & \cdots & 0 \\ \vdots & & & & \vdots \\ 0 & 0 & 0 & \cdots & \lambda_p \end{bmatrix} \quad (6.47)$$

and V_p is the matrix of eigenvectors associated with nonzero eigenvalues of $G^T G$, while U_p is a matrix of eigenvectors associated with the eigenvalues of GG^T. $G^T G$ and GG^T have the same p nonzero eigenvalues, so it is possible to write

$$G_p^{-1} = V_p \Lambda_p^{-1} U_p^T, \quad (6.48)$$

where G_p^{-1} is now the generalized inverse and $G = U\Lambda V^T = U_p \Lambda_p V_p^T$. The restriction to only a limited portion of the model space spanned by the V_p eigenvectors means that there are nonunique parts of the model which cannot be detected by the inversion, while the limitation to the part of the data space spanned by U_p eigenvectors means that there are aspects of the data which cannot be fit by the model. The reduced problem can be solved, but one must be aware of the limitations of the solution.

Equation (6.48) provides a very general means of solving an inverse problem, but we need to step back and evaluate the significance of the inverse. In particular, we should ask, "Is the solution unique,

how well can we determine each of the modal parameters, and how important are the individual observations to our solution?" We can write the model derived from (6.48) as

$$m_p = G_p^{-1} d \quad (6.49)$$

but recall $d = Gm$, where m is the entire model space. Then m_p is related to m by

$$m_p = G_p^{-1} Gm$$

$$= V_p \Lambda_p^{-1} U_p^T U_p \Lambda_p V_p^T m = V_p V_p^T m. \quad (6.50)$$

The matrix $R = V_p V_p^T$ is called the *resolution matrix*. The columns of the resolution matrix indicate how much the true model is smeared into the various parameters of the inversion model. Ideally, one would obtain a diagonal resolution matrix, recovering the full model. Calculation of the resolution matrix is essential for assessing an inversion result.

We make two further definitions. The *information density matrix* is given by

$$D = U_p U_p^T \quad (6.51)$$

and the *covariance matrix* is given by

$$c = V_p \Lambda_p^{-2} V_p^T, \quad (6.52)$$

where the elements of Λ_p^{-2} are $(1/\lambda_1^2, 1/\lambda_2^2, \ldots, 1/\lambda_p^2)$. Equations (6.50), (6.51), and (6.52) are all related and can give us physical insight into the inversion solution. The rank of the matrix G_p is defined as p, or the number of nonzero singular values. Small singular values cause a greater variance in the solution [see (6.52)]. Thus small eigenvalues lower the stability of the inverse. If the smaller eigenvalues are discarded, then the stability increases. However, this decreases the resolution. If all the model parameters are associated with nonzero singular values, then R is an identity matrix, and we have *perfect resolution*. If we decrease the number of singular values, **R** moves away from being an identity matrix. We usually attempt to optimize this trade-off between resolution and stability by using a cutoff on the ratio of a given eigenvalue to the largest eigenvalue. The *condition number* is defined as

$$\gamma = \frac{\lambda_{max}^{-1}}{\lambda_{min}^{-1}}. \qquad (6.53)$$

We choose a cutoff condition number to determine the number of singular values to retain.

Finally, it is clear that errors in measurement of the data will cause errors in the determination of the model parameters. It is usually assumed that the errors associated with the data are random with a *Gaussian distribution*. This means that a given data point d has a 95% probability of falling within $\pm 2\sigma$ of the true value, where σ is the standard deviation. The errors in the data map to the errors in model parameters by the equation

$$\sigma_m^2 = \mathbf{G}^{-1}\sigma^2[\mathbf{G}^{-1}]^{\mathrm{T}}. \qquad (6.54)$$

6.4.2 Example of Generalized Inverse for Earthquake Location

The principles of generalized inversion are best illustrated by an example. Consider the case of an earthquake that occurred in a homogeneous half-space; the *P* waves were recorded at six seismic stations. Table 6.1 gives the location of each station (in grid coordinates) and the *P*-wave arrival times. The arrival times were calculated exactly, and then *white noise* (± 0.01 s) was added to simulate uncorrelated data errors. To start the inversion process, we "guess" a solution ($x_0 = 21$, $y_0 = 21$, $z_0 = 12$, $t_0 = 30.0$) from which we can calculate a data vector. The

TABLE 6.1 Earthquake Location Examples: Station Location

Station	x	y	P-wave time	$d_i - d_o$
1	2.0	31.0	40.02	5.77
2	3.0	−5.0	42.76	6.93
3	50.0	58.0	41.07	2.70
4	55.0	47.0	40.38	2.72
5	81.0	3.0	45.05	4.05
6	−9.0	−18.0	45.75	7.02

differences between the calculated and observed data $(d_i - d_o)$ give the data vector in the form of Eq. (6.12) (given in Table 6.1). Clearly, our initial guess is not very good.

We now need to determine **G**. In this example this is fairly simple since the analytic derivatives of (6.6) evaluated at 0 are easy to obtain

$$\frac{\partial t_i}{\partial x} = \frac{-(x_i - x_0)}{v^2(t_i - t_0)}, \qquad \frac{\partial t_i}{\partial y} = \frac{-(y_i - y_0)}{v^2(t_i - t_0)},$$

$$\frac{\partial t_i}{\partial z} = \frac{-(z_i - z_0)}{v^2(t_i - t_0)}, \qquad \frac{\partial t_i}{\partial t_0} = 1 \qquad (6.55)$$

yielding

$$\mathbf{G} = \begin{bmatrix} 0.1331 & -0.0700 & 0.0841 & 1.000 \\ 0.0917 & 0.1325 & 0.0612 & 1.000 \\ -0.1030 & -0.1315 & 0.0426 & 1.000 \\ -0.1318 & -0.1004 & 0.0465 & 1.000 \\ -0.1621 & 0.0486 & 0.0324 & 1.000 \\ 0.1021 & 0.1327 & 0.0408 & 1.000 \end{bmatrix}.$$

$$(6.56)$$

The eigenvalues of $\mathbf{G}^{\mathrm{T}}\mathbf{G}$, or nonzero singular values, are given by $\boldsymbol{\Lambda}$

$$\boldsymbol{\Lambda} = \begin{bmatrix} 2.452 & 0 & 0 & 0 \\ 0 & 0.342 & 0 & 0 \\ 0 & 0 & 0.2101 & 0 \\ 0 & 0 & 0 & 0.0199 \end{bmatrix}$$

$$(6.57)$$

with the corresponding eigenvectors:

$$
V = \begin{bmatrix}
-0.0117 & 0.7996 & -0.5854 & -0.1327 \\
0.0018 & 0.5981 & 0.7961 & 0.0915 \\
0.0512 & 0.0524 & -0.1528 & 0.9855 \\
0.9986 & 0.0056 & -0.0005 & 0.0522
\end{bmatrix}.
$$

$$(6.58)$$

Even before we obtain a solution, Λ and V provide some valuable insight into how this particular problem is posed. Each element of a particular eigenvector corresponds to a dependence of the eigenvalue on a given model parameter. For example, the largest eigenvalue, 2.452, is associated with the eigenvector in the first column in V. Each element in this vector is related to a model parameter. The fourth model parameter, which is t_0, the origin time, dominates this eigenvector. The vector "points" in the direction of t_0. This eigenvalue is much larger than the other three, and hence the model parameter estimate is most stable. In other words, our inversion estimates for the change in origin time are the most stable. On the other hand, the smallest eigenvalue is 0.0199, with a corresponding eigenvector dominated by the third model parameter, z, focal depth. Thus the estimate of changes to the depth are the least stable part of the inversion process. As we will see shortly, we have difficulty determining true depth given the starting model we chose.

Since there are four nonzero eigenvalues, the model resolution is perfect, but we can look at the data-density matrix to see how important each observation is in constraining the solution:

If we look at the diagonal terms, we can get a relative measure of the importance of each station. The largest values (0.8992 and 0.8120) are associated with stations 1 and 6; these stations constrain the solution the most. Station 4 provides the least constraint, although most of the data are about equally important. The off-diagonal terms give a measure of the influence other stations have on a given value.

The inversion predicts a change to the model m, which is given by

$$
m = \begin{bmatrix}
8.268 \\
9.704 \\
9.063 \\
4.480
\end{bmatrix}.
$$

$$(6.60)$$

The summed squared error for the starting model was 161.8; for the inversion solution it is now 0.0579. Table 6.2 compares the hypocentral guess with the true model value for several iterations. Note that although x, y, and t are very close to the "true" values, z is not. In fact, the inversion initially pushed z in the wrong direction! This is a result of our poor initial guess. After six iterations we obtain a result that is very close to the true location. Remember, we added white noise to the data, so the inversion will not be exact. It is commonplace for inversions that start off with poor starting models to settle into a local minimum, never approaching the actual solution. This is a result of the linearization of the problem, and various strategies are used to move the inversion over a wider range of parameters to seek a global minimum.

$$
D = \begin{bmatrix}
0.8120 & 0.2605 & 0.0987 & 0.1328 & -0.2265 & 0.0775 \\
0.2605 & 0.5887 & -0.2663 & -0.0232 & 0.2741 & 0.1662 \\
0.0987 & -0.2663 & 0.6142 & 0.3450 & 0.0160 & 0.1922 \\
0.1328 & -0.0232 & 0.3450 & 0.3906 & 0.2881 & -0.1333 \\
-0.2265 & 0.2741 & 0.0160 & 0.2881 & 0.6951 & -0.0467 \\
-0.0776 & 0.1662 & 0.1922 & -0.1333 & -0.0467 & 0.8992
\end{bmatrix}.
$$

$$(6.59)$$

Box 6.4 Joint Determination of Hypocenters (JHD)

The example of hypocenter determination in the text assumes that it is possible to accurately predict the travel times for an arbitrary station–source configuration. Unfortunately, this requires the use of an Earth model, and obviously the actual Earth is more complicated, so errors are introduced into the earthquake location process. For a given seismic station, the error in the predicted travel time is due to inaccuracies of the assumed velocity model. These deviations can occur anywhere along the travel path, although in general we divide the errors into three groups: (1) deviations from the velocity structure near the source, (2) deviations near the station, and (3) deviations along the deep travel path. For a single event–station pair it is not possible to isolate the effects of these errors. On the other hand, if a *cluster* of earthquakes occurs (a group of earthquakes with approximately the same location), we can determine something about the errors in the idealized model. Specifically, we can determine a "station correction" that accounts for the inaccuracies of the model structure along the travel path and beneath the station. In this case we can recast the problem to one of determining n station corrections and m earthquake hypocenters. Equation (6.9) can be replaced with

$$r_{ij} = dT_j + \frac{\partial t}{\partial x} dx_j + \frac{\partial t}{\partial y} dy_j + \frac{\partial t}{\partial z} dz_j + \frac{\partial t}{\partial s} ds_i, \qquad (6.4.1)$$

where r_{ij} is the residual, or error, at the ith station for the jth earthquake ($r_{ij} = \hat{t}_{ij} - t_{ij}$, where $\hat{t}_{ij}$ is the observed arrival time and t_{ij} is the computed travel time and station correction). dT_j is the perturbation of the origin time for the jth event. In matrix form, (6.4.1) is just

$$\mathbf{r}_j = A_j \, \mathbf{dx}_j + S_j \, \mathbf{ds}, \qquad (6.4.2)$$

where $\mathbf{r}_j$ is the data change vector and $\mathbf{dx}$ and $\mathbf{ds}$ are separate model change vectors (compare to Eq. (6.12)). The solution of this system of equations is known as *joint hypocentral determination* (JHD) and was first proposed by Douglas (1967).

Numerous authors (Herrmann *et al.*, 1981; Pavlis and Booker, 1983; Pujol, 1988) have proposed efficient inversion schemes for solving (6.4.2); nearly all the schemes involve singular value decomposition. The relative locations obtained by JHD are better than those determined by inversion of more complete and complex velocity models. The resulting hypocentral locations often give a focused picture of the seismicity. Figure 6.B4.1 compares hypocenters determined by conventional inversion and by JHD.

continues

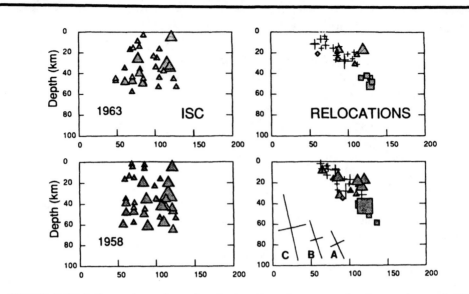

FIGURE 6.B4.1 Comparison of earthquake locations for conventional procedures of ISC (left) and JHD relocations (right). These events are located in the Kurile subduction zone along the rupture zones of large thrust events in 1963 and 1958, and the vertical cross sections traverse the interplate thrust zone from left to right, with the slab dipping toward the right. Note that the JHD relocations reduce scatter and define a dipping plane, which is the main thrust contact. (From Schwartz *et al.*, 1989.)

TABLE 6.2 Comparison of Hypocentral Guess, Solutions After Various Iterations, and True Model Values

Parameter	Initial guess	1st iteration	3rd iteration	6th iteration	True value
x	21.0	29.3	29.9	30.0	30.0
y	21.0	30.7	30.2	30.2	30.0
z	12.0	21.1	9.1	8.9	8.0
t	30.0	34.5	34.9	35.0	35.0

Inversion for source location using more complicated Earth models or travel times obtained from tables proceeds along similar lines, with the only real differences being in the calculation of the partial derivatives needed for each iteration and in the use of spherical geometry. Armed with this cadre of analytic procedures for interpreting seismograms, we can now turn to some of the primary results for Earth structure.

References

Båth, M. (1979). "Introduction to Seismology." 428 pp. Birkhäuser Verlag. Basel.

Bullen, K. E., and Bolt, B. A. (1985). "An Introduction to the Theory of Seismology." Cambridge Univ. Press, New York.

Douglas, A. (1967). Joint hypocenter determination. *Nature* (*London*) **215**, 47–48.

Engdahl, R. R., Flinn, E. A., and Massé, R. P. (1974). Differential *PKiKP* travel times and the radius of the inner core. *Geophys. J. R. Astron. Soc.* **39**, 457–463.

Herrin, E., Arnold, E. P., Bolt, B. A., Clawson, G. E., Engdahl, E. R., Freedman, H. W., Gordon, D. W., Hales, A. L., Lobdell, J. L., Nuttli, O., Romney, C., Taggart, J., and Tucker, W. (1968). 1968 seismological tables for *P* phases. *Bull. Seismol. Soc. Am.* **58**, 1193–1241.

Herrmann, R., Park, S., and Wang, C. (1981). The Denver earthquakes of 1967–1968. *Bull. Seismol. Soc. Am.* **71**, 731–745.

Jeffreys, H., and Bullen, K. E. (1958). "Seismological Tables." Br. Assoc. Adv. Sci., Burlington House, London.

Kafka, A. L., and Ebel, J. E. (1988). Seismic structure of the Earth's crust underlying the state of Maine. *Stud. Maine Geol.* **1**, 137–156.

Kennett, B. L. N. (1991). "IASPEI 1991 Seismological Tables." Research School of Earth Sciences, Australia National University, Canberra.

Kennett, B. L. N., and Engdahl, E. R. (1991). Travel times for global earthquake location and phase identification. *Geophys. J. Int.* **105**, 427–465.

Kulhánek, O. (1990). "Anatomy of Seismograms," pp. 178, Elsevier, Amsterdam/New York.

Menke, W. (1989). "Geophysical Data Analysis: Discrete Inverse Theory," rev. ed., Academic Press, New York.

Pavlis, G., and Booker, J. (1983). Progressive multiple event location (PMEL). *Bull. Seismol. Soc. Am.* **73**, 1753–1777.

Pujol, J. (1988). Comments on the joint determination of hypocenters and station corrections. *Bull. Seismol. Soc. Am.* **78**, 1179–1189.

Schwartz, S. Y., Dewey, J. W., and Lay, T. (1989). Influence of fault plane heterogeneity on the seismic behavior in the southern Kurile Islands arc. *J. Geophys. Res.* **94**, 5637–5649.

Shearer, P. (1991). Imaging global body wave phases by stacking long-period seismograms. *J. Geophys. Res.* **96**, 20,353–20,364.

Simon, R. B. (1981). "Earthquake Interpretations: A Manual for Reading Seismograms." W. M. Kaufmann, Los Altos, CA.

Tarantolla, A. (1987). "Inverse Problem Theory." Elsevier, Amsterdam.

Velasco, A., Lay, T., and Zhang, J. (1994). Long-period surface wave observations of the October 18, 1989 Loma Prieta earthquake, in Loma Prieta. *Geol. Surv. Prof. Pap.* (*U.S.*) (in press).

Additional Reading

Bolt, B. A. (1982). "Inside the Earth: Evidence from Earthquakes." Freeman, San Francisco.

Bullen, K. E., and Bolt, B. A. (1985). "An Introduction to the Theory of Seismology." 4th ed. Cambridge Univ. Press, Cambridge, UK.

Kolman, B. (1970). "Elementary Linear Algebra." Macmillan, London.

Kulhanek, O. (1990). "Anatomy of Seismograms." Elsevier, Amsterdam.

Lancsöz, C. (1957). "Applied Analysis." Prentice-Hall, Englewood Cliffs, NJ.

Menke, W. (1989). "Geophysical Data Analysis: Discrete Inverse Theory." Academic Press, New York.

Tarantolla, A. (1987). "Inverse Problem Theory." Elsevier, Amsterdam.

7

DETERMINATION OF EARTH STRUCTURE

Seismic waves provide a probe of the Earth's deep interior, and their predictable behavior, as set out in Chapters 3 and 4, makes it is possible to obtain high-resolution models of some of the Earth's internal properties. A model is a simplified mathematical representation of the actual three-dimensional material property variations within the planet. Seismology provides primary constraints on the variations of density, rigidity, and incompressibility and secondary constraints on the temperature field at all depths in the Earth. Interpretation of the actual chemistry, physical state, and dynamic behavior associated with the seismological structure requires experimental and modeling results from other disciplines such as mineral physics and geodynamics. Nonetheless, seismology has the primacy of providing our best resolution of the actual structure of the planet.

Not surprisingly, our detailed knowledge of Earth structure generally diminishes with depth. The shallow continental crust has been extensively explored using high-resolution seismic reflection methods in a search for petroleum and mineral resources. The technology developed by the oil industry is now being applied to the deep crust to investigate crustal-scale

faulting and rheological models for crustal evolution, but we still have very limited global sampling of continental deep-crustal structure at high resolution. The processes operating near the surface of the Earth have produced a remarkably complex crustal structure, with the rocks preserving more than 4 billion years of continental evolution and relics of plate tectonic events through the ages. Below the *Moho*, the crust–mantle boundary, our detailed resolution of internal structure diminishes, but surprisingly we obtain increasingly complete global coverage. This is because earthquakes provide the primary seismic sources rather than human-made explosion sources. While both earthquake and seismic station distributions are spatially nonuniform, there are vast numbers of paths through the Earth, yielding fairly complete global coverage. The velocity structure of the upper mantle is very complex, with strong lateral variations associated with the deep structure of plate tectonics processes and depth variations associated with a myriad of high-pressure phase transformations in mantle minerals. The deeper layers of the Earth, the lower mantle and core, are well characterized in their average properties as a function of depth, but only in the past 17 years have seismologists begun to map out what appears to be a modest, few-percent lateral heterogeneity at each depth.

This chapter describes how we use seismic wave recordings to determine large-scale Earth structure. Only a fraction of the methodologies that have been developed can be discussed in this chapter, and new procedures are continually being introduced. Thus, we present a spectrum of methods and the results of their application to various depth ranges in the Earth. Numerous complete texts are devoted to

the techniques and interpretations of shallow crustal reflection seismology (see references for this chapter), so we do not discuss that field here. Our discussion first lays out some basic methodologies common to all applications and then works down from crust to core, considering the seismic waves at the different distance ranges defined in Chapter 6. The regional, upper-mantle, and teleseismic distance ranges (Figure 7.1) provide data that are sensitive to different depth intervals, with very different seismogram characteristics, as revealed in the last chapter. Here we

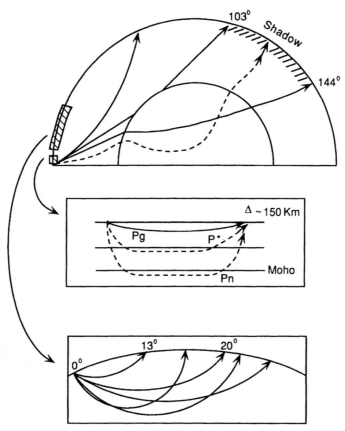

FIGURE 7.1 Schematic diagram of the three characteristic distance ranges used in the study of Earth structure. The range 0–1400 km (0°–13°) is the *near-field* to *regional-distance* range (center), where the seismic wavefield is predominantly crustal phases. The *upper-mantle distance* range (bottom) is from 1400 to 3300 km (13°–30°) and is dominated by upper-mantle triplications. The *teleseismic* range (30°–180°) involves waves that sample the lower mantle and core or reverberate in the upper mantle.

describe how the information in seismograms is transformed into a model of the Earth's interior.

7.1 Earth Structure Inversions

Earlier chapters in this text have provided a general characterization of seismic wave behavior in media that have smooth velocity variations or abrupt discontinuities in material properties. Mathematical expressions predict how the seismic wave amplitudes and travel times should vary as a function of distance from the source. It is logical that comparison of observed travel-time and amplitude behavior with calculations for a suite of model representations of the medium may yield a model or range of models that satisfactorily match the observations. Such a model exploration constitutes *forward modeling* of the data, in which we iteratively perturb model parameters in an effort to predict the observed behavior more accurately. For simple models, with only a few parameters, it is not unreasonable to perform forward modeling to get a "best" model, but even a realistic one-dimensional Earth model may involve many parameters, and a three-dimensional model is guaranteed to. Furthermore, one must address critical issues in defining a best model, such as uniqueness of that representation of the Earth, criteria by which one judges the fit to the data, and resolution of individual model parameters given possible trade-offs between model features.

Although enormous increases in computer speed in recent years enable some "brute force" forward modeling optimization of model parameters by searching over vast suites of models, most current seismological methods employ a different strategy, involving geophysical inverse theory. In Chapter 6 inverse theory was introduced in the context of earthquake location in a medium with a prescribed velocity structure. Clearly, the location depends on the velocity structure, which can at best be an approximation of the actual Earth structure. How, then, can we determine the structure, at least well enough to enable a bootstrapping procedure of iteratively improving both the velocity model and the source location?

The key is to exploit the systematic relationship between seismic wave behavior as a function of distance and the velocity structure encountered along the path. We begin by considering a classic inverse procedure useful for determining a one-dimensional model of velocity variation as a function of depth. Then we will consider a three-dimensional procedure.

7.1.1 Herglotz–Wiechert Inversion

Consider the travel time–epicentral distance behavior for a spherical medium with smoothly varying velocity that is a function only of depth, $v(r)$. In Chapter 3 we developed parametric equations relating travel time and distance

$$T = p\Delta + 2\int_{r_t}^{r_0} \frac{\sqrt{\xi^2 - p^2}}{r}\, dr \quad (3.70)$$

$$\Delta = 2p\int_{r_t}^{r_0} \frac{dr}{r\sqrt{\xi^2 - p^2}}, \quad (3.67)$$

where $\xi = r/v$, r_t is the radius to the turning point of the ray, $p = (r \sin i)/v(r)$, and r_0 is the radius of the sphere. We want to use the observed travel-time curve, $T(\Delta)$, to determine the velocity variation with depth, by a method other than forward modeling. If the $T(\Delta)$ curve is well sampled with a smooth curve fit to the data, we can determine $p(\Delta) = dT(\Delta)/d\Delta$ from the instantaneous slope of the curve. Note that this already implies some smoothing and averaging of the data, which will never lie along a perfectly smooth curve because of both measurement error and three-dimensional heterogeneity in the actual medium. The precise criteria used in fitting a curve to individual measurements $T_j(\Delta_i)$ will affect the $p(\Delta)$ curve and ultimately the model values of $v(r)$. Having determined $T(\Delta)$ and $p(\Delta)$, we proceed by changing the variables of integration in the integral for Δ to

$$\Delta = 2p \int_p^{\xi_0} \frac{1}{r\sqrt{\xi^2 - p^2}} \left(\frac{dr}{d\xi} \right) d\xi, \quad (7.1)$$

where $\xi_t = r_t/v_t = p$ and $\xi_0 = r_0/v_0$. We apply the integral operator

$$\int_{p=p_1}^{p=\xi_0} \frac{dp}{\sqrt{p^2 - \xi_1^2}} \quad (7.2)$$

to (7.1), where $\xi_1 = r_1/v_1 = p_1$ is the ray parameter for a ray bottoming at radius r_1. Thus (7.2) is an integration over all rays from the ray at zero range ($p = \xi_0$, $\Delta = 0$) to a ray with turning point r_1 and ray parameter $\xi_1 = r_1/v_1$:

$$\int_{\xi_1}^{\xi_0} \frac{\Delta \, dp}{\sqrt{p^2 - \xi_1^2}} = \int_{\xi_1}^{\xi_0} dp$$

$$\times \int_p^{\xi_0} \frac{2p}{r\sqrt{\xi^2 - p^2} \sqrt{p^2 - \xi_1^2}} \left(\frac{dr}{d\xi} \right) d\xi.$$

$$(7.3)$$

The left-hand side of (7.3) can be integrated by parts to give

$$\Delta \cosh^{-1}\left(\frac{p}{\xi_1} \right) \Big|_{p=\xi_1}^{p=\xi_0}$$

$$- \int_{\xi_1}^{\xi_0} \frac{d\Delta}{dp} \cosh^{-1}\left(\frac{p}{\xi_1} \right) dp. \quad (7.4)$$

The first term vanishes because $\cosh^{-1}(1) = 0$, and $\Delta = 0$ at $p = \xi_0$ by definition. The second term in (7.4) can be written as an

integral over Δ by

$$\int_0^{\Delta_1} \cosh^{-1}\left(\frac{p}{\xi_1} \right) d\Delta. \quad (7.5)$$

Turning now to the double integral in (7.3), we can change the order of integration to obtain

$$\int_{\xi_1}^{\xi_0} \frac{d\xi}{r} \left(\frac{dr}{d\xi} \right) \int_{\xi_1}^{\xi} \frac{2p \, dp}{\sqrt{p^2 - \xi_1^2} \sqrt{\xi^2 - p^2}}.$$

$$(7.6)$$

The integral on p has a closed form, reducing this to

$$\int_{\xi_1}^{\xi_0} \left(\frac{d\xi}{r} \right) \left(\frac{dr}{d\xi} \right)$$

$$\times \left\{ \sin^{-1}\left[\frac{2p^2 - (\xi^2 + \xi_1^2)}{\xi^2 - \xi_1^2} \right] \right\}_{p=\xi_1}^{p=\xi},$$

$$(7.7)$$

which reduces to

$$\pi \int_{\xi_1}^{\xi_0} \left(\frac{d\xi}{r} \right) \left(\frac{dr}{d\xi} \right) = \pi \int_{r_1}^{r_0} \frac{dr}{r}. \quad (7.8)$$

Integrating (7.8) and combining with (7.5), our Eq. (7.3) becomes

$$\ln\left(\frac{r_0}{r_1} \right) = \frac{1}{\pi} \int_0^{\Delta_1} \cosh^{-1}\left(\frac{p}{\xi_1} \right) d\Delta. \quad (7.9)$$

This is an expression for r_1 in terms of quantities measured from the $T(\Delta)$ plot. The quantity $\xi_1 = p_1 = (dT/d\Delta)_1$ is the slope of $T(\Delta)$ at distance Δ_1. The integral is numerically evaluated with discrete values of $p(\Delta)$ for all Δ from 0 to Δ_1, yielding a value for r_1, with $v(r_1)$ being obtained from $\xi_1 = r_1/v_1$. This procedure is known as the Herglotz–Wiechert formula for inverting a travel-time curve to find velocity as a function of depth. It was used extensively in the development of the earliest *P*-wave and *S*-wave velocity models for deep Earth structure. The procedure is stable as long as $\Delta(p)$ is continuous, with $v(r)/r$ decreasing with r. If a low-velocity zone is present at depth, the formula cannot be used directly, although it is possible to "strip off" layers above the low-velocity zone and then use the contracted travel-time curve to construct smoothly increasing velocities at greater depth. Thus, one could build an Earth model for the mantle, then strip this off before determining the velocity structure of the low-velocity core.

7.1.2 Parameterized Model Inversion

The Herglotz–Wiechert procedure involves an analytic inversion of the integral relationship for distance as a function of ray parameter, using observed travel times to evaluate the inversion numerically. This is actually a rare example in geophysical inversion, in that most Earth structure inversions are constructed as solutions of simultaneous equations giving perturbations of the model in terms of reduced misfit between data and predictions, much like the earthquake location problem in Chapter 6. These are called *discrete inversions*, for a model with a finite set of parameters. This almost always involves solution of a system of equations in the form

$$\mathbf{d} = \mathbf{Gm}, \qquad (7.10)$$

where **d** is a vector of observations or differences between observations and model predictions, **G** is a matrix of partial derivatives $\partial d_i / \partial m_j$, and **m** is the vector of model parameters. If we have n observations and m model unknowns, the solution is *overdetermined* if $n > m$ and **G** has rank m, *underdetermined* if $n < m$, and *exactly determined* if $m = n$. Since the observations have noise, the data may not all be consistent, and one must define the criteria for matching them, such as a least-squares fit. In practice, most seismological inversions are mixed, with both underdetermined and overdetermined aspects, and all solutions and models are intrinsically nonunique.

In some problems, the inversion can be cast in terms of continuous model functions $m(r)$ constrained by discrete, finite data sets that are often *noisy*. Continuous inverse theory explicitly provides an error analysis in the form of trade-offs between resolution and error in the fit to the data over a range of possible model solutions. This complete error analysis is available for assessing any model derived from seismic data but is often computationally demanding.

The choice of inversion method is largely predicated on the nature of the data that are available and the extent of *a priori* constraints on the model. First, let us consider the nature of the data. We have seen that the most reliable travel-time data are generally for first arrivals because later arrivals overlap with multiple reverberations and scattered signal coda. Herglotz–Wiechert inversion is usually applied to first-arrival information alone. However, the information content of first-arrival times may be inadequate to constrain the structure. Figure 7.2a shows an example of a first-arrival travel-time curve that is exactly compatible with an infinite set of structures with markedly different layering, three of which are shown. These different structures could be readily differentiated by the secondary branches in the full wavefield but not by the first arrivals alone.

Another aspect of the limited resolution of seismic data is shown in Figure 7.2b. Four structures are shown that represent different transition structures across a boundary. Each structure would have a different petrological interpretation, but the details of the transition cannot be resolved by signals with wavelengths greater

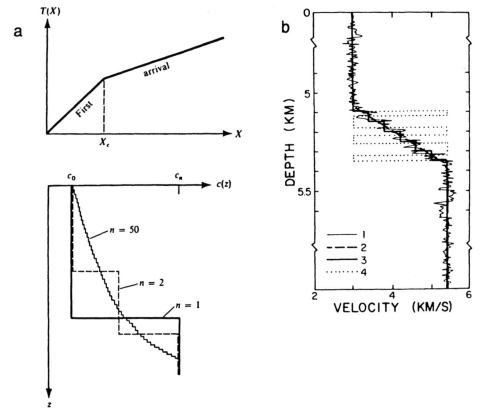

FIGURE 7.2 One of the primary challenges in Earth structure studies is determining the uniqueness of the model. (a) First-arrival travel-time curve that is exactly matched by the three velocity profiles shown below. (b) Four velocity profiles that are indistinguishable when examined using only 1-km-wavelength seismic waves. (Part a is modified from Aki and Richards (1980); part b is modified from Spudich and Orcutt, *Rev. Geophys. Space Phys.* **18**, 627–645, 1980; © copyright by the American Geophysical Union.)

than 1 km. The examples in Figure 7.2 illustrate the need to use both complete wavefield information and broadband seismic data to study deep structure, which has driven the development of all of the procedures discussed later in this chapter.

As an example of the improvement in structural sensitivity offered by waveform and secondary-arrival information, Figure 7.3 shows synthetic seismogram profiles computed using a high-frequency wavelet in the crust for three models of the crust–mantle boundary. The *sharpness*, or depth distribution, of the velocity increase across the boundary clearly affects the amplitude

of both reflected waves near vertical incidence (at close distances) and head waves (P_n) at long distances, which can potentially discriminate between various models. The overall waveform shape can constrain the complexity of the boundary as well.

In the following sections we will see the results of many applications of travel-time and waveform forward modeling and inversion used to determine one-dimensional Earth structures. The applications are wide ranging, from thickness of the crust to velocity increase at the inner core boundary, but the basic principles are the same, exploiting the simple travel time–

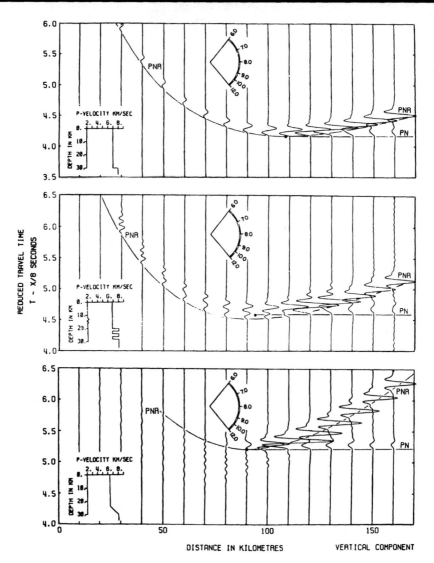

FIGURE 7.3 Synthetic seismogram profiles for three different Moho velocity structures that show how waveform information can potentially distinguish between models of the transition. Reflected arrivals are designated *PNR*; head waves are designated *PN*. One can determine the apparent velocities of arrivals in km/s from the angles shown on the circular scales. (Modified from Braile and Smith, 1975. Reprinted with permission of the Royal Astronomical Society.)

distance and boundary interaction (reflection, refraction, conversion) properties of seismic waves. Although complete description of the methodologies used in each case must be deferred to more advanced seismology texts that develop the mathe-matical procedures for constructing seismograms (i.e., solving the forward problem), the seismology basics in Chapters 2–6 provide sufficient background for one to appreciate the procedure. Before proceeding, we need to describe additional methodologies used for developing three-dimensional models and for studying attenuation structure in the Earth.

7.1.3 Seismic Tomography

To a fairly high level of approximation, the Earth can be viewed as a spherically layered planet, with primary chemical stratification between the crust, mantle, and core. A one-dimensional model such as PREM (Figure 1.20) accurately predicts travel times for waves spreading throughout the planet with an error of less than 1% for teleseismic *P* waves. Thus, seismologists have exerted great effort to develop and refine a one-dimensional model for use in routine earthquake location and geochemical interpretations. The iasp91 model associated with Figures 6.2 and 6.12 represents one of the latest attempts to develop such a refined model. However, we know that some lateral heterogeneity exists at every depth in the Earth. This is directly indicated by the scatter in seismic wave arrival times at all distances (Figure 1.19a). While the lateral variations at a given depth are typically less than 10% for shear velocity (the most variable of elastic parameters), these small fluctuations have great significance as markers of dynamic processes in the Earth's interior, as we shall see. For more than 35 years seismologists have been mapping gross velocity differences near the surface, associated with variations between continental and oceanic crust and upper mantle, and in the past 17 years a concerted effort has emerged to map, or image, the three-dimensional structure everywhere inside the Earth. The process has evolved from localized one-dimensional characterizations of structure beneath a given area to complete three-dimensional modeling using a method called *seismic tomography*.

Seismic tomography was first introduced in the mid-1960s in an earlier form called *regionalization*, in which surface waves traversing mixed oceanic and continental paths were analyzed to determine separate structures beneath each region. This involved calculating a travel-time anomaly relative to a reference symmetric Earth model for a surface-wave phase with a particular period and then partitioning the anomaly with respect to the percentage of path length in a specified tectonic regionalization of the surface. This procedure was necessary because only a few paths occur for which the source, receiver, and entire path length are within a particular region, such as ocean basins less than 20 million years old.

From these early beginnings seismology moved toward a smaller and smaller subdivision of the media, forgoing any regionalization based on surface geology and including both local and global two- and three-dimensional parameterizations of the media for which velocity or *slowness* (reciprocal velocity) perturbations would be found in each region. In every case the principle is that the particular seismic phase has a travel time, T, given by a path integral through the medium of

$$T = \int_s \frac{ds}{v(s)} = \int_s u(s)\,ds, \quad (7.11)$$

where $u(s)$ is the slowness $[1/v(s)]$ along the path. This reflects the localized sampling behavior of all traveling seismic waves, which makes them most sensitive to velocity near the raypath from source to receiver. The *travel-time residual* relative to the reference Earth model may be caused by a velocity or slowness perturbation anywhere along the path (assuming the source location is known; otherwise the location parameters must be included in the problem). A change in velocity along the ray must perturb the raypath, but this effect is often minor or can be addressed by iteratively calculating new raypaths for each model update. Almost all seismic tomography methods involve subdividing the medium into blocks or other spatial functions, such as spherical harmonic expansions, and solving for slowness perturbations that cause predicted times to match observed times better than an initial (usually homogeneous or one-dimensional) model. The idea is that the path integral through the medium perturbations should equal the observed travel-time residual

$$\int_s \Delta u(s)\, ds = \Delta T = T_{\text{obs}} - T_{\text{pred}}, \quad (7.12)$$

where $\Delta u(s)$ is the slowness perturbation to be determined. If the medium is subdivided into blocks, one can calculate the path length l_j in the jth block and discretize (7.12) to give

$$\Delta T = \sum_j l_j \Delta u_j. \quad (7.13)$$

Clearly, a single observation is inadequate for partitioning the slowness perturbations along the path, and the most reasonable choice would be to distribute the anomaly uniformly over the whole path length. But if there are many event–station pairs, each with the ith raypath, we develop a system of i equations

$$\Delta T_i = \sum_j l_{ij} \Delta u_j, \quad (7.14)$$

where raypaths that intersect a common block may require slowness perturbations in that block that are different from the uniformly distributed anomaly along each raypath. The information contained in crossing raypaths provides an integral consistency in the system of equations that can reveal two- or three-dimensional variations in the medium. Equation (7.14) is in the form of a linear system like (7.10), which can be solved by the matrix inversion methods introduced in Chapter 6. In this case the path length of each ray in a block, l_{ij}, is the partial derivative, $\partial T_i/\partial u_j$, of the travel time with respect to the slowness of that block. We usually have many

more raypaths than model parameters, yielding an overdetermined system, but noise in the data and inadequacy of the model parameterization are sure to make the system inconsistent. The generalized inverse solution of (7.14) is provided by

$$\mathbf{m} = [\mathbf{G}^{\text{T}}\mathbf{G}]^{-1}\mathbf{G}^{\text{T}}\mathbf{d}, \quad (6.37)$$

where we let $\mathbf{d} = \Delta T_i$, $\mathbf{G} = l_{ij}$, and $\mathbf{m} = \Delta u_j$. If the generalized inversion is unstable, we resort to damping or singular-value truncation to obtain a solution, as described in Chapter 6. The resolution matrix (6.50) can be computed to reveal how well the model can be reconstructed if the data and model parameterization are perfect. Usually the resolution matrix reveals streaking between adjacent blocks where the ray coverage is inadequate to isolate the anomaly uniquely in each block.

Examples of seismic tomography geometries are shown in Figures 7.4 and 7.5. Figure 7.4a corresponds to the common application of local earthquakes recorded by an array of surface sensors. While it is possible to locate the events with a one-dimensional velocity model, the crossing ray coverage allows us to solve for shallow crustal heterogeneity as well. Sometimes this involves holding the earthquake locations fixed, solving for the three-dimensional structure, and then iterating on both locations and structure, or the problem can be formulated for a simultaneous solution of velocity structure and source locations.

The most widely applied geometry for seismic tomography is illustrated in Figure 7.4b, in which teleseismic waves recorded by a seismic array are used to invert for three-dimensional crust and upper-mantle heterogeneity under the array. This procedure was introduced in the mid-1970s. The basic idea is that in the absence of heterogeneity the incident wavefronts should have a simple plane-wave apparent veloc-

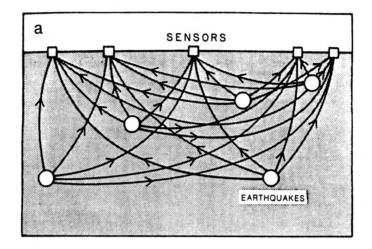

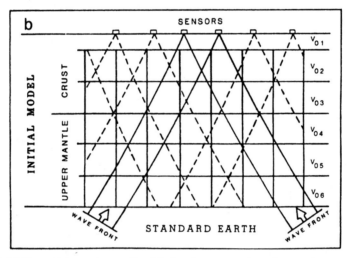

FIGURE 7.4 (a) Raypath coverage utilized in local tomography, where crisscrossing paths from many nearby earthquakes to seismic stations are used in joint inversions to determine both source locations and crustal structure. (b) Tomography that uses teleseismic signals recorded at a densely spaced seismic array. In this case the differential times between arrivals are mapped into velocity perturbations of three-dimensional blocks below the array. (Modified from Iyer, 1989.)

ity across the array. The array data are processed by computing individual station anomalies relative to the best plane-wave fit reduced by the average anomaly across the array for a particular event, giving relative residuals for each raypath. Events at different azimuths are analyzed, providing cones of incident rays under each station, which overlap at depth, giving ray-crossing coverage of each block. A system like (7.14) is determined by calculating raypath segments through the model, and this is solved by damped-least-squares, sin-gular-value decomposition (SVD), or by iterative procedures called *back projection* that iteratively solve the matrix equations without having to invert huge matrices.

The latter techniques became popular in the 1980s as increasingly large data sets and model parameterizations were adopted. One scheme involves solution of (7.14) in the least-squares form

$$\mathbf{G}^{\mathrm{T}}\mathbf{G}(\mathbf{m}) = \mathbf{G}^{\mathrm{T}}\mathbf{d}. \qquad (7.15)$$

Rather than solve for the generalized inverse of $\mathbf{G}^{\mathrm{T}}\mathbf{G}$, one simply uses the diagonal of this matrix to approximate the solution, giving

$$\Delta \tilde{u}_j = \frac{\Sigma_i \, \Delta t_i \, l_{ij}}{\Sigma_i \, l_{ij}^2}. \qquad (7.16)$$

The two sums are computed separately and updated ray by ray. Then each slowness perturbation is calculated, and the model is perturbed. In a procedure called simultaneous iterative reconstruction technique (SIRT), this new model is used to compute new residuals, and the procedure repeats for many iterations. Because no large matrix inversions are performed, huge models can be examined but at the cost of not computing any resolution matrices. This iterative back-projection approach was developed in medical imaging tomography.

The example in Figure 7.5 illustrates an actual data application for a large model with 9360 blocks used in a tomographic inversion of P velocity structure beneath Europe. Both regional and teleseismic earthquakes recorded at European sensors were used in this study by Spakman and Nolet (1988), who analyzed about 500,000 travel times from 25,000 earthquakes. Including earthquake location parameters and combining similar paths by averaging, a total of 20,000 unknowns were extracted from 300,000 equations. An iterative algorithm similar to SIRT was used to solve this problem because the large, sparse matrices were too large to invert. Figure 7.5b shows a contour plot of cell hit counts in one cross section (along the heavy lines) in Figure 7.5a. This indicates the nonuniform distribution of raypaths in the model, which is characteristic of most applications. The darker regions are well sampled and correspond to blocks below stations as

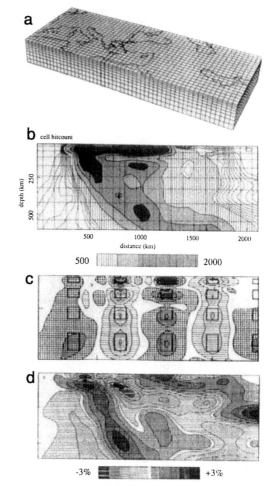

FIGURE 7.5 (a) Example of three-dimensional grid of blocks used in an inversion for structure beneath Europe. Block thickness varies with depth. (b) Cross section along one slice in the grid in (a), showing contours of raypath "hits" of each cell. The dark areas of high cell hits correspond to blocks either below stations or near sources in the dipping ocean slabs subducting in the Mediterranean. (c) Inversion result for known input model in the section shown in (b), showing smearing effects of imperfect ray coverage and inversion instability. (d) Actual data inversion showing velocity perturbations in the section in (b) for data recorded in Europe. A dipping high-velocity structure is seen on the left. (From Spakman and Nolet, 1988.)

well as to earthquake zones in the dipping Aegean slab. The nonuniform raypath distribution affects the model resolution, which is indicated by the results of an inversion with known 5% velocity anomalies for the same ray coverage, as shown in Figure 7.5c. The blocks with dark outlines correspond to locations of anomalous velocities in the simulation, with the smeared-out inversion results indicating the poor vertical resolution and fair horizontal resolution of the particular geometry involved. Only blocks with very high hit counts and favorable raypath geometries are well resolved. The actual model obtained in the same slice is shown in Figure

7.5d. The band of fast-velocity material dipping across the left half of the cross section is inferred to be the high-velocity Aegean slab, dipping toward the northwest, overlain by a low-velocity region under the Aegean Sea. Although the extent of vertical streaking is not easy to assess, the general features of the structure appear to be resolved. This is called an *image* of the interior, being a nonunique, smeared model visualization of the actual Earth structure. Current research in tomography is striving to improve both the image development and the assessment of the reliability of such models. Other tomographic images are shown later in this

---

Box 7.1 Receiver Functions

Although most detailed investigations of crustal structure utilize multiple seismic recordings, a few techniques have been developed to study crustal layering beneath isolated three-component stations. The most widely used is called *receiver function analysis*, which exploits the fact that teleseismic *P* waves that are incident upon the crustal section below a station produce *P* to *S* conversions at crustal boundaries as well as multiple reverberations in the shallow layering. The *P* to *S* conversions have much stronger amplitudes on the longitudinal component (the horizontal component along the great circle from source to receiver) than on the vertical component. By deconvolving the vertical-component signal from the longitudinal component, the obscuring effects of source function and instrument response can be removed, leaving a signal composed of primarily *S*-wave conversions and reverberations below the station. These deconvolved horizontal components are called receiver function traces and can be inverted for a model of the shear velocity layering in the crust (see Figure 7.B1.1).

Receiver function analysis reveals the presence of interfaces at depth, but the absolute velocities and depths of the boundaries are not well resolved, as shown by the excellent fit to the data provided by a suite of models. Using independent determination of the average crustal velocities provided by a surface-wave dispersion analysis can provide a better constrained model. This technique can give stable results if the layers are horizontal, in which case the receiver function is the same for all azimuths, and the tangential component of the *P*-wave motion is very small. In regions of three-dimensional structure, the inversion is usually ill posed, and forward modeling of the azimuthally varying receiver functions is used to constrain the crustal and uppermost mantle structure.

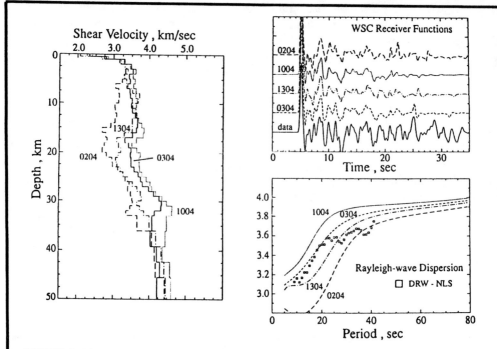

FIGURE 7.B1.1 Receiver function analysis of the crustal velocity structure under Death Valley, California. The left panel shows velocity structures obtained by inversions of the observed receiver function, labeled "data" in the upper-right box. The models differ because of different initial models in the inversion, but all produce reasonable fits to the data. To better resolve the structure, short-period surface-wave dispersion observations are modeled as well. (Courtesy of S. Beck and G. Zandt.)

chapter, and one must keep in mind the caveats about nonuniqueness and limited resolution.

From seismic tomography applications, of which there were well over 100 by the end of the 1980s, it has become well established that the Earth is heterogeneous at all depths at all scale lengths and that we can never achieve a complete deterministic understanding of the full internal spectrum of heterogeneity. However, great progress has already been made toward assessing the strength of heterogeneities in different regions, as shown in Figure 7.6. Perturbations in seismic velocity from 1% to 10% appear to exist throughout the mantle and crust, with smaller perturbations possibly existing in the core. Seismic waves of different ranges and wavelengths

detect this heterogeneity, and any given data set will be able to resolve only a limited portion of the length spectrum. Fortunately, the mantle heterogeneity spectrum appears to be "red," meaning that the longer-wavelength features have more variations. This is a favorable situation for seismic tomography, as much of the important internal structure can be imaged using models with large-scale parameterizations. The heterogeneity spectrum in the lithosphere, and possibly near the core–mantle boundary, appears to be "white," with a more uniform degree of heterogeneity occurring at all spatial scales. This greatly complicates attaining detailed seismic images at shallow depth. At some level there is simply not enough wavefield information to resolve the

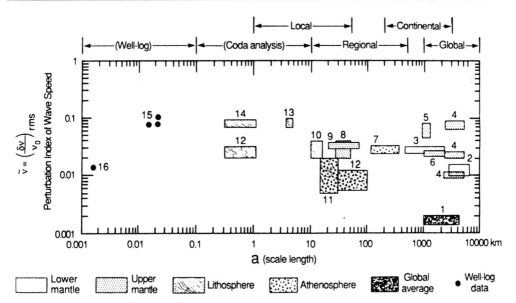

FIGURE 7.6 Summary of many investigations of velocity heterogeneity inside the Earth. The parameter a is the scale length, and $\tilde{v} = (\delta v/v_0)$ rms is the velocity perturbation index of the heterogeneity. (1) Global average, from the analysis of mode splitting of free oscillations. (2) Lower mantle, from body-wave inversion. (3) Lower mantle, from body-wave tomography. (4) Upper mantle, from surface-wave waveform inversion. (5) Upper mantle (Pacific plate), from surface-wave full-wave inversion. (6) Upper mantle (United States), from travel-time inversion. (7) Asthenosphere (central United States, 125–225 km deep), from travel-time inversion. (8) Upper mantle (southern California), from body-wave tomography. (9) Upper mantle, a summary, from travel-time inversion. (10) Lithosphere, from transmission fluctuations at LASA. (11) Lithosphere, from transmission fluctuation at NORSAR. (12) Lithosphere, from transmission fluctuation at NORSAR. (13) Lithosphere, from coda wave analysis. (14) Lithosphere, from coda wave analysis. (15) Crust, from acoustic well log. (16) Crust, from acoustic well log. (From Wu, 1989.)

small-scale heterogeneity in detail, and *statistical tomography* techniques are used to characterize parameters of a random medium representation of the interior. Both statistical and deterministic images of the interior are our main means for studying the dynamic processes presently occurring deep in the Earth, so there are many ongoing efforts to extend and improve both methods.

7.1.4 Attenuation Modeling

As described in Chapter 3, the Earth does not transmit seismic waves with perfect elasticity; small anelastic losses occur that progressively attenuate the wave en-

ergy. This anelasticity causes dispersion, changes pulse shapes, and affects amplitudes of the waves; therefore it can be modeled as well. Unlike the case for seismic velocities, the Earth does not have a simple layered attenuation structure. Instead, lateral variations in attenuation quality factor Q can involve many orders of magnitude at a given depth. In general, the upper mantle has lower Q values (higher attenuation) than the deep mantle, but there are paths through the upper mantle with little attenuation and others with strong attenuation. Thus, it has long been apparent that a three-dimensional model of Q is needed. In the early 1980s it further became widely accepted that attenuation is a function of frequency across the seismic wave frequency range, with the spatial variations having complex frequency dependence as well.

Seismic attenuation is caused by either *intrinsic anelasticity*, associated with small-scale crystal dislocations, friction, and movement of interstitial fluids, or *scattering attenuation* (see Box 3.5), an elastic process of redistributing wave energy by reflection, refraction, and conversion at irregularities in the medium. The latter process is not true anelasticity but has virtually indistinguishable effects that are not accounted for by simple Earth models. At frequencies with wavelengths much larger than the heterogeneities in the medium, intrinsic attenuation dominates. Thus seismic models for attenuation were first constructed for low frequencies.

The main challenge in measuring and developing Earth models for attenuation structure is separating out the anelastic effects from both propagation and source effects. In Chapter 3 we saw that effects of anelasticity on body waves can largely be accounted for by an operator that modifies the amplitude spectrum by a factor of

$$e^{-\pi f t^*(f)}, \qquad (7.17)$$

where

$$t^*(f) = \int_s \frac{ds}{v(s)Q(s,f)}, \qquad (7.18)$$

where the dispersive effect on the velocity, $v(s)$, is neglected. Here the t^* parameter is clearly defined as a path-specific and frequency-dependent function. Because the wave amplitudes are reduced according to (7.17), measurement of t^* involves measuring amplitude reductions beyond that expected from elastic effects. Sometimes this involves assuming a particular source spectrum, following scaling laws like those described in Chapter 9, and then measuring spectral decay of observed body waves relative to the assumed shape. This

can give stable estimates of differences in attenuation between different paths, but it is unreliable for the absolute t^* values. Other procedures give more reliable absolute values by designing an experiment that eliminates uncertainty in the source spectrum, which commonly involves spectral ratios between different phases.

A classic example of a procedure used to estimate stable, absolute attenuation in the Earth is the analysis of multiple ScS (ScS_n) reverberations. Raypaths for these signals and some examples of transverse-component waveforms are shown in Figure 7.7. The core reflections have from one to four transits up and down through the mantle. These arrive as discrete, isolated pulses that decay in amplitude with time. It is straightforward to predict the amplitude decrease induced by geometric spreading expected for these pulses, which predicts a less rapid decrease in amplitude between successive multiples than is observed. The additional amplitude decay is caused by attenuation along the paths through the mantle. The ScS_n arrivals all have similar source radiation, so we can view ScS_{n+1} as a more attenuated version of ScS_n due to its additional transit through the mantle. We can write this as

$$F(\omega, \Delta t)S(\omega) + N(\omega) = S'(\omega),$$

where $S(\omega)$ and $S'(\omega)$ are the spectra of ScS_n and ScS_{n+1}, respectively, and $N(\omega)$ is the noise spectrum. $F(\omega)$ is the attenuation filter $\exp(-\omega \Delta t/2Q_{ScS})$. Computing spectral ratios such as $ScS_3(f)/ScS_2(f)$, $ScS_4(f)/ScS_3(f)$, etc. eliminates the unknown source spectrum to a large extent. Measurement of the slope of the spectral ratios as a function of frequency then reveals the t^* value appropriate for the frequency band of the observations. Less precise resolution of t^* can be obtained by simply making synthetic waveforms and varying the mantle Q model until the synthetics match the data in the time domain,

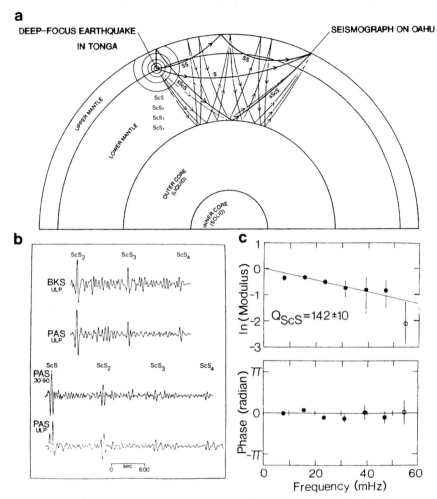

FIGURE 7.7 Analysis of multiple ScS_n reverberations to determine whole-mantle attenuation. (a) Multiple ScS reverberation raypaths. (b) Examples of tangential component ScS_n phases for the October 24, 1980 Huajuapan, Mexico, earthquake. (c) The spectral ratio of successive ScS_{n+1}/ScS_n phases. The slope of the log ratio yields an estimate of Q_{ScS}, in this case appropriate for below Mexico. (Modified from Lay and Wallace, 1988.)

a procedure which emphasizes the period range of the major pulses, which in turn reflects the instrument response characteristics. Since the paths traverse the entire mantle, we cannot assess how the attenuation varies with depth using ScS_n alone; however, comparisons of $sScS_n$ and ScS_n for deep sources allow a separation of attenuation in the extra upper-mantle legs above the source for the surface reflections. Seismologists use many similar comparisons of different phases to isolate the

anelastic losses in signals in an effort to map out the radial and lateral variations of attenuation (see Figure 3.37).

7.2 Earth Structure

All different seismic wave types have been analyzed in determining Earth structure, ranging from free oscillations of the planet to high-frequency body waves reflected from shallow sedimentary layers. The waves reveal aspects of the Earth incorporated in Earth models, functional descriptions of how the material properties vary in the interior. A large number of body-wave travel times, free-oscillation eigenfrequencies, and surface-wave and normal attenuation measurements were modeled in constructing the Preliminary Reference Earth Model (PREM) (Dziewonski and Anderson, 1981) (Figure 1.20). The parameters of this model at a reference period of 1 s are given in Table 7.1, including density, P velocity, S velocity, shear attenuation coefficient (Q_μ), the adiabatic bulk modulus (K_s), rigidity (μ), pressure, and gravity. The PREM model includes anisotropic upper-mantle layers that are not listed here, and the velocities vary with reference period because of anelastic dispersion. Note that the variation of the attenuation coefficient with depth is very simple, with only a few-layer model being resolved. The core has very high values of Q, with almost no seismic wave attenuation, and attenuation of body waves, surface waves, and free oscillations requires a relatively low Q in the upper mantle. Although the PREM Q model is reasonable for globally averaging waves such as free oscillations, it provides only a reference baseline for path-specific attenuation as sampled by body waves and surface waves. Later sections will thus emphasize three-dimensional attenuation variations.

In addition, different PREM models are provided for oceanic and continental lithosphere. As complete a model as PREM is, it still lacks a three-dimensional description of aspherical heterogeneity, and some details of the upper-mantle structure are undergoing revision. Ultimately, seismology will achieve a complete three-dimensional, anelastic (hence, frequency-dependent) anisotropic Earth model, but many aspects of such a complete model are still being resolved. Attaining such a detailed model will be critical for achieving a thorough understanding of the composition and dynamic processes inside the Earth. Since this is an ongoing process, the remainder of this chapter will traverse from the crust to the core, outlining major aspects of what we know about Earth structure and how it is determined by seismic-wave analysis.

7.2.1 Crustal Structure

In terms of relative societal importance, seismological investigations of the structure of the shallow crust unquestionably have the greatest impact and largest effort. Much of that effort involves *reflection seismology*, the collection and processing of multichannel seismic data that record human-made explosive and vibrational sources. Although the principles involved in multichannel seismic processing have basic origins in the behavior of seismic waves and their reflection from boundaries, as described in previous chapters, the processing of the dense and now often two-dimensional recordings of the wavefield involves a multitude of specific procedures beyond the scope of this text. We instead focus our attention on whole-crustal-scale investigations of the shallow crust, which are typically performed with sparser seismic instrumentation than in shallow-crust reflection imaging.

On the larger scale of whole-crustal imaging, the main objective is to determine the basic layered structure of the crust; the P and S velocities as a function of depth, including the depth and contrasts across any internal boundaries; and the overall crustal thickness, or depth to the crust–mantle boundary. The effort to determine crustal thickness dates back to 1910, when Croatian researcher Andrija Mohorovičić first identified an abrupt increase in velocity beneath the shallow rocks under Europe. The boundary separating crustal rocks from mantle rocks is now called the *Moho* and is a ubiquitous

TABLE 7.1 Parameters of the Preliminary Reference Earth Model at a Reference Period of 1 s

Radius (km)	Depth (km)	Density (g/cm^3)	V_p (km/s)	V_S (km/s)	Q_μ	K_s (kbar)	μ (kbar)	Pressure (kbar)	Gravity (cm/s^2)
0	6371.0	13.08	11.26	3.66	85	14253	1761	3638.5	0
200.0	6171.0	13.07	11.25	3.66	85	14231	1755	3628.9	73.1
400.0	5971.0	13.05	11.23	3.65	85	14164	1739	3600.3	146.0
600.0	5771.0	13.01	11.20	3.62	85	14053	1713	3552.7	218.6
800.0	5571.0	12.94	11.16	3.59	85	13898	1676	3486.6	290.6
1000.0	5371.0	12.87	11.10	3.55	85	13701	1630	3402.3	362.0
1200.0	5171.0	12.77	11.03	3.51	85	13462	1574	3300.4	432.5
1221.5	5149.5	12.76	11.02	3.50	85	13434	1567	3288.5	440.0
1221.5	5149.5	12.16	10.35	0	0	13047	0	3288.5	440.0
1400.0	4971.0	12.06	10.24	0	0	12679	0	3187.4	494.1
1600.0	4771.0	11.94	10.12	0	0	12242	0	3061.4	555.4
1800.0	4571.0	11.80	9.98	0	0	11775	0	2922.2	616.6
2000.0	4371.0	11.65	9.83	0	0	11273	0	2770.4	677.1
2200.0	4171.0	11.48	9.66	0	0	10735	0	2606.8	736.4
2400.0	3971.0	11.29	9.48	0	0	10158	0	2432.4	794.2
2600.0	3771.0	11.08	9.27	0	0	9542	0	2248.4	850.2
2800.0	3571.0	10.85	9.05	0	0	8889	0	2055.9	904.1
3000.0	3371.0	10.60	8.79	0	0	8202	0	1856.4	955.7
3200.0	3171.0	10.32	8.51	0	0	7484	0	1651.2	1004.6
3400.0	2971.0	10.02	8.19	0	0	6743	0	1441.9	1050.6
3480.0	2891.0	9.90	8.06	0	0	6441	0	1357.5	1068.2
3480.0	2891.0	5.56	13.71	7.26	312	6556	2938	1357.5	1068.2
3600.0	2771.0	5.50	13.68	7.26	312	6440	2907	1287.0	1052.0
3800.0	2571.0	5.40	13.47	7.18	312	6095	2794	1173.4	1030.9
4000.0	2371.0	5.30	13.24	7.09	312	5744	2675	1063.8	1015.8
4200.0	2171.0	5.20	13.01	7.01	312	5409	2559	957.6	1005.3
4400.0	1971.0	5.10	12.78	6.91	312	5085	2445	854.3	998.5
4600.0	1771.0	5.00	12.54	6.82	312	4766	2331	753.5	994.7
4800.0	1571.0	4.89	12.29	6.72	312	4448	2215	655.2	993.1
5000.0	1371.0	4.78	12.02	6.61	312	4128	2098	558.9	993.2
5200.0	1171.0	4.67	11.73	6.50	312	3803	1979	464.8	994.6
5400.0	971.0	4.56	11.41	6.37	312	3471	1856	372.8	996.9
5600.0	771.0	4.44	11.06	6.24	312	3133	1730	282.9	998.8
5650.0	721.0	4.41	10.91	6.09	312	3067	1639	260.7	1000.6
5701.0	670.0	4.38	10.75	5.94	312	2999	1548	238.3	1001.4
5701.0	670.0	3.99	10.26	5.57	143	2556	1239	238.3	1001.4
5771.0	600.0	3.97	10.15	5.51	143	2489	1210	210.4	1000.3
5871.0	500.0	3.84	9.64	5.22	143	2181	1051	171.3	998.8
5921.0	450.0	3.78	9.38	5.07	143	2037	977	152.2	997.9
5971.0	400.0	3.72	9.13	4.93	143	1899	906	133.5	996.8
5971.0	400.0	3.54	8.90	4.76	143	1735	806	133.5	996.8
6061.0	310.0	3.48	8.73	4.70	143	1630	773	102.0	993.6
6106.0	265.0	3.46	8.64	4.67	143	1579	757	86.4	992.0
6151.0	220.0	3.43	8.55	4.64	143	1529	741	71.1	990.4
6151.0	220.0	3.35	7.98	4.41	80	1270	656	71.1	990.4
6186.0	185.0	3.36	8.01	4.43	80	1278	660	59.4	989.1
6221.0	150.0	3.36	8.03	4.44	80	1287	665	47.8	987.8
6256.0	115.0	3.37	8.05	4.45	80	1295	669	36.1	986.6
6291.0	80.0	3.37	8.07	4.46	80	1303	674	24.5	985.5
6291.0	80.0	3.37	8.07	4.46	600	1303	674	24.5	985.5
6311.0	60.0	3.37	8.08	4.47	600	1307	677	17.8	984.9
6331.0	40.0	3.37	8.10	4.48	600	1311	680	11.2	984.3
6346.6	24.4	3.38	8.11	4.49	600	1315	682	6.0	983.9

(*Continues*)

TABLE 7.1—*Continued*

Radius (km)	Depth (km)	Density (g / cm³)	V_p (km / s)	V_S (km / s)	Q_μ	K_s (kbar)	μ (kbar)	Pressure (kbar)	Gravity (cm / s²)
6346.6	24.4	2.90	6.80	3.90	600	753	441	6.0	983.9
6356.0	15.0	2.90	6.80	3.90	600	753	441	3.3	983.3
6356.0	15.0	2.60	5.80	3.20	600	520	266	3.3	983.3
6368.0	3.0	2.60	5.80	3.20	600	520	266	0.3	982.2
6368.0	3.0	1.02	1.45	0	0	21	0	0.2	982.2
6371.0	0	1.02	1.45	0	0	21	0	0.0	981.5

boundary of highly variable character. Although we generally accept that the crust is chemically distinct from the upper mantle and that the Moho likely involves a chemical contrast, additional contributions to the seismically detectable boundary may arise from transitions in rheological properties, phase transitions in shallow mineral structures, and petrographic fabrics of the rocks. These complexities are combined with the complex tectonic history of the surface to provide a remarkably heterogeneous crustal layer.

While we recognize the complexity of the crust, it is still useful to assess the basic seismological feature common to all crustal environments, which is that the shallow rocks have slower seismic velocities than the deeper rocks, usually approximating a low-velocity layer over a faster mantle. Figure 7.8 shows highly schematic characteristics of crustal structure and

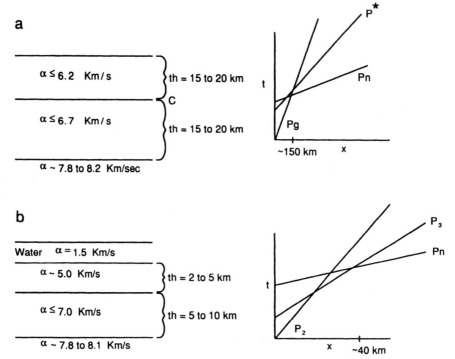

FIGURE 7.8 Schematic travel-time curves and generic continental and oceanic crustal structures. Only primary-wave arrivals are shown, but each interface spawns a set of multiple reflections and conversions that produce many later arrivals.

associated travel-time curves for primary seismic-wave arrivals for continental and oceanic regions. These regions differ primarily in the thickness of the crust, which varies from 20 to 70 km beneath continents and from 5 to 15 km beneath oceans. Both regions have very low seismic velocity surface cover, with the water and mud layers on oceanic crust having a particularly low velocity. Both regions tend to have at least two crustal subdivisions, with the lower-velocity layer below the sediments typically having steep velocity gradients with depth down to a transition or midcrustal discontinuity. At greater depths the average gradients and reflective properties of the deep crust vary substantially.

The travel-time curves for these generic layered crusts are thus composed of distinct primary-wave branches, with direct phases being called P_g in continental environments and P_2 in oceanic crust (layer 1 is the soft sediments; layer 2 is the shallow basaltic layer in the oceanic crust). For continental regions with a midcrustal discontinuity, often called the *Conrad* discontinuity, the head wave from this structure is called P^*, and P_n is the head wave traveling along the Moho. The analogous oceanic arrivals are P_3 and Pn, and both types of crust have PmP reflections from the Moho. As discussed in Chapter 6, S phases have corresponding labels (S_g, S^*, S_n, S_mS, etc.). The crossover distances of the various travel-time branches, the slopes of the branches, and their zero-distance intercepts reveal the layer thickness and velocities using the straightforward wave theory from Chapter 3. The variation in crustal thickness between oceanic and continental regions ensures a very different appearance of the seismograms as a function of distance for the two regions.

A representative continental crustal profile from Globe, Arizona to Silver City, New Mexico is shown in Figure 7.9. The P_g arrival indicates a velocity of 6 km/s, so plotting the profile with a *reduced* time $[T - (X/6)]$, where X is the distance,

causes Pg to arrive along a horizontal reduced time of 0 s. The Pn velocity is 7.9 km/s, so the associated arrival branch slants downward toward the right, as will all arrivals with velocities higher than 6 km/s. The crossover distance is about 140 km, corresponding to a crustal thickness in excess of 30 km. A clear P^* arrival (here labeled P_cR) is apparent at distances of less than 140 km, but as is very often the case, the velocity contrast producing this head wave is not large enough for the P^* branch to be observed as a first arrival at any distance. This makes it difficult to detect and model midcrustal discontinuities, and we can usually study them only by interpreting secondary arrivals.

A data profile of this type, covering several hundred kilometers from the source and allowing the head-wave branches to be identified, is called a *refraction profile*. Refraction seismology is quite straightforward and involves directly identifying travel-time branches; measuring slopes, crossovers, and amplitude behavior along the branches; and relating these to one- or two-dimensional models of the primary layers in the structure. Note that at closer distances, less than 30 km, a weak Moho reflection, P_mP, occurs, as well as other strong arrivals. At close-in distances, direct body and surface waves dominate the seismograms, which display a long sequence of nearly vertically propagating waves composed of single or multiple reflections from crustal layers near the source region. This is the domain of *reflection profiles*. Reflection seismology strives to determine the *reflectivity* of the crust, meaning detailed layering and impedance contrasts below a localized region. The procedures for isolating energy associated with any particular reflector at depth involve unscrambling the many superimposed arrivals generated by the detailed rock layering. Usually, high-precision images of the shallow layering can be extracted from the data using the predictability of seismic-wave interactions with

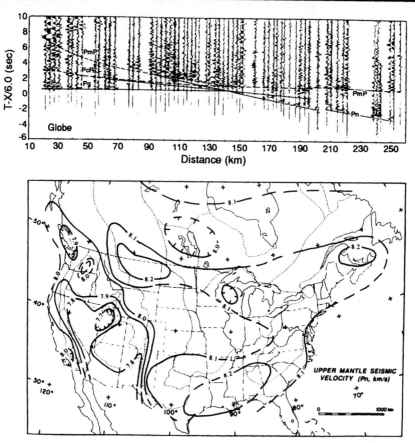

FIGURE 7.9 A typical seismic profile for continental structure is shown at the top, with a reduction velocity of 6 km/s. The first arrival is direct *Pg* out to about 140 km, beyond which the Moho refraction, *Pn*, arrives first. *PmP* is the reflection from the Moho. The later reverberations are multiples and scattered arrivals in the crust. The slope of the *Pn* arrival travel-time curve yields an estimate of the velocity of the uppermost mantle. The map shows a summary of many investigations, with contours of *Pn* velocity under the United States. Note the lower velocities characteristic of the tectonically active western region. (Top from Gish *et al.*, *J. Geophys. Res.* **86**, 6029–6038, 1981; © copyright by the American Geophysical Union. Bottom from Braile *et al.*, 1989.)

layering, just as for refraction work, but the signal environment is much more complex in reflection profiles. Modern whole-crustal imaging actually involves both classes of seismic recording, with refraction and reflection work in the same region giving the best overall model of the crustal structure. Still, the most extensive coverage of the subsurface has involved straightforward refraction modeling, so we emphasize results from those procedures.

One of the earliest, and most important, applications of refraction profiling was the systematic reconnaissance of uppermost mantle velocity variations, as directly measured by *Pn* velocity. Figure 7.9 shows a map of *Pn* velocity contours below the United States, interpolated from many localized *Pn* velocity determinations like that in the Globe profile. Note that in the Basin and Range province the *Pn* velocities are lower than beneath the Colorado

Plateau (under northern Arizona and southeastern Utah) or under the stable continental platform underlying the Midwest. These variations are due to uppermost mantle differences in temperature and petrology, which are directly linked to crustal processes such as ongoing rifting in the Basin and Range. By its very nature, the *Pn* wave travels large horizontal distances, and assigning any single local velocity to it involves a lateral averaging along the path. If the profiles are not reversed, one cannot assess the possible bias due to any dip of the refractor (see Box 3.2), and the Moho boundary itself may not be a sharp boundary or may be underlain by velocity gradients that can bias the *Pn* arrival times (Figure 7.3). Thus, recent investigations of *Pn* velocity variations have attempted to allow for more heterogeneity in *Pn*, using tomographic inversions that explicitly model variations on each path, exploiting path overlap and intersection.

An image of *Pn* velocity variations obtained by seismic tomography for the western United States is shown in Figure 7.10. Many features are found in common with Figure 7.9; however, the tomographic model allows more detailed variations to be detected along any particular path. Both earthquakes and explosions provided the data used in this study, so earthquake locations, crustal thicknesses, and crustal velocity variations must be included in the tomographic model inversion.

The *Pn* crossover distance and *PmP* intercept time provide primary constraints on crustal thickness on any particular refraction profile. Figure 7.11 shows a contour map of crustal thickness variations below North America, which has interesting comparisons with Figure 7.9. Both maps are based on compilations of localized one-dimensional crustal models, which exhibit strong lateral variations between regions. Continental crustal thickness cor-

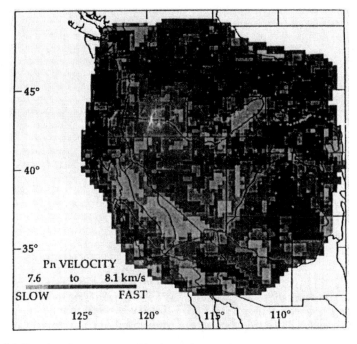

FIGURE 7.10 Result of a tomographic inversion of *Pn* velocity in the western United States, obtained by analysis of many crossing paths. (From Hearn *et al.*, *J. Geophys. Res.* **96**, 16,369–16,381, 1991; © copyright by the American Geophysical Union.)

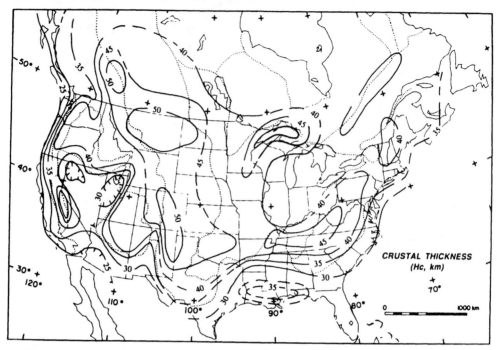

FIGURE 7.11 Contour map of crustal thickness under North America obtained from a great number of studies of regional-distance travel-time curves. Note the thinner crust in the Basin and Range extensional region under Utah and Nevada. (From Braile *et al.*, 1989.)

relates weakly with *Pn* velocity and heat flow; thick crust usually has a high *Pn* velocity and low heat flow. The crust also tends to be thicker under tectonically stable regions and old mountain belts and thinner under actively rifting areas such as the Basin and Range and the Rio Grande Rift of central New Mexico. These gross characteristics of continental thickness and Moho properties, readily determined from refraction profiles, are critical to understanding crustal processes.

Refraction profiling has been extensively performed on several continents, and some gross continental characteristics are summarized for different continental provinces in Figure 7.12. The velocities of the rocks at depth place important bounds on the petrology of the deep crust, but debate continues over the precise composition and state of the deep continental rocks, particularly those overlying the Moho. This has

been fueled by relatively recent applications of reflection seismology approaches to examine details of the *PmP* and midcrustal reflections. Beginning in the 1970s, petroleum industry methods were utilized by university consortiums, such as CO-CORP and CALCRUST in the United States and similar groups in Europe, to examine the deep crust. Using high-frequency, near-vertical-incidence reflections, they have found that the crustal layers and boundaries defined by low-resolution refraction methods have highly variable, small-scale structure. An example of the high-frequency reflected wavefield from a local portion of the Moho beneath Germany is shown in Figure 7.13. This display corresponds to many seismograms running from top to bottom, with darker regions corresponding to larger-amplitude reflections, some of which correlate from trace to trace to define a deep reflector.

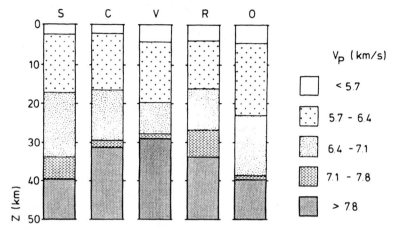

FIGURE 7.12 Idealized velocity–depth distributions in various continental crustal provinces. S, shield areas; C, Caledonian provinces; V, Variscan provinces; R, rifts; O, orogens. (From Meissner and Wever, 1989.)

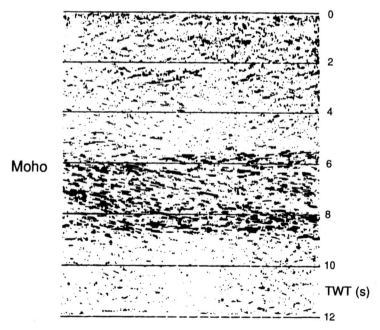

FIGURE 7.13 Interpreted seismic reflection profile showing multiple reflection at two-way travel times (TWT) of 6–9 s below the Black Forest, Germany. This is believed to indicate a layered, or laminated, Moho transition. (After Meissner and Bortfield, 1990.)

The time scale is two-way travel time (from the surface down and back), so the energy between 6 and 8 s corresponds to depths near 24 km, near the Moho in this region. Whereas a 1-s-period wave refracting horizontally along this boundary would reveal little more than the presence of an overall velocity increase, the high-frequency reflections show a complex, laminated Moho boundary with a vertical distribution of reflecting interfaces. In some regions, a comparable strong reflectivity is observed throughout the lower crust, while in other regions the whole crust, or just the shallow crust, has many strong reflections. Seismologists are combining reflection and refraction methods to map out the lateral variations in crustal seismic reflectivity character (Figure 7.14), but no consensus has been achieved about the cause of these variable properties. Likely factors in the seismic response of the crust include rheo-

logical transitions from shallow brittle behavior to deeper ductile behavior, crustal thinning and stretching, fluid concentrations, igneous intrusive layering, and crustal underplating. Seismology will continue to provide the primary probe for interpreting continental crustal fabric.

This variable character of crustal reflections even raises the question of what defines the crust. The Moho is more distinctive than intermittent midcrustal structures like the Conrad discontinuity, but does it strictly represent a chemical transition? Feldspar is the most abundant mineral in the continental crust, followed by quartz and hydrous minerals. The most common minerals in the mantle are ultramafic, such as olivine and pyroxene. A typical K-feldspar rock has a P-wave velocity of about 6 km/s, and dunite (olivine) may have a velocity of 8.5 km/s. The laminated Moho transition suggested by Figure

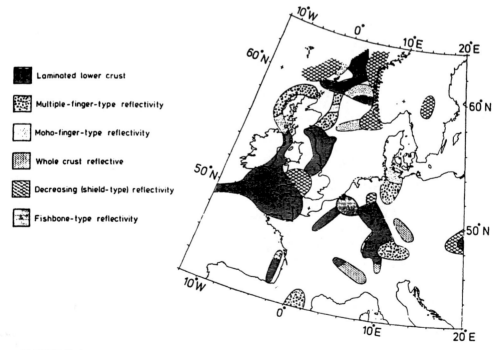

Laminated lower crust

Multiple-finger-type reflectivity

Moho-finger-type reflectivity

Whole crust reflective

Decreasing (shield-type) reflectivity

Fishbone-type reflectivity

FIGURE 7.14 Characterization of crustal reflectivity variations in Europe, showing complex pattern of crustal variations. These are believed to represent the complex tectonic history of the region. (From DEKORP Research Group, 1990.)

7.13 is clearly not a single, sharp chemical boundary, and petrologists have established a complex sequence of mineral reactions with increasing pressure and temperature. A combination of gross chemical stratification, mineralogical phase transformations and reactions, laminated sill injection and partial melting, and rheological variations occur near the crust–mantle boundary, and great caution must be exercised in interpreting limited-resolution seismological models for the transition in terms of associated processes and chemistry. The same caution holds for all models of deeper structure as well.

A key to making progress toward understanding the continental crust is the development of progressively higher-resolution, three-dimensional models. We can achieve this only by merging the methodologies of refraction and reflection seismology with seismic tomography. The *Pn* velocity model in Figure 7.10 is one example of merging refraction methods with tomography to produce a two-dimensional model. Figure 7.15 shows an example of merging reflection seismology with tomography to develop a three-dimensional model for the shallow crust beneath a volcano. In this region, previous reflection work has established the crustal layering surrounding the volcanic caldera, so raypaths from sources at different distances from the recording stations in the caldera were fairly accurately known. This allowed raypaths from different azimuths and distances to thoroughly sample the volume below the array, allowing a block inversion like that in Figure 7.4 to be set up. The east–west profile through the resulting model shows velocity perturbations of $\pm 10\%$ in blocks below the caldera, with the slowest velocities near the central region at a depth of 3–6 km. The bottom figure shows a row of the resolution matrix for the central lowest-velocity block, indicating fair spatial resolution because of the variety of raypaths available. Many comparable seismic tomography experiments are being conducted to analyze other complex regions of continental crust. Figure 1.21 shows a reflection profile in the East African Rift zone, an area where seismologists are using seismic tomography experiments to look at the three-dimensional structure below an active rift.

Investigations of oceanic crust have followed a similar evolution, with primary reconnaissance refraction studies being used to map out gross crustal properties and more recent reflection and tomographic imaging experiments targeting localized regions of particular tectonic interest. Figure 7.16 shows representative one-dimensional seismic-velocity models determined in regions of variable crustal age. The crustal thickness varies less than in continental regions, being on the order of 5–7 km in most places, but anomalous regions occur, such as near fracture zones, where the crust may be as thin as 3 km, and beneath oceanic plateaus, where it is as much as 30 km thick. These anomalous regions account for less than 10% of the oceanic region. For normal oceanic crust the thickness is set by chemical differentiation and intrusion near midocean spreading ridges, and the thickness subsequently varies little. However, the *Pn* velocity does vary strongly with age of the oceanic lithosphere. Young oceanic plate near the ridge has *Pn* velocities of 7.7–7.8 km/s, while the oldest Jurassic crust (~ 200 million years old) may have *Pn* velocities that exceed 8.3 km/s.

The shallowest layers in oceanic crust involve water-saturated muds and ooze (layer 1) that are underlain by layer 2, which has a steep velocity gradient extending over 1 to 2 km. Layer 2 corresponds to fractured, water-saturated basaltic crust that becomes more competent with depth. In some older crust the thickness of layer 2, a "weathered" layer, is large, extending deep into the basaltic crust. The P_2 arrival is the direct wave through this weathered gradient, which may be a smooth increase rather than the stepwise layers seen in

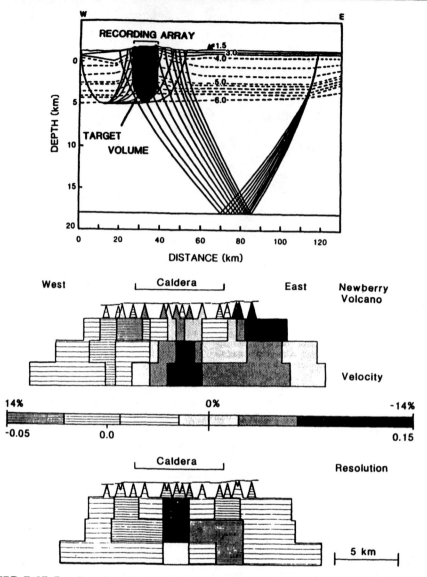

FIGURE 7.15 Results of a high-resolution seismic experiment near Newberry Volcano, Oregon. The top shows typical raypaths used in the tomography, for explosive sources at different distances. An east–west cross section through the center of Newberry caldera (middle) shows velocity perturbations found in the inversion. The resolution matrix for the central block under the caldera is shown at the bottom. (Adapted from Achauer *et al.*, *J. Geophys. Res.* **93**, 10,135–10,147, 1988; © copyright by the American Geophysical Union.)

Figure 7.16. The P_3 head wave along the top of the midcrustal transition to more competent rock is often observable as a first arrival over a limited distance range (Figure 7.8), unlike the Conrad arrival. The typical distance to crossover in oceanic crust is about 40 km; thus refraction pro-

files span a much different distance range in oceanic profiling. As a result, many more joint reflection–refraction analyses of oceanic crust have been done. The thin crust and strong velocity gradients in oceanic regimes motivated many of the early developments of synthetic seismo-

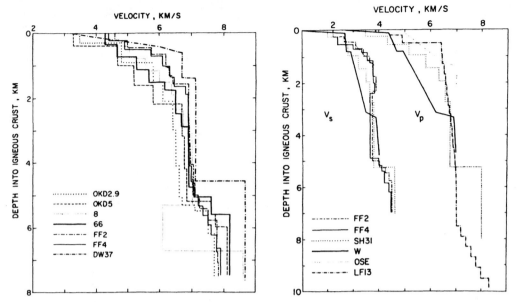

FIGURE 7.16 (Left) Crust *P* velocity profiles for young (< 20 million year) oceanic basin structures found by synthetic modeling. (Right) Crustal *P* and *S* velocity structures from oceanic regions older than 20 million years. (From Spudich and Orcutt, *Rev. Geophys. Space Phys.* **18**, 627–645, 1980; © copyright by the American Geophysical Union.)

gram modeling to extract more information from the seismic wavefield, because crossover and travel-time branch identifications are more difficult in oceanic environments.

Oceanic environments were also the first to provide convincing evidence for uppermost mantle anisotropy. Measurement of *Pn* velocities as a function of azimuth with respect to the strike of spreading-ridge segments has revealed higher velocities perpendicular to the ridge (Figure 7.17). The solid curve in Figure 7.17 is the anisotropic variation measured in laboratory samples of rocks believed to represent uppermost mantle materials under oceanic crust (exposed in upthrust ophiolites). These measurements are very compatible with the spreading-ridge data. The several-percent azimuthal anisotropy is assumed to reflect petrographic fabrics in the residual uppermost mantle material from which the basaltic crust has been extracted. Seismologists are now exten-

sively studying crustal anisotropy in both oceanic and continental regions to infer rock fabric and predominant stress orientations.

Three-dimensional imaging of oceanic crustal regions is concentrating on structure under islands and spreading ridges, and seismic tomography images similar to Figure 7.15 are beginning to map out the volcanic plumbing underlying the rift system. It will require much more extensive mapping before we understand oceanic crustal processes. This is particularly true for transitional structures such as continental margins, oceanic plateaus, and rift systems such as the Afar rift near the Red Sea.

7.2.2 Upper-Mantle Structure

As we delve deeper into the Earth, the information gleaned from seismology plays an increasingly large role in our knowledge

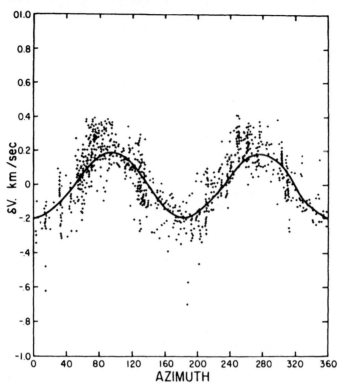

FIGURE 7.17 Azimuthal anisotropy of *Pn* waves in the Pacific upper mantle. Deviations are from a mean of 8.159 km/s. The data are points from seismic refraction results; the curve is a laboratory measurement from an ophiolite structure believed to correspond to oceanic mantle. (Modified from Morris *et al.*, 1969 and Christensen and Salisbury, 1979.)

of the interior. Only a handful of unusual processes have exposed samples of mantle materials at the surface for petrological analysis, and no drill hole has yet penetrated to the Moho anywhere. Thus the material properties revealed by seismology play a dominant role in constraining both the composition and dynamics of the mantle. The disciplines of mineral physics and geodynamics directly utilize seismological observations as boundary conditions or measurements that experimenters strive to explain.

As we stated earlier, the entire Earth can be approximately viewed as a layered, one-dimensionally stratified, chemically differentiated planet composed of crust, mantle, and core. These major layers are separated by boundaries (the Moho and the core–mantle boundary, or CMB) across which seismically detectable material properties have strong contrasts. Additional stratification of the mantle is represented by global seismic-velocity boundaries at depths near 410, 520, and 660 km (Figure 7.18) that define the *transition zone* in the lower part of the upper mantle. These boundaries give rise to reflections and conversions of seismic waves, which reveal the boundaries and allow us to model them in terms of depth and contrasts in velocity, density, and impedance. Once these values are reliably determined by seismology, high-pressure, high-temperature experiments on plausible mantle materials are used to assess the likely cause of the particular structure inside the Earth. Whereas major compositional contrasts

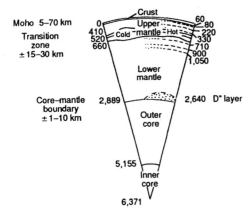

Moho 5–70 km

Transition zone ±15–30 km

Core–mantle boundary ±1–10 km

Crust

Upper mantle

Cold — mantle — Hot

0
410
520
660

60
80
220
330
710
900
1,050

Lower mantle

Core–mantle 2,889

2,640 D″ layer

Outer core

5,155

Inner core

6,371

FIGURE 7.18 Schematic cross section through the Earth, showing depths of globally extensive and intermittent boundaries detected by seismic-wave analysis. Major mantle boundaries exist near depths of 410 and 660 km, bracketing the upper-mantle transition zone. A weak boundary near 520 km may exist globally as well. These are primarily phase boundaries and have minor topographic relief due to thermal variations. The core–mantle boundary is the major chemical boundary inside the Earth, separating the silicate mantle from the molten iron-alloy core. (Reprinted by permission from Lay, *Nature*, vol. 355, pp. 768–769; copyright © 1992 Macmillan Magazines Limited.)

probably underlie the Moho and CMB contrasts, the 410-, 520-, and 660-km discontinuities probably represent mineralogical phase transformations that involve no bulk change in composition but reflect a transition to denser lattice structures with increasing pressure.

The mantle also has localized boundaries at various depths associated with laterally varying thermal and chemical structure (Figure 7.18). The upper 250 km of the mantle is particularly heterogeneous, with strong regional variations associated with surface tectonic provinces. The uppermost mantle just below the Moho is a region with high seismic velocities (*P* velocities of 8.0–8.5 km/s) that is often called the *lid* because it overlies a lower-velocity region. The base of the lid may be anywhere from 60 to 200 km in depth, and it

is thought to represent the rheological transition from the high-viscosity lithosphere to the low-viscosity asthenosphere. The thickness of the lid varies with tectonic environment, generally increasing since the time of the last thermotectonic event. Under very young ocean crust, the lid may actually be absent, but it is commonly observed under old ocean plates. Beneath the lid is a region of reduced velocity, usually referred to as the low-velocity zone (LVZ). Gutenberg first proposed the presence of the LVZ in 1959, and it is thought to correspond to the upper portion of the rheologically defined asthenosphere. The LVZ may be very shallow under ridges, and it deepens as the lid thickness increases. Beneath old continental regions the LVZ may begin at a depth of 200 km, with relatively low velocities usually ending by 330 km, where a seismic discontinuity is intermittently observed. A relatively strong seismic discontinuity is observed under some continental and island-arc regions near a depth of 220 km, which may be associated with LVZ structure or some transition in mantle fabric associated with concentrated LVZ flow structures.

Below 350 km, the transition zone has less pronounced lateral variations, although seismologists have reported intermittent seismological boundaries near depths of 710, 900, and 1050 km. A depth of 710 km or so is reasonable for the boundary between the upper and lower mantle, as known mineralogical phase transformations could persist to this depth. Few, if any, candidate phase transformations are expected in the lower mantle at greater depths; thus intermittent deeper boundaries may involve chemical heterogeneity. The only strong candidate for such a deep intermittent boundary is found about 250 km above the CMB, as discussed later. The only major boundary in the core is the inner core–outer core boundary at 5155 km depth, which is associated with freezing of the inner core as the geotherm crosses the solidus, giving rise to a solid inner core.

Seismological investigation of the mantle and core can be discussed in terms of (1) study of velocity gradients with depth, generally found either by inverting travel-time curves, as in the Herglotz–Wiechert method, or by doing parameterized inversions of model representations of the velocity gradient, and (2) study of seismic discontinuities. A vast array of methods has been developed to study mantle discontinuities, exploiting the many types of interactions with boundaries that seismic waves can have, including refractions, reflections, and conversions. Both travel-time inversions and boundary-interaction analyses can be incorporated in tomographic models to develop three-dimensional models of both smooth and discontinuous variations inside the Earth. The primary motivation for this whole effort is to develop sufficiently accurate representations of the detectable elastic-wave material properties to allow us to interpret composition and

dynamics with some degree of confidence. We will now review some of the methodologies and results about the deep Earth found from systematic application of the principles in Chapters 2 to 4 for body waves, surface waves, and free oscillations.

One of the most important ways of studying upper-mantle structure has been analysis of direct P and S waves at upper-mantle distances. Beyond a distance of 10° to 13°, the first arrivals on seismograms in continental regions correspond to waves that have turned in the upper mantle rather than refracted along the Moho (Figure 7.19). The precise depths to which the waves observed from 10° to 15° have penetrated depend strongly on the velocity structure in the lid, with strong lid-velocity increases refracting energy from relatively shallow depths, while lower velocity gradients or decreases cause the energy to dive deeply into the upper mantle. As early as the 1940s it was clear that at a distance of ~15° the ray parameters associated with the first-arrival travel-time curves of P and

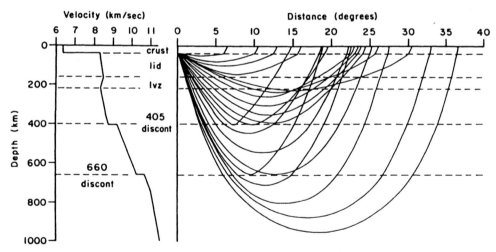

FIGURE 7.19 Complexity of seismic raypaths at upper-mantle distances produced by velocity variations with depth. Raypaths are shown for a velocity structure characteristic of a stable continental region, with a thick high-velocity lid above a weak low-velocity zone (lvz). Oceanic or tectonically active continental structures will have different raypaths to each distance. Multiple arrivals at a given distance correspond to triplications caused by rapid upper-mantle velocity increases. (From LeFevre and Helmberger, *J. Geophys. Res.* **94**, 17,749–17,765, 1989; © copyright by the American Geophysical Union.)

S changed abruptly, but it was not until the 1960s that seismic data were adequate to reveal two triplications in the travel-time curves from 15° to 30°, leading to the discovery of the 410- and 660-km discontinuities (actually, reflections from the boundaries were being detected at around the same time). These triplications dominate the upper-mantle signals, and they arise from the wavefield grazing along the discontinuities with strong refractions, as shown by the raypaths in Figure 7.19.

An example of seismic travel-time and ray-parameter observations at upper-mantle distances is shown in Figure 7.20. Note that the travel times show a particularly clear change in first arrival time near 15°, with a more subtle crossover near 24°. Characterization of the upper-mantle structure in this region, beneath the Gulf of California, depends strongly on identifying the secondary branches of the triplication. When this is possible, as is clearly the case in Figure 7.20b, we can reliably determine the depth and size of the impedance contrasts, with the gradients between the structures being controlled by the curva-

ture of the travel-time curves (measured most directly by the ray-parameter values) and by amplitude variations. Model GCA accurately fits the travel-time and ray-parameter data, constraining the transition zone structure well but any features in the very thin lid and LVZ to a lesser degree. The upper-mantle distance observations are often called *wide-angle reflections*, because they graze along the discontinuities. At wide angles of incidence the reflection coefficients are primarily sensitive to velocity contrasts at the boundaries and are insensitive to density contrasts. The *P* waves constrain the *P* velocity discontinuity only, and *S* waves constrain only the *S* velocity discontinuity.

Modeling the amplitude and waveform information at upper-mantle distances now plays as large a role as fitting the travel times and ray parameters. This is due to the additional sensitivity to sharpness (depth extent) of the discontinuity and adjacent velocity gradients provided by amplitude information. Figure 7.21 displays waveforms corresponding to the travel times in Figure 7.20, with the dense profile

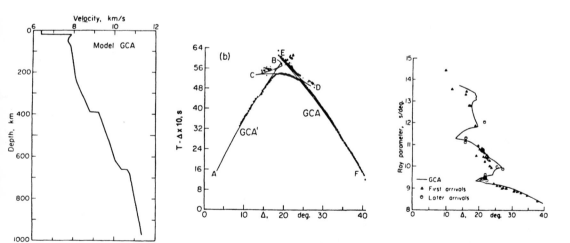

FIGURE 7.20 Left: A *P*-wave velocity model found for the tectonically active upper mantle below the Gulf of California. The observed and predicted travel times for this region are shown in the middle, and observed and predicted ray parameter values are shown on the right. (Modified from Walck, 1984. Reprinted with permission from the Royal Astronomical Society.)

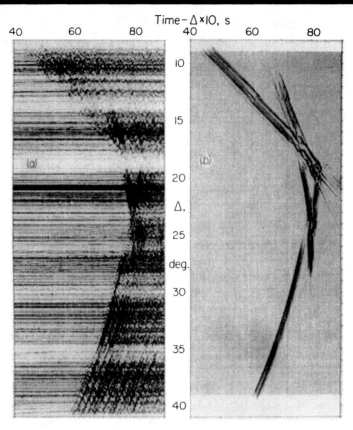

FIGURE 7.21 (a) Data record section combining short-period array recordings for 10 events spanning the range 9°–40°. (b) Synthetic record section for model GCA from Figure 7.20, on the same scale, with source wavelets varying with distance to match the data. (From Walck, 1984. Reprinted with permission from the Royal Astronomical Society.)

being accumulated for multiple sources recorded at a large array of stations in southern California. The profile on the right is for model GCA, which clearly shows the two upper-mantle triplications. Note that near 20° the triplications overlap, giving five arrivals (quintuplication) on a given seismogram. Because any discontinuity near a depth of 520 km would appear at these distances, it has been difficult to identify confidently any such structure with wide-angle reflections. The data in Figure 7.21 are clearly more complex than the synthetics, with many additional incoherent, and some coherent, later arrivals. These correspond to waves reverberating in the structure under the sta-

tions, unmodeled source complexity, scattered arrivals from the surface above the sources, and scattered energy in the volume between the sources and receivers. Although a few clear secondary branches can indeed be observed and timed, it is not surprising that the much sparser data available prior to 1960 failed to clearly reveal upper-mantle triplications. The synthetics provide a helpful guide to the expected relative amplitudes of the later arrivals, and the GCA model was adjusted to match characteristics of the secondary arrivals.

While short-period seismograms, like those in Figure 7.21, or modern broadband seismograms are most useful at upper-mantle distances, extensive modeling of long-period body waves has been performed as well. Probably the most useful long-period data have proved to be the multiple surface reflections *PP* and *SS* observed at distances of 30° to 60°. These phases involve waves that travel twice through the upper-mantle triplications, which doubles the time difference between triplication branches, making them more observable in the long-period signals. Figure 7.22 shows observed and synthetic *SS* phases traversing the tectonically active region of the East Pacific Rise and western North America. The phases are aligned on the direct *S* phases, which are simple because they have bottoming depths in the smooth lower mantle. The large secondary arrivals are *SS*, which show dramatic waveform changes as the signals sweep through the (doubled) triplication distance. Both waveform modeling and timing are used to determine a model, TNA (Figure 7.23), that matches the observed behavior closely.

By modeling *P*, *S*, *PP*, and *SS* waves at different distances in tectonically uniform regions, a suite of upper-mantle velocity models has been developed. Some are shown in Figure 7.23, which illustrates the 10% *S*-wave and 5% *P*-wave velocity variations observed in the lid and LVZ, with the primary transition zone discontinuities at 410 and 660 km having fairly consistent depths and velocity increases. The variable nature of strong velocity gradients near 60, 80, 160, 200, and 300 km is the underlying cause of the intermittence of boundaries near these depths in Figure 7.18. The lower-velocity models, TNA and T7, are for tectonically active western North America, and SNA and K8 are higher-velocity models found under the Canadian and Baltic shields, respectively. The details of the LVZ are extremely hard to determine precisely because the low velocities do not refract energy back to the surface, but some evidence indicates that shear velocity heterogeneity persists to depths of at

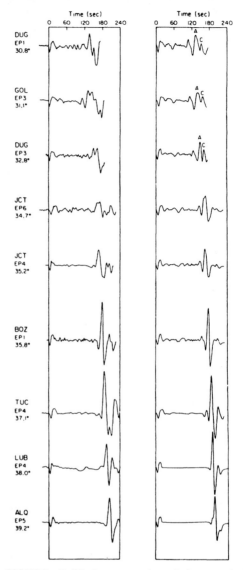

FIGURE 7.22 An example of long-period waveform modeling to determine upper-mantle shear velocity structure. *SH*-component waveforms traversing the mantle under the western United States are shown on the left. The first arrivals are *S* waves turning in the lower mantle. The second arrivals are the *SS* waves, once reflected from the surface, with raypaths encountering upper-mantle triplications. Synthetic waveforms for model TNA (Figure 7.23) are shown on the right. The *SS* waveforms have dramatic waveform changes with distance, caused by upper-mantle triplications, which are well modeled by the synthetics. (From Grand and Helmberger, 1984. Reprinted with permission from the Royal Astronomical Society.)

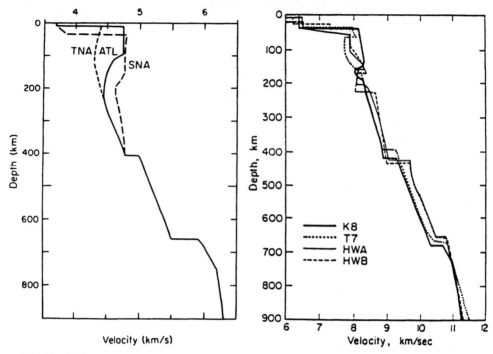

FIGURE 7.23 A variety of *S*-wave (left) and *P*-wave (right) velocity models determined in different regions by waveform modeling of upper-mantle triplication arrivals. Models with higher velocities (SNA, K8) correspond to the upper mantle beneath stable shields. Models with lower velocities (TNA, T7) correspond to tectonically active regions. Note that little variation occurs below 400 km depth.

least 350 km, while *P* velocity heterogeneity may be confined to the upper 250 km. The tectonic associations of these variations, with thick high-velocity regions beneath continental cratonic shields, has led to the idea of a thick root existing beneath ancient continental crust instead of the 100-km-thick lithosphere of classical plate tectonics scenarios.

The velocity discontinuities near 410 and 660 km have contrasts of 4% to 8%, with comparable density increases. Establishing the underlying cause of these changes in material properties has been a major effort of the past 25 years, and the work continues. The primary features that must be established for any structure are (1) its global extent, (2) the velocity and density contrasts, (3) the depth (hence pressure) and variation in depth (topography) of the boundary, (4) the sharpness of the con-

trast, and (5) behavior of the contrast in regions of known temperature structure, such as near downwelling slabs.

The 410- and 660-km discontinuities are observed rather extensively and appear to be global features, but they do vary in depth. Depths as shallow as 360 km and as deep as 430 km have been reported for the "410," with 20- to 30-km variations reported for the "660." The precision of depth estimates from wide-angle studies is not that great because of their limited resolution of lid and LVZ structure, which can baseline-shift the transition zone structure. Thus, depth estimates are usually based on near-vertical-incidence reflections from both above and below the boundary.

One of the classical procedures used for detecting and constraining the depth of mantle discontinuities is illustrated in

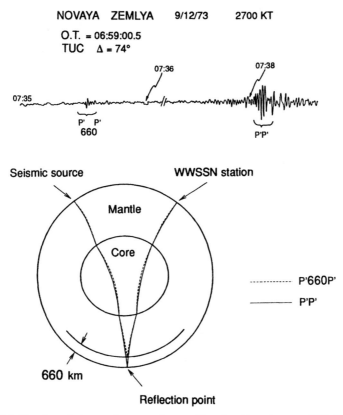

NOVAYA   ZEMLYA   9/12/73   2700 KT

O.T. = 06:59:00.5
TUC   Δ = 74°

Seismic source   WWSSN station

Mantle

Core

----------- P'660P'

————— P'P'

660 km

Reflection point

FIGURE 7.24 Short-period waveform and raypaths for the phases *P'P'* (*PKPPKP*) and *P'₆₆₀P'*, the underside reflection from a discontinuity 660 km deep below the *P'P'* surface reflection point. The difference in arrival times of these phases reveals the depth of the mantle reflector, while the amplitude ratio indicates the impedance contrast.

Figure 7.24. Underside reflections arrive as precursors to *PKPPKP* (*P'P'* for short), observed at a distance near 70° from the source. The time difference between *P'P'* and precursors such as $P'_{660}P'$ is proportional to the depth of the discontinuity. The depth is determined using a model for seismic velocities above the reflector. The amplitude ratio and frequency content of the precursor provide a measure of the impedance contrast. We must account for the frequency-dependent propagation effects caused by focusing due to core velocity structure, but this technique, first developed in the 1960s, is reliable if the signals are rich in high-frequency energy. The sample record shown, which has an unusually large, isolated $P'_{660}P'$ arrival, is from an underground nuclear explosion, a good source of high-frequency energy.

Unfortunately, the *P'P'* precursor approach is limited by the narrow distance range of favorable focusing by the core, but many analogous discontinuity interactions have been utilized to bound the depth, magnitude, and sharpness of upper-mantle discontinuites. Example raypaths and associated nomenclature are shown in Figure 7.25. A variety of topside and bottomside reflections and conversions can be detected, each with a distance range favorable for observing isolated arrivals. Many localized observations of the mantle-discontinuity phases have been examined;

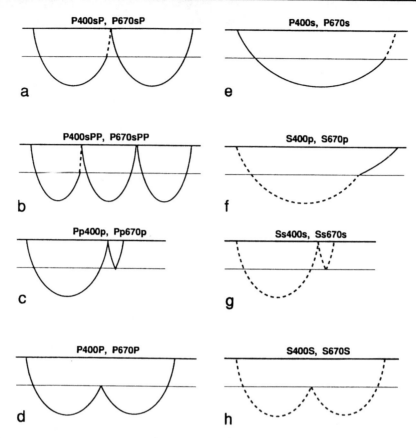

FIGURE 7.25 Examples of the multitude of raypath reflections and conversions produced by all mantle discontinuities. Many of the corresponding arrivals are very weak, but they can be detected by stacking waveforms at appropriate distances. (From Shearer, *J. Geophys. Res.* **96**, 18,147–18,182, 1991; © copyright by the American Geophysical Union.)

however, the availability of large digital data sets has created the opportunity to seek these phases on a global scale. Using a variation of the waveform-stacking procedure described in Box 6.1, large numbers of seismograms have been combined to make the impressive displays in Figure 7.26. Digitally recorded long-period body-wave seismograms from globally distributed shallow events were combined (stacked by summation after alignment on a reference phase) in 0.5° windows throughout the teleseismic distance range where mantle discontinuity interaction phases are observed. The stacks aligned on

P (Figure 7.26a), SH (Figure 7.26b), and SSH (Figure 7.26c) are shown along with computed travel-time curves for major and minor seismic phases for model PREM, which has discontinuities at 410- and 660-km depths. The first two stacks exhibit clear topside reflection arrivals ($Pp_{410}P$, $Pp_{660}P$, $Ss_{410}s$, $Ss_{660}s$) as well as strong converted phases ($P_{410}sP$, $P_{660}sP$, $P_{410}sPP$, $P_{660}sPP$) from the boundaries. The SS stacks show prominent underside reflections ($S_{410}S$, $S_{660}S$). The clarity of these low-amplitude arrivals convincingly demonstrates the global existence of these discontinuities, and the coherence of the

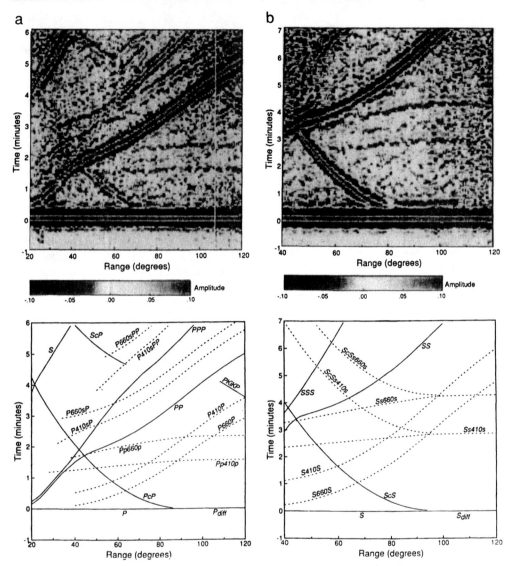

FIGURE 7.26 Observed seismogram profiles comprising thousands of stacked waveforms at different distances for (a) vertical components aligned on the *P* waves, (b) horizontal transverse components aligned on the *S* waves, and (c) horizontal transverse components aligned on the *SS* waves. In each case a travel-time curve is shown below for model PREM, with arrival times of major seismic phases (solid lines) and minor phases generated by interactions with upper-mantle discontinuities near 410 and 660 km depth. (From Shearer, *J. Geophys. Res.* **96**, 18,147–18,182, 1991; © copyright by the American Geophysical Union.)

C

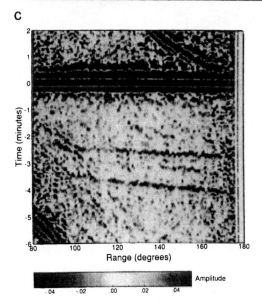

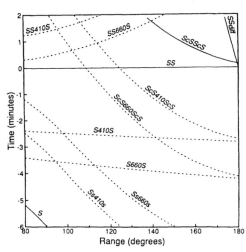

FIGURE 7.26—*Continued*

stacked images bounds the topographic variability of the boundaries to a few tens of kilometers.

The *SS* underside precursors are exceptionally clear (Figure 7.26c) and allow isolated observations to be investigated. The time difference between $S_{660}S$ and *SS* can be used in a fashion similar to the *P'P'* precursors to map the depth variation of the boundary on a much more extensive scale. The $S_{660}S$–*SS* travel-time differ-

ences can be corrected for shallow heterogeneity by using models that correct for *SS* midpoint travel-time anomalies caused mainly by structure in the upper 400 km under the *SS* bounce point, allowing a mapping of depth to the "660"-km boundary. The first such global map of an internal mantle discontinuity is shown in Figure 7.27, derived from the $S_{660}S$ observations. This is a smoothed version of the isolated observations but shows that the depth varies globally by 20 to 30 km, with the boundary tending to be deeper in regions of subducting cold material. The magnitude of the topography reveals physical characteristics of the boundary, such as the magnitude of thermal heterogeneity that must exist to cause a phase boundary to vary in depth by this amount.

We have seen a few of the ways in which seismology determines characteristics of the major transition zone mantle boundaries. This information is used by mineral physicists to constrain the mineralogy of the mantle materials. Since the 1950s it has been known that common upper-mantle minerals such as olivine $(Mg,Fe)_2SiO_4$ and enstatite $(Mg,Fe)SiO_3$ undergo phase transformations with increasing pressure. In particular, the low-pressure olivine crystal converts to the β-spinel structure at pressures and temperatures expected near 410 km depth, the β-spinel structure converts to γ-spinel structure near 500 km depth, and then the γ-spinel structure converts to perovskite $(Mg,Fe)SiO_3$ and magnesiowüstite $(Mg,Fe)O$ at conditions near 660 km depth. These mineralogical phase transitions compact the crystal lattice (Figure 7.28) and cause an increase in seismic-wave velocity. Because the phase transformations can occur over a fairly narrow range in depth (pressure), it is plausible that upper-mantle discontinuities may indicate internal phase boundaries.

The experimental and observational agreement in depth, sharpness, and topographic variability of transition zone boundaries favors their interpretation as

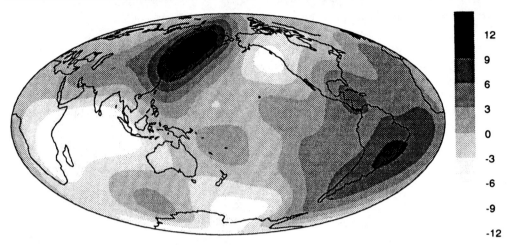

FIGURE 7.27 A smooth fit to the apparent depth of the 660-km discontinuity from *SS* precursor $S_{660}S$ observations. The depths have been corrected for an upper-mantle shear velocity model and are plotted relative to the mean depth of 659 km. (Adapted from Shearer and Masters, 1992. Reprinted by permission from *Nature* **355**, 791–796; © copyright 1992 Macmillan Magazines Limited.)

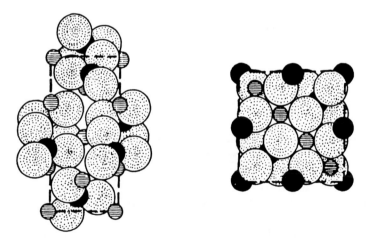

FIGURE 7.28 One of the major causes of rapid upper-mantle seismic-velocity increases is phase transformation, in which material with uniform composition collapses to a denser crystal structure with increasing pressure. An example is shown here for olivine, a common mineral in the upper mantle. On the left is shown the low-pressure form of the olivine crystal, in which large atoms are oxygen, horizontally striped ones are silicon, and black atoms are magnesium. On the right is shown the high-pressure β-spinel phase of the same mineral, in which the atoms are closer together. This transformation occurs at pressures near a depth of 410 km and probably accounts for the seismic velocity increase near that depth. (Modified from Press and Siever, 1978.)

primarily the result of phase transitions. The olivine–β-spinel transition is widely accepted as the cause of the 410-km discontinuity, and this boundary should occur at shallower depth in cold regions, such as descending slabs, and at greater depth in warm upwellings. The 660-km discontinuity is most likely due to the dissociative phase transformations of olivine (γ-spinel) and enstatite to perovskite and magnesiowüstite, with a similar transition in garnet perhaps occurring a bit deeper (~ 710 km). These endothermic transitions have pressure–temperature slopes opposite to those of the exothermic olivine–β-spinel

transition; thus cold downwellings should depress the boundary (Figure 7.18). This is consistent with the pattern of 660-km discontinuity topography seen in Figure 7.27.

If these major transition zone discontinuities are predictable phase boundaries, does it follow that we know the upper-mantle composition? Unfortunately, the problem is complex, and many plausible mantle constituents are consistent with observed seismic models. The main seismic model parameters are compared with experimental results for a variety of plausible Earth materials in Figure 7.29. Comparisons such as these are used to construct

---

Box 7.2 S-Wave Splitting and Upper-Mantle Anisotropy

The upper mantle has laterally variable anisotropy, and we can use body waves to determine local anisotropic properties beneath three-component broadband stations. One technique that has been widely applied is analysis of teleseismic shear-wave splitting. Anisotropy causes shear waves to split into two pulses, one traveling faster than the other, with the differential time between the two pulses accumulating with the path length traversed in the anisotropic region. The two pulses can most readily be observed for phases that have a known initial polarization before they enter the anisotropic region. One such phase is *SKS*, which has purely *SV* polarization at the conversion from *P* to *SV* at the core–mantle boundary. If the entire mantle along the upgoing path from the core to the surface is isotropic, then the *SKS* phases will have no transverse component. If anisotropy is encountered, the *SKS* phases split into two polarized shear waves traveling at different velocities, and this generally results in a tangential component of motion when the ground motion is rotated to the great-circle coordinate frame. An example of *SKS* arrivals on horizontal broadband seismograms is shown in Figure 7.B2.1a, with a clear, distorted tangential component being observed. By searching over all possible back azimuths, it is possible to find the polarization direction of the anisotropic splitting, which has fast and slow waves with the same shape (Figure 7.B2.1b). Note that the ground motion has a complex nonlinear polarization.

By shifting the fast and slow waves to eliminate the anisotropic splitting, a linear polarization is retrieved. Rotation of the shifted traces into the great-circle coordinate system then eliminates the transverse component of the *SKS* phase. This procedure reveals the polarization direction of the fast and slow shear waves and the magnitude of the splitting. If paths from different azimuths give uniform anisotropic properties, the region where splitting occurs is likely to be in the shallow structure under the station.

continues

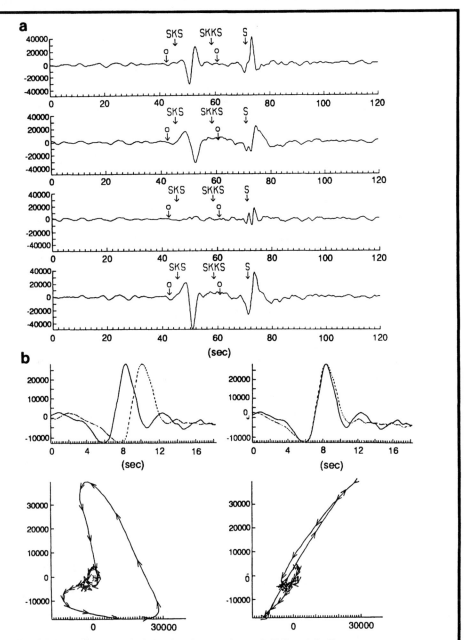

FIGURE 7.B2.1 (a) Teleseismic broadband recordings of *SKS* and *S*. The top two traces are the original transverse and radial components, showing strong *SKS* energy on the transverse component. The bottom two traces are corrected for anisotropic splitting by finding the fast and slow directions and shifting the signals for the splitting prior to rotation. (b) Top traces: Superposition of fast and slow *SKS* arrivals with observed splitting (left) and after correction (right). Bottom: Particle motion plots in the horizontal plane for uncorrected (left) and corrected (right) observations, showing how correction restores the linear polarization of the wave. (Modified from Silver and Chan, *J. Geophys. Res.* **96**, 16,429–16,454, 1991; © copyright by the American Geophysical Union.)

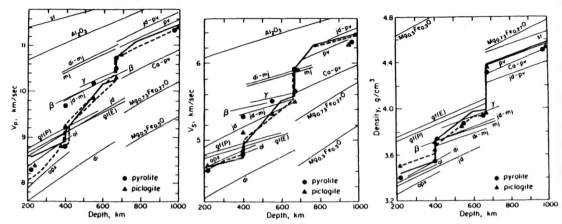

FIGURE 7.29 Comparison of globally averaged models for upper-mantle *P* velocity (left), *S* velocity (middle), and density (right) with various experimental results for minerals and mineral assemblages. Such comparisons are used to interpret the velocity discontinuities and gradients found by seismology. (From Bass and Anderson, *Geophys. Res. Lett.* **11**, 237–240, 1984; © copyright by the American Geophysical Union.)

models of bulk-mantle composition compatible with the observed density and seismic velocities; however, this is a nonunique process. At this time it is still possible to conclude that either upper and lower mantles must be chemically stratified, even allowing for phase transformations in the transition zone, or that the upper and lower mantles have a uniform composition. We must refine both seismological models and mineral physics experiments before the extent of mantle chemical stratification can be resolved.

7.2.3 Upper-Mantle Tomography

The variation in one-dimensional velocity models for different tectonic regions indicates upper-mantle heterogeneity, but tomographic techniques are required to determine the complete structure. Numerous seismic wave data sets have been used to investigate upper-mantle heterogeneity on a variety of scale lengths. Dense arrays of seismometers provide sufficient raypath coverage of the underlying lithosphere to develop tomographic images of the upper mantle to depths of several hundred kilometers. Figure 7.30 shows one example of

P velocity heterogeneity beneath the southern California region, developed using teleseismic travel-time relative residuals. The tomographic image, obtained using the SIRT algorithm, shows a high-velocity tabular structure extending into the upper mantle beneath the Transverse Ranges. This is believed to represent deep crustal material forced downward by convergent flux near the big bend of the San Andreas fault in the San Gabriel Mountains.

A continental-scale tomographic image, developed from *S*- and *SS*-wave travel-time residuals beneath North America, is shown in Figure 7.31. The shear velocity variations range from the slow-velocity model TNA beneath the western tectonic region to the high-velocity structure SNA (Figure 7.23) beneath the Canadian shield, but the tomographic model shows the full three-dimensional variation of the continental root. This model also suggests that low-velocity material extends quite deep below the Mid-Atlantic Ridge and near segments of the East Pacific Rise.

Even larger-scale tomographic models have been developed using large data sets of digitally recorded surface waves. A

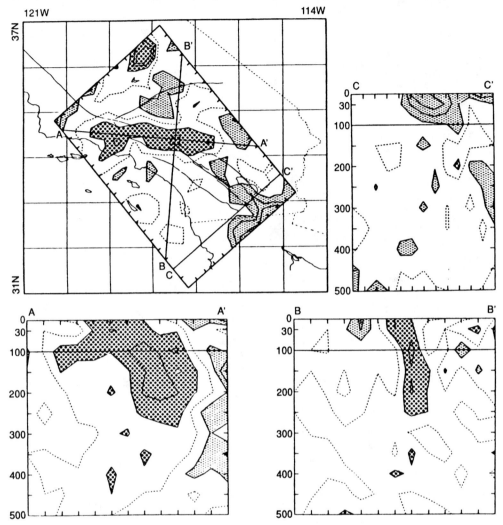

FIGURE 7.30 Lithospheric heterogeneity under southern California revealed by seismic tomography using *P*-wave travel times. The dark, stippled region is a fast-velocity body that extends vertically into the mantle below the Transverse Ranges. (From Humphreys *et al.*, *Geophys. Res. Lett.* **11**, 625–627, 1984; © copyright by the American Geophysical Union.)

model for the entire Pacific Ocean was developed by tomographic modeling of crossing Love- and Rayleigh-wave paths, with the phase velocity for Love waves with different periods for lithosphere of different ages being shown in Figure 7.32. Along with the data, a simple model with a thickening seismic lid overlying an LVZ is shown, with predicted variation of phase velocity. This simple model, based on a thermally growing plate structure, fits the data well except for the longest-period waves, which suggest less pronounced thickening of the plate with age. Thus, modeling seismic velocities under oceanic regions can constrain the thermal evolution of the oceanic lithosphere. An additional factor influencing the plate thickness is the spreading rate of the ridge system. The variation in lithospheric age

Depth = 140 to 235 km   Depth = 235 to 320 km

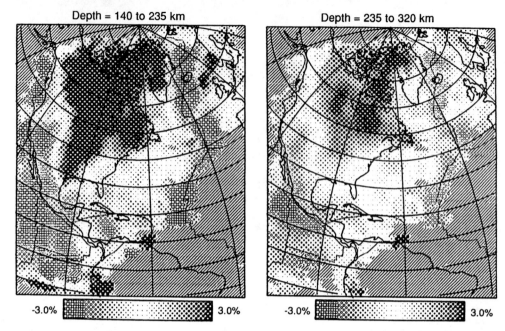

-3.0%   3.0%   -3.0%   3.0%

FIGURE 7.31 Shear velocity variations with respect to average velocity in two depth ranges below North America. Darker areas correspond to fast velocities below the stable shield region of the continent. The tectonically active area of western North America has slow upper-mantle shear velocities. (Modified from Grand, *J. Geophys. Res.* **92**, 14,065–14,090, 1987; © copyright by the American Geophysical Union.)

of oceanic plates is shown in Figure 7.33, and it is clear that the Pacific plate has long had the fastest spreading rate. This is manifested in differences between Love-wave dispersion in oceanic lithosphere of the same age, with the relatively slow-spreading Atlantic and Indian plates having higher velocities by a given age than the Pacific plate.

In addition to characterizing the gross structure of oceanic upper mantle, the to-mographic models reveal detailed struc-tures in the upper mantle under important tectonic elements such as ridges and hot spots. Figure 7.34 shows cross sections through a relatively high-resolution tomo-graphic shear-wave model derived from surface-wave dispersion observations, for profiles across the midocean ridges, and across major hot spots. The profiles across ridges reveal broad low-velocity regions in the upper 100 km extending laterally over

500 to 1500 km on either side of the ridge axis. The width of the low-velocity regions found under the ridges is proportional to the spreading rate. The ridges overlie broad regions of slower than average man-tle, but there is not clear evidence of con-centrated deep anomalies. This suggests a passive upwelling process with plates pulling apart and the decrease in pressure of upwelling material leading to partial melting. This velocity structure appears to be very different from the deep velocity structure beneath hot spots, where the to-mographic images reveal concentrated low-velocity regions from 100 to 200 km under the surface volcanic centers, sug-gesting a deeper source and perhaps a more active upwelling mantle flow under hot spots. Future efforts will improve the resolution of such images, which at present have a horizontal scale of 1000 km and a vertical resolution of 60 km.

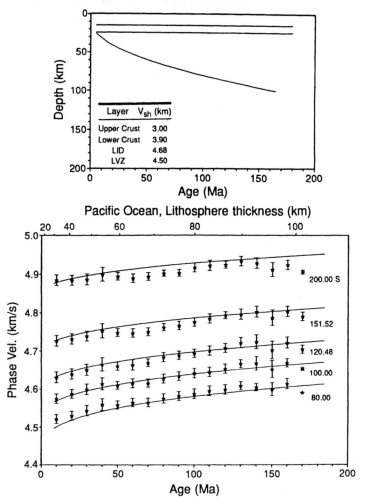

FIGURE 7.32 (Top) A model for lithospheric evolution of shear velocities under the Pacific Ocean. (Bottom) Comparison of the model predictions (solid lines) with observed Love-wave phase velocities with periods from 80 to 200 s plotted as a function of plate age. The observations were obtained by a tomographic inversion of Love waves traversing the Pacific. (From Zhang, 1991.)

In addition to being applied to regions of upwelling flow, seismic tomography has been extensively applied to determine upper-mantle structure beneath subduction zones, where old oceanic lithosphere sinks into the mantle. For over 25 years, seismic observations have revealed that the sinking tabular structure of the plate provides a high-velocity, low-attenuation (high-Q) zone through which waves can propagate upward or downward through the mantle, and seismologists have expended extensive effort to determine properties of both the slab and the overlying mantle wedge. The arc volcanism overlying the cold slab is clearly associated with anomalous properties of the wedge, as has been known for several decades (Figure 7.35). Localized regions of very slow shear velocities and strong attenuation indicate partial melting in the mantle above the slab. The locations of anomalous zones have been constrained

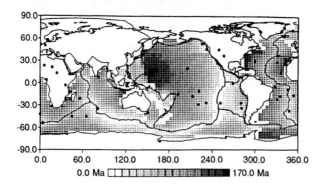

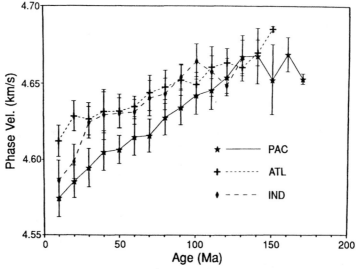

FIGURE 7.33 (Top) Map of oceanic lithosphere age variations. (Bottom) Comparison of Love-wave phase velocities for 100-s-period waves in different age plate beneath the Pacific (PAC), Atlantic (ATL), and Indian (IND) oceans. (From Zhang, 1991.)

mainly by using phases from intermediate- and deep-focus events in the slab recorded either at stations in the overlying plate or at teleseismic distances, where surface-reflected phases such as pP and sS are observed.

Up until the mid-1970s subducting slabs were always shown as cartoons, geometrically constrained by the intraplate earthquake locations in the Benioff zone, but as vast numbers of body-wave travel times began to accumulate, it became possible to develop tomographic images of the subducting high-velocity slab without *a priori*

constraints. One such image, shown in a cross section through the model, is displayed in Figure 7.36. This plot shows the P velocity anomaly relative to the average velocity at each depth for the Aleutian slab with 2% contour intervals. The locations of shallow and intermediate-depth earthquakes are superimposed on the velocity anomaly structure. On average, the slab appears to be about 5% faster than surrounding mantle in the upper 400 km, with local perturbations as large as 10–12% at depths near the low-velocity zone. Earthquakes deeper than 40 km ap-

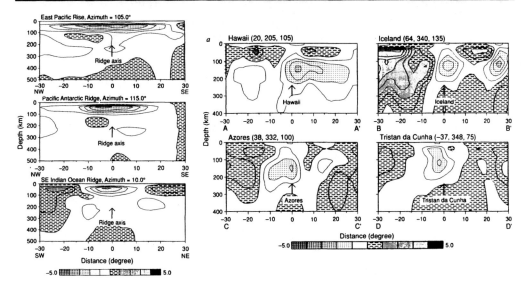

FIGURE 7.34 Cross sections into a three-dimensional shear velocity structure determined from Love- and Rayleigh-wave tomography on transects perpendicular to mid-ocean ridges (left) and hotspots (right). Dotted regions correspond to more than 1% slow shear velocities and indicate shallow low velocities under the ridges and deeper low velocities under the hot spots. (Reprinted with permission from *Nature* **355**, 45–49; © copyright 1992 Macmillan Magazine Limited.)

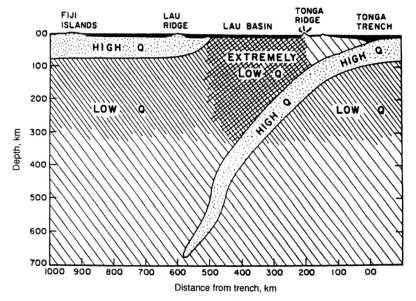

FIGURE 7.35 Cartoon of subducting oceanic lithosphere showing variations in attenuation below an island arc. The slab has high seismic velocities and low attenuation (high Q) relative to the surrounding mantle. (From Barazangi and Isacks, *J. Geophys. Res.* **76**, 8493–8516, 1971; © copyright by the American Geophysical Union.)

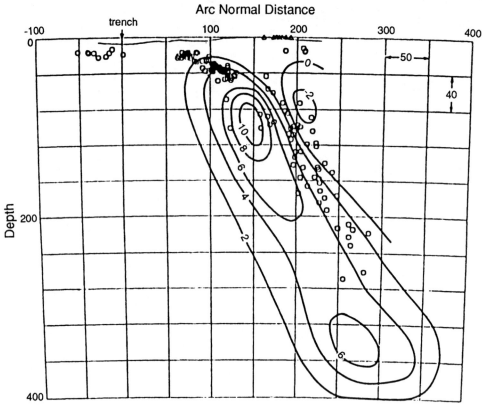

FIGURE 7.36 Cross section through a seismic tomography model determined for beneath the Aleutian island arc. The contours show velocity perturbations relative to the average velocity at each depth and define a fast-velocity, dipping lithospheric slab. The squares are earthquake locations near the cross section, showing that most intermediate-depth events are located near the upper-mantle boundary of the subducting slab. Most events shallower than 50 km are thrust events on the contact between the two plates. (From Engdahl and Gubbins, *J. Geophys. Res.* **92**, 13,855–13,862, 1987; © copyright by the American Geophysical Union.)

pear to be located within the upper portion of the subducting slab. Other tomographic images of subducting slabs are shown in Figure 1.22. Developing velocity models like these requires relocation of the sources using three-dimensional ray tracing because the strong velocity anomalies can significantly deflect the raypath. Stable solution of the nonlinear inversion for structure and source location when raypath perturbations must be included is a current area of active research in seismic tomography.

Lower-resolution images, but of a global extent, have been developed using huge sets of long-period body waves, surface waves, and free oscillations. The surface-wave inversions often involve a two-step procedure of first extracting tomographic models for the Rayleigh- and Love-wave dispersion, using either block models or spherical harmonic expansions, and then doing a second inversion to find an *S* velocity model (usually with constrained *P* velocity and density models) that fits the dispersion. Examples of surface-wave dis-

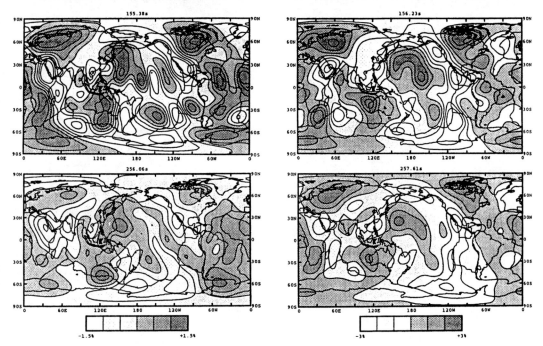

FIGURE 7.37 Global models of Rayleigh-wave (left) and Love-wave (right) phase-velocity variations relative to the average velocity at periods near 155 and 256 s. Fast regions are darker, with the contour interval being 0.5% for Rayleigh waves and 1% for Love waves. (Modified from Wong, 1989.)

persion models represented in spherical-harmonic expansions up to degree 12 are shown in Figure 7.37. The first such models were obtained in 1984 (see Figure 1.23), following several years of accumulating global digital seismic data from the ASRO, SRO, and IDA networks (Chapter 5). The surface-wave dispersion models have varying depth sensitivity depending on how deeply each period wave samples. Since the shorter-period fundamental modes sense the lithosphere, a strong correspondence in Rayleigh- and Love-wave phase-velocity variations with tectonic region is found. Note that ocean ridges tend to overlie slow-velocity material, while continental cratons and older ocean regions tend to overlie fast-velocity material. The heterogeneity in phase velocity is stronger at short periods, and the patterns for longer-period waves lose any clear relationship with shallow structure.

When surface-wave dispersion and body-wave travel-time observations are inverted for shear velocity, models like those in Figure 7.38 are obtained, with shear velocity variations at shallow depths of 50 to 150 km having patterns that are clearly related to surface tectonics. However, variations at greater depths have weaker, tectonically unrelated patterns. These heterogeneities reflect both chemical and thermal variations in the mantle, with shear velocities varying moderately by $\pm 5\%$. The associated thermal variations should be on the order of several hundred degrees, perhaps stronger in small-scale regions that are not resolved by the low-resolution models. Fluid-dynamics calculations show that such large thermal variations have density variations that drive solid-state convective motions of the upper mantle over long time scales. Thus, to the degree that the velocity variations map thermal

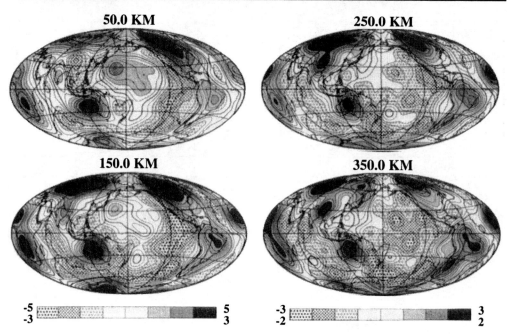

FIGURE 7.38 Maps of global *S*-wave velocity heterogeneity at four depths in the upper mantle, determined from Love- and Rayleigh-wave dispersion observations. The model is expanded to spherical harmonic degree and order 12. Faster regions are darker, and the scale differs between two columns. (From Su *et al.*, 1994. Figure prepared by Y. Zheng).

heterogeneity, these images can be associated with dynamic flow in the interior. However, one must isolate the thermal variations from possible petrological variations, which might, for example, contribute to the high-velocity deep roots of continents, before inferring dynamics from any seismic model.

One procedure for obtaining independent constraints on thermal variations in the mantle is to map the aspherical structure of seismic-wave attenuation. As described in Chapter 3, seismic attenuation structure in the Earth is only crudely approximated by any radially layered structure. With order-of-magnitude lateral variations in the quality factor, Q, at the upper-mantle depths, a simple stratified Q model like that for PREM (Table 7.1) or SL8 (Figure 4.33) is useful only to provide a baseline for body-wave attenuation or for global averages of surface-wave paths or free oscillations. Seismic attenuation is

path specific in the Earth, and some general relationships with surface tectonics have been observed. Typically, paths through the mantle under stable cratons, which tend to be relatively high-velocity regions, have much lower attenuation, or lower t^* values, than paths under tectonically active regions such as the western United States or midocean ridges. The lower mantle appears to have very little attenuation everywhere except possibly near the base of the mantle, so most of the regional variations in attenuation occur in the upper-mantle low-attenuation region, from 50 to 350 km depth. This has been demonstrated by comparing high-frequency attenuation between shallow and deep-focus earthquakes. Because most attenuation mechanisms at upper-mantle depths are expected to involve thermally activated processes, developing aspherical upper-mantle attenuation models can provide a mantle thermometer to complement

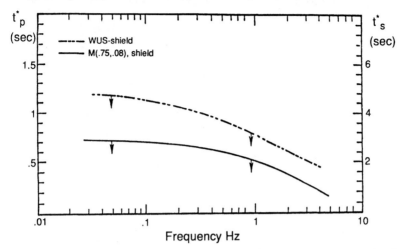

FIGURE 7.39 The frequency dependence of t_P^* and t_S^* (the path-integrated effects of intrinsic attenuation on P and S waves) for paths under North America. The solid curve is for paths entirely within the mantle under the stable continental shield. The dashed curve is for a path that traverses from the tectonically active area to the shield. It is assumed that $t_P^* = 0.25 t_S^*$. (From Der *et al.*, 1982. Reprinted with permission from the Royal Astronomical Society.)

velocity-variation and boundary-topography models.

In the 1980s much progress was made in determining high-frequency body-wave attenuation models and crude tectonic regionalizations. An important complication that was demonstrated is that attenuation does vary with frequency in the short-period body-wave band, leading to frequency-dependent t^* models like those shown in Figure 7.39. Two models are shown, both of which have an S-wave attenuation factor that is four times the P-wave attenuation factor, $t_\beta^* = 4t_\alpha^*$. The upper curve is for a body-wave path with one leg through the upper mantle under a tectonically active region and a second leg through a stable shield, the lower curve for source and receiver paths that both pass through a shield structure. A purely tectonic path would have systematically higher t^* values. For the mixed path the t_α^* at a period of 1 s is 0.8 s, with models such as PREM predicting values of 0.8–1.2 s over the body-wave frequency band for teleseismic raypaths. Purely shield paths are much

less attenuating than average Earth models or mixed paths, and it is remarkable to see how small the t^* values become at frequencies above 1 Hz. The implication is that 4- to 5-Hz waves are transmitted through the Earth extremely efficiently, with almost no anelastic loss. Recalling that t^* affects body-wave spectral amplitudes by a factor proportional to $e^{-\pi f t^*}$, the decrease in t^* with increasing frequency results in a high-frequency content of teleseismic waves that is many orders of magnitude larger than would be the case if t^* had constant values near 1 s, as was commonly assumed in the 1970s.

Both the frequency dependence and regional variation of t^* near a period of 1 s contribute to variability of short-period body-wave magnitudes, m_b. This has actually had great political significance, since most nuclear test sites in the Soviet Union overlie shield-like mantle and the U.S. test site in Nevada (NTS) overlies a tectonically active region. As a result, explosions of the same yield have smaller teleseismic magnitudes for NTS explosions than for

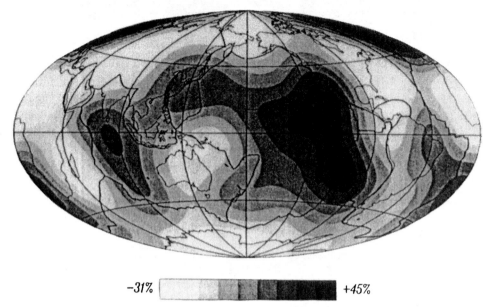

−31% +45%

FIGURE 7.40 Lateral variations of attenuation of seismic waves at a depth of 160 km in the mantle, obtained by analysis of surface wave amplitudes. The attenuation is very high in the eastern Pacific near the East Pacific Rise, as well as along the Mid-Atlantic Ridge. (Courtesy of B. Romanowicz.)

Soviet tests. This has complicated the verification of nuclear test yield limitation treaties, and this concern motivated many of the early studies of regional variations in seismic attenuation.

The high-frequency t^* determinations have relied mainly on comparison of teleseismic ground-motion spectra with theoretical source models or near-source recordings of the seismic radiation. At longer periods, many studies of body-wave attenuation have been conducted to determine average levels and lateral variations in t^* near a period of 10 s. Analysis of multiple ScS_n phases has been the most extensively applied procedure because of its intrinsic stability and source-effect cancellation properties. It is well documented that t^* varies regionally in the long-period body-wave band, but frequency dependence is not reliably resolved. Surface-wave attenuation coefficients can also be determined to develop regionalized and frequency-dependent models, although reli-

able separation of dispersive effects from attenuation effects is a complex procedure. Surface-wave analyses have confirmed the strong regional variations in upper-mantle attenuation across North America and Eurasia, and they constrain the vertical distribution of attenuation variations, which is difficult to attain with most body-wave studies.

Although much progress has been made, development of an aspherical attenuation model is in a state of infancy compared with elastic velocity models. This is in part because it is more difficult to isolate effects of attenuation from propagation and source effects. In the case of surface-wave analyses, it is important to have a detailed velocity model first to account for focusing and dispersive effects, as discussed in Section 4.7. When this is done, the residual amplitude variations can be tomographically mapped into an aspherical model, as shown in Figure 7.40. This is a first-generation, low-resolution model obtained from global surface-wave data, but it is a harbinger of increasingly refined aspherical models yet to come. As the reliability of these models improves, comparisons with the elastic velocity models will enable an improved separation of thermal and chemical causes of the heterogeneity.

Another property of the upper mantle that is still in very early stages of being mapped out is the global distribution of anisotropy. As for attenuation, any simple stratified model of anisotropic properties in the Earth is grossly inaccurate for describing path-specific behavior. This is largely due to systematic regional differences in anisotropic orientation, apparently associated with large-scale lithospheric motions. While teleseismic body waves largely provide point examples of upper-mantle anisotropic character (see Box 7.2), surface waves provide global path coverage required to constrain large-scale anisotropic properties. As might be expected given the oceanic *Pn* velocity anisotropy shown in Figure 7.17, a component of the surface-wave anisotropy appears to correlate with plate motion directions. This is illustrated in Figure 7.41, which plots the orientation of horizontally fast directions of *azimuthal anisotropy* affecting surface waves. Note that the fast directions tend to parallel plate spreading directions in the ocean basins, indicating lithospheric and asthenospheric shear flow–induced fabrics that result in the directional dependence of surface-wave velocities. Improving azimuthal coverage of the surface with increasing numbers of surface-wave observations should improve these azimuthal anisotropy models in the future.

Perhaps an even more fundamental aspect of upper-mantle anisotropy is that it is generally required in order to explain simultaneously Love- and Rayleigh-wave observations when fitting the data to high precision. In other words, isotropic models fail to explain jointly Love- and Rayleigh-

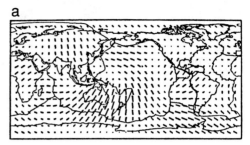

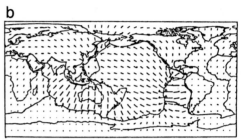

FIGURE 7.41 (a) A map showing the fast direction for Rayleigh-wave phase velocities with a period of 200 s for a model with azimuthal anisotropy. The length of the lines is proportional to the extent of anisotropy. (b) Flow lines at a depth of 260 km for a kinematic plate tectonics model with a low-viscosity channel in the upper mantle. Note that there is some correspondence with azimuthal anisotropy. ((a) From Tanimoto and Anderson, 1984; (b) from the model of Hager and O'Connell, 1979; reproduced from Anderson, 1989.)

wave dispersion on many paths, but simplified anisotropic parameterizations, such as transverse isotropy, provide sufficient parameters to reconcile the Love/Rayleigh-wave discrepancy. Unfortunately, introduction of additional parameters into the large-scale tomographic inversions is often not stable. This issue is important because the best isotropic models that tolerate misfit of the joint Love- and Rayleigh-wave data may have significantly different features than alternative (albeit poorly resolved) anisotropic models. An example is provided by oceanic lithospheric models, for which isotropic inversions imply plate thicknesses that increase to more than 100 km with increasing age, overlying strong low-velocity zones. Anisotropic inversions of the same data can result in maximum lid thicknesses of only 60–70 km and much less pronounced low-velocity zones. Anisotropy introduces many additional parameters and attendant resolution problems into mantle imaging, but it is clearly critical for future efforts to include anisotropic models.

7.2.4 Lower-Mantle Structure

The standard seismological procedure for studying the lower mantle has involved inversion of travel-time curves for smoothly varying structure, with array measurements of ray parameters as a function of distance playing a major role. This is due to the absence of significant boundaries throughout much of the vast region from 710 to 2600 km depth (Figure 7.42). Some observations indicate possible impedance boundaries near 900 and 1050 km (Figure 7.18), but the extent and nature of any structures at these depths are unresolved. The extremal bounds shown in Figure 7.42 constrain any associated impedance contrasts to less than a few

percent for a spherically averaged model. No deeper impedance contrasts appear to exist except locally in the lowermost 250 km of the mantle, which is a region of anomalous velocity gradients called the D'' region.

Both Herglotz–Wiechert and parameterized model inversions for lower-mantle structure suggest smooth variations in properties compatible with self-compression of a homogeneous medium throughout the bulk of the lower mantle. Because all common upper-mantle materials have undergone phase transitions to the perovskite structure at depths from 650 to 720 km, it is generally believed that the lower mantle is primarily composed of $(Mg_{0.9}Fe_{0.1})SiO_3$ perovskite plus $(MgFe)O$, with additional SiO_2 in the high-pressure stishovite form and uncertain amounts of calcium perovskite. No high-pressure phase transitions in the perovskite, stishovite, and MgO (rock salt) structures are expected over the pressure range spanned by the lower mantle, so the absence of structure is compatible with uniform composition. However, bulk com-

---

Box 7.3 Absorption Bands and Frequency Dependence of *Q* in the Mantle

The lower mantle transmits teleseismic body waves with very little attenuation for periods shorter than 0.5 s, but long-period body waves are more attenuated, and free oscillations require even stronger lower-mantle attenuation. This frequency dependence can be interpreted in the context of an absorption band model (Chapter 3) in which the absorption band shifts to lower frequencies in the deep mantle and to higher frequencies in the upper mantle, although with strong lateral variations. There is some evidence that the D'' region has stronger short-period attenuation as well. One model of depth-varying frequency-dependent attenuation throughout the mantle is shown in Figure 7.B3.1. This model is very simplified, and no one-dimensional structure can describe the attenuation for specific paths, but this type of model provides a useful baseline behavior for deep-mantle thermally activated processes. Complete mapping of three-dimensional frequency-dependent anisotropic properties of the Earth is a long-term goal of seismology that will improve our understanding of the Earth's thermal structure at depth.

continues

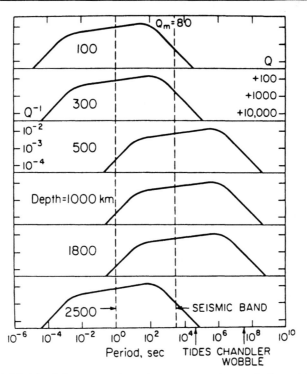

FIGURE 7.B3.1 Model of frequency-dependent attenuation variation with depth in the mantle. Attenuation is parameterized as an absorption band that represents a continuum of relaxation mechanisms, which vary with depth. When Q^{-1} is large, attenuation is high. This model allows short-period waves to travel through the lower mantle with little attenuation, except in the D'' region. (From Anderson and Given, *J. Geophys. Res.* **87**, 3893–3904, 1982; © copyright by the American Geophysical Union.)

position is still uncertain, in particular the Fe and Si component relative to the upper mantle, so it remains unresolved whether a contrast in overall chemistry occurs that would favor stratified rather than whole-mantle convection. A critical parameter in this issue is the density structure of the lower mantle. Body-wave observations do not constrain the density structure in the lower mantle because it lacks detectable impedance contrasts. Instead, it is low-frequency free oscillations, which involve deep-mantle motions of large volumes, that we use to constrain the lower-mantle density structure. The free oscillations are not sensitive to the detailed density structure, but in combination with gross mass con-

straints from the Earth's moments of inertia, they can resolve the average density structure of the lower mantle rather well. The average density, along with density contrasts in the transition zone boundaries, provide the primary data modeled by mineral physicists in their effort to constrain the bulk composition of the lower mantle.

Although no major internal boundaries occur in the lower mantle above the D'' region, lateral heterogeneities occur throughout the region that can be modeled using seismic tomography. Several investigators have used body-wave travel times from thousands of events in three-dimensional inversions for lower-mantle

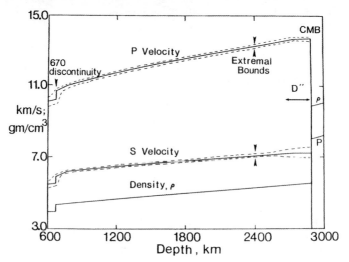

FIGURE 7.42 Average lower-mantle seismic velocity and density structures. No major radial structures appear to occur below 1000 km depth (see Figure 7.18) down to 2600 km depth. The 200- to 300-km-thick *D"* region at the base of the mantle has strong lateral and radial heterogeneity. The extremal bound curves bracket the range of average model parameters consistent with travel-time data. Local structure may extend beyond these bounds. (From Lay, 1989.)

structure. The data are the millions of travel times assembled over the past 25 years by the ISC bulletin. The scatter in these times relative to any homogeneous model (see Figure 1.19) constitutes the evidence for lower-mantle heterogeneity. The models differ in parameterization and inversion method, but they are generally consistent in their large-scale structure. Constant-depth slices through one of the *P*-wave tomography models are shown in Figure 7.43. From 700 to 2500 km depth the lateral variations tend to be less than ±1%, with large slow-velocity regions under Africa and the central Pacific extending throughout the lower mantle. In the *D"* region the heterogeneity is stronger, ±2% for *P* waves and ±3% for *S* waves. The inversions yielding such models involve huge matrices that must be inverted either on the largest supercomputers or by using back-projection methods that lack a resolution analysis. Up to several million travel times are used in these inversions, although redundant data are averaged to

weight the inversion evenly and to enhance signal stability. Since the models result in small heterogeneities, raypath perturbations are ignored, which is reasonable for all depths other than for *D"*, where grazing rays may be sensitive to even a few-percent heterogeneity.

Large data sets of long-period waveforms have also been inverted for lower-mantle tomographic images, sometimes in conjunction with free-oscillation inversions. An example of a shear-wave velocity model for the entire lower mantle is shown in Figure 7.44. The body-wave inversion involves constructing synthetic waveforms for a heterogeneous model that match the data better than waveforms from a starting model, mainly by perturbing the travel time of the wave. The free-oscillation inversions involve calculating multiplet splitting for the heterogeneous structure that matches observed spectral peaks. Although these inversions are based on far fewer observations than the ISC inversions, the data quality is much higher and is free of subtle

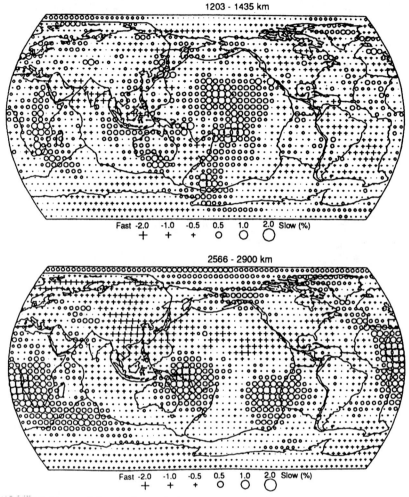

FIGURE 7.43 Maps of *P* velocity variations at two depths in the lower mantle obtained by tomography. Circles indicate slow-velocity regions. (From Inoue *et al.*, 1990.)

biases and noise processes that afflict the ISC data base. Nonetheless, some of the large-scale features from the travel-time and waveform inversions are very similar. The model in Figure 7.44 shows several-percent shear velocity variations throughout the lower mantle, with stronger heterogeneity in the *D″* region. Relatively high velocities in the lower mantle tend to extend downward around the Pacific rim, with relatively slow velocities occurring under the central Pacific and Africa.

The large-scale patterns of lower-mantle heterogeneity are not yet reliably resolved, but they strongly suggest a dynamic convective system. If the velocity heterogeneity is caused by temperature variations, then temperature fluctuations of several hundred degrees occur in the deep mantle. This is sufficient to induce density heterogeneity that will in turn drive solid-state convection. Thus, slow-velocity regions are probably hot, relatively buoyant regions that are slowly rising, while fast-velocity

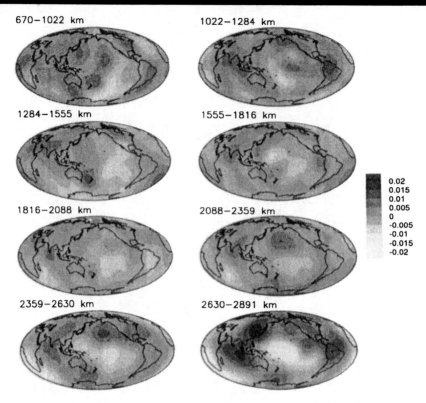

670–1022 km   1022–1284 km

1284–1555 km   1555–1816 km

1816–2088 km   2088–2359 km

2359–2630 km   2630–2891 km

0.02
0.015
0.01
0.005
0
-0.005
-0.01
-0.015
-0.02

FIGURE 7.44 Maps of *S* velocity variations in the lower mantle obtained from tomography using *S* waves and free oscillations. Darker regions are faster velocities. Note that the most pronounced heterogeneity is near the base of the mantle. (Courtesy of G. Masters, 1992.)

regions are colder, sinking regions. If the seismic images indeed reveal the convective flow of the interior, can they resolve the dynamic configuration of the whole mantle? Clearly, higher-resolution studies are needed to resolve this issue, so seismologists have concentrated on two regions in particular: (1) the upper-mantle downwellings with deep seismicity and (2) the *D″* region at the base of the mantle.

One of the most straightforward ways to determine whether upper-mantle material penetrates into the lower mantle or is deflected into a stratified convection system is to determine the fate of subducting oceanic lithosphere. Several methods have been developed to assess the extent of

lower-mantle slab penetration by detecting the high-velocity structure associated with the cold slab material. Earthquakes occur no deeper than 700 km in the mantle; presumably the mechanism by which they occur is terminated by the 660-km phase transition, but they do mark the downwelling flow to at least the base of the upper mantle. If the slab continues to sink unimpeded into the lower mantle, one would expect to observe a tabular high-velocity structure along the downdip extension of the seismic zone. One approach to imaging such a structure is to construct a *residual sphere*, which is a stereographic plot of travel-time anomalies of *P* or *S* waves from a deep-focus event, with the anomalies projected to the azimuth and takeoff angle at the source for each associated raypath. A tabular high-velocity structure below the source should then produce a systematic pattern of fast and slow anomalies, which can be reproduced by modeling. This procedure has been applied over several decades with somewhat mixed results and interpretations. Figure 7.45 shows one of the most convincing cases favoring a high-velocity slab ex-

tension into the lower mantle at least a few hundred kilometers. Residual spheres for events at various depths in the vertically plunging Marianas slab are compared with predictions obtained by ray tracing through a fast slab structure extending to a depth of 1300 km. A cross section perpendicular to the slab model is shown on the right. The intermediate-depth events clearly show a regular pattern of positive

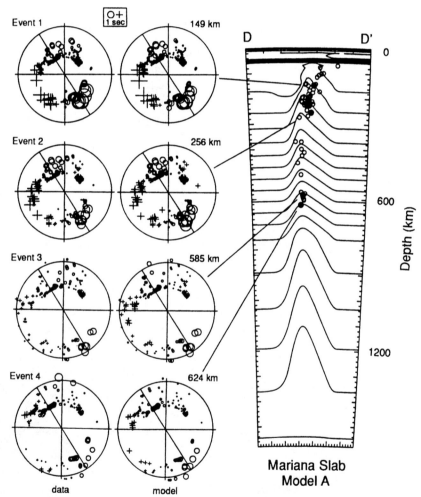

FIGURE 7.45 Residual sphere analysis for intermediate- and deep-focus earthquakes in the Marianas slab. Data for individual events at different depths are shown on the left, synthetic predictions for a slab model are shown in the middle. Seismicity and contours of the velocity model are shown in the cross section on the right, which is in a plane perpendicular to the strike of the slab. (From Creager and Jordan, *J. Geophys. Res.* **91**, 3573–3589, 1986; © copyright by the American Geophysical Union.)

and negative anomalies, with early arrival times observed along the strike of the slab. This is predicted by the model because raypaths at these azimuths have a relatively long path length in the high-velocity material. A similar, although weaker, pattern is observed for very deep focus events. The lower-mantle extension of the slab is introduced to explain the patterns of residuals from the deep event. This procedure is greatly complicated by the fact that the earthquake locations are not known independently and must be determined in the modeling. Also, corrections for deep-mantle and near-receiver velocity heterogeneity must be made before the near-source contribution to the patterns can be reliably modeled. Differences among seismologists as to the magnitude of these corrections have prevented a clear consensus on the depth of slab penetration using residual-sphere analysis.

Rather than model specific events, one can use tomographic imaging to detect any aseismic extension of the slab velocity heterogeneity. Figure 7.36 shows one example for which the velocity heterogeneity of the slab appears to extend about 150 km below the deepest earthquakes. Several studies have produced corresponding images of velocity heterogeneity near deep-focus source regions. One example is shown in Figure 7.46a in a cross section through the Kuril subduction zone. Note the broad region of slightly fast material that extends the high-velocity slab structure into the lower mantle. In order to assess the reliability of this image, the same raypaths were used for the input slab model in Figure 7.46b, and the inversion was performed on

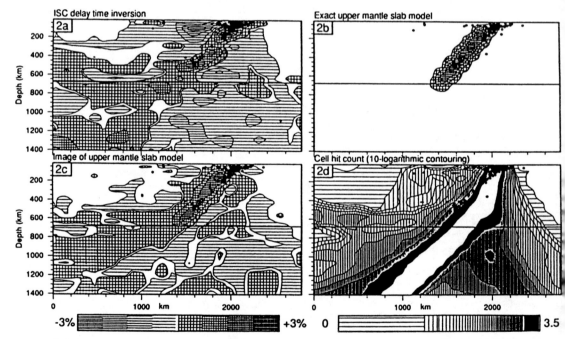

FIGURE 7.46 Cross section through a three-dimensional velocity model in the vicinity of a subducting slab for (a) actual data inversion, (b) an input synthetic mode, and (c) inversion of the synthetic data set. (d) Cell hit count for blocks of the model. The lack of crossing coverage causes downward smearing of the synthetic slab anomaly (c), giving artifacts similar to the patterns in the data. (From Spakman *et al.*, *Geophys. Res. Lett.* **16**, 1097–1100, 1989; © copyright by the American Geophysical Union.)

the synthetic data set, giving the image in Figure 7.46c. Clearly the inversion tends to smear the known anomaly into the lower mantle, giving an erroneous impression of slab penetration. The cell-hit-count plot shows the variable ray coverage of the model that leads to this streaking effect. In order to overcome these contaminating effects, the data sets used in deep-slab imaging have been augmented with more raypaths, including upgoing phases that help to stabilize the depth determination, and with procedures that remove common path effects from the data. This is resulting in much clearer slab images (see Figure 1.22), yet these still provide an ambiguous resolution of the slab-penetration question. In many cases, as in Figure 1.22, the high-velocity material appears to broaden and deflect horizontally near 600–800 km depth, without a tabular extension into the lower mantle. This is compatible with the

large-scale downward deflection of the 660-km phase transition landward of subduction zones, as seen in Figure 7.27. However, it is unclear whether this broadened high-velocity region continues to sink, causing the lower mantle ring of high-velocity material beneath the Pacific rim, or whether thermal coupling may occur across a boundary in a layered convecting system, by which the cool slab material induces downwellings in the deeper layer. As seismic tomography improves, it is hoped that this issue will be resolved.

The D'' region at the base of the mantle has received much attention as well, since the core–mantle boundary is likely to be a major thermal boundary layer. If significant heat is coming out of the core, the mantle will be at least partially heated from below, and D'' may be a source of boundary-layer instabilities. It is often proposed that hot spots are caused by thermal

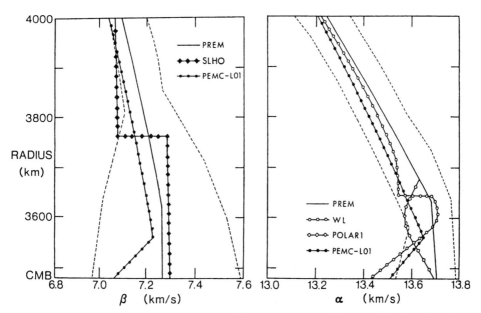

FIGURE 7.47 Various shear-wave (left) and P-wave (right) velocity models that have been proposed for the D'' region at the base of the mantle. A variety of travel-time, waveform-modeling, triplication-modeling, and diffracted-wave analyses were used to obtain these models, and there is no clear, best-average structure. (Reproduced, with permission, from Young and Lay, *Annual Review of Earth and Planetary Sciences* **15**, 25–46, © 1987 by Annual Reviews Inc.)

plumes rising from a hot internal boundary layer, with D'' being a plausible candidate.

Seismological investigations of D'' have not yet resulted in a complete characterization of the region, although it is clearly a more laterally heterogeneous region than the overlying bulk of the lower mantle. A small sample of the remarkable diversity of seismic models for the D'' region is shown in Figure 7.47. These models have been obtained by travel-time studies, free-oscillation inversions, analysis of reflected phases, and studies of waves diffracted into the core-shadow zone. The low-resolution travel-time and free-oscillation inversions tend to give smooth models like the PREM structures, which simply show a tapering of velocity gradients 175 km above the core–mantle boundary. Diffracted-wave studies (Figure 7.48) tend to give models with stronger velocity gradient reductions, producing weak low-velocity zones just above the core (e.g., PEMC-LOI), while other studies have detected abrupt velocity discontinuities near the top of D''. The variation among models is reminiscent of the variety of lithospheric models seen in Figure 7.23, and no single best radial model can be reliably interpreted in terms of boundary-layer structure. Thus, there have been many efforts to determine the three-dimensional structure of the D'' region. This includes tomographic inversions, such as those in Figures 7.43 and

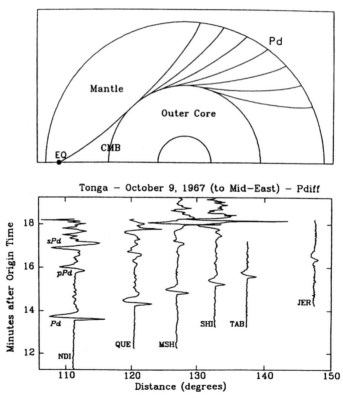

FIGURE 7.48 Raypaths for *P* waves diffracted by the outer core, which arrive in the core-shadow zone, along with a profile of observations. The timing and waveforms of the diffracted waves are most sensitive to velocity structure at the base of the mantle. (Modified from Wysession *et al.*, *J. Geophys. Res.* **97**, 8749–8764, 1992; © copyright by the American Geophysical Union.)

7.44, which resolve large-scale heterogeneities, and analyses of travel times and waveforms of diffracted waves that traverse long distances in the D'' region, as shown in Figure 7.48. Those large-scale studies indicate the presence of $\pm 2\%$ P velocity and $\pm 3\%$ S velocity heterogeneity.

Determination of more localized D'' structure associated with these large lateral variations is performed using vertical reflections and wide-angle grazing body waves. Examples of the latter are shown in Figure 7.49, which shows profiles of S and sS phases and the core reflections ScS and $sScS$. An intermediate arrival between these phases is observed at distances from 72° to 95°, which appears to be a triplication caused by an abrupt velocity increase at the top of D'' in regions that are faster than average. The travel times show a clear triplication. Predictions for a model with a 2.75% shear velocity increase 250 km above the core are shown in Figure 7.49. A similar triplication is intermittently observed in broadband P waves. Figure 7.50 shows a map of regions of D'' that have been studied with wide-angle body waves and the shear velocity models found in each case. The models are similar except beneath the central Pacific, where the velocities vary from slow in the southern portion of the SGHE region to fast in the north. Precursors to near-vertical-incident ScS phases indicate intermittent occurrence of a D'' impedance contrast under the western Pacific. The cause of such a laterally variable boundary is uncertain, but it may plausibly represent a variable chemical boundary layer, somewhat analogous to the continental structures embedded in the surface lithosphere. Much further work is needed before the precise nature and dynamics of the D'' region can be reliably determined.

7.2.5 Structure of the Core

The Earth's core was first discovered in 1906 when Oldham found a rapid decay of P waves beyond distances of 100°, and he postulated that a low-velocity region in the interior produced a shadow zone. Gutenberg accurately estimated a depth to the core of 2900 km in 1912, and by 1926 Jeffreys showed that the absence of S waves traversing the core required it to be fluid. The core extends over half the radius of the planet and contains 30% of its mass. The boundary between the mantle and core is very sharp and is the largest compositional contrast in the interior, separating the molten core alloy from the silicate crystalline mantle. Seismologists have used reflections from the top side of the core–mantle boundary (PcP), underside reflections ($PKKP$), and transmitted and converted waves (SKS, PKP, and $PKIKP$) to determine topography on the boundary, which appears to be less than a few kilometers. The contrast in density across the boundary is larger than that at the surface of the Earth; thus it is not surprising that little if any topography exists. The strong density contrast may be responsible for a concentration of chemical heterogeneities in the D'' region composed of materials that are denser than average mantle but less dense than the core. The material properties of the core are quite uniform (Figure 7.51), with a smoothly increasing velocity structure down to a depth of 5150 km, where a sharp boundary separates the outer core from the solid inner core. This 7–9% velocity increase boundary was discovered by the presence of refracted energy in the core shadow zone by Lehmann in 1936. It was not until the early 1970s that solidity of the inner core was demonstrated by the existence of finite rigidity affecting normal modes that sense the deep structure of the core.

The decrease in P velocity from values near 13.7 km/s at the base of the mantle to around 8 km/s at the top of the outer core profoundly affects seismic raypaths through the deep Earth, as discussed in Chapter 6. The principal P waves with

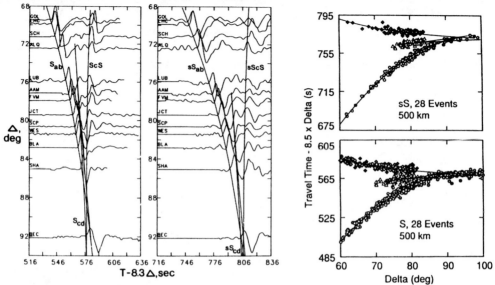

FIGURE 7.49 Lower-mantle shear-wave triplications in *S* and *sS* phases. Data profiles from a single deep-focus event are shown on the left, with superimposed travel-time curves identifying the *S* triplication and *ScS* arrivals in these horizontally transverse, long-period signals. The travel-time curves show observed triplication arrivals (triangles) that require a complex structure in *D″*. The curves are for model SYLO, shown in Figure 7.50. (Left from Lay, 1986; right from Young and Lay, 1990.)

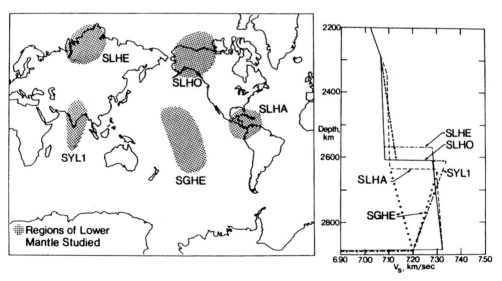

FIGURE 7.50 Map indicating regions of *D″* for which detailed shear velocity structures have been determined. All regions shown except for that under the central Pacific (SGHE), have been found to have a 2.5–3% shear velocity increase from 250 to 300 km above the core–mantle boundary. (From Young and Lay, *J. Geophys. Res.* **95**, 17,385–17,402, 1990; © copyright by the American Geophysical Union.)

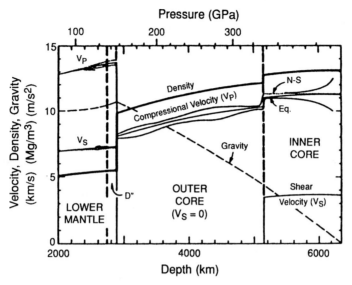

FIGURE 7.51 Seismologically measured density $(1 \times 10^3$ kg/m^3, bold solid curve), elastic-wave velocities (compressional (V_P) and shear (V_S) velocities in km/s, thin solid curves), and gravitational acceleration (in m/s^2, thin dashed curve) through the core and lowermost mantle are shown as functions of depth and corresponding pressure (top scale). Extremal bounds on the V_P profile through the core are included for comparison (thin solid lines about the average V_P curve). The differences between the polar (N–S) and equatorial (Eq.) compressional velocities through the inner core, as proposed by Morelli *et al.* (1986) and Woodhouse *et al.* (1986), are indicated by dotted lines. Heterogeneity in the lowermost mantle (the *D″* region) is illustrated by the variations in V_P and V_S profiles (dotted curves) observed at different locations above the core. (Reproduced with permission, from Jeanloz, *Annual Review of Earth and Planetary Sciences* **18**, 357–386, © 1990 by Annual Reviews Inc.)

paths through the core are shown in Figure 7.52. As the takeoff angle from the source decreases from that of rays that just graze the core and diffract into the shadow zone, the *PKP* waves are deflected downward by the low velocities, being observed at distances of 188° to 143° (PKP_{AB}) and then again at 143° to 170° (PKP_{BC}) as the takeoff angle continues to decrease. Reflections from the inner-core boundary define the *PKiKP* (*CD*) branch (Figure 7.53), after which the *P* wave penetrates the inner core as *PKIKP*, which is observed from 110° to 180°. Note that most of this complexity of the core travel-time curve stems from the velocity decrease, the spherical geometry of the core, and the increase in velocity in the inner core.

These core phases are readily identified on teleseismic short-period recordings, and vast numbers of travel times are reported by the ISC (Figure 7.53). The travel times, ray parameters, and positions of the cusps of the travel-time data are used to invert for models of the core. Since the outer core does not transmit *S* waves, we believe it to be fluid, but the fact that the *S* velocity at the base of the mantle is slightly lower than the *P* velocity at the top of the core (Figure 7.51) means that the core is not a low-velocity zone for the converted phase *SKS*. As a result, the *SKS* phase and attendant underside reflections off the core–mantle boundary, such as *SKKS*, are also used to determine the velocity structure of the outer core, particularly in the

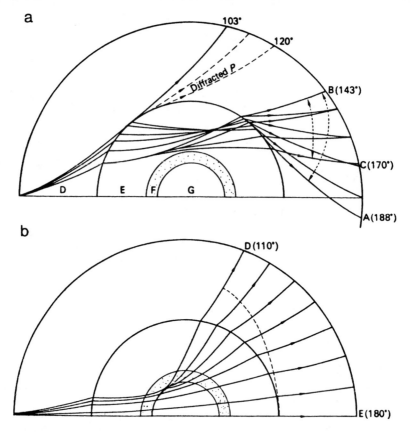

FIGURE 7.52 (a) Raypaths for *PKP* waves traversing the outer core. (b) Raypaths for *PKIKP* waves. (After B. Gutenburg.)

outermost 800 km where *PKP* phases do not have turning points. Some evidence from *SKKS* phases suggests that the outermost 100 km of the core may have reduced seismic velocities relative to a uniform core, possibly representing a chemical boundary layer.

The *P*-wave impedance contrast at the core–mantle boundary is small (which causes *PcP* to be weak), but the boundary is very sharp, so it can effectively reflect short-period energy several times. Phases such as *PKKKKP* (*PK₄P*) and *PK₇P* are thus observed (Figure 7.54), and they provide useful measures of the core velocity structure as well as bounds on the topography on the boundary. The observation of these phases indicates that seismic attenu-

ation in the outer core is negligible, allowing waves to ring on for long durations.

The inner core–outer core boundary has been extensively investigated for more than 50 years, in part due to the scattered arrivals preceding the B cusp, or caustic (Figure 7.53). These arrivals were originally attributed to a complex transition zone at the top of the inner core, and a wide variety of models have been developed for this boundary (Figure 7.55). However, seismic arrays established that the *PKP* precursors are caused by scattering, probably in the *D″* region or at the core–mantle boundary. The currently preferred models for the inner-core boundary involve a simple discontinuity, perhaps with a slightly reduced outer-core velocity gra-

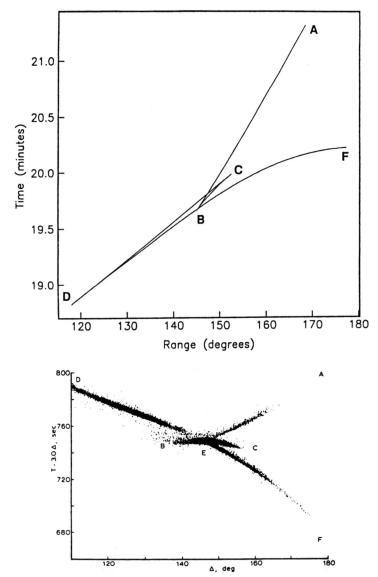

FIGURE 7.53 (Top) Theoretical travel-time curves for *PKP* (*ABC*), *PKiKP* (*CD*), and *PKIKP* (*DF*) branches for PREM. (Bottom) Reduced travel times of about 60,000 *P* waves in the core showing the various branches of the core phases. (Top from Shearer and Toy, 1991; *J. Geophys. Res.* **96**, 2233–2247, 1991; © copyright by the American Geophysical Union. Bottom from Johnson and Lee, 1985.)

dient just above the boundary. No topography has been detected on the inner-core boundary, but it appears that attenuation strongly increases in the inner core.

Both the velocity and density contrasts and attenuation structure of the inner core have been studied using *PKiKP* reflections and *PKIKP* refractions. An example of waveform comparisons used to determine inner-core properties is shown in Figure 7.56. Here the core phases *PKP* (*BC*) and *PKIKP* are seen on a single recording. When the two arrivals are rescaled and superimposed, it is clear that the *DF*

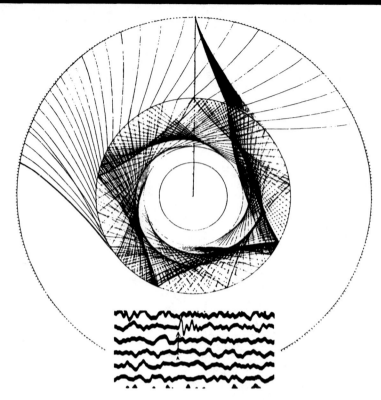

FIGURE 7.54 Multiple-reflection raypaths for PK_nP waves in the outer core. The inset shows the short-period recording of the *PKKKKP* phase from an underground nuclear explosion in the Soviet Union. (Modified from Bolt, 1973.)

branch arrival is broadened relative to the *BC* arrival, indicating a lower Q in the inner core. Waveform modeling is used to match observations like these by considering a suite of Earth models and finding models that match the relative timing, amplitude, and frequency content of the core phases.

Mineral physics experiments demonstrate that the seismologically determined density of the core is lower than expected for pure iron, so the outer core is believed to have about 10% of a light alloying component such as Si, O, C, or S. The inner core may be almost pure iron, with the freezing process brought about by the geotherm dipping below the alloy melting temperature. The freezing process tends to concentrate the lighter component in the fluid. Rise of this buoyant material provides compositionally driven convection in the core, which is believed to sustain core dynamics that produce the Earth's magnetic field. Proximity to the solidus is implied by the existence of the inner core; thus the outer core may actually contain suspended particles, up to 30% by volume. It is not known whether these impart any effective rigidity to the outer core, but anomalous, unexplained core modes (Chapter 4) may require a complex mechanism. Because the core rotates, the polar regions of the outer core may have a separate flow regime from the spherical annulus of material along the equator. Thus, cylindrical symmetry may play a role in

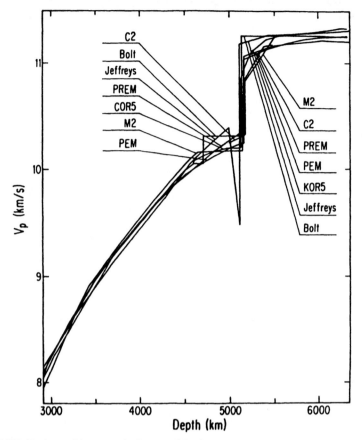

FIGURE 7.55 Various *P*-wave velocity models for the inner core–outer core boundary. Analysis of *PKiKP*, *PKIKP/PKP* waveforms and travel-time behavior underlies most of these models. Most recent models favor a relatively simple boundary, with a sharp velocity increase of 0.8–1.0 km/s, possibly overlain by a zone of reduced *P* velocity gradient at the base of the outer core. (From Song and Helmberger, *J. Geophys. Res.* **97**, 6573–6586, 1992; © copyright by the American Geophysical Union.)

Box 7.4 Structure of the Inner Core

The inner core is a very small region inside the Earth but appears to have surprising internal structure. Seismic waves traversing the inner core on paths parallel to the spin axis travel faster than waves in the equatorial plane, indicating the existence of inner-core anisotropy. This has been detected by travel times of *PKIKP* waves (Figure 7.B4.1) as well as by innercore–sensitive normal modes. The travel-time variations have ~1-s systematic differences with angle from the north–south axis. This axial symmetry may result from convective flow in the inner core that induces an alignment in weakly anisotropic crystals of solid iron. Thus, seismological measurements can reveal dynamic processes as deep as 6000 km into the Earth.

continues

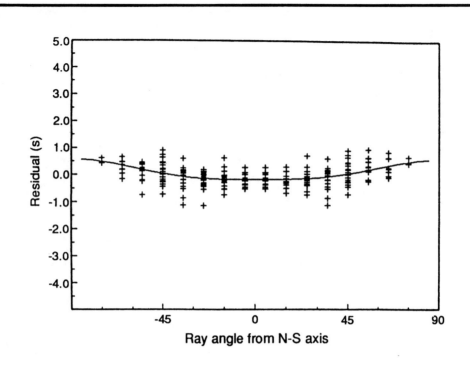

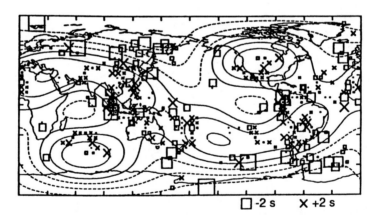

FIGURE 7.B4.1 Two separate analyses of *PKIKP* travel-time anomalies, revealing a pattern of more negative anomalies (faster velocities) for paths along the polar direction. The top plot shows the residuals as a function of azimuth from the axis. The lower plot is a map of the anomalies at the source and receiver locations, with a low-order, degree-4 expansion of the pattern. (Top from Shearer and Toy, *J. Geophys. Res.* **96**, 2233–2247, 1991; ©copyright by the American Geophysical Union. Bottom from Morelli *et al.*, *Geophys. Res. Lett.* **13**, 1545–1548, 1986; © Copyright by the American Geophysical Union.)

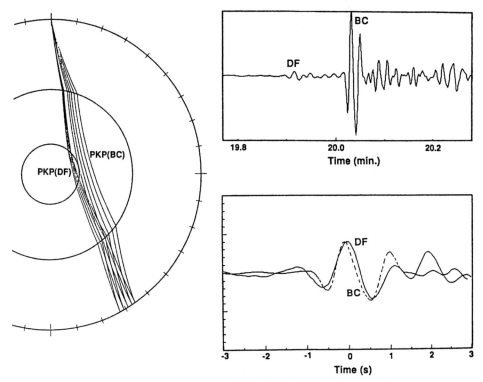

FIGURE 7.56 Comparison of *PKP* (*BC*) and *PKIKP* [*PKP* (*DF*)] arrivals on the same seismogram. Raypaths for these phases are shown on the left. When the *DF* and *BC* branches are overlain, it is clear that *PKIKP* is broadened by its transit through the inner core. (From Shearer and Toy, 1991.)

core characteristics, perhaps with varying degrees of suspended particles in the polar regions. Improved seismic models are needed to detect any such aspherical structure of the outer core, but inner-core asphericity has already been detected (Box 7.4).

References

Achauer, U., Evans, J. R., and Stauber, D. A. (1988). High-resolution tomography of compressional wave velocity structure at Newberry volcano, Oregon, Cascade Range. *J. Geophys. Res.* **93**, 10,135–10,147.

Aki, K., and Richards, P. G. (1980). "Quantitative Seismology." Freeman, San Francisco.

Anderson, D. L., (1989). "Theory of the Earth." Blackwell, Boston.

Anderson, D. L., and Given, J. W. (1982). Absorption band *Q* model for the Earth. *J. Geophys. Res.* **87**, 3893–3904.

Barazangi, M., and Isacks, B. (1971). Lateral variations of seismic-wave attenuation in the upper mantle above the inclined earthquake zone of the Tonga Island arc: Deep anomaly in the upper mantle. *J. Geophys. Res.* **76**, 8493–8516.

Bass, J. D., and Anderson, D. L. (1984). Composition of the upper mantle: Geophysical tests of two petrological models. *Geophys. Res. Lett.* **11**, 237–240.

Bolt, B. A. (1973). The fine structure of the earth's interior. *Sci. Amer.* **228**, 24–33.

Braile, L. W., and Smith, R. (1975). Guide to the interpretation of crustal refraction profiles. *Geophys. J. R. Astron. Soc.* **40**, 145–176.

Braile, L. W., Hinze, W. J., von Freses, R. R. B., and Keller, G. R. (1989). Seismic properties of the crust and uppermost mantle of the conterminous United States and adjacent Canada. *Mem.—Geol. Soc. Am.* **172**, 655–680.

Christensen, N. I., and Salisbury, M. H. (1979). Seismic anisotropy in the upper mantle: Evidence from the Bay of Islands ophiolite complex. *J. Geophys. Res.* **84**, 4601–4610.

Creager, K. C., and Jordan, T. H. (1986). Slab penetration into the lower mantle beneath the

Mariana and other island arcs of the northwest Pacific. *J. Geophys. Res.* **91,** 3573–3589.

DEKORP Research Group (1990). *Tectonophysics* **173,** 361-378.

Der, Z. A., McElfresh, T. W., and A. O'Donnell, A. (1982). An investigation of regional variations and frequency dependence of anelastic attenuation in the United States in the 0.5–4 Hz band. *Geophys. J. R. Astron. Soc.* **69,** 67–100.

Dziewonski, A. M., and Anderson, D. L. (1981). Preliminary Reference Earth Model. *Phys. Earth Planet. Inter.* **25,** 297–356.

Engdahl, E. R., and Gubbins, D. (1987). Simultaneous travel time inversion for earthquake location and subduction zone structure in the central Aleutian Islands. *J. Geophys. Res.* **92,** 13,855–13,862.

Gish, D. M., Keller, G. R., and Sbar, M. L. (1981). A refraction study of deep crustal structure in the Basin and Range: Colorado Plateau of eastern Arizona. *J. Geophys. Res.* **86,** 6029–6038.

Grand, S. P. (1987). Tomographic inversion for shear velocity beneath the North American plate. *J. Geophys. Res.* **92,** 14,065–14,090.

Grand, S. P., and Helmberger, D. V. (1984). Upper mantle shear structure of North America. *Geophys. J. R. Astron. Soc.* **76,** 399–438.

Hager, B. H., and O'Connell, R. (1979). Kinematic models of large-scale flow in the Earth's mantle. *J. Geophys. Res.* **84,** 1031–1048.

Hearn, T., Beghoul, N., and Barazangi, M. (1991). Tomography of the western United States from regional arrival times. *J. Geophys. Res.* **96,** 16,369–16,381.

Humphreys, E., Clayton, R. W., and Hager, B. H. (1984). A tomographic image of mantle structure beneath southern California. *Geophys. Res. Lett.* **11,** 625–627.

Inoue, H., Fukao, Y., Tanabe, K., and Ogata, Y. (1990). Whole mantle *P* wave travel time tomography. *Phys. Earth Planet. Inter.* **59,** 294–328.

Iyer, H. M. (1989). Seismic tomography. *In* "The Encyclopedia of Solid Earth Geophysics" (D. James, ed.), pp. 1133–1151. Van Nostrand–Reinhold, New York.

Jeanloz, R. (1990). The nature of the earth's core. *Annu. Rev. Earth Planet. Sci.* **18,** 357–386.

Johnson, L. R., and Lee, R. C. (1985). Extremal bounds on the *P* velocity in the Earth' core. *Bull. Seismol. Soc. Am.* **75,** 115–130.

Lay, T. (1986). Evidence of a lower mantle shear velocity discontinuity in *S* and *sS* phases. *Geophys. Res. Lett.* **13,** 1493–1496.

Lay, T. (1989). Mantle, lower structure. *In* "The Encyclopedia of Solid Earth Geophysics" (D. James, ed.), pp. 770–775. Van Nostrand–Reinhold, New York.

Lay, T. (1992). Wrinkles on the inside. *Nature* (*London*) **355,** 768–769.

Lay, T., and Wallace, T. C. (1988). Multiple *ScS* attenuation and travel times beneath western North America. *Bull. Seismol. Soc. Am.* **78,** 2041–2061.

LeFevre, L. V., and Helmberger, D. V. (1989). Upper mantle *P* velocity structure of the Canadian shield. *J. Geophys. Res.* **94,** 17,749–17,765.

Meissner, R., and Bortfield, (1990). "DEKORP Atlas." Springer-Verlag, Berlin.

Meissner, R., and Wever, T. (1989). Continental crustal structure. *In* "The Encyclopedia of Solid Earth Geophysics" (D. James, ed.), pp. 75–89. Van Nostrand–Reinhold, New York.

Morelli, A., Dziewonski, A. M., and Woodhouse, J. H. (1986). Anisotropy of the inner core inferred from *PKIKP* travel times. *Geophys. Res. Lett.* **13,** 1545–1548.

Morris, E. M., Raitt, R. W., and Shor, G. G. (1969). Velocity anisotropy and delay time maps of the mantle near Hawaii, *J. Geophys. Res.* **74,** 4300–4316.

Press, F., and Siever, R. (1978). "Earth." Freeman, San Francisco.

Shearer, P. M. (1991). Constraints on upper mantle discontinuities from observations of long-period reflected and converted phases. *J. Geophys. Res.* **96,** 18,147–18,182.

Shearer, P. M., and Masters, T. G. (1992). Global mapping of topography on the 660-km discontinuity. *Nature* (*London*) **355,** 791–796.

Shearer, P. M., and Toy, K. M. (1991). *PKP*(*BC*) versus *PKP*(*DF*) differential travel times and aspherical structure in the Earth's inner core. *J. Geophys. Res.* **96,** 2233–2247.

Silver, P., and Chan, W. W. (1991). Shear wave splitting and subcontinental mantle deformation. *J. Geophys. Res.* **96,** 16,429–16,454.

Song, X., and Helmberger, D. V. (1992). Velocity structure near the inner core boundary from waveform modeling. *J. Geophys. Res.* **97,** 6573–6586.

Spakman, W., and Nolet, G. (1988). Imaging algorithms, accuracy and resolution in time delay tomography. *In* "Mathematical Geophysics" (N. J. Vlaar, G. Nolet, M. J. R. Wortel, and S. A. P. L. Cloetingh, eds.), pp. 155–187. Reidel Publ., Dordrecht, The Netherlands.

Spakman, W., Stein, S., Van der Hilst, R., and Wortel, R. (1989). Resolution experiments for NW Pacific subduction zone tomography. *Geophys. Res. Lett.* **16,** 1097–1100.

Spudich, P., and Orcutt, J. (1980). A new look at the velocity structure of the crust. *Rev. Geophys. Space Phys.* **18,** 627–645.

Su, W.-J., Woodward, R. L., and Dziewonski, A. M. (1994). Degree 12 model of shear velocity heterogeneity in the mantle. *J. Geophys. Res.* **99,** 6945–6980.

Tanimoto, T., and Anderson, D. L. (1984). Mapping convection in the mantle. *Geophys. Res. Lett.* **11**, 287–290.

Walck, M. C. (1984). The *P* wave upper mantle structure beneath an active spreading centre: The Gulf of California. *Geophys. J. R. Astron. Soc.* **76**, 697–723.

Wong, Y. K. (1989). Upper mantle heterogeneity from phase and amplitude data on mantle waves. Ph.D. Thesis, Harvard University, Cambridge, MA.

Woodhouse, J., Giardini, D., Li, X. D. (1986). Evidence for inner core anisotropy from free oscillations. *Geophys. Res. Lett.* **13**, 1549–1552.

Wu, R. S. (1989). Seismic wave scattering. *In* "The Encyclopedia of Solid Earth Geophysics" (D. James, ed.), pp. 1166–1187. Van Nostrand–Reinhold, New York.

Wysession, M. E., Okal, E. A., and Bina, C. R. (1992). The structure of the core–mantle boundary from diffracted waves. *J. Geophys. Res.* **97**, 8749–8764.

Young, C. J., and Lay, T. (1987). The core–mantle boundary. *Annu. Rev. Earth Planet. Sci.* **15**, 25–46.

Young, C. J., and Lay, T. (1990). Multiple phase analysis of the shear velocity structure in the D'' region beneath Alaska. *J. Geophys. Res.* **95**, 17,385–17,402.

Zhang, Y.-S. (1991). Three-dimensional modeling of upper mantle structure and its significance to tectonics. Ph.D. Thesis, California Institute of Technology, Pasadena.

Zhang, Y.-S., and Tanimoto, T. (1992). Ridges, hotspots and their interaction as observed in seismic velocity maps. *Nature* **355**, 45–49.

Additional Reading

Anderson, D. L. (1989). "Theory of the Earth." Blackwell, Boston.

Bolt, B. A. (1982). " Inside the Earth: Evidence from Earthquakes." Freeman, San Francisco.

Brown, G. C., and Mussett, A. E. (1981). "The Inaccessible Earth." Allen & Unwin, London.

Claerbout, J. F. (1985a). "Fundamentals of Geophysical Data Processing with Applications to Petroleum Prospecting." Blackwell, Oxford.

Claerbout, J. F. (1985b). "Imaging the Earth's Interior." Blackwell, Oxford.

Dziewonski, A. M., and Anderson, D. L. (1984). Seismic tomography of the Earth's interior. *Am. Sci.* **72**, 483–494.

Fuchs, K., and Froidevaux, C., eds. (1987). "Composition, Structure and Dynamics of the Lithosphere–Asthenosphere System," Geodyn. Ser., Vol. 16. Am. Geophys. Union, Washington, DC.

Jacobs, J. A. (1987). "The Earth's Core." Academic Press, London.

James, D. E., ed. (1989). "The Encyclopedia of Solid Earth Geophysics." Van Nostrand–Reinhold, New York.

Jeffreys, H. (1976). "The Earth." Cambridge Univ. Press, Cambridge, UK.

Nolet, G., ed. (1987). "Seismic Tomography–With Applications in Global Seismology and Exploration Geophysics." Reidel Publ., Dordrecht, The Netherlands.

Press, F., and Siever, R. (1978). "Earth." Freeman, San Francisco.

Ringwood, A. E. (1975). "Composition and Petrology of the Earth's Mantle." McGraw–Hill, New York.

Sheriff, R. E., and Geldart, L. P. (1982). "Exploration Seismology," Vol. 1. Cambridge Univ. Press, Cambridge, UK.

Sheriff, R. E., and Geldart, L. P. (1983). "Exploration Seismology," Vol. 2. Cambridge Univ. Press, Cambridge, UK.

Taylor, S. R., and McLennan, S. M. (1985). "The Continental Crust: Its Composition and Evolution." Blackwell, Boston.

Waters, K. H. (1981). "Reflection Seismology—A Tool for Energy Resource Exploration," 2nd ed. Wiley, New York.

8

SEISMIC SOURCES

Having investigated the various types of seismic waves that propagate in the Earth, and having surveyed the interior structure of the planet which controls the evolution of the seismic wavefields, it is now time that we address the wave sources and their quantitative description. Although many investigations of Earth structure can be designed to minimize the effects of uncertainty in the source, in general, structural investigations do require detailed knowledge of the seismic source. Of course, the sources themselves are also of great interest, because they commonly represent important dynamic processes. The observed seismogram is a complex marriage of the signature of the source and the effects of propagation. Knowledge of the propagation effects allows us to constrain the physical process of the source. This is a startling proposition: to use the limited sampling of seismic wavefields provided by seismometers located sparsely on the surface to deduce complex transient phenomena that have taken place thousands of kilometers away, perhaps at great depth, in a medium as complicated as the Earth! Certainly, a student of the Earth's gravitational, thermal, magnetic, or chemical fields could not hope to analyze remotely a distant ephemeral process inside the planet with

any claim of uniqueness. It is again the properties of seismic waves that prove to be the seismologist's ally in this endeavor, with detailed information about the localized, remote source process being conveyed to the seismometer with relatively little loss of resolution, provided the propagation effects can be accounted for. The last chapter demonstrated the advanced state of current capabilities for doing the latter. In the next few chapters we will see how this capability enables remarkable characterization of seismic sources.

The vast majority of important seismic sources involve faulting, or shearing motions on surfaces inside the Earth. However, let us first consider source representations of human-made underground explosions. These, too, are common sources, having a vast range of energy scales. Small explosions are used for mining, quarrying, road excavating, and other constructional applications, as well as in natural resource exploration and crustal studies. Larger sources include underground nuclear tests, which produce waves strong enough to be observed on the opposite side of the Earth. Naturally occurring explosive or implosive sources are rare, but some may occur with metastable mineralogical phase transitions or magmatic

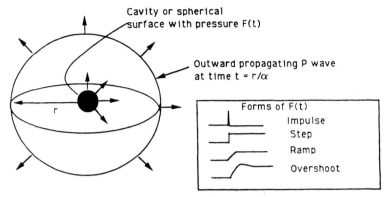

FIGURE 8.1 An underground explosion source is conceptualized as a small cavity or spherical surface within which a spherically symmetric pressure pulse is suddenly applied. At the instant of application a P wave is initiated in the surrounding elastic medium, which has P-wave velocity α. This wave spreads spherically outward from the source. The shape of the P-wave potential will directly reflect the time history of the pressure pulse, examples of which are shown in the inset, with the displacements being given by Eq. (8.3).

processes at depth. The most general aspect that we intuitively associate with all explosions is spherical symmetry.

Consider an explosion in a whole space (Figure 8.1). The explosion can be idealized as a sudden application of a pressure pulse to the inside of a small cavity with spherical symmetry. In detail, of course, the explosion itself may produce the cavity by melting, vaporizing, and deforming the surrounding rock (underground nuclear explosions can produce 0.5-km-radius cavities in preexisting rock). Nonetheless, whatever nonlinear process has occurred immediately after the energy release, at some distance from the source called the *elastic radius*, r_e, a spherical surface exists beyond which infinitesimal strain theory is valid and on which we can predict the elastic displacements and strains due to the effective pressure force, $F(t)$, produced by the inelastically deformed interior. The time history of the effective pressure force can range from an impulse (the pressure quickly drops back to its initial state) to a step (permanent strains in the inelastically distorted medium serve as an effective permanent pressure). Nuclear explosions often involve a combination of

these, with an overshoot of pressure that decays to a static permanent step as gas pressure in the newly created cavity dissipates. The elastic radius for an underground nuclear explosion may be 1 km or larger.

Beyond the elastic radius the equation of motion reduces to a one-dimensional inhomogeneous wave equation

$$\frac{\partial^2 \phi}{\partial r^2} - \frac{1}{\alpha^2}\frac{\partial^2 \phi}{\partial t^2} = -4\pi F(t)\delta(r_e), \quad (8.1)$$

which is in terms of the P-wave displacement potential, $\phi(r,t)$, and the effective force function, $F(t)$, applied at the elastic radius. The solution for the displacement potential has the form

$$\phi(r,t) = \frac{-F(t-(r/\alpha))}{r}, \quad (8.2)$$

where $F(t)$ is called the reduced displacement potential, and r is the distance from the elastic radius (often the latter is negligible and the source is treated as a point source). Thus, seismic waves propagate outward with equal amplitude in all directions on a spherical wavefront. It is a common characteristic of particular solutions of inhomogeneous wave equations that the wave potential has a D'Alembert solution of this type, with the functional shape being the same as the force–time history. Since the P-wave energy is spread over a spherical wavefront, the wave potential decays as r^{-1}.

The spherically symmetric displacement field $u(r, t)$ is given by

$$u(r,t) = \frac{\partial \phi(r,t)}{\partial r} = \left(\frac{1}{r^2}\right)F\left(t - \frac{r}{\alpha}\right)$$

$$+ \left(\frac{1}{r\alpha}\right)\frac{\partial F(t - (r/\alpha))}{\partial \tau},$$

(8.3)

where $\tau = t - (r/\alpha)$ is the *retarded time*. The surrounding medium is at rest until time $t = (r/\alpha)$ ($\tau = 0$) after the explosion, which corresponds to the time that it takes for the P wave to travel to the observing point. The first term in (8.3) involves displacements that are directly proportional to the reduced displacement potential and that decay rapidly (proportional to $1/r^2$) with distance from the source. This is called the *near-field term*, and if any step in the effective pressure occurs, a static (permanent) deformation of the surrounding elastic medium will result. The second term decays more slowly, thus dominating displacement at large distances, and corresponds to the *far-field term*, which is proportional to the time derivative of the reduced displacement potential. Thus, a step in effective pressure at the source produces an impulsive far-field ground motion. This is a characteristic that we will find for far-field motions from other sources, and our ability to infer the time history of the forces acting at the source hinges on the sensitivity of far-field displacements to the temporal derivative of the source time history.

It turns out that three mutually perpendicular *dipoles* (pairs of forces acting along

the same line in opposite directions) produce waves identical to those from an explosive point source. The strength of a dipole, or force couple, is given by its moment, $M = f\,dx$, where f is the strength of the force and dx is the distance separating the forces. For our explosion source, the time history of $F(t)$ is related to $M(t)$ by $M(t) = -4\pi\rho\alpha^2 F(t)$ [where $F(t)$ is defined in terms of force per unit mass]. Thus, an explosive source can be represented by *equivalent body forces*, which produce the same motions. We will find wave solutions for equivalent body-force representations of other seismic sources later in this chapter, also using solutions of the inhomogeneous wave equation with body-force source terms.

The solution in (8.3) indicates that the particle motions produced by an explosion in a whole space are outward along radial directions (corresponding to radial static deformations, if any, and to transient P-wave motions). This is actually observed in the Earth, which is clearly not a whole space, with the direct P waves from a large explosion having *compressional* first arrivals at all stations, at all azimuths. Compressional P-wave motion is defined as P-wave particle motion away from the source, after allowing for wavefront curvature effects along the seismic raypath (Figure 8.2). Since elastic-wave propagation does not modify the shape of the wave initiated at the source, except by predictable transmission effects, the near-source compressions are observable at large distance. An idealized explosion involves no shear deformation at the source, and hence no S wave will be directly generated or *excited* by the explosion. However, even if the Earth had perfect radial symmetry, we do expect to see SV waves for an explosion, and hence Rayleigh waves, due to P to SV conversions at boundaries such as the free surface (pS, PS, PPS, etc.) or the core–mantle boundary (PcS, PKS, PKIKS, etc.). Theoretically we would not expect to see either SH or

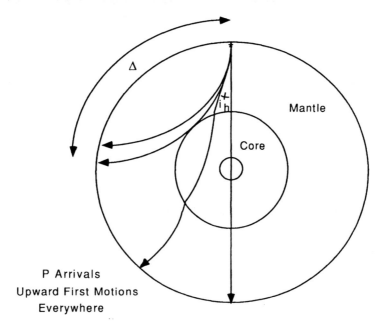

P Arrivals
Upward First Motions
Everywhere

FIGURE 8.2 Initial *P*-wave motions from an underground explosion are compressional, defined as being in the direction of wave propagation, or "away" from the source, allowing for the raypath perturbations caused by Earth structure. Thus, explosions produce upward ground motions for the first *P* or *PKP* arrivals everywhere on the surface of the Earth.

Love waves from an explosion in such a radially symmetric Earth. Actually, we do commonly observe *SH* and Love waves from both large and small explosions. The transverse-component energy is produced by departure from spherical symmetry of the source, triggering of an earthquake or deviatoric strain release near the explosion, and/or conversion of *P* and *SV* energy to *SH* energy by heterogeneous structures and nonspherical asymmetry of the Earth (Box 8.1).

In this chapter we will pursue more complex earthquake-faulting source representations, demonstrating both intuitive and mathematical representations of these internal sources. We will see that elastic-wave theory provides simplified force systems that represent complex physical processes, revealing many features about the source, but not a first-principles theory for the source phenomena.

8.1 Faulting Sources

Unlike an underground explosion, most seismic sources lack spherical symmetry. For example, faulting involves shear dislocation on a planar surface, which is certainly not spherically symmetric but does have some low-order symmetry. The outgoing waves are influenced by the strain distribution near the source and hence convey the asymmetry to distant locations. This leads to the concept of a *radiation pattern*, which is a geometric description of the amplitude and sense of initial motion distributed over the *P* and *S* wavefronts in the vicinity of a source. The low-order symmetry of shear dislocations provides predictable relationships between the radiation pattern of detectable wave motions and the fault-plane orientation, which allows remote determination of the faulting processes.

Box 8.1 Explosion SH Waves?

One of the sociologically important applications of modern seismology is the monitoring of global underground nuclear testing. The seismic waves generated by such explosions reveal the occurrence of the event as well as provide an estimate of the size of the explosion, mainly by empirical calibration of *P*- and Rayleigh-wave amplitudes with explosions of known *yield*, or energy release, in equivalent kilotons of TNT. But first an event must be identified as an explosion rather than a natural source. Usually this discrimination of explosion events is accomplished by examining a variety of waveform characteristics that may distinguish earthquakes from explosions. It would seem reasonable to rely mainly on whether or not *SH*-wave energy is observed, for an explosion source theoretically will not generate significant transverse-component radiation at the source. However, Figure 8.B1.1 compares *SH* and Love waves from an earthquake with the same component recorded for an explosion.

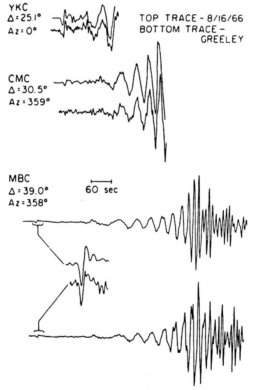

FIGURE 8.B1.1 Comparison between *SH* and Love waves for the nuclear explosion GREE-LEY (December 20, 1966) and an earthquake in eastern Nevada (August 16, 1966). The seismograms are lined up on the *S* arrival, and the amplitude scale is the same for both the earthquake and explosion. (From Wallace *et al.*, 1983.)

continues

The *SH* components are virtually identical, and this holds for the full azimuthal range surrounding the event. For this large explosion, fault motion was presumably triggered in the surrounding crust, a process called *tectonic release*, which accounts for the *SH*-wave energy. Although not all explosions produce such clear *SH* radiation, it is clear that other waveform attributes have to be used to discriminate explosions. The most successful discriminants prove to be location (source depths greater than humans can drill to, or in underwater locations where other means of detecting explosions are very reliable) and the ratio of short-period (1-s) *P*-wave energy (m_b) to 20-s-period Rayleigh-wave energy (M_S), which is higher for explosions than for earthquakes larger than magnitude 3.5.

In Chapter 1 the elastic rebound theory was introduced (Figure 1.4), which hypothesizes that shearing motions on a fault occur when the elastic strain accumulation in the vicinity of the fault overcomes the static frictional stress that resists motion. Sliding motions initiate at a point (the earthquake hypocenter), and a slip front expands outward over the fault, separating regions that are slipping (or perhaps have slipped and then come to rest) from regions that have not yet slipped (Figure 8.3). The expansion of the rupture area is thus a function of space and time, $A(\mathbf{x}, t)$, as is the corresponding slip function, $\mathbf{D}(\mathbf{x}, t)$, which gives the actual vector sliding motions on the fault. The stored elastic strain energy in the source region is liberated as heat and seismic waves, and eventually the fault slippage ends. For the long-wavelength waves excited by the source motions, the rupture area and source volume that release strain energy are relatively small and can be approxi-

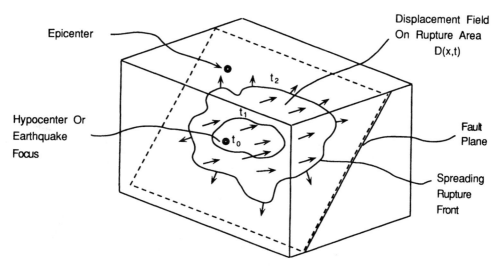

FIGURE 8.3 A schematic diagram of rupture on a fault spreading from the hypocenter, or earthquake focus, over the fault plane. All regions that are sliding continually radiate outgoing *P*- and *S*-wave energy. The displacement field, $\mathbf{D}(\mathbf{x}, t)$, varies over the surface of the fault. Note that the direction of rupture propagation does not generally parallel the slip direction. (Modified from Bolt, 1988.)

mated as a *point source* of the seismic waves, concentrated in space. Shorter-wavelength waves are sensitive to the finite extent and detailed variation of slip process on the fault, and they require *finite source* models. In the next chapter we will address the nature of the frictional sliding of faults and the associated stress regime in the rock; in this chapter we will concentrate on the geometric nature of faulting and its mathematical representation.

A standard nomenclature has evolved for describing fault orientation and slip direction that is useful to introduce at this point. We visualize faulting as slippage between two blocks of material (Figure 8.4a), where the slip is constrained to lie in the plane connecting the two blocks. This constraint precludes extensional crack opening or closing, but we will discuss such sources later. To describe the orientation of the fault plane in geographic coordinates, we require two angular parameters. These are the *strike* of the fault, ϕ_f, the azimuth of the fault's projection onto the surface measured from north, and the *dip* of the fault, δ, the angle measured downward from the surface to the fault plane in the vertical plane perpendicular to the strike. The strike direction is defined such that the dip is $< 90°$

(i.e., so that if you orient the thumb on your right hand along the strike and rotate your hand downward from the horizontal to the fault plane, your hand will pivot through an angle less than 90°). The strike direction is arbitrarily either orientation for a vertically dipping fault ($\delta = 90°$). The actual motion of the two blocks on either side of the fault is defined by a slip vector, which can have any orientation on the fault plane. The direction of the slip vector is given by the angle of *slip*, or *rake* (λ), measured in the plane of the fault from the strike direction to the slip vector showing the motion of the hanging wall relative to the footwall. The magnitude of the slip vector is given by D, the total displacement of the two blocks. In general, ϕ_f, δ, λ, and D can vary over the finite fault surface, with *average* values being used for simple models.

Three basic categories of fault slip are commonly used to characterize motions on faults relative to the Earth's surface (Figure 8.5). When the two sides of a fault slip horizontally relative to each other, the motion is called *pure strike slip* ($\lambda = 0°, 180°$), and if the dip is 90° the geometry is called *vertical strike slip*. For $\lambda = 0°$, the hanging wall (or near side of a vertical fault) moves to the right, so that a point on the other

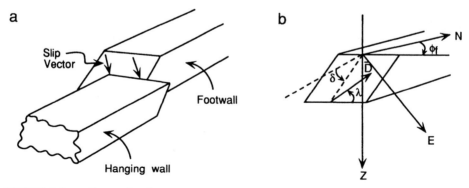

FIGURE 8.4 (a) Convention for naming the two blocks on either side of a nonvertical fault. The block above the fault is the *hanging wall*; that below is the *footwall*. (b) Standard definition of fault-plane and slip-vector orientation parameters. The figure also defines strike (ϕ_f), dip (δ), and rake or slip (λ), as discussed in the text.

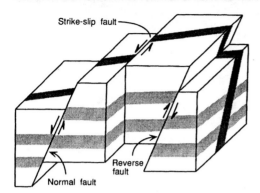

FIGURE 8.5 Examples of the end-member styles of faulting for different slip vectors. (From Bolt, 1988. Copyright © 1988 by W. H. Freeman and Co. Reprinted with permission.)

side of the fault moves to the left; this type of fault movement is called *left-lateral slip*. If $\lambda = 180°$, the fault movement is called *right-lateral slip*. The other end-member types of motion involve slip with a vertical component of relative displacement of the two blocks, which is called *dip–slip* faulting because the slip vector parallels the dip direction. For $\lambda = 90°$, the hanging wall moves upward, causing *thrust faulting*, and for $\lambda = 270°$ it moves downward, causing *normal faulting*. In general, λ will have a value different than these special cases, and the motion is then called *oblique slip*, with the predominant character being described by stringing together appropriate modifiers (e.g., *right-lateral oblique normal faulting*, for $180° < \lambda < 270°$).

Now, let us apply our intuition to consider the pattern of *P*-wave displacement motions that we expect for an arbitrarily oriented shear dislocation. Referring to Figure 8.6, we expect that alternating quadrants will exist in which static motions will occur and that the initial motions of the *P* arrival will be *compressional* (motion away from the source) or *dilatational* (motion toward the source). This is inferred from considering the respective pushing and pulling of the quadrants surrounding the fault when sudden slip oc-

curs. The changes in static deformations and initial *P*-wave polarity on the spreading wavefront do not occur abruptly, but rather a smooth, three-dimensional radiation pattern results that we can describe using polar coordinates with simple trigonometric functions. Thus, we introduce a local Cartesian coordinate system at the source, with the x_1 axis along the slip direction, x_2 also in the plane of the fault, and x_3 perpendicular to the fault. We also introduce spherical coordinates r, θ, ϕ for this source reference frame, as defined in Figure 8.7a. In this geometry, the *P*-wave displacements can be described by $u_r \propto \sin 2\theta \cos^2 \phi$, as will be derived later. When $\phi = 0°$ (i.e., in the $x_1 x_3$ plane) $u_r \propto \sin 2\theta$ (Figure 8.7b), which is a simple four-lobed azimuthal pattern reflecting the alternating quadrants in Figure 8.6. The smoothly varying function makes intuitive sense, with reversals in polarity occurring where amplitudes have smoothly gone to zero. Thus, no tearing of the ground occurs outside the slip zone or on the *P* wavefront, just a smooth transition from outward initial motions to inward initial motions. The strongest *P*-wave motions are thus expected in the middle of the four quadrants, at 45° angles to the fault plane $(x_1 x_2)$ (Figure 8.7c). We will see that such simple geometric patterns are characteristic of radiation from faults and that most of the complexity of inferring the source geometry comes from mapping from the Earth's geographic reference frame in which we make wave mea-

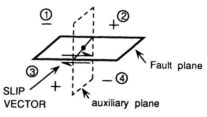

FIGURE 8.6 Sense of initial *P*-wave motion with respect to the fault plane and auxiliary plane.

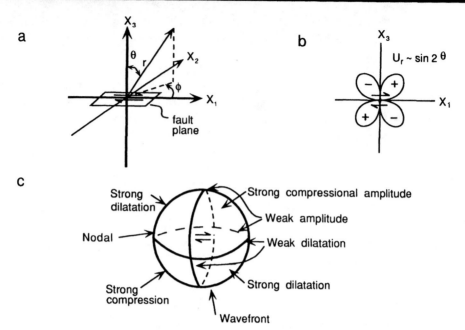

FIGURE 8.7 (a) The local source coordinate system (x_1, x_2, x_3) with coordinates parallel or perpendicular to the fault plane with one axis along the slip direction. (b) The radiation-pattern variation over the P wavefront in the $x_1 x_3$ plane. (c) The three-dimensional variation of P-wave amplitude and polarity on a spreading wavefront from a shear dislocation.

surements back to the source coordinate system, where radiation patterns have simple low-order symmetry.

The amplitude and polarity of the P-wave displacements will be preserved along the associated path to any receiver because the wave transmits the initial motion from particle to particle without modification. If sufficient observations of the static deformations near the fault or of the first P-wave motions are made and if their initial orientation relative to the source is determined by "propagating" the wave back to the source, the orientation of the fault plane can be determined. If static deformations or actual surface rupture is measured, the fault plane can be determined uniquely; however, the symmetry of the pattern of alternating compressional and dilatational quadrants prevents the fault plane from being uniquely determined by P-wave first motions alone. In

this case an orthogonal *auxiliary plane* exists, perpendicular to the fault plane (Figure 8.6), that could just as well have experienced faulting dislocation, with an opposing sense of slip (i.e., right-lateral motion on the true fault is indistinguishable from left-lateral motion on the auxiliary plane). In the Earth, we must account for the curvature of the seismic-wave raypaths (e.g., Figure 8.2) when relating initial motions to the source orientation. The situation is relatively straightforward near the event, where raypaths are simple and the quadrantal distribution of motions is readily apparent for a vertical strike–slip fault (Figure 8.8). Procedures for determining the fault orientation for arbitrary orientations and more complex raypaths are described later.

We now consider how these simple ideas of faulting can be represented in a form suitable for incorporation in the elastic

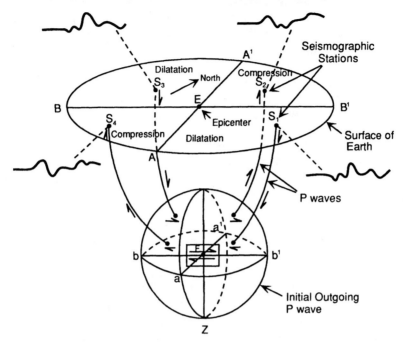

FIGURE 8.8 First motions of P waves at seismometers located in various directions about an earthquake allow determination of the fault orientation. P waves, for which upward motions are compressional (away from the source) and downward motions are dilatational (toward the source), exhibit a simple alternation with quadrants for a vertical strike–slip fault, as shown. (Modified from Bolt, 1988. Copyright © 1988 by W. H. Freeman and Co. Reprinted with permission.)

equations of equilibrium or motion. Then we proceed to relate the source radiation from shear-dislocation sources to the ground-motion observations provided by seismograms.

8.2 Equivalent Body Forces

We have considered intuitively the geometry of P-wave displacement radiation from a shear dislocation, and we will now proceed to quantify this. We recall the full equation of motion for an elastic medium

$$\rho \ddot{\mathbf{u}} = \mathbf{f} + (\lambda + 2\mu)\nabla(\nabla \cdot \mathbf{u}) - \mu\nabla \times \nabla \times \mathbf{u}.$$

$$(2.52)$$

In the earlier chapters of this book we set $\mathbf{f} = 0$ (no body forces) and let $\mathbf{u} = \nabla\phi +$

$\nabla \times \Psi$ to separate P- and S-wave disturbances. Now, we want to ask the question: Are there body forces, $\mathbf{f}$, that are equivalent to earthquake-faulting dislocations, and if so, can we still solve this equation to find the wave disturbances they produce?

At first thought it seems unlikely that any simple body-force system could possibly be an adequate representation of a finite shear-faulting failure. First of all, there is the finite extent of the faulting. As described in the last section, the rupture initiates at the hypocenter and spreads over the fault plane as a rupture front, or spreading dislocation surface (Figure 8.3). As the rupture front expands, all enclosed sliding regions of the fault continually radiate energy in the form of P and S waves. Individual particles on the fault may have smooth or irregular time histories of particle dislocation. Each point on the rupture surface may also have a slightly different slip vector. In general, these slip vectors are expected to be nearly parallel, but the amount of slip can vary spatially within the rupture zone and must vary at the edges of the final rupture surface (where the displacement goes to zero). While fault slippage results from the earthquake process, we ignore many other phenomena, such as local heating and perhaps melting of rock, hydrologic pressure variations, and rock fracturing, to conceptualize the kinematic rupture history. Our goal is to replace this kinematic rupture process with a useful force system that produces equivalent seismic-wave radiation.

We proceed by "standing back" from the fault and considering the average properties of the rupture. We are mainly interested in gross characteristics such as the total rupture area, A, the average displacement over the fault, $\overline{D}$, and the average velocity and direction of rupture propagation, V_r. For seismic waves with periods longer than or comparable to the duration of rupture and for wavelengths that are large relative to the fault dimensions, we can visualize replacing the complex fault-

ing by a simple dislocation representation (Figure 8.9). In its simplest form the dislocation model idealization will involve a point source (i.e., no spatial extent), with a simple dislocation time history to approximate the process of seismic-wave radiation during particle dislocation and expansion of the rupture area. More complex models of spatial distributions of dislocations can be constructed from this end-member case. Model complexity increases as the ratio of seismic-energy wavelength to fault length decreases.

The average dislocation model in Figure 8.9 now looks like a simple enough system to be replaced by a force system that would be *dynamically equivalent*, meaning one that produces equivalent seismic-wave radiation. Indeed, it would appear that we simply need a *time-varying force couple* applied within the elastic medium to simulate the dislocation. The level of approximation implied in Figure 8.9 clearly depends on the sensitivity of the seismic waves to the details of the faulting complexity, which is frequency and wavelength dependent, and on the extent to which one wants to determine actual stresses on the fault. Both dislocation and equivalent

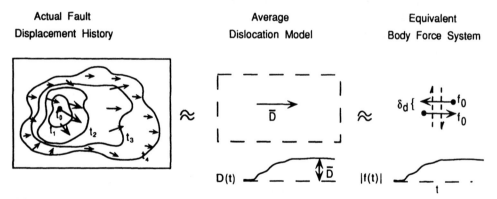

Actual Fault Displacement History	Average Dislocation Model	Equivalent Body Force System

FIGURE 8.9 Concepts underlying equivalent body forces. Actual faulting involves complex cracking and frictional sliding over a surface in a short time that results in a space–time history of slipping motion. The finite spatial–temporal faulting process can be approximated by a dislocation model with dislocation time history $D(t)$. In turn, this dislocation model can be idealized by an equivalent force system that can be directly incorporated in the equations of motion.

body-force models must fail at some point to express the actual physics of faulting, but point-source and finite-source models of fault kinematics have revealed many gross characteristics of faulting. The main failing of such kinematic models is that they do not explicitly include physics for initiating or terminating rupture, and hence they leave many fundamental issues unresolved.

A clear problem with the *single-couple* model of faulting is that a force couple has an associated moment, $M_0 = |f|\delta d$ (as the separation between the force arms, $\delta d \to 0$ and $|f| \to \infty$, giving a constant M_0). Thus, we would have a net unbalanced moment introduced into the medium when the dislocation occurred. Recalling the ambiguity between fault plane and auxiliary plane for *P*-wave radiation patterns, it is reasonable to expect that we need a second force couple to balance the moment of the force system (Figure 8.9) so that no net moment is added to the medium and so that the force couple will be directed along the auxiliary plane. This is the *double-couple* model, which is the currently preferred model of equivalent body forces for a dislocation source. Later in this chapter we will introduce a more general system of body forces called a *seismic moment tensor*, which includes double couples as a special case.

There was much debate about the suitability of the single-couple versus the double-couple model for faulting that lasted from the 1920s until the mid-1960s. Although the single-couple model makes less sense physically, the main reason for not ruling it out was that the *P*-wave radiation from both models is indistinguishable. However, the *S*-wave and surface-wave radiation is not, and when sufficient data became available, it was unambiguously demonstrated that the double-couple model was more appropriate. In addition, elastodynamic solutions for actual stress and displacement discontinuities in the

medium (an alternate way of modeling shear faults, which we do not consider in this text) confirm the equivalence of double-couple body forces and shear dislocations.

If we consider the geometry of the double-couple force system, we can anticipate some of the results of the next section. Figure 8.10 shows that a double-couple force system can be equivalently represented by a pair of orthogonal dipoles without shear, or what are called the *principal axes*. These axes are at 45° angles to the $x_1 x_2$ fault plane, in the orthogonal plane $x_1 x_3$. The dipole directed toward the source is the *compressional* or *P axis* and lies in the quadrants of dilatational *P*-wave first motions *toward* the source (you can avoid confusion over the nomenclature by thinking of the *P* axis as pushing the ground toward the source, which clearly implies *P*-wave dilatation). The dipole directed outward from the source is the *tensional* or *T axis*, and it lies in the quadrants of compressional initial *P*-wave motion (think of these as regions where the ground is pulled away from the source, hence giving compressional *P* waves). *S*-wave radiation patterns follow the geometry shown in Figure 8.10, with the sense of shearing motion on the wavefront reflecting the orientation of the force on the near side of the source double couple. The *P* and *S* radiation patterns are rotated by 45° from one another, but a $\sin 2\theta$ symmetry still precludes distinguishing the fault plane and the auxiliary plane using only first motions. The *S*-wave radiation for a single couple has only a two-lobed radiation pattern, which provides the basis for distinguishing it from the double-couple mechanism. While the double-couple force system has no net moment, the strength of the two couples can be represented by the seismic moment, M_0, which will be shown to equal $\mu A \overline{D}$, where μ is the rigidity, A is the fault area, and $\overline{D}$ is the average displacement on the fault.

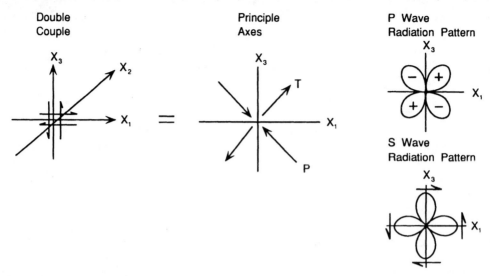

FIGURE 8.10 The double-couple force system in the $x_1 x_3$ plane for a shear dislocation in the $x_1 x_2$ plane. An equivalent set of point forces composed of two dipoles without shear, or the principal axes, is shown in the center. On the right are the patterns of P- and S-wave radiation distributed over the respective wavefronts in the $x_1 x_3$ plane.

The key to constructing solutions of the equations of motion for a complex set of body forces such as a double couple is to solve first for the displacement field due to a single point force and then to use the linearity of elastic solutions to superimpose the solutions for several forces to produce a displacement field for force couples. Thus, it is important to gain physical insight into the most elementary elastic solutions for a point force. We do so by considering the solution of the static problem of a force, **F**, applied at a point in a homogeneous elastic medium, as in Figure 8.11. Let us consider the displacement field, **u**, on a spherical surface S of radius r centered on the point source. The magnitude of the displacement can be approximated as follows. For the system to be in equilibrium, the body force, **F**, must be balanced by the stresses acting on S. The stress will be compressional at point C, purely shear at Q, and mixed shear and compressional or dilatational at intermediate points on S. If we generically represent the stress on S by σ, a first-order force balance gives $|\mathbf{F}| \approx 4\pi r^2 \sigma$. For linear elasticity we know that $\sigma = E\varepsilon = E(\nabla \mathbf{u})$, where E is some general elastic modulus and ε is the strain. If we let $u(r)$ be the magnitude of the displacement at r,

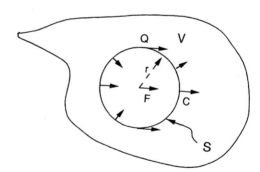

FIGURE 8.11 A planar cut through a three-dimensional volume, V, in which a point force, **F**, is acting. We consider the nature of the displacement field, **u**, on a spherical surface of radius r centered at the point of application of the force.

then du/dr gives the magnitude of σ:

$$-E\frac{du}{dr} \approx \sigma \approx \frac{F}{4\pi r^2}.$$

Thus

$$du \approx \frac{-F\,dr}{4\pi r^2 E} \Rightarrow U = \frac{F}{4\pi Er}. \quad (8.4)$$

Thus the static displacement field from a point force is expected to diminish with distance from the source, and the displacements are directly proportional to the force divided by an elastic modulus. This approximation lacks the directionality information of the displacement field, which involves a combination of radial and shear components, but the logic sequence is the same, as we will now follow to derive complete solutions. We will introduce a point force, use the equations of equilibrium or motion to produce a force balance, and then find a representation of the observable ground motions (both static and transient) in terms of the force magnitude and time history.

8.3 Elastostatics

We want to determine the static displacement $\mathbf{u}$ at point P in an isotropic, infinite, homogeneous elastic medium with density ρ and elastic constants λ and μ due to a force at point O. At large distances from the source, $\mathbf{u} = 0$. We define the point force $\mathbf{F}$ by

$$\mathbf{F} = \lim_{\delta V \to 0} \rho\mathbf{f}\,\delta V, \quad (8.5)$$

where $\mathbf{f}$ is the force per unit mass, $\rho\mathbf{f}$ is the body force per unit volume, and δV is a small volume element being acted on. We introduce the three-dimensional delta function $\delta(r)$

$$\delta(r) = \begin{cases} 0 & r \neq 0 \\ \int_V \delta(r)\,dV = 1. \end{cases} \quad (8.6)$$

Using Gauss' theorem (Box 2.3) we can find that

$$\delta(r) = \frac{-1}{4\pi}\nabla^2\left(\frac{1}{r}\right), \quad (8.7)$$

which allows us to represent the delta function by spatial derivatives of the radial coordinate r^{-1}.

8.3.1 Static Displacement Field Due to a Single Force

We now introduce this mathematical representation of a point force into our basic elastic equations for equilibrium (2.52), with $\ddot{\mathbf{u}} = 0$:

$$\mathbf{F} + (\lambda + 2\mu)\nabla(\nabla \cdot \mathbf{u}) - \mu\nabla \times \nabla \times \mathbf{u} = 0. \quad (8.8)$$

Consider a point force of magnitude $\mathbf{F}$ at the origin.

$$\mathbf{F} = \rho\mathbf{f} = F\mathbf{a}\delta(r) = -F\nabla^2\left(\frac{\mathbf{a}}{4\pi r}\right)$$

$$= -F\left[\nabla\left(\nabla \cdot \frac{\mathbf{a}}{4\pi r}\right) - \nabla \times \nabla \times \left(\frac{\mathbf{a}}{4\pi r}\right)\right], \quad (8.9)$$

where $\mathbf{a}$ is a unit vector in the direction of the force. We have used the vector identity (2.51) $[\nabla^2\mathbf{u} = \nabla(\nabla \cdot \mathbf{u}) - \nabla \times \nabla \times \mathbf{u}]$. The equation of equilibrium becomes

$$-F\nabla^2\left(\frac{\mathbf{a}}{4\pi r}\right)$$

$$= -F\left[\nabla\left(\nabla\cdot\frac{\mathbf{a}}{4\pi r}\right) - \nabla\times\nabla\times\left(\frac{\mathbf{a}}{4\pi r}\right)\right]$$

$$= -(\lambda + 2\mu)\nabla(\nabla\cdot\mathbf{u}) + \mu\nabla\times\nabla\times\mathbf{u}.$$

$$(8.10)$$

We look for a solution of the form

$$\mathbf{u} = \nabla(\nabla\cdot\mathbf{A}_P) - \nabla\times\nabla\times\mathbf{A}_S \quad \text{where}$$

$$\begin{cases} \nabla\times\mathbf{A}_P = 0 & \therefore \nabla^2\mathbf{A}_P = \nabla(\nabla\cdot\mathbf{A}_P) \\ \nabla\cdot\mathbf{A}_S = 0 & \therefore \nabla^2\mathbf{A}_S = -\nabla\times\nabla\times\mathbf{A}_S \end{cases}$$

$$(8.11)$$

based on our knowledge (Box 2.3) that any displacement field can be represented by a sum of solenoidal and irrotational fields, as the source has been represented. Substitution of this solution leads to

$$\nabla\left\{\nabla\cdot\left[\frac{-F\mathbf{a}}{4\pi r} + (\lambda + 2\mu)\nabla^2\mathbf{A}_P\right]\right\}$$

$$+ \nabla\times\nabla\times\left(\frac{F\mathbf{a}}{4\pi r} - \mu\nabla^2\mathbf{A}_S\right) = 0,$$

$$(8.12)$$

which can be satisfied by having

$$(\lambda + 2\mu)\nabla^2\mathbf{A}_P = \frac{F\mathbf{a}}{4\pi r}$$

$$\mu\nabla^2\mathbf{A}_S = \frac{F\mathbf{a}}{4\pi r}. \quad (8.13)$$

If we put $\mathbf{A}_P = A_P\mathbf{a}$ and $\mathbf{A}_S = A_S\mathbf{a}$, we

obtain Poisson's equations

$$\nabla^2 A_P = \frac{F}{4\pi(\lambda + 2\mu)r}$$

$$\nabla^2 A_S = \frac{F}{4\pi\mu r}. \quad (8.14)$$

Since $\nabla^2 r = 2/r$, we can integrate these to obtain

$$A_P = \frac{Fr}{8\pi(\lambda + 2\mu)}$$

$$A_S = \frac{Fr}{8\pi\mu}. \quad (8.15)$$

These solutions are potentials that solve the inhomogeneous equations (8.13). We compute the displacements by inserting them into (8.11). Plugging in our potentials A_P and A_S and expressing the vector operations with indicial notation, (8.11) and (8.15) yield *the ith component of displacement for a unit force* $(F = 1)$ *in the jth direction, u_i^j*:

$$u_i^j = \frac{1}{8\pi(\lambda + 2\mu)}\frac{\partial}{\partial x_i}\frac{\partial r}{\partial x_j}$$

$$- \frac{1}{8\pi\mu}\frac{\partial}{\partial x_i}\frac{\partial r}{\partial x_j} + \delta_{ij}\frac{1}{8\pi\mu}\nabla^2 r$$

$$= \frac{1}{8\pi\mu}\left(\delta_{ij}\nabla^2 r - \frac{\lambda + \mu}{\lambda + 2\mu}\frac{\partial^2 r}{\partial x_i \partial x_j}\right)$$

or

$$u_i^j = \frac{1}{8\pi\mu}(\delta_{ij}\, r_{,kk} - \Gamma r_{,ij}) \quad (8.16)$$

where

$$\Gamma = \frac{\lambda + \mu}{\lambda + 2\mu}. \quad (8.17)$$

For $\lambda \approx \mu$, a Poisson solid, $\Gamma \approx 2/3$. We have oriented our Cartesian coordinate system in the reference frame of the source

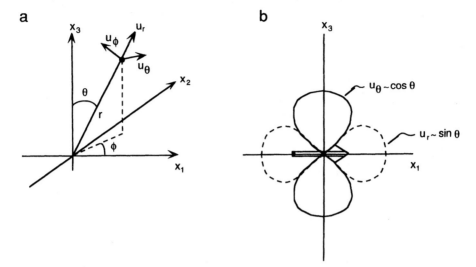

FIGURE 8.12 (a) The unit vectors in the polar coordinate system centered on the source. (b) The azimuthal pattern of radial displacements, u_r, and tangential displacements, u_θ, in the $x_1 x_3$ plane for a point force in the x_1 direction. These combine to give the deformation field shown in Figure 8.11.

(i.e., the force acts along one of the axes), and we determine displacements relative to this local source reference frame. Equation (8.16) is the *Somigliana tensor*. Note that it is symmetric, $u_i^j = u_j^i$. For a force, **F**, the six independent permutations are

$$u_1^1 = \frac{F}{8\pi\mu}\left[\frac{2}{r} - \Gamma\left(\frac{1}{r} - \frac{x_1^2}{r^3}\right)\right]$$

$$u_1^2 = \frac{F}{8\pi\mu}\left(\Gamma\frac{x_1 x_2}{r^3}\right)$$

$$u_1^3 = \frac{F}{8\pi\mu}\left(\Gamma\frac{x_1 x_3}{r^3}\right)$$

$$u_2^2 = \frac{F}{8\pi\mu}\left[\frac{2}{r} - \Gamma\left(\frac{1}{r} - \frac{x_2^2}{r^3}\right)\right]$$

$$u_2^3 = \frac{F}{8\pi\mu}\left(\Gamma\frac{x_2 x_3}{r^3}\right)$$

$$u_3^3 = \frac{F}{8\pi\mu}\left[\frac{2}{r} - \Gamma\left(\frac{1}{r} - \frac{x_3^2}{r^3}\right)\right]. \quad (8.18)$$

We consider the case in which the force is applied in the x_1 direction and find the displacements given in polar coordinates (r, θ, ϕ) for the source reference frame (Figure 8.12), using the Jacobian coordinate transformation:

$$\begin{bmatrix} u_r \\ u_\theta \\ u_\phi \end{bmatrix}$$

$$= \begin{bmatrix} \sin\theta\cos\phi & \sin\theta\sin\phi & \cos\theta \\ \cos\theta\cos\phi & \cos\theta\sin\phi & -\sin\phi \\ -\sin\phi & \cos\phi & 0 \end{bmatrix}\begin{bmatrix} u_1^1 \\ u_2^1 \\ u_3^1 \end{bmatrix}.$$

$$(8.19)$$

In the $x_1 x_3$ plane, $\phi = 0$; thus we find

$$u_r = \sin\theta\, u_1^1 + \cos\theta\, u_3^1 = \frac{F}{4\pi\mu r}\sin\theta$$

$$u_\theta = \cos\theta\, u_1^1 - \sin\theta\, u_3^1$$

$$= \frac{F}{4\pi\mu r}\left(1 - \frac{\Gamma}{2}\right)\cos\theta. \quad (8.20)$$

These solutions, with simple trigonometric patterns of static deformation as illustrated in Figure 8.12, are very similar to the intuitive derivation at the end of the preceding section. The radial motions are static deformations that have a simple two-lobed sinusoidal distribution around the source, with maxima toward and away from the point force in the direction of the force. The shearing deformations in the $x_1 x_3$ plane also have a two-lobed pattern of shearing parallel to the direction of the force. Axial symmetry requires that these patterns be rotated around the x_1 axis to obtain the full, three-dimensional patterns.

8.3.2 Static Displacement Field Due to a Force Couple

If we apply a force at position (ξ_1, ξ_2, ξ_3) instead of at the origin, the displacement at $P(x_1, x_2, x_3)$ will still be given by the Somigliana tensor, with all distances corrected by the offset of the source location, e.g., $r^2 = (x_1 - \xi_1)^2 + (x_2 - \xi_2)^2 + (x_3 - \xi_3)^2$. If we apply a force $\mathbf{F}$ in the x_1 direction at $\xi_2 + \frac{1}{2}d\xi_2$ and another force $\mathbf{F}$ in the $-x_1$ direction at $\xi_2 - \frac{1}{2}d\xi_2$, we construct a single force couple as shown in Figure 8.13. The displacement at $P(x_1, x_2, x_3)$ is the sum of the displacements from the two individual forces:

$$u_i^1(\xi_1, \xi_2 + \tfrac{1}{2}d\xi_2, \xi_3 : x_1, x_2, x_3)$$

$$- u_i^1(\xi_1, \xi_2 - \tfrac{1}{2}d\xi_2, \xi_3 : x_1, x_2, x_3)$$

$$= \frac{\partial u_i^1}{\partial \xi_2} d\xi_2 + O(d\xi_2)^2. \qquad (8.21)$$

Here the arguments give the source location and then the observing position. Thus, we calculate the difference between the displacement fields due to the single forces, allowing for the slight spatial offset $d\xi_2$, to

find the couple response. Since

$$r^2 = (x_1 - \xi_1)^2 + (x_2 - \xi_2)^2 + (x_3 - \xi_3)^2$$

we see that

$$\frac{\partial r}{\partial \xi_i} = -\frac{\partial r}{\partial x_i}.$$

Therefore

$$\frac{\partial u_i^j}{\partial \xi_k} = -\frac{\partial u_i^j}{\partial x_k} \qquad (8.22)$$

and the displacement for the force couple in Figure 8.13 is given by

$$-\frac{\partial u_i^1}{\partial x_2} d\xi_2 + O(d\xi_2)^2. \qquad (8.23)$$

If we let $d\xi_2 \to 0$ and $F \to \infty$ so that $F d\xi_2 \to M$, a finite moment, this result gives the static displacement field for the single couple with moment M. This is a direct consequence of linear superposition of solutions for elastic materials. Thus, the displacement (u_i) due to a single couple at the origin with forces acting in the x_1 direction offset in the x_2 direction can be obtained by replacing F by M in the Somigliana tensor, taking the derivative of all the terms with respect to x_2, and changing the sign:

$$u_1 = -\frac{M}{8\pi\mu}\left[\frac{-2x_2}{r^3} - \Gamma\left(\frac{-x_2}{r^3} + 3\frac{x_1^2 x_2}{r^5} \right) \right]$$

$$u_2 = -\frac{M}{8\pi\mu}\Gamma\left(\frac{x_1}{r^3} - 3\frac{x_1 x_2^2}{r^5} \right)$$

$$u_3 = -\frac{M}{8\pi\mu}\Gamma\left(-3\frac{x_1 x_2 x_3}{r^5} \right). \qquad (8.24)$$

Similarly, if the single couple is oriented along the x_2 axis with offset arm along the

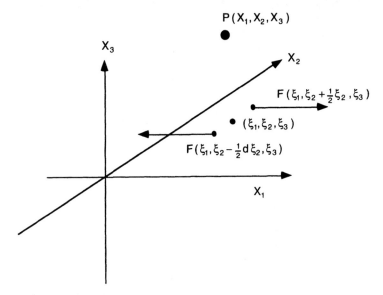

FIGURE 8.13 A force couple acting at position (ξ_1, ξ_2, ξ_3) parallel to the $x_1 x_2$ plane.

x_1 direction, we find

$$u_1 = -\frac{M}{8\pi\mu}\Gamma\left(\frac{x_2}{r^3} - 3\frac{x_1^2 x_2}{r^5}\right)$$

$$u_2 = -\frac{M}{8\pi\mu}\left[\frac{-2x_1}{r^3} - \Gamma\left(\frac{-x_1}{r^3} + 3\frac{x_2^2 x_1}{r^5}\right)\right]$$

$$u_3 = -\frac{M}{8\pi\mu}\Gamma\left(-3\frac{x_1 x_2 x_3}{r^5}\right). \tag{8.25}$$

In general, for a couple oriented with forces in the jth direction with offset in the kth direction, the displacements are $u_i = -u_{i,k}^j$, where moment is defined to be positive for clockwise rotation and negative for counterclockwise rotation around the perpendicular axis.

8.3.3 Static Displacement Field Due to a Double Couple

In the previous section, it was argued that a double-couple model is an appropriate equivalent body-force system for an earthquake dislocation. To determine the displacement field due to the double cou-

ple, all we do is sum the displacements for the individual couples! This is because the principle of superposition that holds for point forces must also hold for couples. Thus, for the double couple in the $x_1 x_2$ plane shown in Figure 8.14a, the displacements are given by

$$u_i = -u_{i,2}^1 + u_{i,1}^2 \tag{8.26}$$

$$u_1 = \frac{M}{8\pi\mu}\left(\frac{2x_2}{r^3}\right) - \frac{2M\Gamma}{8\pi\mu}\left(\frac{x_2}{r^3} - \frac{3x_2 x_1^2}{r^5}\right)$$

$$= \frac{M}{4\pi\mu r^2}\frac{x_2}{r}\left[1 - \Gamma\left(1 - \frac{3x_1^2}{r^2}\right)\right]$$

$$u_2 = \frac{M}{8\pi\mu}\left(\frac{2x_1}{r^3}\right) - \frac{2M\Gamma}{8\pi\mu}\left(\frac{x_1}{r^3} - \frac{3x_2^2 x_1}{r^5}\right)$$

$$= \frac{M}{4\pi\mu r^2}\frac{x_1}{r}\left[1 - \Gamma\left(1 - \frac{3x_2^2}{r^2}\right)\right]$$

$$u_3 = \frac{M}{8\pi\mu}\left(6\Gamma\frac{x_1 x_2 x_3}{r^5}\right)$$

$$= \frac{M}{4\pi\mu r^2}\left(3\Gamma\frac{x_1 x_2 x_3}{r^3}\right). \tag{8.27}$$

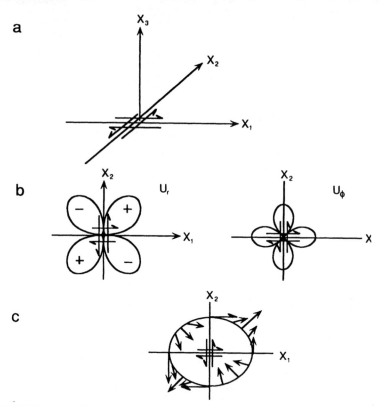

FIGURE 8.14 (a) A double couple in the x_1x_2 plane. (b) Azimuthal pattern of radial (u_r) and tangential (u_ϕ) displacements in the x_1x_2 plane. (c) The total displacement pattern in the x_1x_2 plane on a circle around the source, involving a combination of u_r and u_ϕ components.

Using the polar coordinates (r, θ, ϕ) in Figure 8.12, the displacements can be written in terms of u_r, u_θ, and u_ϕ:

$$u_r = \frac{M}{4\pi\mu r^2}\left(1 + \frac{\Gamma}{2}\right) \sin^2\theta \sin 2\phi$$

$$u_\theta = \frac{M}{4\pi\mu r^2}\left(\frac{1}{2} - \frac{\Gamma}{2}\right) \sin 2\theta \sin 2\phi$$

$$u_\phi = \frac{M}{4\pi\mu r^2}(1 - \Gamma) \sin\theta \cos 2\phi. \quad (8.28)$$

Note that the displacement fields for the double couple drop off with distance from the source much more rapidly than for the point force. On the x_1x_2 plane, $\theta = \pi/2$,

$u_\theta = 0$, and

$$u_r \approx \left(1 + \frac{\Gamma}{2}\right) \sin 2\phi$$

$$u_\phi \approx (1 - \Gamma) \cos 2\phi. \quad (8.29)$$

These are the azimuthal variation of u_r, u_ϕ displacements in the x_1x_2 plane and are shown in Figure 8.14b. If we consider the static displacements on a circle, we have the pattern in Figure 8.14c, involving both shearing and radial components.

So are these results useful for anything? Indeed they are, as Eqs. (8.27), or the similar terms for any other double-couple orientation, can be numerically computed to determine static deformations of the Earth around a slipped fault. The basic

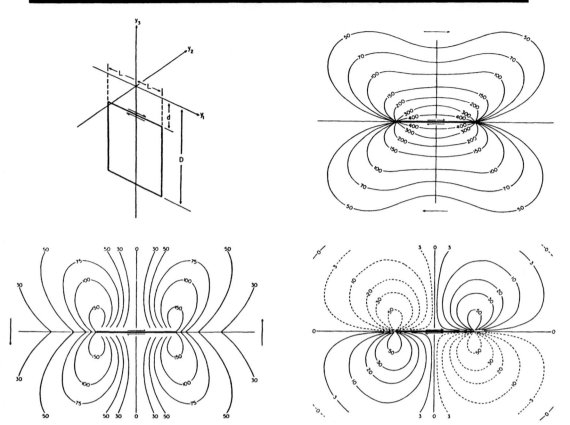

FIGURE 8.15 Ground displacements on the surface of a half-space computed for slip on a vertical strike slip, with a length (2L) equal to the downdip width (D). Fault parallel motions are shown on the upper right, fault perpendicular motions are shown on the lower left, and vertical motions are shown on the lower right. Contour values in units of $10^{-3} U$, where U is the uniform dislocation on the fault. Solid contours in the lower right indicate uplift, dashed indicate depression. (From Chinnery, 1961.)

idea is that *the displacement field due to a shear dislocation can be given by the displacement field due to a distribution of equivalent double couples that are placed in a medium without any dislocation.* This is one of the most important concepts in seismology, and it underlies both static and dynamic displacement modeling.

Since static deformations decay rapidly with distance from the source, most modeling involves vertical and horizontal ground deformations that are observed near the fault, so a point-source approximation is almost never valid. Thus, a finite fault with a numerically discretized distri-

bution of double couples is usually used. Calculations of surface horizontal and vertical displacements for a vertical strike–slip fault are shown in Figure 8.15. Note that the finite dimensions of the fault produce complex patterns of displacement at the ends of the fault, even for this uniform-slip model. Observed patterns can be modeled to constrain the faulting parameters. An actual data application is shown in Figure 8.16, where surface vertical ground motions measured by leveling before and after the 1989 Loma Prieta earthquake are used to derive an average fault model for the event. Oblique right-lateral thrust

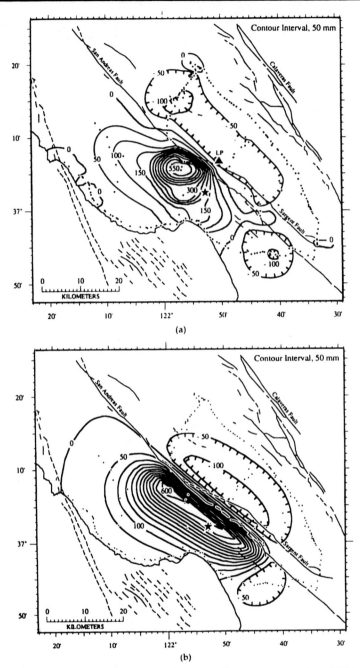

FIGURE 8.16 Observed vertical ground motions (in millimeters) near the 1989 Loma Prieta rupture zone compared to predictions of a dislocation model. The orientation of the fault zone is a 30-km-long fault parallel to the San Andreas dipping about 70° to the south. The star shows the epicenter of the event; the triangle labeled LP is the location of Loma Prieta Mountain, for which the event is named. (From Marshall *et al.*, 1991.)

faulting produced the 0.6-m uplift as well as horizontal motions of about 1.5 m. Typically, some constraints can be placed on the total slip and fault area (and hence on the seismic moment) of well-recorded events. Ongoing research in geodetic modeling of faulting includes incorporating viscoelastic effects of the deeper crust and upper mantle, adding layering and elastic parameter heterogeneity in the Earth model, and inverting for variable slip functions or changing fault mechanisms.

We proceed now to allow for time dependence of the source force system, which gives rise to both transient waves and permanent static deformations of the medium.

8.4 Elastodynamics

We now consider the application of a time-varying force system to a homogeneous elastic whole space. Mathematically this is a much more difficult problem than the statics case. The temporal dependence means that we will have to perform an extra set of integrations, with respect to both time and space. The elastostatics provided the logical framework for elastodynamics, and we will follow the same basic procedures. Because some of the mathematics is beyond the scope of this text, we have left out some of the mathematical steps. More detailed analysis can be found in advanced texts such as Aki and Richards (1980) or Kasahara (1981). As for the static problem, we construct the full solution for the double-couple force system by summing up solutions for the single force response. The elastodynamic equations are

$$\rho \ddot{\mathbf{u}} = \rho \mathbf{f} + (\lambda + 2\mu)\nabla(\nabla \cdot \mathbf{u}) - \mu \nabla \times \nabla \times \mathbf{u}.$$
$$(2.52)$$

We let the body force per unit volume ($\rho \mathbf{f}$) have the time-dependent form $\rho \mathbf{f}(t) = F(t)\delta(\mathbf{r})\mathbf{a}$, where $F(t)$ is the time history of the applied force. Common force–time

histories to consider are delta functions, $\delta(t)$, step functions, $H(t)$, and ramp functions, $R(t)$, like those in Figure 8.1. As before, we use a vector identity to write

$$\rho \mathbf{f} = F(t)\delta(\mathbf{r})\mathbf{a} = -F(t)\nabla^2 \left(\frac{\mathbf{a}}{4\pi r} \right)$$

$$= -F(t) \left\{ \nabla \left[\nabla \cdot \left(\frac{\mathbf{a}}{4\pi r} \right) \right] \right.$$

$$\left. -\nabla \times \nabla \times \left(\frac{\mathbf{a}}{4\pi r} \right) \right\}. \quad (8.30)$$

So we seek a solution of the form [see (8.11)]

$$\mathbf{u}(t) = \nabla(\nabla \cdot \mathbf{A}_P) - \nabla \times \nabla \times \nabla \mathbf{A}_S$$

$$\text{where} \begin{cases} \nabla \times \mathbf{A}_P = 0 \\ \nabla \cdot \mathbf{A}_S = 0. \end{cases} \quad (8.31)$$

The elastodynamic equation separates into

$$(\lambda + 2\mu)\nabla^2 \mathbf{A}_P = \frac{F(t)}{4\pi r}\mathbf{a} + \rho \frac{\partial^2 \mathbf{A}_P}{\partial t^2}$$

$$\mu \nabla^2 \mathbf{A}_S = \frac{F(t)}{4\pi r}\mathbf{a} + \rho \frac{\partial^2 \mathbf{A}_S}{\partial t^2}.$$

$$(8.32)$$

Putting $\mathbf{A}_P = A_P \mathbf{a}$, $\mathbf{A}_S = A_S \mathbf{a}$, these reduce to scalar equations:

$$\nabla^2 A_P = \frac{F(t)}{4\pi(\lambda + 2\mu)r} + \frac{1}{\alpha^2}\frac{\partial^2 A_P}{\partial t^2}$$

$$\text{where } \alpha = P \text{ velocity} = \sqrt{\frac{\lambda + 2\mu}{\rho}}$$

$$\nabla^2 A_S = \frac{F(t)}{4\pi\mu r} + \frac{1}{\beta^2}\frac{\partial^2 A_S}{\partial t^2}$$

$$\text{where } \beta = S \text{ velocity} = \sqrt{\frac{\mu}{\rho}}. \quad (8.33)$$

Before proceeding, we want to consider the form of solutions to the inhomogeneous wave equation

$$\nabla^2\phi(x_1, x_2, x_3, t) - \frac{1}{c^2}\frac{\partial^2\phi}{\partial t^2}(x_1, x_2, x_3, t)$$

$$= g(x_1, x_2, x_3, t). \qquad (8.34)$$

A specific form of g that is useful to know the solution for is

$$g(\mathbf{x}, t) = -\delta(\mathbf{x})\delta(t) = -\delta(\mathbf{r})\delta(t)$$

$$(8.35)$$

or a point force in space and time. The solution is

$$\phi(\mathbf{x}, t) = \frac{1}{4\pi}\frac{\delta(t - (r/c))}{r}. \qquad (8.36)$$

This solution is very important. It states that the solution to a symmetric point-source impulse is an outward-propagating wave that decays in amplitude as $1/r$. A corresponding solution was invoked in the initial discussion of explosion sources at the start of this chapter. From Chapter 2 (Box 2.5), we recall that the form of spherically symmetric waves that solve the homogeneous wave equation is

$$\nabla^2\phi - \frac{1}{c^2}\ddot{\phi} = 0$$

$$\phi = \frac{1}{r}f\left(t - \frac{r}{c}\right) + \frac{1}{r}g\left(t + \frac{r}{c}\right),$$

$$(8.37)$$

a standard D'Alembert-type solution. The $1/r$ term stems from the amplitude decrease required to keep the total energy on the spreading wavefront constant. We also see that the delta function source shape is of the same form as the far-field displacement time history. Given (8.36), we can readily construct additional solu-

tions:

$$\nabla^2\phi - \frac{1}{c^2}\frac{\partial^2\phi}{\partial t^2} = -\delta(\mathbf{x} - \boldsymbol{\xi})\delta(t - \tau)$$

$$\phi(\mathbf{r}, t) = \frac{1}{4\pi}\frac{\delta(t - \tau - |\mathbf{x} - \boldsymbol{\xi}|/c)}{|\mathbf{x} - \boldsymbol{\xi}|},$$

$$(8.38)$$

which is the solution for a point force at position $x = (\xi_1, \xi_2, \xi_3)$ applied at time $t = \tau$. Another basic solution is

$$\nabla^2\phi - \frac{1}{c^2}\frac{\partial^2\phi}{\partial t^2} = -\delta(\mathbf{x} - \boldsymbol{\xi})f(t)$$

$$\phi = \frac{1}{4\pi}\frac{f(t - |\mathbf{x} - \boldsymbol{\xi}|/c)}{|\mathbf{x} - \boldsymbol{\xi}|}.$$

$$(8.39)$$

If the source is extended throughout a volume, V, as well as in time

$$\nabla^2\phi - \frac{1}{c^2}\frac{\partial^2\phi}{\partial t^2} = -\Phi(\mathbf{x}, t)$$

$$\phi(\mathbf{x}, t) = \frac{1}{4\pi}\iiint_V \frac{\Phi(\boldsymbol{\xi}, t - |\mathbf{x} - \boldsymbol{\xi}|/c)}{|\mathbf{x} - \boldsymbol{\xi}|} dV.$$

$$(8.40)$$

This states that the field at $(\mathbf{x}, t)$ is sensitive to source activity in the element dV (at $\boldsymbol{\xi}$) only at the *retarded time* $t - (|\mathbf{x} - \boldsymbol{\xi}|/c)$. We can thus write down solutions to Eq. (8.33) as

$$A_P = \frac{1}{4\pi}\iiint_V \frac{-F(t - |\mathbf{x} - \boldsymbol{\xi}|/\alpha)}{4\pi(\lambda + 2\mu)r|\mathbf{x} - \boldsymbol{\xi}|} dV$$

$$A_S = \frac{1}{4\pi}\iiint_V \frac{-F(t - |\mathbf{x} - \boldsymbol{\xi}|/\beta)}{4\pi\mu r|\mathbf{x} - \boldsymbol{\xi}|} dV,$$

$$(8.41)$$

where $\boldsymbol{\xi} = 0$ for a point force at the origin.

The next steps are rather messy. It is necessary to integrate over the volume around position **x**. Let the distance $|\mathbf{x} - \boldsymbol{\xi}| = \alpha\tau$, where τ is the transit time. Then it can be shown that

$$A_P(r,t)$$

$$= \frac{1}{4\rho\pi r}\left(\int_0^\infty F(t - (r/\alpha) - \tau)\tau\, d\tau\right.$$

$$\left. - \int_0^\infty F(t - \tau)\tau\, d\tau\right)$$

$$A_S(r,t) = \frac{1}{4\rho\pi r}\left(\int_0^\infty F(t - (r/\beta) - \tau)\tau\, d\tau\right.$$

$$\left. - \int_0^\infty F(t - \tau)\tau\, d\tau\right). \quad (8.42)$$

The actual displacement field is obtained by computing

$$\mathbf{u} = \nabla(\nabla \cdot \mathbf{A}_P) - \nabla \times \nabla \times \mathbf{A}_S. \quad (8.43)$$

For a body force $F(t)$ applied in the x_1 direction at the origin the total displacement field is

$$u_i(\mathbf{x},t)$$

$$= \frac{1}{4\pi\rho}\left(\frac{\partial^2}{\partial x_i \partial x_1}\frac{1}{r}\right)\int_{r/\alpha}^{r/\beta} \tau F(t - \tau)\, d\tau$$

$$+ \frac{1}{4\pi\rho\alpha^2 r}\left(\frac{\partial r}{\partial x_i}\frac{\partial r}{\partial x_1}\right)F\left(t - \frac{r}{\alpha}\right)$$

$$+ \frac{1}{4\pi\rho\beta^2 r}\left(\delta_{i1}\frac{-\partial r}{\partial x_i}\frac{\partial r}{\partial x_1}\right)F\left(t - \frac{r}{\beta}\right).$$

$$(8.44)$$

We can repeat this process by letting each ∂x_1 be ∂x_2 or ∂x_3, thus easily obtaining the solutions for point forces in the x_2 or x_3 direction.

For a point force $F(t)$ in the x_j direction, located at the origin, we have the classic *Stokes* solution:

$$u_i(\mathbf{x},t)$$

$$= \frac{1}{4\pi\rho}(3\gamma_i\gamma_j - \delta_{ij})\frac{1}{r^3}\int_{r/\alpha}^{r/\beta}\tau F(t - \tau)\, d\tau$$

$$+ \frac{1}{4\pi\rho\alpha^2}\gamma_i\gamma_j\frac{1}{r}F\left(t - \frac{r}{\alpha}\right)$$

$$- \frac{1}{4\pi\rho\beta^2}(\gamma_i\gamma_j - \delta_{ij})\frac{1}{r}F\left(t - \frac{r}{\beta}\right),$$

$$(8.45)$$

where γ_i is the direction cosine; $\gamma_i = (x_i/r) = \partial r/\partial x_i$. The first integral term behaves like $1/r^2$ for short-duration sources, and the other terms behave like $1/r$. Thus the first term is called the *near-field* term, and the latter terms are *far-field* terms. The first far-field term is a P wave:

$$u_i^P = \frac{1}{4\pi\rho\alpha^2}\gamma_i\gamma_j\frac{1}{r}F\left(t - \frac{r}{\alpha}\right) \quad (8.46)$$

with the following properties: (1) it attenuates as $1/r$, (2) the wave propagates with velocity $\alpha = [(\lambda + 2\mu)/\rho]^{1/2}$, (3) the waveform is proportional to the applied force, and (4) the displacement is parallel to the direction from the source. Figure 8.17 indicates the sense of P radial motions associated with this term. The second far-field term is an S wave

$$u_i^S = \frac{1}{4\pi\rho\beta^2}(\delta_{ij} - \gamma_i\gamma_j)\frac{1}{r}F\left(t - \frac{r}{\beta}\right)$$

$$(8.47)$$

with the following properties: (1) it attenuates as $1/r$, (2) the wave propagates with velocity $\beta = (\mu/\rho)^{1/2}$, (3) the displacement waveform is proportional to the force, and (4) the direction of displacement is perpendicular to the direction from the source. The sense of shearing motions is shown in Figure 8.17. The near-field displacements comprise contributions from

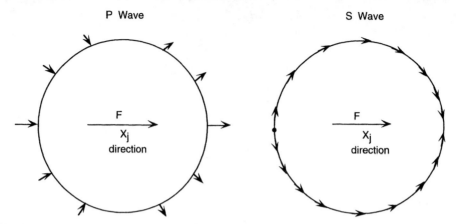

FIGURE 8.17 Sense of far-field displacements on P and S wavefronts produced by a single force in the x_j direction in an infinite, homogeneous, isotropic medium.

both the P and S wavefields, and these cannot be easily separated.

The displacement field for single couples and double couples can be obtained by differentiating the single-force results with respect to appropriate coordinates. This is the same as we saw for the static fields.

Differentiating the complete Stokes solution with respect to each coordinate direction gives the result for a couple paralleling each direction. The complete solution including near-field terms is given on pages 78–83 of Aki and Richards (1980). The major results can be obtained from considering the far-field P wave alone, with the other results following from similar analysis.

P waves in the far field from a force couple (Figure 8.18) with time history $h(t)$ are given by summing the motions due to individual point forces:

$$u_i^F = \frac{1}{4\pi\rho\alpha^2}\gamma_i\gamma_1 \frac{h(t - (r/\alpha))}{r}$$

$$u_i^{F_2} = \frac{1}{4\pi\rho\alpha^2}\gamma_i^{F_2}\gamma_{-1}^{F_2} \frac{h(t - (r_2/\alpha))}{r_2}.$$

$$(8.48)$$

The sum of these gives the total displacement due to the force couple:

$$u_i^c = \frac{1}{4\pi\rho\alpha^2}\left[\gamma_i\gamma_1 \frac{h(t - (r/\alpha))}{r} \right.$$

$$\left. + \gamma_i^{F_2}\gamma_{-1}^{F_2} \frac{h(t - (r_2/\alpha))}{r_2}\right]. \quad (8.49)$$

Let's consider $\gamma_i^{F_2}, \gamma_{-1}^{F_2}, r_2$ for $|\Delta\mathbf{r}| = |\mathbf{r}_2 - \mathbf{r}| \ll r$

$$\gamma_i = \frac{x_i}{r}$$

$$\gamma_i^{F_2} = \frac{x_i^{F_2}}{r + \Delta r} \approx \frac{x_i^{F_2}}{r} \approx \frac{x_i}{r} \quad (\text{for } j \neq 2)$$

$$= \frac{x_2 - \Delta x_2}{r} \quad (\text{for } j = 2)$$

$$\gamma_i^{F_2} = \gamma_i - \frac{\Delta x_2}{r}\delta_{2j}. \quad (8.50)$$

Note that although $\Delta r \ll r$, Δx_2 is not necessarily $\ll x_2$. Referring to Figure 8.18,

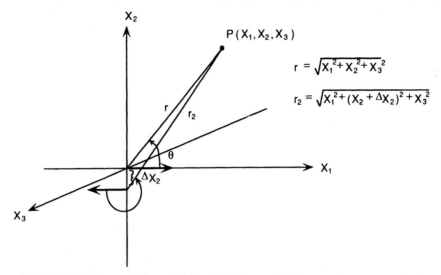

X₂

P(X₁,X₂,X₃)

$$r = \sqrt{X_1^2 + X_2^2 + X_3^2}$$

$$r_2 = \sqrt{X_1^2 + (X_2 + \Delta X_2)^2 + X_3^2}$$

FIGURE 8.18 A couple for which far-field P-wave displacements are computed.

we approximate the direction cosines:

$$\gamma_i = \frac{x_1}{r} = \cos\theta$$

$$\gamma_{-1}^{F_2} = \cos\psi = \frac{-x_1}{r + \Delta r}$$

$$\psi \approx \pi + \theta$$

$$\cos(\pi + \theta) \approx -\cos\theta$$

$$\gamma_{-1}^{F_2} \approx -\gamma_i. \tag{8.51}$$

So

$$u_i^c = \frac{\gamma_1}{4\pi\rho\alpha^2}\left[\gamma_i \frac{h(t - (r/\alpha))}{r}\right.$$

$$\left. -\left(\gamma_i - \frac{\Delta x_2}{r}\delta_{2j}\right)\frac{h(t - (r_2/\alpha))}{r_2}\right]. \tag{8.52}$$

We can approximate $1/r_2$ as $1/r$ for $\Delta r \ll r$, but what about $h(t - (r/\alpha))$ com-

pared with $h(t - (r_2/\alpha))$?

$$h(t - (r_2/\alpha)) = h(t - (r/\alpha) - (\Delta r/\alpha)). \tag{8.53}$$

$\Delta r/\alpha$ may not be small relative to $t - (r/\alpha) = t_0 = $ arrival time. We need a Taylor series expansion

$$h(t) = h(t_0) + \frac{\partial h}{\partial t}(t_0)(t - t_0) + \cdots$$

$$h(t - (r/\alpha) - (\Delta r/\alpha))$$

$$= h(t - (r/\alpha)) + \dot{h}(t - (r/\alpha))$$

$$\times \left[(t - (r/\alpha) - (\Delta r/\alpha))\right.$$

$$\left. -(t - (r/\alpha))\right] + \cdots$$

$$= h(t - (r/\alpha))$$

$$-(\Delta r/\alpha)\dot{h}(t - (r/\alpha)) + \cdots \tag{8.54}$$

Then, to first order

$$u_i^c = \frac{\gamma_1}{4\pi\rho\alpha^2} \left\{ \gamma_i \left[\frac{h(t-(r/\alpha))}{r} \right] \right.$$

$$-\gamma_i \left[\frac{h(t-(r/\alpha))}{r} \right.$$

$$\left. -\frac{\Delta r}{\alpha} \frac{\dot{h}(t-(r/\alpha))}{r} \right]$$

$$+\frac{\Delta x_2}{r} \delta_{2j} \left[\frac{h(t-(r/\alpha))}{r} \right.$$

$$\left. \left. -\frac{\Delta r}{\alpha} \frac{\dot{h}(t-(r/\alpha))}{r} \right] \right\}$$

$$u_i^c = \frac{\gamma_1}{4\pi\rho\alpha^2} \left\{ \gamma_i \frac{\Delta r \dot{h}(t-(r/\alpha))}{\alpha r} \right.$$

$$+\frac{\Delta x_2 \delta_{2j}}{r^2} \left[h\left(t-\frac{r}{\alpha}\right) \right.$$

$$\left. \left. -\frac{\Delta r}{\alpha} \dot{h}\left(t-\frac{r}{\alpha}\right) \right] \right\}. \quad (8.55)$$

The second group of terms decays as $1/r^2$; therefore these are near-field terms that we can dismiss relative to the far-field terms, which decay as $1/r$. This leaves the far-field displacements

$$u_i^c = \frac{\gamma_1}{4\pi\rho\alpha^3} \left[\frac{\gamma_i \Delta r \dot{h}(t-(r/\alpha))}{r} \right]. \quad (8.56)$$

It is critical to note that this differencing process has yielded a temporal differentiation of the source time history. The spatial offset of the forces leads to far-field displacement sensitivity to particle velocities at the source rather than to particle displacements. This was also found for the

explosion case described at the start of this chapter. We now need to consider $\Delta r = r_2 - r$:

$$\Delta r \approx \frac{\partial r}{\partial x_2} \Delta x_2 = \gamma_2 \Delta x_2 \quad (8.57)$$

(recall direction cosines $\gamma_i = x_i/r = \partial r/\partial x_i$)

$$u_i^c = \frac{\gamma_1\gamma_2\gamma_i}{4\pi\rho\alpha^3} \left[\frac{\Delta x_2}{r} \dot{h}\left(t-\frac{r}{\alpha}\right) \right]. \quad (8.58)$$

Now, we must consider the limit as $\Delta x_2 \to 0$ and $h \to \infty$ such that $\Delta x_2 h \to M$, which is the moment of the couple:

$$M\left(t-\frac{r}{\alpha}\right) = \lim_{\substack{\Delta x_2 \to 0 \\ h \to \infty}} \Delta x_2 h\left(t-\frac{r}{\alpha}\right)$$

$$\dot{M}\left(t-\frac{r}{\alpha}\right) = \lim_{\substack{\Delta x_2 \to 0 \\ h \to \infty}} \Delta x_2 \dot{h}\left(t-\frac{r}{\alpha}\right).$$

$$(8.59)$$

So the solution for the couple is given by

$$u_i^c = \frac{\gamma_1\gamma_2\gamma_i}{4\pi\rho\alpha^3} \frac{\dot{M}(t-(r/\alpha))}{r}. \quad (8.60)$$

If we did the same analysis for a couple with an orientation parallel to the x_2 axis in the x_1x_2 plane and if we offset F_2 by Δx_1 in the negative x_2 direction, we would get the identical results because of symmetry of the products of the direction cosines. Summation of these two couples gives the double couple shown in Figure 8.14.

Thus, the total response to the double-couple system in far-field P waves is

$$u_i^{Dc} = 2 \left[\frac{\gamma_1\gamma_2\gamma_i}{4\pi\rho\alpha^3} \frac{\dot{M}(t-(r/\alpha))}{r} \right]. \quad (8.61)$$

The general form of the far-field displacements for a couple in the pq plane is given by

$$u_n^c = M_{pq} * G_{np,q}$$

$$= \frac{\gamma_n \gamma_p \gamma_q}{4\pi\rho\alpha^3} \frac{\dot{M}_{pq}(t - (r/\alpha))}{r}, \quad (8.62)$$

where M_{pq} is any of the nine possible force couples (or dipoles) for three-dimensional geometries. These nine couples compose the *seismic moment tensor*. The next section will discuss this general force system in greater detail.

The general form of the far-field S-wave displacements for a couple in the pq plane is given by

$$u_n^c = \frac{-(\gamma_n \gamma_p - \delta_{np})\gamma_q}{4\pi\rho\beta^3} \frac{\dot{M}_{pq}(t - (r/\beta))}{r}.$$

$$(8.63)$$

If the displacements are considered in a spherical coordinate system (Figure 8.12),

we find that the far-field displacements are

$$\mathbf{u}(\mathbf{x}, t)$$

$$= \frac{1}{4\pi\rho\alpha^3}(\sin 2\theta \cos \phi \hat{r}) \frac{\dot{M}_0(t - (r/\alpha))}{r}$$

$$+ \frac{1}{4\pi\rho\beta^3}(\cos 2\theta \cos\phi \, \hat{\theta} - \cos\theta \sin\phi\hat{\phi})$$

$$\times \frac{\dot{M}_0(t - (r/\beta))}{r}, \quad (8.64)$$

where $M_0(t) = \mu A(t)D(t)$ is the time-dependent moment function. The first term is the P-wave radiation, and the second is the S-wave radiation. These expressions have the four-lobed patterns shown in Figure 8.19. Note that the azimuthal patterns are the same as found in the static case (Figure 8.14). Also note that the far-field displacements are proportional to the time derivative of the moment function, $\dot{M}(t)$, which is called the *moment rate* function.

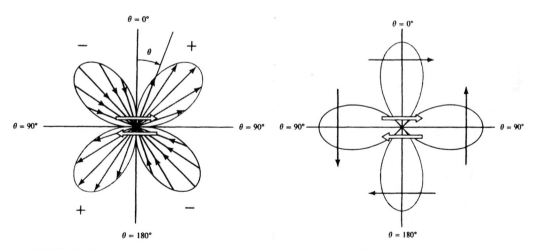

FIGURE 8.19 Far-field radiation patterns in the $x_1 x_3$ plane ($\phi = 0$) for radial components of displacement (left) and transverse components of displacement (right), for a double couple in the $x_1 x_3$ plane. The full vector displacements are given by Eq. (8.64). (Modified from Aki and Richards, 1980. Copyright ©1980 by W. H. Freeman and Co. Reprinted with Permission.)

Box 8.2 Point-Force Sources

Our theoretical development of equivalent body forces used point-force solutions mainly to build couples and double couples, and one might ask whether point-force sources exist. Several natural phenomena have been found to be well explained by equivalent point-force sources, on the basis of observing compatible radiation patterns. The simplest examples include vertical volcanic eruptions, which can be modeled as a point force representing the counterforce of the eruption. Figure 8.B2.1 shows the theoretical horizontal and vertical ground motions for a vertical point force on the surface, which primarily excites a large Rayleigh wave. The solution for this system was first provided by Lamb (1904) and represents the ground-motion calculation of a transient wave. The figure on the right shows observed and synthetic ground motion (mainly Rayleigh wave) for a station 67 km from an eruption of Mt. St. Helens, on June 13, 1980 (a minor eruption after the main blast). Kanamori and Given (1983) estimated a force strength of 5.5×10^{15} dyne by matching the observed amplitude.

Another process that can be modeled by a point force quite successfully is a nearly horizontal landslide. In this case the point represents the reaction force on the surface due to laterally moving the slide mass off the hillside. Kanamori and Given (1982) modeled the massive landslide of the May 18, 1980 Mt. St. Helens eruption with a horizontal force, finding that this was the best way to model long-period Love and Rayleigh waves from the event. Another example is the 1975 Kalapana, Hawaii event, which involved either slip on a very shallow dipping fault

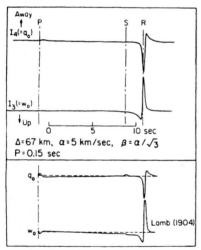
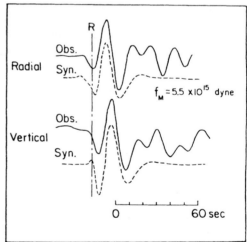

FIGURE 8.B2.1 (Left) Theoretical radial and vertical ground motions for a point force on the surface of a half-space. (Right) Comparison of observed and synthetic ground motions for the June 13, 1980 eruption of Mt. St. Helens. The synthetics are computed for a vertical downward point force at the source, with strength $f_m = 5.5 \times 10^{15}$ dyne. (From Kanamori and Given, *Geophys. Res. Lett.* **10**, 373–376, 1983; © Copyright by the American Geophysical Union.)

continues

or a large slump (Figure 8-B2.2). The observed radiation pattern of Love waves from the event is two-lobed, consistent with the single-force model as shown. From the strength of the point force one can estimate the peak acceleration of the slide block as $0.1–1.0$ m/s^2.

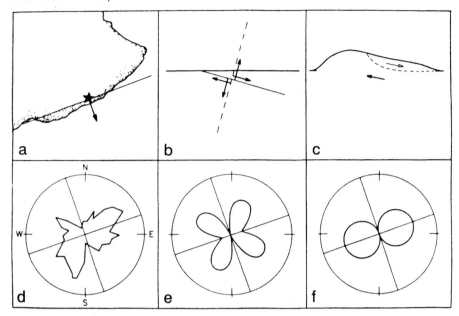

FIGURE 8.B2.2 Observations and interpretations of the source mechanism for the 1975 Kalapana, Hawaii event. Observations are shown in (a) and (d), where subhorizontal surface ground motions were observed southeast of the hypocenter and the teleseismic Love-wave radiation has the observed pattern in (d). Interpretation of the source as a shallow double couple (b) predicts a four-lobed Love wave radiation (e); interpretation as a reaction point force (c) due to land sliding produces a Love-wave radiation pattern (f) in better agreement with the data. (From Eissler and Kanamori, *J. Geophys. Res.* **92**, 4827–4836, 1987; © Copyright by the American Geophysical Union.)

It is usually much more convenient to express the radiation pattern in terms of a geographical coordinate system rather than the coordinate system defined by the double couple. This requires an algebraic mapping between the two systems, as shown in Figure 8.20. We follow Aki and Richards (1980) in defining a ray coordinate system with directions $\hat{l}, \hat{p}, \hat{\phi}$ along the P, SV, and SH directions at the source. The equations for the far-field P and S waves from a point double-couple source with standard fault orientation parameters ϕ_f, δ, and λ and source takeoff angle i_h can be written as:

$$U_P(r,t) = \frac{1}{4\pi\rho\alpha^3} R^P \dot{M}\left(t - \frac{r}{\alpha}\right)$$

$$U_{SV}(r,t) = \frac{1}{4\pi\rho r\beta^3} R^{SV} \dot{M}\left(t - \frac{r}{\beta}\right)$$

$$U_{SH}(r,t) = \frac{1}{4\pi\rho r\beta^3} R^{SH} \dot{M}\left(t - \frac{r}{\beta}\right)$$

$$R^P = \cos\lambda \sin\delta \sin^2 i_h \sin 2\phi$$

$$- \cos\lambda \cos\delta \sin 2i_h \cos\phi$$

$$+ \sin\lambda \sin 2\delta(\cos^2 i_h - \sin^2 i_h \sin^2\phi)$$

$$+ \sin\lambda \cos 2\delta \sin 2i_h \sin\phi$$

$$R^{SV} = \sin\lambda \cos 2\delta \cos 2i_h \sin\phi$$

$$- \cos\lambda \cos\delta \cos 2i_h \cos\phi$$

$$+ \tfrac{1}{2}\cos\lambda \sin\delta \sin 2i_h \sin 2\phi$$

$$- \tfrac{1}{2}\sin\lambda \sin 2\delta \sin 2i_h(1+\sin^2\phi)$$

$$R^{SH} = \cos\lambda \cos\delta \cos i_h \sin\phi$$

$$+ \cos\lambda \sin\delta \sin i_h \cos 2\phi$$

$$+ \sin\lambda \cos 2\delta \cos i_h \cos\phi$$

$$- \tfrac{1}{2}\sin\lambda \sin 2\delta \sin i_h \sin 2\phi,$$

$$(8.65)$$

where $\phi = \phi_f - \phi_s$, with ϕ_f the fault strike

and ϕ_s the station azimuth. The $1/r$ term accounts for geometric spreading in a whole space, and we need to modify this for actual geometric spreading in the Earth. To estimate the actual geometric spreading to the far field, we use expression (3.72) for the decrease in energy per unit area on the wavefront as a function of distance

$$E(\Delta) = \frac{E_0 \sin i_h v_0}{\cos i_0 \sin\Delta r_0^3 \cos i_0}\left|\frac{\partial^2 T}{\partial\Delta^2}\right|$$

$$= \frac{E_0 \sin i_h}{r_0^2 \sin\Delta \cos i_0}\left|\frac{di_h}{d\Delta}\right|, \quad (8.66)$$

where E_0 is the energy emitted per unit solid angle at the source, i_h is the takeoff angle at the source, and i_0 is the incident angle at the receiver.

We will consider geometric spreading for a P wave. If we take the Fourier transform of the P-wave time-domain sig-

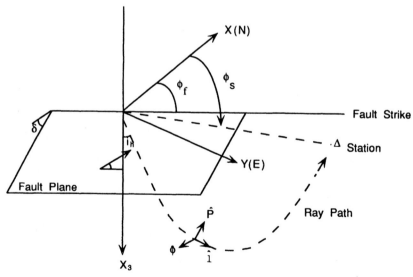

FIGURE 8.20 Definition of a geographic coordinate system, with positive x_3 being downward. The fault strike, ϕ_f, is measured from north; the dip, δ, is measured in the plane of the fault. The observing station is at azimuth ϕ_s, and the raypath to that station has a takeoff angle, i_h, relative to the x_3 axis.

nal and consider the spectral amplitude, we have

$$\hat{u}_r(r,\omega) = \frac{1}{4\pi\rho r\alpha^3} R^P \omega |\hat{M}(\omega)|. \quad (8.67)$$

Recall that the Fourier transform of a time derivative of a function is given by ω times the transform of the function. Consider a small sphere of radius $r = \varepsilon_h$ at the source. Since the energy passing through a unit area of wavefront per unit time is proportional to $\rho\alpha\omega^2|\hat{u}(\omega)|^2$, then

$$E_0 = \varepsilon_h^2 \rho_h \alpha_h \omega^2 |\hat{u}_r(r,\omega)|^2$$

$$= \rho_h \alpha_h \omega^2 \left(\frac{1}{4\pi\rho_h\alpha_h^3} R_P \omega |\hat{M}(\omega)| \right)^2.$$

$$(8.68)$$

If we let $|\hat{u}_r(\omega)|$ be the amplitude spectral density at the station, we have

$$E(\Delta) = \rho_0 \alpha_0 \omega^2 |\hat{u}_r(\omega)|^2. \quad (8.69)$$

Then we have

$$|\hat{u}_r(\omega)| = \frac{1}{4\pi\rho_h\alpha_h^3} \frac{g(\Delta,h)}{r_0} R^P \omega |\hat{M}(\omega)|,$$

$$(8.70)$$

where

$$g(\Delta,h) = \sqrt{\frac{\rho_h\alpha_h}{\rho_0\alpha_0} \frac{\sin i_h}{\sin \Delta} \frac{1}{\cos i_0} \left| \frac{di_h}{d\Delta} \right|}.$$

$$(8.71)$$

Inverting this to the time domain, we find

$$|\hat{u}_r(\Delta,t)|$$

$$= \frac{1}{4\pi\rho_h\alpha_h^3} \frac{g(\Delta,h)}{r_0} R^P \dot{M}(t-(r/\alpha)).$$

$$(8.72)$$

Similar expressions are found for SH and SV waves. The term $g(\Delta,h)/r_0$ is a *geometric spreading* factor, which replaces $1/r$ in (8.65).

We will consider the nature of the moment rate function $\dot{M}(t-(r/\alpha))$ in detail in later chapters. For the present we simply recognize that $\dot{M}$ is the time derivative of the moment function. We thus need to consider the time history of $M(t)$. Earlier we defined the static moment, $M_0 = \mu A\bar{D}$, where $\bar{D}$ is the average displacement over the fault area A after motion ends. $M(t) = \mu A(t)D(t)$. Also, $\dot{M}(t) = \mu[\partial(A(t)D(t))/\partial t]$.

If all the displacement occurred instantaneously in time, then

$$M(t) = H(t), \quad (8.73)$$

where $H(t)$ is a step function and

$$\dot{M}(t) = \delta(t), \quad (8.74)$$

as shown in Figure 8.21. More realistically, it takes a finite length of time for any given particle to achieve its total offset, even for a point source for which $A(t)$ is a step function. In this case, a ramp function can

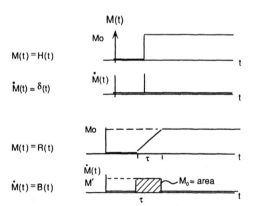

FIGURE 8.21 Far-field *P*- and *S*-wave displacements are proportional to $\dot{M}(t)$, the time derivative of the moment function $M(t) = \mu A(t)D(t)$. Simple step and ramp moment functions generate far-field impulses or boxcar ground motions.

represent the moment function (Figure 8.21)

$$M(t) = R(t) \qquad (8.75)$$

and

$$\dot{M}(t) = B(t), \qquad (8.76)$$

where $B(t)$ is a boxcar function. The area A under the boxcar function is equal to M_0:

$$A = \int_{-\infty}^{\infty} \dot{M}(t)\, dt = M'\tau = M_0. \qquad (8.77)$$

Thus, the details of the *particle-dislocation time history* affect the far-field body-wave signals. In the next chapter we will consider a few simple rupture time histories, including both finite particle dislocation and areal expansion histories.

8.5 The Seismic Moment Tensor

In the previous section we derived a general form for the far-field P- and S-wave displacements for any of the nine possible couples in the local-source Cartesian coordinate system [Eqs. (8.62) and (8.63)]. The full set of couples, M_{ij}, is shown in Figure 8.22. We summed the displacement fields produced by M_{12} and M_{21} to produce double-couple solutions (8.61), but we could just as well sum $M_{13} + M_{31}$ or $M_{23} + M_{32}$ to produce double couples acting in orthogonal coordinate planes of the local-source coordinate system. The explosion source described early in this chapter can be modeled by the sum of the three dipole terms, $M_{11} + M_{22} + M_{33}$, with each having equal moment. The full set of couples can clearly be summed to produce a very wide range of effective source deformations and time histories. In many cases we can assume that the force couples all have the same time dependence, $s(t)$, so we can define a second-order tensor $\mathbf{M}$ with M_{pq} compo-

nents, giving the excitation of the pq couple:

$$\mathbf{M} = \begin{bmatrix} M_{11} & M_{12} & M_{13} \\ M_{21} & M_{22} & M_{23} \\ M_{31} & M_{32} & M_{33} \end{bmatrix}. \qquad (8.78)$$

The double-couple solution given by (8.61) has the corresponding moment tensor

$$\mathbf{M} = M_0 \begin{bmatrix} 0 & M_{12} & 0 \\ M_{21} & 0 & 0 \\ 0 & 0 & 0 \end{bmatrix}, \qquad (8.79)$$

where the scalar factor, M_0, is taken outside.

The seismic moment tensor is always symmetric. Because it is a tensor, coordinate transformations must exist that relate the terms of the moment tensor in the source coordinate system to the more useful geographic coordinate system in which the couples are not necessarily aligned with the axes. Thus, we want to find expressions analogous to (8.65) for the full moment tensor.

We initially consider double-couple moment tensors. The arbitrarily oriented double couple will act on a fault plane with a slip vector, $\mathbf{D}$, and a normal to the plane, $\mathbf{v}$. The elements of M_{ij} for the geographic (or any other) reference frame are

$$M_{kj} = \mu A (D_k v_j + D_j v_k). \qquad (8.80)$$

Note the symmetry of the slip vector $\mathbf{D}$ and the fault normal $\mathbf{v}$, which gives rise to the ambiguity of the fault plane and auxiliary plane for a point double couple. If we adopt the coordinate system in Figure 8.20 and if we express the coordinates of the slip vector and the fault normal in terms of ϕ_f, δ, and λ, we find

$$\mathbf{D} = \overline{D}(\cos \lambda \cos \phi_f + \cos \delta \sin \lambda \sin \phi_f)\hat{\mathbf{x}}_1$$
$$+ \overline{D}(\cos \lambda \sin \phi_f$$
$$- \cos \delta \sin \lambda \cos \phi_f)\hat{\mathbf{x}}_2$$
$$- \overline{D} \sin \delta \sin \lambda\, \hat{\mathbf{x}}_3, \qquad (8.81)$$

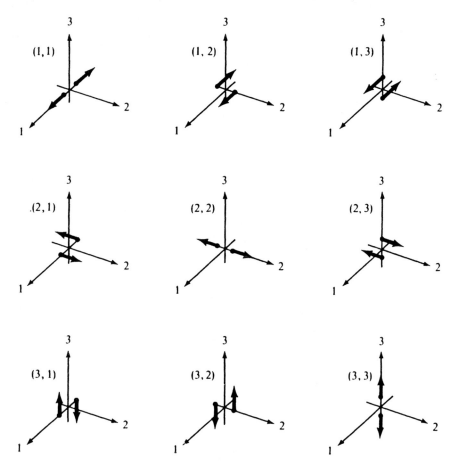

FIGURE 8.22 The nine couples composing the seismic moment tensor. (From Aki and Richards, 1980. Copyright ©1980 by W. H. Freeman and Co. Reprinted with permission.)

where $\overline{D}$ is the average slip;

$$\mathbf{v} = -\sin\delta\sin\phi_f\,\hat{\mathbf{x}}_1$$
$$+ \sin\delta\cos\phi_f\,\hat{\mathbf{x}}_2 - \cos\delta\hat{\mathbf{x}}_3 \quad (8.82)$$

and

$$M_{11} = -M_0\big(\sin\delta\cos\lambda\sin 2\phi_f$$
$$+ \sin 2\delta\sin\lambda\sin^2\phi_f\big)$$

$$M_{22} = M_0\big(\sin\delta\cos\lambda\sin 2\phi_f$$
$$- \sin 2\delta\sin\lambda\cos^2\phi_f\big)$$

$$M_{33} = M_0(\sin 2\delta\sin\lambda) = -(M_{11} + M_{22})$$

$$M_{12} = M_0\big(\sin\delta\cos\lambda\cos 2\phi_f$$
$$+ \tfrac{1}{2}\sin 2\delta\sin\lambda\sin 2\phi_f\big)$$

$$M_{13} = -M_0(\cos\delta\cos\lambda\cos\phi_f$$
$$+ \cos 2\delta\sin\lambda\sin\phi_f)$$

$$M_{23} = -M_0(\cos\delta\cos\lambda\sin\phi_f$$
$$- \cos 2\delta\sin\lambda\cos\phi_f). \quad (8.83)$$

It is possible to construct the P, SV, or SH motion for a moment tensor by summing the moment tensor weighted Green's functions using:

$$u_n(\mathbf{x}, t) = \sum_{i=1}^{5} m_i * G_{in}, \quad (8.84)$$

where $m_1 = M_{11}$, $m_2 = M_{22}$, $m_3 = M_{12}$, $m_4 = M_{13}$, and $m_5 = M_{23}$, and G_{in} are the Green's functions corresponding to each of the respective moment tensor elements. This is attractive for formulating waveform inversions directly for the moment tensor elements. Aki and Richards (1980) show that a double couple can also be represented in terms of four combinations of elementary moment tensors, the simplest possible combination, where the elementary moment tensors correspond to shear dislocations. A double-couple displacement can actually be represented by weighted summation of three *fundamental fault* Green's functions, corresponding to a vertical strike–slip fault ($\delta = \pi/2$, $\lambda = 0$), a vertical dip–slip fault ($\delta = \pi/2$, $\lambda = \pi/2$), and a fault dipping 45° with slip being purely updip ($\delta = \pi/4$, $\lambda = \pi/2$) evaluated at a 45° azimuth. The first two of these correspond to moment tensor Green's functions, but the third differs slightly from any of the M_{ij} Green's functions because it is evaluated at a particular azimuth. This formulation is useful for inverting for fault parameters directly rather than moment tensor elements. The ability to express the displacement for an arbitrary double-couple fault mechanism as the sum of five moment tensor Green's functions or three fundamental faults is the basis for many synthetic seismogram programs and waveform inversions, which will be described in Chapter 10.

Since the moment tensor is symmetric, it can be rotated into a principal-axis system. For example, the source described by (8.79) can be diagonalized to

$$\mathbf{M}' = \begin{bmatrix} M_0 & 0 & 0 \\ 0 & -M_0 & 0 \\ 0 & 0 & 0 \end{bmatrix}, \quad (8.85)$$

where the new elements are along axes defined by the principal dipole axes, P (maximum compressional deformation), T (minimum compressional deformation),

and B (intermediate or null axis). This shows that a double-couple force system is equivalent to two orthogonal force dipoles, as illustrated in Figure 8.10.

In general, a seismic moment tensor need not correspond to a pure double couple, but the symmetric tensor can still be diagonalized, with linear combination of three orthogonal dipoles completely describing the moment tensor excitation. The diagonalized values then correspond to eigenvalues of the moment tensor, with associated orthonormal eigenvectors $\mathbf{a}_i = (a_{ix_1}, a_{ix_2}, a_{ix_3})^{\mathrm{T}}$. For a double couple, the eigenvector corresponding to the positive eigenvalue gives the tension axis, T; the eigenvector for the zero eigenvalue gives the intermediate stress axis, B; and the eigenvector for the negative value gives the compressional axis, P. In general, we can decompose the diagonalized moment tensor

$$\mathbf{M} = \begin{bmatrix} M_1 & 0 & 0 \\ 0 & M_2 & 0 \\ 0 & 0 & M_3 \end{bmatrix}$$

$$= \frac{1}{3} \begin{bmatrix} \mathrm{tr}(\mathbf{M}) & 0 & 0 \\ 0 & \mathrm{tr}(\mathbf{M}) & 0 \\ 0 & 0 & \mathrm{tr}(\mathbf{M}) \end{bmatrix}$$

$$+ \begin{bmatrix} M_1^1 & 0 & 0 \\ 0 & M_2^1 & 0 \\ 0 & 0 & M_3^1 \end{bmatrix}, \quad (8.86)$$

where $\mathrm{tr}(\mathbf{M}) = M_1 + M_2 + M_3$ is the trace of $\mathbf{M}$, and the remaining terms M_i^1 are the deviatoric eigenvalues of $\mathbf{M}$. The isotropic terms given by the trace correspond to volume changes in the medium due to either explosion or implosion. Most shearing sources appear to have little isotropic component, and moment tensors for faulting events are often determined with the constraint $\mathrm{tr}(\mathbf{M}) = 0$.

The diagonalized deviatoric moment tensor can be decomposed into a variety of eigenvalue combinations that have somewhat different physical implications. One decomposition is into three vector dipoles

$$
\begin{bmatrix} M_1^1 & 0 & 0 \\ 0 & M_2^1 & 0 \\ 0 & 0 & M_3^1 \end{bmatrix} = \begin{bmatrix} M_1^1 & 0 & 0 \\ 0 & 0 & 0 \\ 0 & 0 & 0 \end{bmatrix}
$$

$$
+ \begin{bmatrix} 0 & 0 & 0 \\ 0 & M_2^1 & 0 \\ 0 & 0 & 0 \end{bmatrix} + \begin{bmatrix} 0 & 0 & 0 \\ 0 & 0 & 0 \\ 0 & 0 & M_3^1 \end{bmatrix},
$$

$$(8.87)$$

for which the dipoles act in the directions of the eigenvectors of **M**. In this case, the null (B) axis is nonzero if the moment tensor is not a double couple.

Alternatively, a moment tensor can be decomposed into an isotropic part and three double couples

$$
\begin{bmatrix} M_1 & 0 & 0 \\ 0 & M_2 & 0 \\ 0 & 0 & M_3 \end{bmatrix}
$$

$$
= \frac{1}{3} \begin{bmatrix} \mathrm{tr}(\mathbf{M}) & 0 & 0 \\ 0 & \mathrm{tr}(\mathbf{M}) & 0 \\ 0 & 0 & \mathrm{tr}(\mathbf{M}) \end{bmatrix}
$$

$$
+ \frac{1}{3} \begin{bmatrix} M_1 - M_2 & 0 & 0 \\ 0 & -(M_1 - M_2) & 0 \\ 0 & 0 & 0 \end{bmatrix}
$$

$$
+ \frac{1}{3} \begin{bmatrix} 0 & 0 & 0 \\ 0 & M_2 - M_3 & 0 \\ 0 & 0 & -(M_2 - M_3) \end{bmatrix}
$$

$$
+ \frac{1}{3} \begin{bmatrix} M_1 - M_3 & 0 & 0 \\ 0 & 0 & 0 \\ 0 & 0 & -(M_1 - M_3) \end{bmatrix},
$$

$$(8.88)$$

where each deviatoric term is a double couple.

Other decompositions involve compensated linear vector dipoles (CLVDs), which have one dipole of strength 2 in the direction of one eigenvector and two dipoles of unit strength in the directions of the other eigenvectors. A moment tensor can be represented by an isotropic term and three CLVDs

$$
\begin{bmatrix} M_1 & 0 & 0 \\ 0 & M_2 & 0 \\ 0 & 0 & M_3 \end{bmatrix}
$$

$$
= \frac{1}{3} \begin{bmatrix} \mathrm{tr}(\mathbf{M}) & 0 & 0 \\ 0 & \mathrm{tr}(\mathbf{M}) & 0 \\ 0 & 0 & \mathrm{tr}(\mathbf{M}) \end{bmatrix}
$$

$$
+ \frac{1}{3} \begin{bmatrix} 2M_1 & 0 & 0 \\ 0 & -M_1 & 0 \\ 0 & 0 & -M_1 \end{bmatrix}
$$

$$
+ \frac{1}{3} \begin{bmatrix} -M_2 & 0 & 0 \\ 0 & 2M_2 & 0 \\ 0 & 0 & -M_2 \end{bmatrix}
$$

$$
+ \frac{1}{3} \begin{bmatrix} -M_3 & 0 & 0 \\ 0 & -M_3 & 0 \\ 0 & 0 & 2M_3 \end{bmatrix}.
$$

$$(8.89)$$

A common decomposition is in terms of a major and minor double couple, where the major double couple is the best approximation of the moment tensor by a double couple with the same principal axes. Since the trace of the deviatoric part of the moment tensor is $M_1^1 + M_2^1 + M_3^1 = 0$

and if $|M_1^1| \geq |M_2^1| \geq |M_3^1|$, we can write

$$
\begin{bmatrix} M_1 & 0 & 0 \\ 0 & M_2 & 0 \\ 0 & 0 & M_3 \end{bmatrix}
$$

$$
= \frac{1}{3} \begin{bmatrix} \text{tr}(\mathbf{M}) & 0 & 0 \\ 0 & \text{tr}(\mathbf{M}) & 0 \\ 0 & 0 & \text{tr}(\mathbf{M}) \end{bmatrix}
$$

$$
+ \begin{bmatrix} M_1^1 & 0 & 0 \\ 0 & -M_1^1 & 0 \\ 0 & 0 & 0 \end{bmatrix} + \begin{bmatrix} 0 & 0 & 0 \\ 0 & -M_3^1 & 0 \\ 0 & 0 & M_3^1 \end{bmatrix},
$$

$$(8.90)$$

where the middle term is the major double couple, comprising the largest eigenvalue, and the other double couple is the minor double couple.

Yet another approach involves decomposition into an isotropic part, a double couple, and a CLVD. For $|M_1^1| \geq |M_2^1| \geq |M_3^1|$, we compute $\varepsilon = -M_2^1/M_3^1$, and then we have

$$
\begin{bmatrix} M_1 & 0 & 0 \\ 0 & M_2 & 0 \\ 0 & 0 & M_3 \end{bmatrix}
$$

$$
= \frac{1}{3} \begin{bmatrix} \text{tr}(\mathbf{M}) & 0 & 0 \\ 0 & \text{tr}(\mathbf{M}) & 0 \\ 0 & 0 & \text{tr}(\mathbf{M}) \end{bmatrix}
$$

$$
+ (1 - 2\varepsilon) \begin{bmatrix} 0 & 0 & 0 \\ 0 & -M_3 & 0 \\ 0 & 0 & M_3 \end{bmatrix}
$$

$$
+ \varepsilon \begin{bmatrix} -M_3 & 0 & 0 \\ 0 & -M_3 & 0 \\ 0 & 0 & 2M_3 \end{bmatrix},
$$

$$(8.91)$$

where ε is a measure of the size of the CLVD component relative to the double couple. For a pure double couple, $\varepsilon = 0$; for a pure CLVD, $\varepsilon = \pm 0.5$.

Seismologists use all of these moment tensor decompositions to characterize how well an equivalent body-force system of the moment tensor explains any particular source. The great utility of moment tensors is that once Earth responses are computed for the various couple orientations, actual ground motions are simply a linear sum of the couple responses, with weighting factors being the constant (or time-dependent) terms of the moment tensor. This enables linear *moment tensor inversion*, the process of retrieving an estimate of the moment tensor from actual data. Chapter 10 will discuss this further.

8.6 Determination of Faulting Orientation

We conclude this chapter with a brief discussion of how fault orientations are determined from seismic-wave observations. At this point we will restrict our attention to methods based on first-motion polarity and amplitude patterns for body waves and surface waves. After further developing finite-fault theory in the next chapter, we will return to complete waveform analysis in Chapter 10, showing how both the fault orientation and slip function are determined by both fault parameter and moment tensor inversions.

8.6.1 Stereographic Projections

The arbitrary orientation of faulting requires a procedure for handling the three-dimensional character of seismic-wave radiation from a source. A convenient approach is to imagine a small sphere around the source called the *focal sphere* (Figure 8.23). The focal sphere is imagined to be in a homogeneous medium, so raypaths to it from the source are simple radial spokes. This is useful because it is the curvature of raypaths in the Earth that

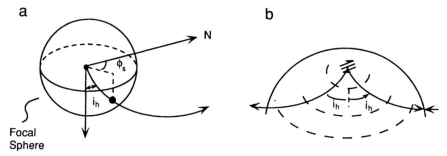

FIGURE 8.23 (a) The small focal sphere near the source, which can be thought of as the initial outgoing P (or S) wavefront. The raypath to a point on the Earth's surface (b) will have an associated takeoff angle and azimuth.

complicates observing the simple double-couple radiation patterns. Any P-wave raypath leaving the source can be identified by two parameters: the azimuth from the source, ϕ_s, and the ray parameter or takeoff angle, i_h (Figure 8.23). Each ϕ_s, i_h combination prescribes a unique path through the Earth to a point on the surface, and a corresponding portion of the outgoing wavefront is destined to reach that distant point, conveying the initial motion in the associated region of the outgoing wave.

Stereographic or *equal-area projections* are used to project the focal sphere onto a single plane as shown in Figure 8.24. Rays that take off upward will intersect the upper hemisphere of the focal sphere, but these are simply projected back to the lower hemisphere by adding 180° to the station azimuth (this exploits the low-order symmetry of fault mechanisms). The fault plane and auxiliary plane intersect the focal sphere, and the intersections project to the equatorial plane as curves that separate regions of compressional and dilata-

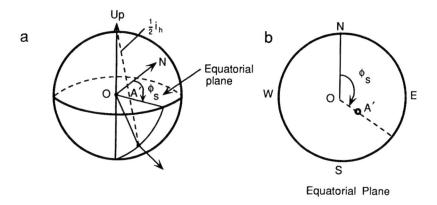

Stereographic $\quad OA' = \tan\left(\frac{1}{2}i_h\right)$

Equal area $\quad\quad OA' = \sqrt{2}\,\sin\left(\frac{1}{2}i_h\right)$

FIGURE 8.24 Projections for mapping spherical surfaces onto a plane. Both stereographic and equal-area projections are used, with the difference being the radial point A' used to represent the chord from the top of the focal sphere to the point intersected by the outgoing raypath. A ray going straight down intersects the center of the equatorial plane. Azimuth is preserved in the projection.

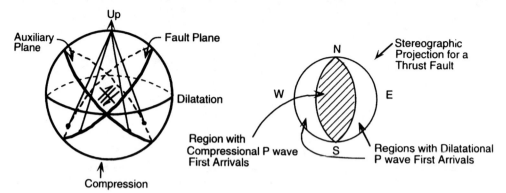

FIGURE 8.25 Projection of the fault plane and auxiliary plane onto the focal-sphere equatorial plane.

tional *P*-wave motions (Figure 8.25). The curves follow great circles on the equatorial plane, and their orthogonality produces characteristic appearance of the different faulting types. Examples of *focal mechanisms*, as these projections are called, for common faults are shown in Figure 8.26.

If we project observed *P*-wave first motions to the correct ϕ_s and i_h in the focal mechanism, we should see systematic quadrants of compressional or dilatational arrivals, and given adequate data these can be used to find the strike and dip of the two *P*-wave nodal planes (i.e., the auxiliary and fault planes). The same is true for *S*-wave polarizations. Figure 8.27 shows the expected patterns of *P*-wave polarities and amplitudes; it also shows *S*-wave polarizations and amplitudes projected onto focal mechanisms for the same fault orientation. An equal-area projection is used for the *P* waves, a stereographic projection for the *S* waves. Note the regular patterns arising from the smooth *P* and *S* radiation patterns and the systematic tendency for *S*-wave polarizations to converge on the projection of the *T* axis. The *B* axis lies on the intersection of the two planes, and this is also the "pole" (perpendicular) to the plane containing the *P* and *T* axes. The intersection of that plane with the fault plane is the slip-vector direction.

All that we need to do to determine the fault orientation from data is to project observed *P* and *S* polarities and amplitudes onto the equatorial plane and then determine the orthogonal planes. The following procedures are used to determine a focal mechanism:

1. For known earthquake and receiver locations we can determine the azimuth, ϕ, and the distance, Δ, for each station from the source.
2. Since we know $T(\Delta)$ for the Earth, we know $p(\Delta)$, and for each station we have a value of $p = (r_s \sin i_h)/V_s$, where r_s and V_s are values at the source. We can thus determine $i_h(\Delta)$ for any given source depth. Table 8.1 gives a table of i_h for *P* waves for shallow sources.

TABLE 8.1 Takeoff Angles for Surface Source

Δ	i_h	Δ	i_h	Δ	i_h
15°	45°	37°	27°	79°	17°
17	43	41	26	83	16
18	39	45	25	87	15
21	35	49	24	91	14
23	32	51	23	95	14
25	30	55	22		
27	29	59	22		
29	29	63	21		
31	29	67	20		
33	28	71	19		
35	28	75	18		

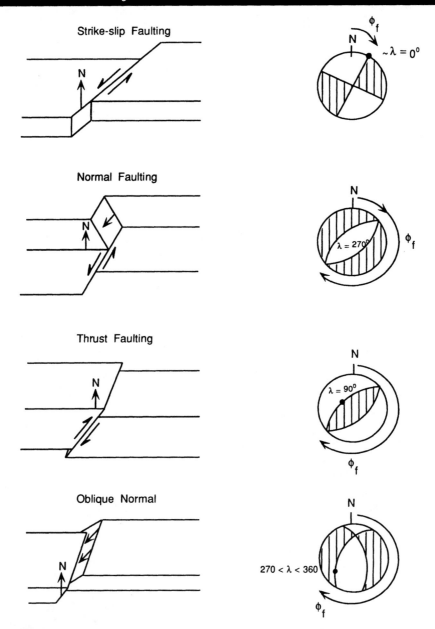

FIGURE 8.26 Basic fault types and their appearance in focal mechanism projections. Dark regions indicate compressional *P*-wave motions.

3. Using ϕ and i_h, we project the ray position to each station on a lower-hemisphere stereographic projection. We use different symbols to indicate whether compressional or dilatational first *P* arrivals were observed at each

station (usually open circles for dilatations and filled circles for compressions).

4. The first-motion data are rotated on the stereonet to find a meridian line that separates compressions from di-

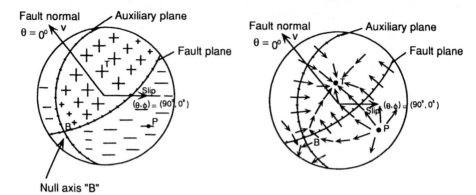

FIGURE 8.27 Focal mechanisms for an oblique-slip event showing *P*-wave polarities and relative amplitudes (left) and *S*-wave polarizations and amplitudes (right). Plus signs (+) indicate compressions. The fault and auxiliary planes are shown as well as projections of the *P, T,* and *B* axes. (Modified from Aki and Richards, 1980. Copyright ©1980 by W. H. Freeman and Co. Reprinted with permission.)

latations. This plane is drawn in along with the pole to the plane, which projects at 90° from the plane.

5. The data plot is rotated to find a second meridian that separates dilatations and compressions, but which also passes through the pole of the first plane.

6. The maximum compressive strain axis, *P*, is a pole lying on the plane containing the poles of the two "fault" planes (one is really the auxiliary plane). It lies in the dilatational quadrant 45° from the two planes.

7. The minimum compressive stress axis, *T*, lies in the compressional quadrant 45° from the two planes.

8. The intermediate compressive stress axis, *B*, is the intersection of the two fault planes.

9. The slip vector is defined as the pole of the auxiliary plane.

Well-constrained focal mechanisms for oblique–slip and strike–slip events are shown in Figure 8.28. Regional *P* waves provide good coverage of the focal sphere, as shown by the solution for the Loma Prieta event, while for teleseismic signals, the data cluster in the center of the mechanism. This reflects the narrow cone of

takeoff angles spanned by teleseismic *P* and *SH*; many other examples of focal mechanisms are shown in Chapters 10 and 11. Often one plane is well constrained, and the other can lie within a large range of possible orthogonal orientations. Joint use of *P* and *S* polarities reduces the ambiguity. To discriminate which plane is the actual fault plane, one must either observe ground breakage, see aftershocks preferentially distributed on one plane, or analyze short-period seismic signals to resolve any source finiteness effects.

8.6.2 Focal Mechanisms from Surface Waves

Surface waves can also be used to determine fault orientations because their azimuthal amplitude and phase patterns contain information about the fault orientation. Let us briefly consider a surface-wave displacement

$$x(t) = \frac{1}{2\pi} \int_{-\infty}^{\infty} \chi(\omega) e^{i\omega t} \, d\omega, \quad (8.92)$$

where $\chi(\omega)$ is the Fourier spectrum of $x(t)$. At a given frequency, ω, for a Love-

---

Box 8.3 Moment Tensor Conventions

Unfortunately, not all seismologists use the same coordinate system to define the general moment tensor terms. Alternative choices of reference frame to that in Figure 8.20 are shown in Figure 8.B.3.1.

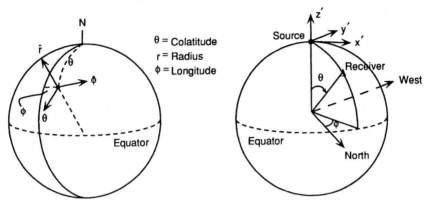

FIGURE 8.B3.1 In the case on the left, which is a convention commonly used in free-oscillation analyses, the unit vectors $\hat{r}, \hat{\theta}, \hat{\phi}$ point upward, southward, and eastward, respectively. The terms in Eq. (8.83) transform to $M_{rr} = M_{33}$, $M_{\theta\theta} = M_{11}$, $M_{\phi\phi} = M_{22}$, $M_{r\theta} = M_{13}$, $M_{r\phi} = -M_{23}$, and $M_{\theta\phi} = -M_{12}$. The (r, θ, ϕ) geographic coordinate system is used in the routinely reported moment tensors obtained by the centroid moment tensor inversions performed at Harvard. The $x'y'z'$ system has been extensively used by Hiroo Kanamori in his surface-wave inversion procedures. In this case

$$M_{11} = M_{x'x'}, \quad M_{22} = M_{y'y'}, \quad M_{33} = M_{z'z'},$$
$$M_{13} = -M_{x'z'}, \quad M_{23} = M_{y'z'}, \quad M_{12} = -M_{x'y'}.$$

Care in choice and consistency of the reference frame is critical.

---

wave seismogram observed at a distance Δ (in degrees) from the earthquake source, we have

$$\chi_L(\omega)$$

$$= \frac{1}{\sqrt{\sin \Delta}} e^{-i(\pi/4)} e^{i\omega a(\Delta/c)} e^{-\omega\Delta(a/2QU)}$$

$$\times \left(p_L P_L^1 + i q_L Q_L^1 \right), \qquad (8.93)$$

while for a Rayleigh wave we have

$$\chi_R(\omega)$$

$$= \frac{1}{\sqrt{\sin \Delta}} e^{-i(\pi/4)} e^{i\omega a(\Delta/c)} e^{-\omega\Delta(a/2QU)}$$

$$\times \left(s_R S_R^1 + p_R P_R^1 + i q_R Q_R^1 \right), \qquad (8.94)$$

where a is the Earth's radius, c and U are

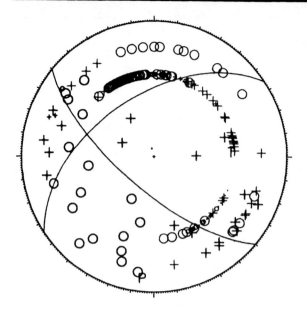 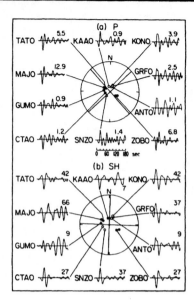

FIGURE 8.28 Examples of well-constrained focal mechanisms. On the left, *P*-wave first motions for the 1989 Loma Prieta earthquake from regional-distance stations are shown in an equal-area lower-hemisphere projection. Compressional motions are indicated by (+) and dilatations by (O). In this case $\phi_f = 130°$, $\delta = 70°$, and $\lambda = 140°$. On the right, teleseismic *P*-wave and *SH*-wave first motions are shown with *P*- and *SH*-radiation nodal planes for the November 8, 1980 Eureka, California earthquake. This left-lateral strike–slip event has $\phi_f = 48°$, $\delta = 90°$, and $\lambda = 0°$. Upward motions of *P* waves correspond to compressions (solid dots), while upward motion of *SH* corresponds to counterclockwise motion at the source. First-arrival amplitudes are shown for an equalized instrument gain. (Left from Oppenheimer, *Geophys. Res. Lett.* **17**, 1199–1202, 1990; © Copyright by the American Geophysical Union. Right from Lay *et al.*, 1982.)

phase and group velocity at ω, and Q is the corresponding attenuation quality factor. The phase function is $\exp[i\omega a(\Delta/c)]$, and $1/(\sin \Delta)^{1/2}$ describes the geometric spreading. The other parameters are as follows:

$$p_L = \cos \delta \sin \lambda \sin \delta \sin 2\phi$$

$$+ \cos \lambda \sin \delta \cos 2\phi$$

$$q_L = -\cos \lambda \cos \delta \sin \phi + \sin \lambda \cos 2\delta \cos \phi$$

$$s_R = \sin \lambda \sin \delta \cos \delta$$

$$q_R = \sin \lambda \cos 2\delta \sin \phi + \cos \lambda \cos \delta \cos \phi$$

$$p_R = \cos \lambda \sin \delta \sin 2\phi$$

$$- \sin \lambda \sin \delta \cos \delta \cos 2\phi, \qquad (8.95)$$

where $\phi = \phi_f - \phi_s$, and P_L^1, Q_L^1, S_R^1, P_R^1, and Q_R^1 are the surface-wave excitation functions, which are complicated expressions of the elastic constants and source depth. Therefore we can write $\chi_L(\omega) = F_L(\Delta)(p_L P_L^1 + iq_L Q_L^1)$ or $\chi_R(\omega) = F_R(\Delta)(s_R S_R^1 + p_R P_R^1 + iq_R Q_R^1)$. This implies that we can compute surface-wave amplitude radiation patterns at a given frequency as

$$A_R(\phi)$$

$$= \sqrt{(s_R S_R^1)^2 + (p_R P_R^1)^2 + (q_R Q_R^1)^2}$$

$$A_L(\phi) = \sqrt{(p_L P_L^1)^2 + (q_L Q_L^1)^2}. \qquad (8.96)$$

Figure 8.29 shows the surface-wave radia-

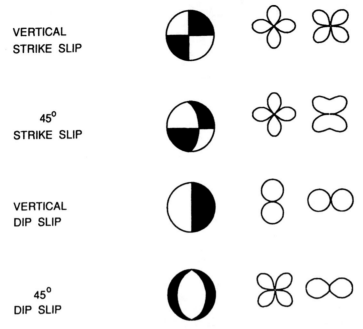

FIGURE 8.29 *P*-wave focal-mechanism projections and surface-wave radiation patterns for four basic fault orientations. The relative amplitude of the Love-wave (left) and Rayleigh-wave (right) radiation patterns is not drawn to scale. (Adapted from Kanamori, *J. Geophys. Res.* **75**, 5011–5027, 1970; © Copyright by the American Geophysical Union.)

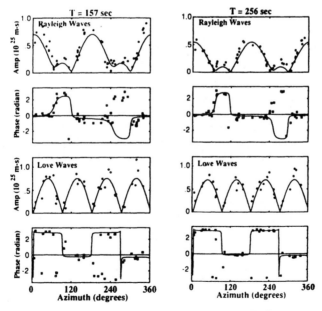

FIGURE 8.30 Azimuthal amplitude and phase variations from Rayleigh-wave and Love-wave observations for the 1989 Loma Prieta earthquake. The spectra have been equalized back to the source. The solid curve is the fit of a theoretical moment tensor source with a major double couple that is oblique right-lateral thrust. (Modified from Velasco *et al.*, 1993. Reprinted with permission.)

tion patterns for several fault orientations. The azimuthal patterns can be used to determine the fault orientation and moment by either inversion or forward modeling. In the actual analysis of surface waves, the seismograms are Fourier analyzed and equalized to the same distance from the epicenter (including phase shifts and amplitude effects) before they are inverted for a mechanism. Figure 8.30 shows equalized Rayleigh-wave and Love-wave spectra for the 1989 Loma Prieta earthquake, which exhibit azimuthal radiation patterns that constrain the fault geometry. In Chapter 10 we will expand on the ac-

tual inversion for fault parameters or moment tensor terms.

References

Aki, K., and Richards, P. G. (1980). "Quantitative Seismology." Freeman, San Francisco.

Bolt, B. A. (1988). "Earthquakes." Freeman, New York.

Chinnery, M. A. (1961). The deformation of the ground around surface faults. *Bull. Seismol. Soc. Am.* **51**, 355–372.

Eissler, H., and Kanamori, H. (1987). A single-force model for the 1975 Kalapana, Hawaii, earthquake. *J. Geophys. Res.* **92**, 4827–4836.

Box 8.4 A Non-Double-Couple Source

Moment tensor inversions are routinely performed using a variety of seismic phases for each earthquake. Most natural events appear to be primarily a double-couple shearing source, with any minor double-couple or CLVD component [measured by ε as defined in the text for Equation (8.91)] being attributed to noise in the inversion arising from inaccurate propagation corrections. However, some events have ε values larger than expected due to noise. An example is the January 1, 1984 deep earthquake under Japan. Figure 8.B4.1 shows the non-double-couple moment tensor solutions found using waves with different frequencies. Note that these projections, which show the P-wave nodal surfaces separating compressional (dark regions) and tensional P-wave motions, do not have the simple orthogonal planes expected for a double couple (e.g., Figure 8.28). The ε values range from -0.2 to -0.33, indicating that the B axis has a positive value, contributing to the tensional character of the source (essentially making the P-wave compressional motion zone larger). The question is then, What is the significance of this non-double-couple component?

Kuge and Kawakatsu (1990) analyzed the broadband P-wave motions and found evidence for complex rupture. This can be well modeled assuming two double couples shifted in space and time (procedures for this modeling are described in Chapter 10). The sum of the two double couples equals the total moment tensor determined at long periods. Thus, in this case the deep event involved faulting on two separate fault planes, with only high-frequency body-wave signals having sufficient resolution to determine the detailed nature of the source.

continues

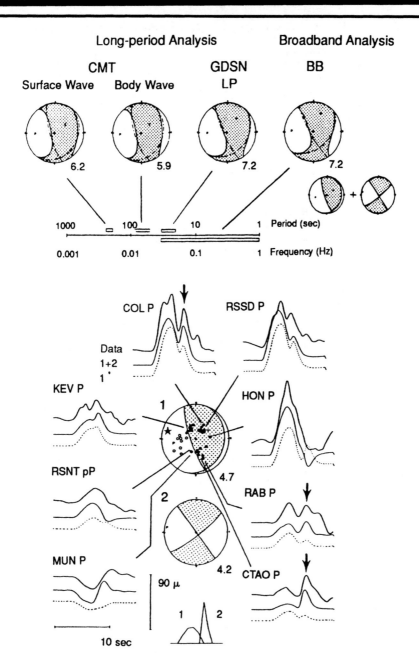

FIGURE 8.B4.1 (Top) Projection of *P*-wave radiation nodes of moment tensor solutions for the source of a deep earthquake on January 1, 1984. (Bottom) Observed *P* waves (top traces) and predictions for two subevent models with different mechanisms (middle) or the same mechanism. (Top from Kuge and Kawakatsu, *Geophys. Res. Lett.* **17**, 227–230, 1990; © copyright by the American Geophysical Union.)

Kanamori, H. (1970). Synthesis of long-period surface waves and its application to earthquake source studies—Kurile Islands earthquake of October 13, 1963. *J. Geophys. Res.* **75**, 5011–5027.

Kanamori, H., and Given, J. W. (1982). Analysis of long-period seismic waves excited by the May 18, 1980 eruption of Mt. St. Helens—A terrestrial monopole? *J. Geophys. Res.* **87**, 5422–5432.

Kanamori, H., and Given, J. W. (1983). Lamb pulse observed in nature. *Geophys. Res. Lett.* **10**, 373–376.

Kasahara, K. (1981). "Earthquake Mechanics." Cambridge Univ. Press, Cambridge, UK.

Kuge, K., and Kawakatsu, H. (1990). Analysis of a deep "non double couple" earthquake using very broadband data. *Geophys. Res. Lett.* **17**, 227–230.

Lamb, H. (1904). On the propagation of tremors over the surface of an elastic solid. *Philos. Trans. R. Soc. London, Ser. A* **203**, 1–42.

Lay, T., Given, J. W., and Kanamori, H. (1982). Long period mechanism of the 8 November 1980 Eureka, California earthquake. *Bull. Seismol. Soc. Am.* **72**, 439–456.

Marshall, G. A., Stein, R. S., and Thatcher, W. (1991). Faulting geometry and slip from coseismic elevation changes: The October 18, 1989 Loma Prieta, California, earthquake. *Bull. Seismol. Soc. Am.* **81**, 1660–1698.

Oppenheimer, D. H. (1990). Aftershock slip behavior of the 1989 Loma Prieta, California earthquake. *Geophys. Res. Lett.* **17**, 1199–1202.

Velasco, A. A., Lay, T., and Zhang, J. (1993). Long-period surface wave inversion for source parameters of the October 18, 1989 Loma Prieta earthquake. *Phys. Earth Planet. Inter.* **76**, 43–66.

Wallace, T. C., Helmberger, D. V., and Engen, G. R. (1983). Evidence of tectonic release from underground nuclear explosions in long-period *P* waves. *Bull. Seismol. Soc. Am.* **73**, 593–613.

Additional Reading

Aki, K., and Richards, P. G. (1980). "Quantitative Seismology." Freeman, San Francisco.

Ben-Menahem, A., and Singh, S. J. (1981). "Seismic Waves and Sources." Springer-Verlag, New York.

Burridge, R., and Knopoff, L. (1964). Body force equivalents for seismic dislocations. *Bull. Seismol. Soc. Am.* **54**, 1875–1888.

Chinnery, M. A. (1969). Theoretical fault models. *Publ. Dom. Obs.* (*Ottawa*) **17**, 211–223.

De Hoop, A. T. (1958). Representation theorems for the displacement in an elastic solid and their application to elastodynamic theory. Ph.D. Thesis, Technische Hogeschool, Delft.

Herrmann, R. B. (1975). A student's guide to the use of *P* and *S* wave data for focal mechanism determination. *Earthquake Notes* **46**, 29–39.

Jost, M. O., and Herrmann, R. B. (1989). A student's guide to and review of moment tensors. *Seismol. Res. Lett.* **60**, 37–52.

Kasahara, K. (1981). "Earthquake Mechanics." Cambridge Univ. Press, Cambridge, UK.

Knopoff, L., and Gilbert, F. (1960). First motions from seismic sources. *Bull. Seismol. Soc. Am.* **50**, 117–134.

Love, A. E. H. (1944). "Mathematical Theory of Elasticity." Dover, New York.

Marayama, T. (1963). On the force equivalents of dynamical elastic dislocations with reference to the earthquake mechanism. *Bull. Earthquake Res. Inst. Univ. Tokyo* **41**, 467–486.

Minster, B. J. (1985). Twenty-five years of source theory. *In* "The VELA Program" (A. U. Kerr, ed.), pp. 67–116. DARPA, Washington, DC.

Stekefee, J. A. (1958). On Volterra's dislocations in a semi-infinite medium. *Can. J. Phys.* **36**, 192–205.

9

EARTHQUAKE KINEMATICS AND DYNAMICS

In the preceding chapter we developed equations that govern the far-field P and S waves generated from a double-couple source. These equations, (8.65) and (8.72), describe the size of the body-wave pulses and give the displacement pulse shape as a function of time in terms of the seismic moment rate, $\dot{M}(t)$. The equivalence of the double-couple force system to shear displacement on a fault plane means that the moment rate maps the spatiotemporal history of the fault slip. We will consider this mapping in this chapter. Fault slip involves three main stages: (1) initiation of the fault sliding (or formation of a crack), (2) growth of the slip zone, or rupture front expansion, (3) and termination of the rupture process. The details of the fault slip impose a signature on the body waveforms, providing an empirical means for studying the faulting process using seismology. The complementary laboratory and theoretical study of rock failure and the growth of fractures is called *fault mechanics*. A simple review of the basics of fault mechanics will serve as a framework for discussing the dynamics of earthquake sources.

Let us consider the most basic stages involved in faulting: (1) initiation of rupture and (2) frictional sliding during the rupture process. These processes clearly involve phenomena that linear elasticity cannot describe. The earliest roots of fault mechanics can be traced to the work of Amonton in 1699 and Coulomb in 1773. Coulomb introduced a simple theory for rock failure that states that the shear strength of a rock is equal to the initial strength of the rock, plus a constant times the normal stress, σ_n, on the plane of failure:

$$|\tau|_{\text{failure}} = c + \mu_i \sigma_n, \qquad (9.1)$$

where c is the strength of the rock, sometimes called *cohesion*, and μ_i is a constant called the *coefficient of internal friction*. Numerous variations on (9.1) have been proposed, but this simple equation predicts failure surprisingly well. The equation is usually referred to as the *Coulomb failure criterion*.

A similar relationship, known as Amonton's second law (Amonton's first law states that frictional forces are independent of the size of the fault surface), is used to describe frictional sliding on an existing crack:

$$\tau = \mu_s \sigma. \qquad (9.2)$$

Note that this equation has the same form as (9.1) without the cohesion term. In (9.2) μ_s is called the *coefficient of friction*, and it does not have the same value, μ_i, required for failure of unfaulted rock. The coefficient of friction, μ_s, generally has a larger value before sliding takes place, called the *static friction coefficient*, and a smaller value once sliding commences, called the dynamic or *kinetic friction coefficient*. Amonton proposed that μ_s was related to roughness or protrusions on the fault surface. These protrusions, known as *asperities*, are welded contacts between the two sides of a fault. These welds must be overcome to allow sliding, and once these welds are broken, sliding proceeds at a reduced friction level.

Much experimental work has been done on rock friction, and in general it has been shown that Amonton's law is applicable over a wide range of normal stress values. Byerlee (1978) compiled data from a large number of rock-friction experiments and found that maximum friction was nearly independent of rock type. For normal stresses > 200 MPa, the relationship is

$$|\tau| = 50 + 0.6\sigma_n \text{ MPa} \qquad (9.3)$$

and for normal stresses less than 200 MPa the relationship is

$$|\tau| = 0.85\sigma_n. \qquad (9.4)$$

If we assume that normal stress is approximately equal to the overburden pressure, then Eq. (9.3) is valid for all depths greater than about 6 km. This is known as *Byerlee's law*. It is rather remarkable that Byerlee's law is so independent of rock type, which suggests that the details of fault roughness are relatively unimportant in the frictional behavior associated with earthquake rupture.

The dynamics of frictional sliding are more complex than Eq. (9.3) would seem to indicate. In the laboratory the relationship between sliding displacement and applied shear stress is not smooth. In general, no slip occurs on the fault surface until the critical value of τ is reached, and then sudden slip occurs followed by a drop in stress. This causes a time interval of "no slip" during which the stress builds up again to the critical value, and then the sudden-slip episode is repeated. This type of frictional behavior is known as *stick–slip*, or unstable sliding. For extremely smooth fault surfaces the slip may be continuous, or nonepisodic. This is referred to as *stable sliding*. Brace and Byerlee (1966) proposed stick–slip behavior as a possible explanation for shallow earthquakes, building on ideas developed by Bridgman in the 1930s and 1940s. Earthquakes are generally thought to be recurring slip episodes on preexisting faults followed by periods of no slip and increasing strain. This was a profound change in the understanding of faults; the emphasis is not on the *strength* of the rock but rather on the stress–stability cycle. The difference between the shear stress just before the slip episode and just after the slipping has ceased is known as the *stress drop*. The stress drop observed in an earthquake may represent only a fraction of the total stress supported by the rock.

In detail, a number of factors control friction. These include temperature, slip rate, and slip history. Many materials become weaker with repeated slip and eventually enter the stable sliding mode. This behavior is known as *slip weakening*. Also, most Earth materials exhibit an inverse dependence of friction on slip velocity. This type of behavior is known as *velocity weakening*. Finally, for most materials, stick–slip behavior is observed only at temperatures below about 300°C. Almost all seismogenic faults involve a shallow surficial region where stable sliding normally occurs, with occasional coseismic slip during ruptures that nucleate in a deeper unstable sliding regime that may extend throughout much of the crust. At yet greater depth, a transition to stable sliding again occurs, grading into a zone of continuous ductile deformation. Slip-rate dependence of friction can cause stable sliding regimes to fail episodically, so it is frictional behavior rather than internal strength that controls rupture.

The application of Eq. (9.1) to geologic materials and stresses in the Earth to predict newly created fault orientations is known as *Anderson's theory of faulting*. Equation (9.1) can be displayed graphically in a Mohr diagram, as shown in Figure 9.1. The upper straight line gives the condition for failure, or the *failure envelope*. Failure in an unfaulted medium occurs only when the difference between the maximum and minimum compressive stresses intersects the envelope. The direction of failure, or orientation of the fault, is defined by the perpendicular to the envelope (angle 2θ in Figure 9.1). The slope of the failure envelope is related to the internal friction by $\mu_i = \tan \phi$. A fault will form at an angle $(90° - \theta)$ from the axis of maximum compressive stress σ_1:

$$\theta = \pm \left(45° + \frac{\phi}{2} \right). \qquad (9.5)$$

For most rocks, the angle of internal friction, ϕ, is $\sim 30°$, and thus failure will occur on either fault of a conjugate pair oriented at $\pm 30°$ to σ_1. Note that in this case the P and T axes of earthquake focal mechanisms will *not* correspond to principal stress directions, and only for $\mu = 0$ will this be the case.

Near the surface of the Earth one of the principal stresses is almost always vertical; thus we can predict the orientations of faults for a particular environment. In a compressive regime, σ_1 and σ_2 are horizontal and σ_3 is vertical, resulting in 30°-dipping thrust faults that strike parallel to σ_2. In an extensional environment, σ_1 is vertical and σ_3 is horizontal, resulting in 60°-dipping normal faults. In regions where σ_3 and σ_1 are both horizontal, vertical strike–slip faults are formed, oriented at

30° from σ_1. Anderson's theory has been very successful for describing basic characteristics of observed faults. One shortcoming of the theory is that it only relates the orientation of the faults to the stress field at the time of fault formation. Because the frictional strength of faults is less than the strength of unbroken rock, once they form, faults represent a plane of weakness that will continue to move even though the stress field is not optimally aligned. The line through the origin in Figure 9.1 gives the condition for frictional sliding of preexisting faults. If the medium contains preexisting faults with normals that lie between the angles defined by 2α and 2γ, slip on those fractures will occur, probably preventing failure on a new fault at angles θ. This makes it difficult to infer principal stress orientations from isolated seismic focal mechanisms. It is possible to infer a regional stress orientation based on analysis of many focal mechanisms in a region, since the suite of fault mechanisms activated by a given stress regime will be bounded.

Coulomb failure criteria and the Anderson theory for faulting provide a *static* framework for understanding the faulting process, but the *dynamics* of the sudden slip are much more difficult to understand. The earthquake rupture can be described as a two-step process: (1) formation of a crack and (2) propagation, or growth, of the crack. The crack tip serves as a stress concentrator; if the stress at the crack tip exceeds some critical value, then the crack grows unstably (sudden slip). Figure 9.2 shows the stress at a point, P, along a fault. Before the rupture reaches the neighborhood of P, the stress is τ_0, which is less than the strength of the fault, τ_s. As the rupture approaches, the stress at P rises due to the stress concentration ahead of the crack tip; when the stress exceeds τ_s (at time t_0), slip at point P begins. As point P slips, stress diminishes and drops to the dynamic frictional value τ_f. After the slip has stopped, the stress level ad-

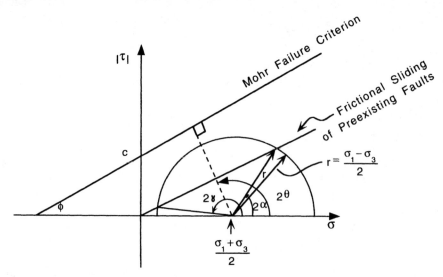

FIGURE 9.1 Mohr–Coulomb failure envelope and its relation to the difference between σ_1 and σ_3. c is the cohesion.

justs up or down slightly to τ_1, the final stress, depending on whether velocity weakening or hardening occurred. The *stress drop* is equal to $\Delta\sigma = \tau_0 - \tau_1$.

The connection to the observable seismic radiation is through energy released as the fault surface moves. In Figure 9.2 the fault at point P slipped from zero to its final value at time t_1 during a finite period of time. Figure 9.3 shows a hypothetical slip curve for point P and the corresponding time derivative, which is directly proportional to $\dot{M}(t)$. The slip function may have substantial irregularity, but it can often be approximated by a simple ramp function with the derivative being boxcar

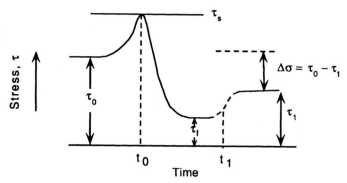

FIGURE 9.2 Stress at a point on a fault surface. As the rupture front approaches the point, stress increases to a value of τ_s, after which failure occurs at the point. The point slips to a displacement D, and stress is reduced to some value τ_f. The difference between the initial stress and the final stress, $\Delta\sigma$, is defined as the stress drop. (After Yamashita, 1976.)

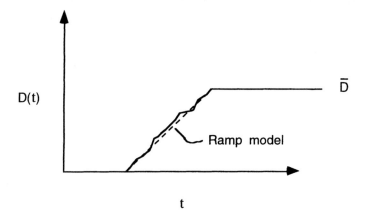

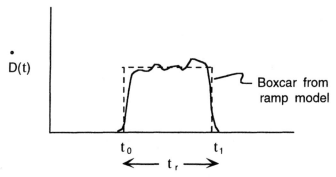

FIGURE 9.3 The relation between the displacement history of a particle on the fault and the far-field source time function.

shaped. The history of the slip will map directly into the moment rate, which in turn will give the shape of the *P*- or *S*-wave energy radiated from the earthquake. The time derivative of slip at a point, called the *particle velocity*, depends on the frictional constitutive law, the determination of which is an active area of research. The seismic moment of an entire fault is the sum of the slip of all the particles on the fault; thus the moment rate is actually a convolution of the slip history at a point and the history of propagation of the rupture front along the fault.

One of the basic unresolved issues of earthquake rupture dynamics is the duration of particle displacement at each point on a finite fault. Most dynamic crack mod-els predict that after a rupture front passes a point, that point will continue to slide until information is received that the rupture front has stopped. That information comes in the form of a *healing front* initiated from the outermost extent of rupture, which sweeps inward across the faulted surface, bringing slip to a halt. This idea requires, for example, that for a circular, radially outward growing crack, the center of the fault is the first and last place to be sliding. Figure 9.4 illustrates the continued slip and changing slip velocity on an ex-panding crack surface for such a model. Each point on the fault experiences a stress history like that in Figure 9.2 as the crack tip passes by and produces far-field radia-tion like that in Figure 9.3, but with the

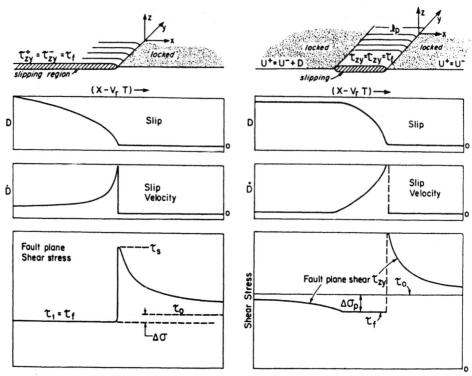

FIGURE 9.4 Two idealized models of rupture. On the left is the crack model of Kostrov (1966) for a crack propagating at rupture velocity v_r in the x direction. Fracture initiates when the stress at the crack tip exceeds the static strength of the fault, τ_s, and the stress on the slipping fault is a constant frictional level, τ_f. The ambient stress before rupture is τ_0, and the stress drop, $\Delta\sigma$, is $\tau_0 - \tau_f$. Every point on the rupture surface continues to slide until the rupture front stops and sends a "healing pulse" back across the fault. On the right is the self-healing slip pulse model of Heaton (1990) in which slip-velocity-dependent friction heals the rupture surface behind the slip pulse. The sliding friction is very low and returns to a high level after the rupture. The static stress drop, $\tau_0 - \tau_1$, may be very low (~ 10 bars), but the average stress drop within the rupture pulse may be an order of magnitude larger. (From Heaton, 1990.)

particle time history having a prolonged tail of "slow slip" following the passage of the slip front, terminating only when the healing phase occurs. The average slip duration of a point for such crack models is approximately $t_r \approx 2\sqrt{s}/3v_r$, where s is the slip area and v_r is the rupture velocity. This clearly is on the same order as the time required for the entire fault to rupture, $t_c \approx \sqrt{s}/v_r$.

Alternatively, it is possible that rupture surfaces heal before the crack expansion terminates. Heaton (1990) has proposed a

model of slip pulses, in which the fault dynamic friction is inversely proportional to the slip velocity, greatly reducing friction just as the crack tip passes by, followed by a rapid return to high friction as slip velocity diminishes (Figure 9.4). In such a case, the particle slip duration, t_r, may be only a small fraction of the total rupture time. Because seismic-wave radiation is sensitive to the derivative of the particle dislocation history as well as to growth of the rupture area, $\dot{M}(t) = \mu(\partial/\partial t)[A(t) \cdot D(t)]$, far-field seismic sig-

Box 9.1 Failure Modes

The brittle failure of rocks has long been studied both experimentally and theoretically, and the field of modern *fracture mechanics* grew out of a major discrepancy between the two approaches. In general, the theoretical strength of rock, the stress required to break the atomic bonds in a crystal lattice, is several orders of magnitude greater than the stress required to break rock in the laboratory. In the first quarter of this century, A. A. Griffith suggested an explanation based on the fact that all materials contain defects or microscopic cracks. *Griffith's theory* is based on the theorem of minimum energy and involves a balance of energy in the system. Creating a new crack requires *work* in the breaking of molecular bonds, which increases the potential energy of the system. This increase is balanced by a reduction in the strain energy. If the rate of strain energy supplied to a crack tip equals the energy required to extend the crack, stable crack growth occurs.

The growth of a crack depends on the tip of the crack serving as a stress concentrator. The ability of a crack to concentrate stress depends on the type of displacement at the crack tip. There are three fundamental types of cracks. The simplest, referred to as mode I, or tensile mode, is a crack that opens normal to the direction of crack propagation. The other two types of cracks have shear displacements. Mode II has displacements in the plane of the crack and along the direction of crack propagation (referred to as the *sliding mode*). Mode III has displacements in the plane of the crack but normal to the direction of crack growth (referred to as the *tearing mode*). A crack could be any combination of the three fundamental types, but in general, earthquake rupture is modeled with mode II cracks. The crack-tip stress depends on the mode type and is proportional to $K_n r^{-1/2}$, where r is the distance from the crack tip. K_n is called the *stress intensity factor* (n is the mode type). Griffith's theory requires an energy balance for cracks to grow, which can be used to define the crack extension force, G

$$G = K_n^2 (1 - \nu^2)/E \tag{9.1.1}$$

for mode II cracks. When G exceeds some critical value G_c, unstable fracturing occurs.

The dynamics of fracture growth is an area of active research. A number of modifications to Griffith's theory are now used and are being applied to seismology to predict rupture velocity and to model the way faults grow with time.

nals have limited resolution of any small prolonged tails on the displacement time history for each particle. Many observations given later in this chapter and in Chapter 10 tend to favor short particle slip durations, but the question of which model is superior is still undergoing active research.

In this chapter we will further explore the factors contributing to the shape of body-wave pulses due to the faulting process. The "shape" is often referred to as the *source time function*. Since moment rate controls the amplitude of seismic pulses, it has a strong frequency dependence. Hence, the "size" of an earthquake depends on the frequency of the waves being analyzed. We will use this to develop the concept of magnitude scales.

9.1 The One-Dimensional Haskell Source

Four gross faulting source parameters primarily affect the seismic radiation. These are (1) the final (average) displacement on the fault ($\overline{D}$), (2) the dimensions of the fault (L, length, and w, width), (3) the rupture velocity (v_r), and (4) the particle velocity (the rate at which an individual particle on the fault travels from its initial to its final position). We will consider how each of these affects the source time function. In the last chapter we found that the displacement signature associated with a shear dislocation had two distinct terms: near field and far field. At distances beyond a few fault lengths, far-field effects dominate, and it is in the far field where we will examine the source time function.

From Eq. (8.65) let us consider the far-field P-wave displacements associated with a double couple with its orientation specified in a geographic coordinate system:

$$u_r(r,t) = \frac{1}{4\pi\rho\alpha^3}\frac{R^P}{r}\dot{M}\left(t - \frac{r}{\alpha}\right). \quad (9.6)$$

In the simplest case, the fault can be considered a single point source. Then the seismic moment rate is just the displacement history of that particle. If the displacement occurred instantaneously as a step, the moment rate would be a delta function. More realistically, it will take a finite length of time for the particle to achieve its total offset. The simplest case would be represented by a ramp history, as in Figure 9.3. The source time function

which arises from a ramp history is a boxcar of length τ_r, where τ_r is called the *rise time*. The rise time is the time it takes a single particle on the fault to achieve its final displacement. Thus if a fault could be described by a single point with a ramp particle time history, the far-field P- and S-wave displacements would be shaped like boxcars. Their amplitudes would vary with azimuth depending on the radiation pattern, but the pulse shape would be identical everywhere. A simple ramp displacement history describes faults as viewed in the far field surprisingly well.

Clearly, all faults involve more than a single point displacement. We will now consider a simple kinematic fault model to assess the role of rupture expansion. Many points along the finite fault follow a similar displacement history, but at different times, as the rupture front expands. All the point sources need to be summed with the appropriate time lags to give the complete time function. We can see how to do this by considering a simple "ribbon" fault (see Figure 9.5), where the fault ruptures initially at one end and the rupture propagates with a finite velocity to the other end. The fault is long and narrow and can be treated as a series of small segments that individually approximate point sources. We use the principle of linear

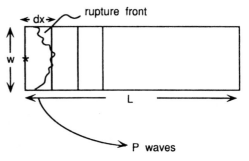

FIGURE 9.5 Geometry of a one-dimensional fault of width w and length L. The individual segments of the fault are of length dx, and the moment of a segment is $m\,dx$. The fault ruptures with velocity v_r.

superposition to determine the far-field displacement; it is just the summation of the subevent point-source displacements:

$$u(r,t) = \sum_{i=1}^{N} u_i\left(r_i, t - \frac{r_i}{\alpha} - \Delta t_i\right), \quad (9.7)$$

where Δt_i is the lag between subevents. We will proceed with it being understood that time for the displacement field always involves the delay time r/α, which is the P-wave propagation time. Using (9.6), for the far-field P wave each $u_{ri} = R_i^P(1/4\pi\rho\alpha^3)(\dot{M}_i/r_i)$, but $\dot{M}_i$ is just $\mu A_i \dot{D}_i(t) = \mu w\, dx\, \dot{D}_i(t)$. Therefore, we can rewrite (9.7) as

$$u_r(r,t) = \frac{R_i^P\mu}{4\pi\rho\alpha^3} w \sum_{i=1}^{N} \frac{\dot{D}_i}{r_i}(t - \Delta t_i)\, dx.$$

$$(9.8)$$

Let us consider a simplified geometry. If the station is at a large distance perpendicular to the fault, then r_i is approximately constant, as is R_i^P. If the rupture front proceeds with a constant rupture velocity, v_r, and the displacement history is the same everywhere on the fault, then Δt_i is the distance along the fault divided by the rupture velocity:

$$u_r(r,t) = \frac{R^P\mu}{4\pi\rho\alpha^3} \frac{w}{r} \sum_{i=1}^{N} \dot{D}\left(t - \frac{x}{v_r}\right) dx.$$

$$(9.9)$$

We can rewrite (9.9) in a more useful form by using the shift property of the delta function

$$\dot{D}\left(t - \frac{x}{v_r}\right) = \dot{D}(t) * \delta\left(t - \frac{x}{v_r}\right), \quad (9.10)$$

where $*$ denotes convolution, and the particle velocity is everywhere the same on the fault. Note that this assumption is not fully consistent with a dynamic crack model as discussed earlier, but it could be valid

for the slip pulse model or for high-frequency radiation from a crack model in which the particle time histories are close to ramp functions. We will analyze this simple model primarily to gain insight into the effects of finiteness, which will hold qualitatively for all rupture models. If we substitute (9.10) into (9.9) and take the limit of the sum as $dx \to 0$, we obtain an integral equation

$$u_r(r,t)$$

$$= \frac{R^P\mu}{4\pi\rho\alpha^3} \frac{w}{r} \int_0^x \dot{D}(t) * \delta\left(t - \frac{x}{v_r}\right) dx,$$

$$(9.11)$$

where x is the length of the fault. Note that $\dot{D}(t)$ is independent of x, so it can be taken outside the integral

$$u_r(r,t)$$

$$= \frac{R^P\mu}{4\pi\rho\alpha^3} \frac{w}{r} \dot{D}(t) * \int_0^x \delta\left(t - \frac{x}{v_r}\right) dx.$$

$$(9.12)$$

The solution requires the integration of the delta function. Let $z = t - (x/v_r)$, $x = tv_r - zv_r$, and $dx = (dx/dz)\, dz = -v_r\, dz$. Thus

$$\int_0^x \delta\left(t - \frac{x}{v_r}\right) dx = \int_t^{t-(x/v_r)} -v_r\, \delta(z)\, dz.$$

$$(9.13)$$

The integral of $\delta(z)$ is the Heaviside step function, $H(t)$, where

$$H(t) = 0 \quad \text{before time } t$$

$$H(t) = 1 \quad \text{after time } t. \quad (9.14)$$

Thus

$$u_r(r,t) = \frac{R^P \mu w}{4\pi\rho\alpha^3 r} \dot{D}(t) * v_r H(z) \Big|_{t-(x/v_r)}^{t}$$

$$= \frac{R^P \mu w}{4\pi\rho\alpha^3 r} v_r \dot{D}(t)$$

$$* \left[H(t) - H\left(t - \frac{x}{v_r}\right) \right]$$

$$= \frac{R^P \mu w}{4\pi\rho\alpha^3 r} v_r \dot{D}(t) * B(t;\tau_c),$$

$$\text{(9.15)}$$

where $B(t;\tau_c)$ is a boxcar of duration τ_c, the rupture time ($\tau_c = x/v_r$), which starts at time t. Thus the far-field displacement pulse shape is defined by the convolution of two boxcars, one representing the displacement history of a single particle and the second representing the effects of a finite fault. The convolution results in a time function which is a trapezoid (see Box 9.2). This trapezoid has two fundamental dimensions: (1) a duration equal to the sum of the two boxcars and (2) the *rise* and *fall* of the trapezoid, which are equal to the duration of the shortest boxcar, usually assumed to be the particle velocity (see Figure 9.6).

Box 9.2 Convolution

A seismogram can be considered the output of a series of filters that represent different processes such as propagation (reflection and transmission at various boundaries), attenuation, and recording on a frequency-band-limited instrument. Each filter distorts an input signal based on prescribed rules, which can be thought of as a *transfer function*. The mathematical link between an input signal, the transfer function, and the output signal is known as a *convolution*. Mathematically the convolution is written

$$g(t) = S(t) * I(t) = \int_{-\infty}^{\infty} S(\tau) I(t - \tau) \, d\tau, \qquad (9.2.1)$$

where $S(t)$ is the input signal and $I(t)$ is the filter ($*$ denotes the convolution operator). The integral is simple to understand if you think of $S(t)$ as a collection of single time point amplitudes that are passed through the filter. For the ith element in $S(t)$, the entire signal $I(t)$ is multiplied by the amplitude of the ith point. The $(i + 1)$th point in $S(t)$ also serves as a multiplier of $I(t)$; this new series is shifted by dt, the spacing between the ith and $(i + 1)$th point in $S(t)$. The two series are summed, and the resultant signal is the convolution. Consider an example of $S(t)$ being a delta function and $I(t)$ being a boxcar:

$S(t) = 0$ everywhere except at $t = 10$, where $S = 1$

$I(t) = 0$ between 0 and 5, 1 between 5 and 10, and 0 elsewhere.

$S(t) * I(t)$ can be seen graphically in Figure 9.B2.1a. The convolution of a time series with a delta function is just the same time series. Now consider the convolution of two boxcars (see Figure 9.B2.1b). The resulting output function is a trapezoid.

continues

It is far easier to perform a convolution in the frequency domain than in the time domain. It turns out that the convolution operator is just multiplication in the frequency domain:

$$\mathscr{F}(g(t)) = \mathscr{F}(S(t) * I(t)) = \hat{S}(\omega)\hat{I}(\omega). \qquad (9.2.2)$$

This can be proved by considering the definition of the Fourier transform (Box 5.1)

$$\int_{-\infty}^{\infty} g(t)e^{-i\omega t}\,dt = \hat{g}(\omega) = \int_{-\infty}^{\infty}\left[\int_{-\infty}^{\infty} S(\tau)I(t-\tau)\,d\tau\right]e^{-i\omega t}\,dt. \qquad (9.2.3)$$

Let $z = (t - \tau)$, and change the variable of integration from dt to dz; $t = (z + \tau)$, $dt = dz$:

$$\hat{g}(\omega) = \int_{-\infty}^{\infty}\left[\int_{-\infty}^{\infty} S(\tau)I(z)\,d\tau\right]e^{-i\omega z}e^{-i\omega\tau}\,dz \qquad (9.2.4)$$

$$= \int_{-\infty}^{\infty}\left[\int_{-\infty}^{\infty} S(\tau)e^{-i\omega\tau}\,d\tau\right]I(z)e^{-i\omega z}\,dz \qquad (9.2.5)$$

$$= \int_{-\infty}^{\infty}\hat{S}(\omega)I(z)e^{-i\omega z}\,dz = \hat{S}(\omega)\hat{I}(\omega). \qquad (9.2.6)$$

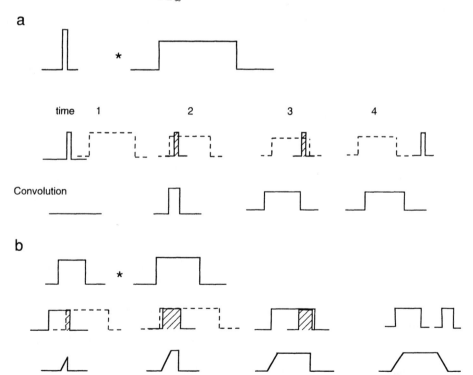

FIGURE 9.B2.1 Graphical representation of convolution. (a) Convolution of a delta function and a boxcar gives the same boxcar. (b) Convolution of the boxcars gives a trapezoid.

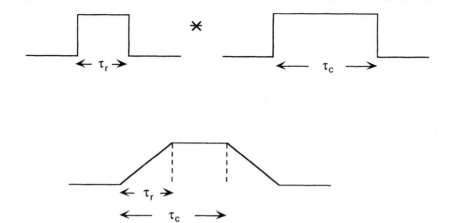

FIGURE 9.6 The convolution of two boxcars, one of length τ_r and the other of length τ_c ($\tau_c > \tau_r$). The result is a trapezoid with a rise time of τ_r, a top of length $\tau_c - \tau_r$, and a fall of width τ_r.

For this simple line source, or *Haskell fault model* (Haskell, 1964), the far-field *P*- and *S*-wave displacements should be trapezoidally shaped. Now consider the area under a far-field *P* wave:

$$\int_{-\infty}^{\infty} u_r(r,t)\, dt$$

$$= \int_{-\infty}^{\infty} \frac{R^P \mu}{4\pi\rho\alpha^3} v_r \frac{w}{r} \dot{D}(t) * B(t;\tau_c)\, dt$$

$$(9.16)$$

or, rearranging terms,

$$\frac{4\pi r\rho\alpha^3}{R^P} \int_{-\infty}^{\infty} u_r(r,t)\, dt$$

$$= \int_{-\infty}^{\infty} \dot{D}(t)\mu w v_r B(t;\tau_c)\, dt. \quad (9.17)$$

The right-hand side of (9.17) is the area of $\dot{D}(t)$ (D) multiplied by the area of $\mu w v_r \cdot B(t;\tau_c)$ ($\mu w L$). Thus, the right-hand side is equal to the seismic moment, $M_0 = \mu DA$. The left-hand side is the area under the displacement pulse corrected for spreading, the radiation pattern, and the source material constants. This equality provides a procedure for determining the seismic moment from far-field displacements. Figure 9.7 shows the *SH* displacement waveform from an earthquake near Parkfield, California. Note that its shape is roughly trapezoidal.

9.1.1 Directivity

In the simple Haskell source model, the boxcar associated with the propagation of the rupture had a length τ_c for a station at an azimuth perpendicular to the strike of the ribbon fault. Obviously, τ_c depends on the dimensions of the fault and on v_r, but it also depends on the orientation of the observer relative to the fault. In general, the rupture velocity is less than the *S*-wave velocity of the faulted material; the body waves generated from a breaking segment of the fault will arrive at a station before the body waves arrive from a segment that ruptures later. On the other hand, when the path to the station is not perpendicular to the fault, the body waves generated from different segments of the fault will have different travel path lengths to the recording station and thus unequal travel times. Figure 9.8 shows a fault of length

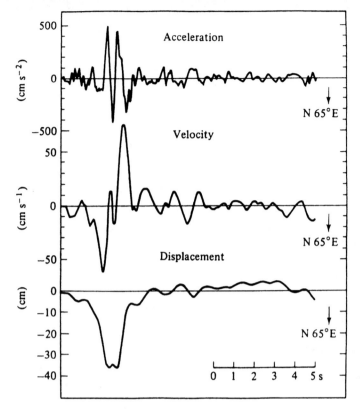

FIGURE 9.7 A recording of the ground motion near the epicenter of an earthquake at Parkfield, California. The station is located on a node for *P* waves and a maximum for *SH*. The displacement pulse is the *SH* wave. Note the trapezoidal shape. (From Aki, *J. Geophys. Res.* 73, 5359–5375, 1968; © copyright by the American Geophysical Union.)

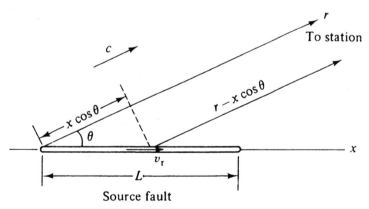

FIGURE 9.8 Geometry of a rupturing fault and the path to a remote recording station. (From Kasahara, 1981.)

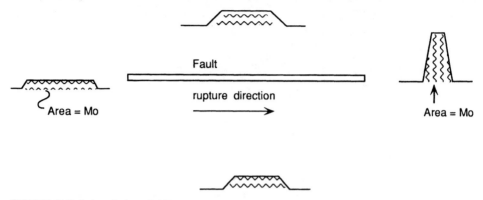

FIGURE 9.9 Azimuthal variability of the source time function for a unilaterally rupturing fault. The duration changes, but the area of the source time function is the seismic moment and is independent of azimuth.

L, rupturing from left to right. If the distance to the recording station is r $(r \gg L)$, then the arrival time of a ray from the beginning of the fault is $t = (r/c)$, where c is the velocity of the wave type. The arrival time of waves from a faulting segment at point x on the fault is given by

$$t_x = \frac{x}{v_r} + \frac{(r - x \cos \theta)}{c}. \quad (9.18)$$

Thus the difference in time between energy arriving from the end of the fault, at position L and that arriving from the beginning of the fault can be used to define τ_c:

$$\tau_c = \left[\frac{L}{v_r} + \left(\frac{r - L \cos \theta}{c} \right) \right] - \left(\frac{r}{c} \right) \quad (9.19)$$

$$\tau_c = \frac{L}{v_r} - \left(\frac{L \cos \theta}{c} \right). \quad (9.20)$$

The component of the time function associated with the fault finiteness is still a boxcar, but its length, or *rupture time*, depends on the viewing azimuth. Thus, for the Haskell model the source time function is a trapezoid at all stations, but its overall length varies. This azimuth dependence due to fault propagation is called

directivity. If a seismic station is located along the direction of rupture propagation $(\theta = 0)$, the trapezoid is very narrow and has a high amplitude. If the seismic station is located such that the fault is rupturing away from it, the source time function will be spread out and have a small amplitude (see Figure 9.9).

The area under the time function is directly proportional to the seismic moment, which must be independent of azimuth. Thus the ratio of the rupture and phase velocities will strongly affect the amplitude of a particular phase. As the rupture velocity approaches the phase velocity of interest, the directivity effects become more pronounced. Figure 9.10 illustrates the effect on the radiation patterns of P and S waves for two different rupture velocities, with the relative effect on the S waves being more dramatic. Figure 9.11 shows some actual observations of azimuthal variation in the width of P-wave source functions for the 1976 Haicheng, China, earthquake. Recall from Chapter 8 that amplitude radiation patterns could give the orientation of two double-couple nodal planes but could not distinguish *between* the fault and auxiliary planes. Directivity provides a technique for doing so, since the quadrupolar symmetry is broken. For example, for the Haicheng event, the

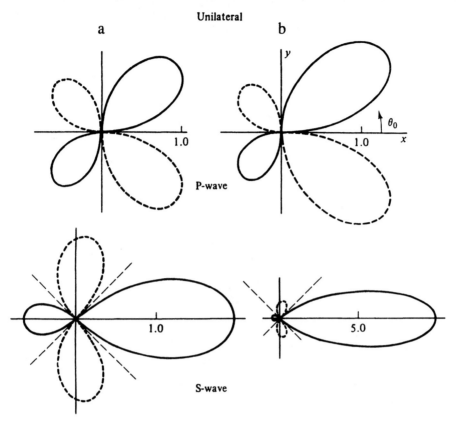

FIGURE 9.10 The variability of *P*- and *SH*-wave amplitude for a propagating fault (from left to right). For the column on the left $v_r/v_s = 0.5$, while for the column on the right $v_r/v_s = 0.9$. Note that the effects are amplified as rupture velocity approaches the propagation velocity. (From Kasahara, 1981.)

fault plane must be the east–west trending plane, for there is no way to produce the observed variations in pulse width by rupture on the north–south plane. One of the most dramatic examples of directivity was associated with the great 1960 Chilean earthquake. The fault was more than 1000 km long, allowing τ_c to vary tremendously. Figure 9.12 shows the successive passages of *R* and *G* waves recorded at PAS (Pasadena, California) due north of the fault. Note that even-order arrivals (R_4, G_4, etc.) are larger than odd-order arrivals (R_3, G_3, etc.) that have propagated shorter distances. This is because the rupture expanded southward, enhancing the major-arc arrivals.

The simple Haskell line source representation that we have considered involved *unilateral rupture*, or rupture in only one direction. For some earthquakes, unilateral rupture is a sufficient model of the faulting process, but many earthquakes nucleate in the center of a fault segment and spread in both directions. This is known as *bilateral rupture*. The source time function for bilateral rupture varies much less with azimuth, and it is often impossible to distinguish bilateral rupture from a point source. Some faults appear to expand radially, as *circular ruptures*. Generally, it is much easier to observe the horizontal component of source finiteness than any vertical component because the tem-

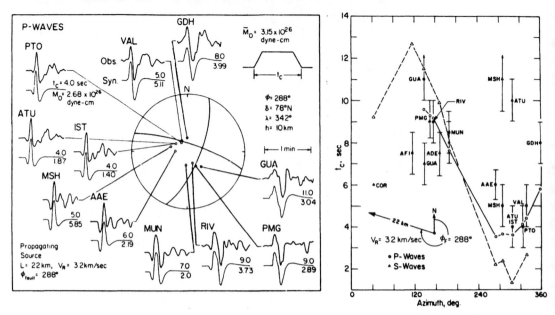

FIGURE 9.11 Effects of directivity on *P*-wave pulse width. The observed long-period WWSSN waveforms shown in the left part of the figure (upper traces) are from the 1976 Haicheng earthquake. Compare the pulse width at PTO, which the fault is rupturing toward, with that at GUA, which the fault is rupturing away from. The azimuthal variation of the pulse width is summarized at the right. This indicates that the fault plane is the plane striking 288°. Synthetic seismograms that include the directivity effect (lower traces on the left) match the data well. (From Cipar, 1979.)

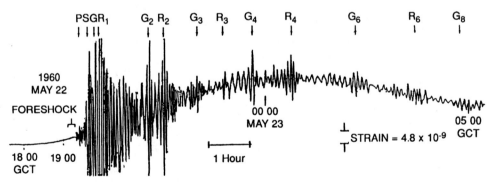

FIGURE 9.12 An ultra-long-period seismic recording of the 1960 Chilean earthquake. The station is situated due north of the epicenter, and the fault ruptured more than 1000 km in the direction away from the station. Note that G_4 is much larger than G_3 even though it traveled much farther.

poral delays are more pronounced for horizontal offsets. However, in some rare cases, the effects of *vertical directivity* are observed. This is possible when the length of the fault is short compared with its width and if the rupture is mainly up or down dip. In this case, the source time function for direct arrivals (P and S) will differ from those for the depth phases (pP, sS, sP, pS). Directivity can also be detected for complex ruptures with multiple subevents that are spatially and temporally offset. Chapter 10 will describe directivity effects for nonuniform faulting, which allow variable-displacement models to be determined for some earthquakes.

9.2 The Source Spectrum

The equivalence of time- and frequency-domain representations of a seismogram can be used to provide valuable insight into source characteristics such as magnitude and scaling, which will be discussed later in this chapter. The source time function (9.17) for the faulting model developed in the last section can be written

$$u(t) = M_0(B(t;\tau_r) * B(t;\tau_c)), \quad (9.21)$$

where the boxcar of width τ_r represents the particle dislocation history, and the boxcar of width τ_c represents the effects of fault finiteness. The heights of the boxcars are normalized to $1/\tau_r$ and $1/\tau_c$, respectively. The Fourier transform of a boxcar is given by

$$F(B(t;\tau_r)) = \hat{B}(\omega) = \frac{\sin(\omega\tau_r/2)}{\omega\tau_r/2}. \quad (9.22)$$

Recall (Box 9.2) that convolution of two functions in the time domain has a frequency-domain representation equal to the

multiplication of the Fourier transforms of the two functions. Thus we can write the spectral density of the source time function as

$$\hat{u}(\omega) = M_0 \left| \frac{\sin(\omega\tau_r/2)}{\omega\tau_r/2} \right| \left| \frac{\sin(\omega\tau_c/2)}{\omega\tau_c/2} \right|. \quad (9.23)$$

It is clear from Eq. (9.23) that the displacement amplitude decreases with increasing frequency. Figure 9.13a shows that we can approximate the boxcar amplitude spectrum (for normalized area, $A_0 = 1$) as

$$\left| \frac{\sin(\omega\tau_r/2)}{\omega\tau_r/2} \right| \approx \begin{cases} 1 & \omega < \dfrac{2}{\tau_r} \\ \dfrac{1}{\omega\tau_r/2} & \omega > \dfrac{2}{\tau_r} \end{cases}. \quad (9.24)$$

This implies that the spectrum of a boxcar has a plateau at frequencies less than $2/\tau_r$ and then decays in proportion to $1/\omega$. The crossover frequency between the plateau and ω^{-1} behavior, defined by the intersection of the asymptotes to the low- and high-frequency spectra, is called a *corner frequency*. For a convolution of two boxcars, as in Eq. (9.23), and assuming $\tau_r < \tau_c$, the peak-amplitude spectra will have three distinct trends:

$$u(\omega) \approx \begin{cases} M_0 & \omega < \dfrac{2}{\tau_c} \\ \dfrac{M_0}{\omega\tau_c/2} & \dfrac{2}{\tau_c} < \omega < \dfrac{2}{\tau_r} \\ \dfrac{M_0}{\omega^2(\tau_r\tau_c/4)} & \omega > \dfrac{2}{\tau_r} \end{cases} \quad (9.25)$$

The amplitude spectrum content of a seismic pulse should be flat at periods longer

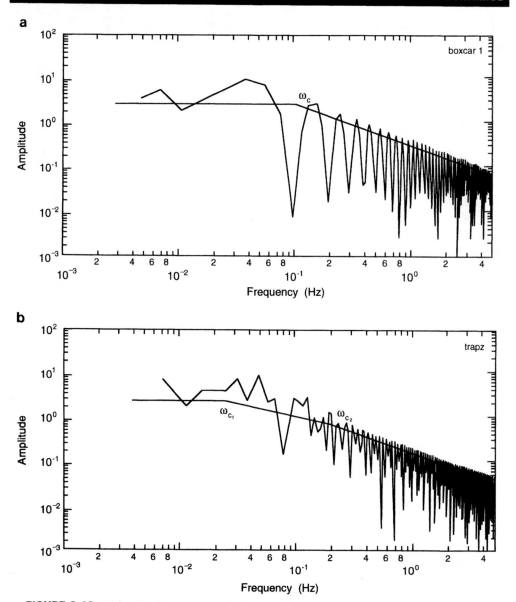

FIGURE 9.13 (a) Amplitude spectrum of a boxcar. The spectrum has two distinct regions, one where the spectral density is flat, and a second where the spectral density decays as ω^{-1}. (b) Seismic spectrum of a trapezoid (two boxcars convolved together). The intersection of the asymptotes to the low-frequency and high-frequency portions of the spectrum defines the corner frequency, ω_c, for a boxcar, and two corner frequencies, ω_{c1} and ω_{c2}, for a trapezoid.

than the rupture time of the fault. At periods between the rise time and rupture time, the spectra will decay as $1/\omega$, and at high frequencies the spectra will decay in proportion to $1/\omega^2$ (see Figure 9.13b). This is called the ω^2 source model. In practice we usually identify only one corner, which is defined by the intersection of the asymptote to the plateau and the asymptote of the $1/\omega^2$ decay (see Figure 9.13b). Figure 9.14 shows the amplitude spectrum for a P wave recorded at station ARU for the Loma Prieta earthquake. The data exhibit the expected behavior, although interfer-

ence of depth phases prevents the low-frequency spectra from being flat. The decay of far-field displacement spectra with increasing frequency is thus a natural consequence of interference between high-frequency waves caused by temporal and spatial finiteness of the source. Signals with periods less than the rupture duration tend to interfere destructively. While the Haskell model is only a kinematic model for a finite source, any physical finite source model will give rise to qualitatively comparable behavior of the far-field radiation.

Box 9.3 Frequency-Domain Representation of Seismic Signals: Aliasing and the Nyquist Frequency

In Chapter 5 we presented the concept of Fourier transforms and the equivalence of the frequency and time domains. The Fourier transform integral provides a simple procedure to convert $F(t)$ to $\hat{F}(\omega)$. But digital seismograms are not *continuous* functions in the time domain. In fact, they are measured values of ground displacement or velocity at regular time intervals. We refer to this as a *sampled* time series, and instead of integral transforms we use *discrete* transforms. The principles are similar to those already developed, although there are a number of variations and limitations. It is possible to think of a digital seismic signal as a collection of N individual data points that happen to be spaced at a discrete time interval Δt. If the seismic signal is an impulse ($N = 1$) with amplitude A, then the Fourier transform is given by

$$F(t) = \begin{cases} A & \text{at } t = t_1 \\ 0 & \text{elsewhere} \end{cases} \qquad \hat{F}(\omega) = \int_{-\infty}^{\infty} F(t)e^{-i\omega t}\, dt = Ae^{-i\omega t_1}. \quad (9.3.1)$$

This is the Fourier transform of a *delta function*, $\delta(t_1)$, multiplied by an amplitude A. The amplitude spectrum of this time series is $|Ae^{-i\omega t_1}| = A$, and the phase spectrum is $\phi(\omega) = \omega t_1$. We can generalize this to a longer time series. If $N > 1$ and the amplitude is unity, then we can consider a discrete time series

$$S(t, \Delta t) = \begin{cases} 1 & \text{at } t = n\,\Delta t, \ n = 1, \dots, N \\ 0 & \text{elsewhere} \end{cases} \qquad \hat{S}(\omega) = \sum_{n=-N}^{N} e^{-i\omega n\,\Delta t}. \quad (9.3.2)$$

The time series $S(t, \Delta t)$ is sometimes called the *Shah* or *sampling function*. The Fourier series of $S(t, \Delta t)$ is given in terms of discrete frequencies, $\omega_m = 2\pi m/T$,

continues

by

$$S(t, \Delta t) = \sum_{m = -\infty}^{\infty} F_m e^{i\omega_m t} \tag{9.3.3}$$

$$F_m = \frac{1}{T} \int_{-\tau/2}^{\tau/2} S(t, \Delta t) e^{-i\omega_m t} \, dt. \tag{9.3.4}$$

For $T = \Delta t$, evaluation of (9.3.4) gives $F_m = 1/(\Delta t)$. Then Eq. (9.3.3) can be rewritten

$$S(t, \Delta t) = \sum_{n = -\infty}^{\infty} e^{i2m\pi t/\Delta t} \cdot 1/(\Delta t). \tag{9.3.5}$$

Equation (9.3.5) is $2\pi/\Delta t$ times the inverse Fourier transform of a Shaw function in the frequency domain spaced at $(2\pi/\Delta t)$; $S(\omega; 2\pi/\Delta t)$. Thus the transform of a Shah function gives a Shaw function with sampling in time of Δt giving sampling in angular frequency of $2\pi/\Delta t$. We can use this to determine the spectrum of any sampled time signal:

$$x(t, \Delta t) = x(t) S(t, \Delta t). \tag{9.3.6}$$

$$\hat{x}(\omega, \Delta\omega) = \hat{x}(\omega) * (2\pi/\Delta t) S(\omega, (2\pi/\Delta t)). \tag{9.3.7}$$

Thus, the spectrum of the sampled time series is periodic in the frequency domain; every $2\pi/\Delta t$ the spectrum repeats itself.

The discretization of a time series introduces the concept of *bandwidth*. If $\hat{x}(\omega)$ is zero for $|\omega| > \pi/\Delta t$, then the spectra of successive frequency points, $\Delta\omega$, do not overlap. However, if $\hat{x}(\omega)$ is *nonzero* for $|\omega| > \pi/\Delta t$, the spectra of adjacent frequency points overlap, and we cannot decipher the individual contributions. This results in a phenomenon called *aliasing*. The only way we can avoid aliasing is to decrease the sampling interval, Δt, such that $\pi/\Delta t$ is a higher frequency than the highest angular frequency content of the signal. The frequency $f_n = 1/2\Delta t$ is called the *Nyquist frequency*.

When spectra are presented for digital data, the highest frequency shown is the Nyquist frequency. For IRIS broadband seismic stations, $\Delta t = 0.05$ s, so the Nyquist frequency is 10 Hz. The lowest frequency in a spectrum is given by the inverse of the length of the time window being investigated.

9.3 Stress Drop, Particle Velocity, and Rupture Velocity

The most important parameter in the dynamics of the faulting process that can be determined seismically is the stress drop. In Figure 9.2 we defined stress drop as the difference between the state of stress at a point on the fault before and after rupture. This can vary over the fault. For a finite faulting event we define the *static stress drop* as the stress drop integrated over the fault area, divided by the fault area. Figure 9.15 shows a simplified fault of length L and width w with an average displacement $\overline{D}$. Intuitively, the strain associated with $\overline{D}$, the slip during the earthquake, will be proportional to $\overline{D}/w$ or $\overline{D}/L$. We can relate this strain change to

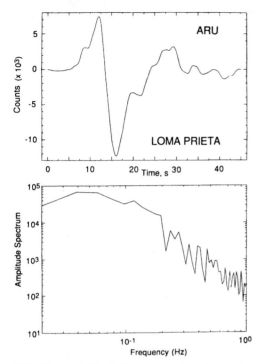

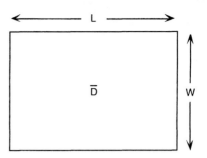

FIGURE 9.15 Dimensional analysis of a fault. The length is nearly always larger than the fault width. *L* scales in proportion to *w* up to the dimensions of the seismogenic zone, and then *L* increases while *w* does not.

FIGURE 9.14 (Top) *P*-wave arrival from the Loma Prieta earthquake recorded at the IRIS station ARU located in Russia. (Bottom) The spectrum of the *P*-wave arrival.

the static stress drop, $\Delta\sigma$, by Hooke's law:

$$\Delta\sigma = C\mu\left(\frac{\overline{D}}{\tilde{L}}\right), \qquad (9.26)$$

where $\tilde{L}$ is a *characteristic rupture dimension* (either L or w) and C is a nondimensional constant that depends on the fault geometry. Table 9.1 prescribes the constant C for various fault geometries in terms of stress drop, $\Delta\sigma$, and complemen-

tary relations in terms of seismic moment, M_0. Note that in the equations for $\Delta\sigma$, $\mu\overline{D}$ is just M_0/A. Thus, although moment can be accurately determined, the stress drop depends inversely on a much more poorly resolved length scale, $\tilde{L}^3$.

It is possible to evaluate the scaling of stress drop of earthquakes by analyzing the dependence of seismic moment on fault area. However, measuring fault area is difficult for many events. If we assume that the fault width scales approximately as the fault length (a good assumption for small-magnitude earthquakes), then we can relate the square root of the fault area, $\sqrt{A}$, to τ_c, the rupture time, which can be determined for many events. Figure 9.16 shows τ_c determinations for a global distribution of earthquakes as a function of moment. Although the scatter is substantial, the cube of duration ($\propto A^{3/2}$) scales with M_0. Referring to the equations for M_0 in Table 9.1, we see that this implies

TABLE 9.1 Stress Drop and Moment Relations for Three Fault Types

	Circular (radius = a)	Strike slip	Dip slip
$\Delta\sigma$	$\dfrac{7\pi}{16}\mu\left(\dfrac{\overline{D}}{a}\right)$	$\dfrac{2}{\pi}\mu\left(\dfrac{\overline{D}}{w}\right)$	$\dfrac{4(\lambda+\mu)}{\pi(\lambda+2\mu)}\mu\left(\dfrac{\overline{D}}{w}\right)$
M_0	$\dfrac{16}{7}\Delta\sigma a^3$	$\dfrac{\pi}{2}\Delta\sigma w^2 L$	$\dfrac{\pi(\lambda+2\mu)}{4(\lambda+\mu)}\Delta\sigma w^2 L$

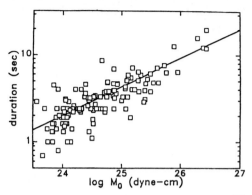

FIGURE 9.16 Time function duration $(\alpha \sqrt{A})$ versus seismic moment for a global distribution of earthquakes. The linear trend is consistent with a constant stress drop, independent of earthquake size. (Courtesy of G. Ekström.)

σ_0, is applied to the surface of Σ. At time $t = 0^+$ this traction relaxes to a new value, σ_1. Thus $\Delta\sigma = \sigma_0 - \sigma_1$. It takes a time τ_r to achieve the final stress state. Figure 9.17 shows one side of Σ and a point P some distance from the fault. The relaxation of σ_0 at time $t = 0^+$ is equivalent to applying a negative shear stress to the fault surface. This shear stress creates a stress imbalance which propagates through the material with a velocity β. At any given time, the shear-wave pulse will have propagated a distance βt into the medium. Therefore, we can calculate an instantaneous strain

$$\varepsilon = \frac{u(t)}{\beta t} = \left(\frac{\Delta\sigma}{\mu}\right) \qquad (9.27)$$

or

$$u(t) = \frac{\Delta\sigma\beta t}{\mu}. \qquad (9.28)$$

that stress drop is essentially independent of M_0. This observation is the basis for the idea of self-similarity of earthquakes. It is a remarkable observation that we will explore more completely when we discuss seismic scaling.

Stress drop also scales with the particle velocity. If we manipulate Eq. (9.26), then $\bar{D} = (\bar{L}\Delta\sigma)/\mu C$. Since $D(t)$ is a function of time, $\Delta\sigma$ must have an identical time rate history. Brune (1970) developed a very simple model to explain the temporal history of stress drop. He considered a crack with surface area Σ in a homogeneous material (Figure 9.17). An initial traction,

Thus the particle velocity, $\dot{u}(t)$, is directly proportional to the stress drop: $\dot{u}(t) = \Delta\sigma(\beta/\mu)$. This simple model predicts that the particle displacement history on a fault should be ramp-like as in Figure 9.3, and when $\Delta\sigma$ is nearly constant, $\dot{u}(t)$ should also be relatively earthquake independent. In general, $\dot{u}(t)$ appears to be on the order of 100 cm/s.

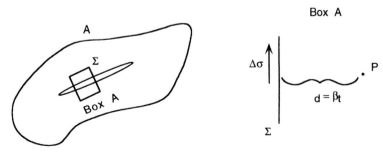

FIGURE 9.17 Geometry of a fault, Σ, which cuts a continuous medium. One surface of Σ is subjected to a stress that produces a *SH* wave that travels to a point *P*.

Equation (9.28) assumes that the stress relaxes along the entire fault at some time t; in actuality, the fault propagates from one end of the fault to the other at a velocity v_r. Therefore, the point P will actually "see" a stress drop from the ruptured segment of the fault before the rupture front reaches the fault adjacent to P, as long as $v_r < \beta$ (this is usually the case). This results in a reduced particle velocity

$$\dot{u}(t) = \frac{\Delta \sigma}{2\mu} \beta e^{-(t/\tau)}, \qquad (9.29)$$

where τ is a time constant given by L/v_r. This form of particle velocity does not differ much from the boxcar form assumed earlier, except that the source time function is now a boxcar with a tail. In practice, we usually assume the simpler form because the data cannot resolve greater detail.

Stress drop has no role in controlling the rupture velocity, v_r. Fracture energy conditions usually require that v_r starts small and then increases to a final value when the fault rupture front exceeds some critical dimension. In general, theoretical work has shown that mode II and mode III cracks rupture in different ways, but in both cases the main part of the rupture cannot exceed the shear-wave velocity. Because most of our seismic observations can see only the main part of the slip, a general rule of thumb is $v_r = 0.8\beta$, where β is the shear velocity at the fault.

9.4 Magnitude Scales

The best way to quantify the size of an earthquake is to determine its seismic moment, M_0, and the shape of the overall source spectrum. This can be done by recovering the source time function from either body or surface waves, but this requires relatively complete modeling of the waveform in question. It is desirable to have a measure of earthquake size that is much simpler to make, for example, using the amplitude of a single seismic phase, such as the P wave. Unfortunately, as we have seen, the amplitude and waveform character of a far-field P or S wave are proportional to the moment rate; thus different fault dislocation histories with the same seismic moment can produce very different amplitude signals. Further, the effects of the time function will depend on the frequency band of observation, and thus the amplitude of various phases will vary greatly from instrument to instrument. These limitations aside, measurements based on wave amplitude are still very useful because of their simplicity and because high-frequency shaking in a narrow frequency band is often responsible for damage from earthquakes. The concept of *earthquake magnitude*, a relative-size scale based on measurements of seismic phase amplitudes, was developed by K. Wadati and C. Richter in the 1930s, over 30 yr before the first seismic moment was calculated in 1964.

Magnitude scales are based on two simple assumptions. The first is that given the same source–receiver geometry and two earthquakes of different size, the "larger" event will on average produce larger-amplitude arrivals. The second is that the amplitudes of arrivals behave in a "predictable" fashion. In other words, the effects of geometric spreading and attenuation are known in a statistical fashion. The general form of all magnitude scales is given by

$$M = \log(A/T) + f(\Delta, h) + C_s + C_r,$$

$$(9.30)$$

where A is the ground displacement of the phase on which the amplitude scale is based, T is the period of the signal, f is a correction for epicentral distance (Δ) and focal depth (h), C_s is a correction for the siting of a station (e.g., variability in amplification due to rock type), and C_r is a source region correction. The logarithmic scale is used because the seismic-wave amplitudes of earthquakes vary enormously. A unit increase in magnitude corresponds to a 10-fold increase in amplitude of ground displacement. Magnitudes are obtained from multiple stations to overcome amplitude biases caused by radiation pattern, directivity, and anomalous path properties. Four basic magnitude scales are in use today: M_L, m_b, M_s, and M_w.

9.4.1 Local Magnitude (M_L)

The first seismic magnitude scale was developed by C. Richter in the early 1930s and was motivated by his desire to issue the first catálogue of California earthquakes. This catalogue contained several hundred events, whose size ranged from barely perceptible to large, and Richter felt that an earthquake description must include some objective size measurement to assess its significance. Richter observed that the logarithm of maximum ground motion decayed with distance along parallel curves for many earthquakes. All the

observations were from the same type of seismometer, a simple Wood–Anderson torsion instrument. Figure 9.18 shows some of the original data from Richter. The relative size of events is calculated by comparison to a *reference event*:

$$\log A - \log A_0 = M_L, \qquad (9.31)$$

where A and A_0 are the displacements of the earthquake and a reference event at a prescribed distance, respectively. Richter chose his reference earthquake, with $M_L = 0$, such that A_0 was 1×10^{-3} m at an epicentral distance of 100 km. By using the reference event to define a curve, we can rewrite Eq. (9.31) as

$$M_L = \log A - 2.48 + 2.76 \log \Delta. \quad (9.32)$$

A graphical form of this relation is shown in Figure 1.9. At first glance this equation is not in the form of (9.30), but Richter made a number of restrictions that can be factored out of (9.30). First, all the instruments used were narrowband and identi-

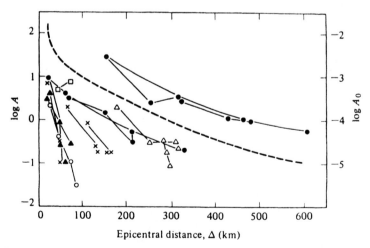

FIGURE 9.18 Origin of the local magnitude scale, based on the systematic decrease of seismic amplitudes with distance. The data are for Southern California earthquakes in January, 1932. (From Richter, 1958. Copyright © 1958 by W. H. Freeman and Co. Reprinted with permission.)

cal, and thus the maximum-amplitude phase was always of a single dominant period, T. Second, all the seismicity was shallow (less than 15 km deep), and the travel paths were confined to southern California. Thus the corrections for regional dependence and focal depth are approximately constant, and Eq. (9.32) is actually a particular subset of (9.30).

Earthquakes with $M_L \leq 2.5$ are called *microearthquakes* and are rarely felt. The smallest events that are recorded have magnitudes less than zero, and the largest M_L recorded is about 7, which gives seven orders of magnitude in ground displacement. In practice, M_L is usually a measure of the regional-distance S wave. The magnitude for each of the horizontal seismometers is averaged in a least-squares sense to give an M_L for a given station. The values of M_L from each station are averaged to give the "magnitude." M_L may vary considerably from station to station, due not only to station corrections but also to variability in the radiation pattern.

M_L in its original form is rarely used today because Wood–Anderson torsion instruments are uncommon and, of course, because most earthquakes do not occur in southern California. However, M_L remains a very important magnitude scale because it was the first widely used "size measure," and all other magnitude scales are tied to M_L. Further, M_L is a very useful scale for engineering. Many structures have natural periods close to that of a Wood–Anderson instrument (0.8 s), and the extent of earthquake damage is closely related to M_L.

9.4.2 Body-Wave Magnitude (m_b)

Although the local magnitude is useful, the limitations imposed by instrument type and distance range make it impractical for global characterization of earthquake size. Beyond regional distances, where direct P becomes a distinct phase, it is convenient to define a magnitude based on the amplitude of the P wave, which is termed m_b. This magnitude is based on the first few cycles of the P-wave arrival and is given by

$$m_b = \log(A/T) + Q(h, \Delta), \quad (9.33)$$

where A is the actual ground-motion amplitude in micrometers and T is the corresponding period in seconds. The reason for using the first few swings of the P wave is that the effects of radiation pattern and depth phases can result in a complicated waveform signature. In practice, the period at which m_b is usually determined is 1 s (the WWSSN and many regional network short-period instruments have a "peaked" response near 1 Hz). It is not unusual to have scatter of the order of ± 0.3 for individual m_b measurements for a given event, requiring extensive averaging. Occasionally, long-period instruments are used to determine body-wave magnitude for periods from 5 to 15 s, and these are usually referred to as m_B. When m_B is measured, it is usually for the *largest* body wave (P, PP, etc.).

The correction for distance and depth $Q(h, \Delta)$ is determined empirically. Figure 9.19 shows values for $Q(h, \Delta)$; note that the corrections are fairly uniform beyond 30° but are complex at upper-mantle distances. This reflects the complexity of the body waves in this epicentral distance range. The correction dramatically *decreases* at 20° because the upper-mantle triplications result in very large amplitude arrivals.

9.4.3 Surface-Wave Magnitude (M_s)

Beyond about 600 km the long-period seismograms of shallow earthquakes are dominated by surface waves, usually with a period of approximately 20 s (recall the *Airy phase* discussed in Chapter 4). The amplitude of these waves depends on dis-

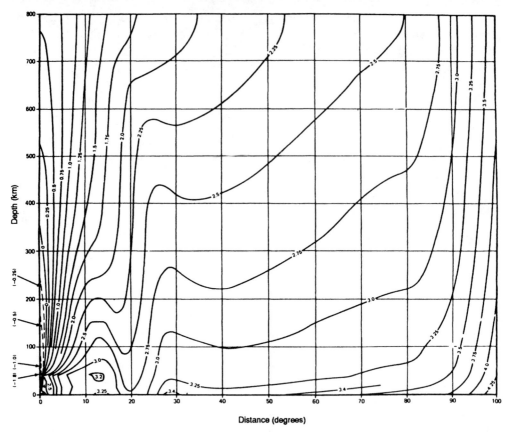

FIGURE 9.19 The correction $Q(h, \Delta)$ that is applied to determine the body-wave magnitude. The correction is read off the contour for the appropriate depth and epicentral distance. (From Veith and Clawson, *Bull. Seismol. Soc. Am.* **62**, 435–452, 1972. Reprinted with permission.)

tance differently than the amplitude of body waves, and surface-wave amplitudes are strongly affected by the source depth. Deep earthquakes do not generate much surface-wave amplitude, and thus there is no appropriate correction for source depth. The equation for surface-wave magnitude is given by

$$M_s = \log A_{20} + 1.66 \log \Delta + 2.0, \quad (9.34)$$

where A_{20} is the amplitude of the 20-s-period surface wave in micrometers. In general, the amplitude of the Rayleigh wave on the vertical component is used in Eq. (9.34).

Both M_s and m_b were designed to be as compatible as possible with M_L; thus at times all three magnitudes give the same value for an earthquake. Unfortunately, this is rarely the case. In all three magnitudes we are making a frequency-dependent measurement of amplitudes at about 1.2, 1.0, and 0.05 Hz for M_L, m_b, and M_s, respectively. If we consider the source spectrum of a seismic pulse (e.g., Figure 9.14), it is easy to see that only for small earthquakes (very short fault lengths) with corner frequencies well above 1 Hz will the amplitude be the same for all three frequencies. For earthquakes above a certain size, the frequency at which we mea-

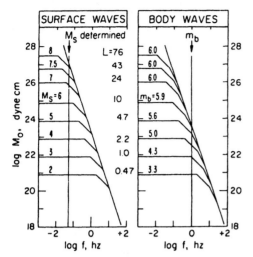

FIGURE 9.20 Spectra for different-sized earthquakes and the relationship of these spectra to the frequencies at which M_S and m_b are determined. (From Geller, 1976.)

sure m_b will be located on the ω^{-2} decay slope, and thus all earthquakes above this size will have a constant m_b. This is called *magnitude saturation*. Figure 9.20 illustrates one model of the variation of source spectra for different size earthquakes, and

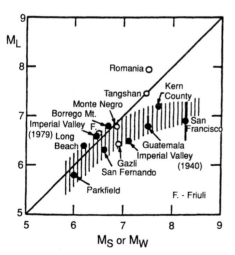

FIGURE 9.21 Effects of magnitude saturation for the high-frequency magnitude M_L versus the lower-frequency magnitudes M_S and M_w. (Courtesy of H. Kanamori.)

it is apparent that m_b begins to saturate at approximately $m_b = 5.5$ and is fully saturated by $m_b = 6.0$, whereas M_S does not saturate until approximately $M_S = 7.25$ and is fully saturated by $M_S = 8.0$. There are many examples of reported m_b larger than 6.0, which implies that these particular source spectra are not valid for all events, something we will discuss later. Figure 9.21 shows measured values of M_L and M_s for several different earthquakes. It is clear that M_L begins to saturate at about magnitude 6.5, but for rare examples such as Tangshan and Romania, ML does not saturate. It is desirable to have a magnitude measure that does not suffer from this saturation deficiency.

9.5 Seismic Energy and Magnitude

Although the magnitude scales discussed above give a means of comparing earthquakes, the total size of an earthquake is best represented by the seismic moment, M_0. An alternative measure of earthquake size would be energy released. We calculate energy by considering the history of a particle as it responds to a transient seismic wavefield. As a wave passes, the particle, which has a potential energy, will have a velocity and thus a kinetic energy. The sum of the potential and kinetic energies integrated over time will yield the work, or the energy expended. As an example, consider a seismic station situated directly above a monochromatic source of seismic energy. The displacement of the ground at the seismic station will be given by

$$x = A \cos\left(\frac{2\pi t}{T}\right), \qquad (9.35)$$

where A is the amplitude of the wave of

period T. The ground velocity is given by

$$v = -\left(\frac{2\pi A}{T}\right)\sin\left(\frac{2\pi t}{T}\right). \quad (9.36)$$

The kinetic energy of a unit mass at a recording station is just given by $1/2\rho v^2$. If we average this over one complete cycle, we obtain the *kinetic energy density*:

$$e = \frac{1}{2}\frac{\rho}{T}\int_0^T v^2\, dt$$

$$= \left(\frac{\rho}{2T}\right)\left(\frac{2\pi A}{T}\right)^2\int_0^T \sin^2\left(\frac{2\pi t}{T}\right) dt$$

$$= \rho\pi^2\frac{A^2}{T^2}. \quad (9.37)$$

Note that the energy density is proportional to A^2, which is the expected result. Since the mean potential and kinetic energies are equal, we can write $E = 2e$. If we integrate over the spherical wavefront to correct for geometric spreading, we obtain an equation of the form

$$E = F(r,\rho,c)\left(\frac{A}{T}\right)^2, \quad (9.38)$$

where r is the distance traveled, ρ is the density, and c is the velocity of the wave type. This equation can be recast in a form similar to the general equation for magnitude scales:

$$\log E = \log F(r,\rho,c) + 2\log\left(\frac{A}{T}\right).$$
$$(9.39)$$

Thus it is possible to relate energy to magnitude if $F(r,\rho,c)$ is known. Gutenberg and Richter found empirical relationships for m_b and M_s:

$$\log E = 5.8 + 2.4m_b \quad (9.40)$$

$$\log E = 11.8 + 1.5M_s. \quad (9.41)$$

Obviously, the energy calculation in Eqs. (9.40) and (9.41) suffers from all the problems of the magnitude determination. In particular, since m_b saturates, the estimate of E using (9.40) for any earthquake larger than about magnitude 6.5 is probably low. Equation (9.41) is fairly robust, since M_s does not saturate until very large magnitudes are reached. Equation (9.41) gives an interesting insight into the tremendous range of earthquake size. The difference between the energy released in an $M_s = 6.0$ and an $M_s = 7.0$ earthquake is a factor of $10^{1.5}$, or ~ 32. In other words, the seismic energy released in a magnitude 7.0 earthquake is over 30 times greater than that released in a magnitude 6.0 earthquake, and it is three orders of magnitude greater than that released in a $M_s = 5.0$ earthquake.

It is also possible to relate seismic moment to the seismic energy. Kostrov (1974) showed that the radiated seismic energy is proportional to the stress drop:

$$E_s \approx \tfrac{1}{2}\Delta\sigma\overline{D}A \quad (9.42)$$

or, rearranging terms using the definition of M_0,

$$E_s \approx \frac{\Delta\sigma}{2\mu}M_0. \quad (9.43)$$

We can use this expression to relate M_0 to magnitude through Eq. (9.41). If we assume that stress drop is constant and equal to about 30 bars, this yields the relation

$$\log M_0 = 1.5M_s + 16.1. \quad (9.44)$$

This equation gives a simple way to relate magnitude to seismic moment and, in fact, can be used to define a new magnitude scale, M_w, called the *moment magnitude*:

$$M_w = \left(\frac{\log M_0}{1.5}\right) - 10.73. \quad (9.45)$$

This scale, derived by Kanamori (1977), is tied to M_s but will not saturate because

M_0 does not saturate. Generally, determination of M_0 is much more complicated than magnitude measurement, although modern seismic analyses are routinely providing M_0 for all global events larger than $M_w = 5.0$. The largest earthquake recorded this century was the 1960 Chilean earthquake, with $M_w = 9.5$. Table 1.6 lists M_s and M_w values for large events this century.

9.6 Aftershocks and Fault Area

In our discussion of earthquake rupture and stick–slip mechanics, we defined the sudden release of seismic energy in terms of a single seismic event. However, nearly all large earthquakes are followed by a sequence of smaller earthquakes, known as *aftershocks*, which are apparently related to the fault plane that slipped during the event. The large earthquake, known as the *mainshock*, introduces a major stress adjustment to a complex system by its sudden slip. Regions within the rupture zone, or adjacent to it, may require readjustment to the new stress state in the source volume, thus generating aftershocks. Aftershocks typically begin immediately after a mainshock and are distributed throughout the source volume. For a typical $M_s = 7.0$ earthquake thousands of small aftershocks may occur. In general, the largest aftershock is usually more than a magnitude unit smaller than the mainshock (aftershocks can still be quite dangerous due to the damage to structures caused by the mainshock). The total seismic moment release of an aftershock series rarely exceeds 10% of the moment of the mainshock.

Typically, the frequency of occurrence of aftershocks decays rapidly. Omori studied aftershocks in Japan in the 1930s and developed an empirical formula for the aftershock activity (*Omori's law*)

$$n = \frac{C}{(K+t)^P}, \qquad (9.46)$$

where n is the frequency of aftershocks at time t after the mainshock. K, C, and P are constants that depend on the size of the earthquake, and the P value is usually close to 1.0–1.4. Figure 9.22 shows the time history of the aftershocks of the 1974 Friuli, Italy, earthquake.

The distribution of aftershocks is often used to infer the fault area. For most

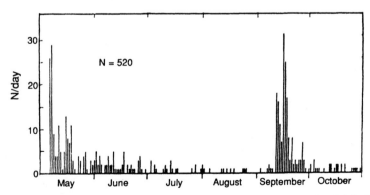

FIGURE 9.22 The number of aftershocks as a function of time for the Friuli, Italy, earthquake of May 6, 1976. The aftershocks decay according to Omori's law until early September, when a second event occurred. (From Cagnetti and Pasquale, *Bull. Seismol. Soc. Am.* **69**, 1797–1818, 1979. Reprinted with permission.)

earthquakes, the fault area (or aftershock area) scales with magnitude. Utsu and Seki (1954) developed the empirical formula

$$\log A = 1.02 M_s + 6.0, \qquad (9.47)$$

where A is measured in cm^2. In general, the fault area is estimated from the extent of the aftershock zone after 1 to 2 days. This limit is imposed because sometimes aftershock zones grow continually for a month or more, presumably involving outward expansion of the mainshock rupture zone.

Figure 9.23 shows the aftershocks of the 1983 Borah Peak, Idaho, earthquake. The aftershocks define a zone approximately 70 km long and a dipping plane whose orientation agrees closely with that of the southwest-dipping nodal plane of the focal mechanism. Note that the mainshock hypocenter is located at the southeast end of the fault, near the bottom of the fault plane. Apparently the Borah Peak earthquake rupture mode was unilateral, rupturing to the northwest. It is quite common for the hypocenter of a mainshock to

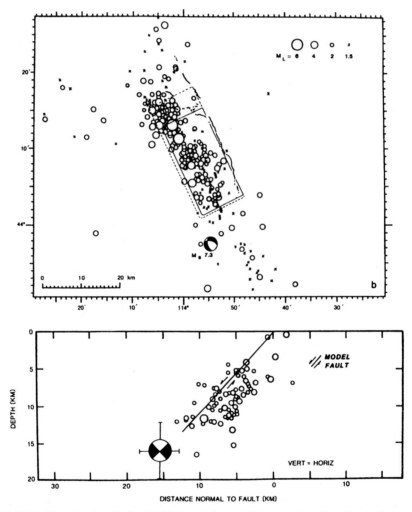

FIGURE 9.23 Aftershocks of the 1983 Borah Peak, Idaho, earthquake. The fault ruptured the surface for about 40 km. Note that the aftershocks define a plane that is consistent with the N22°W plane of the focal mechanism. (From Stein and Barrientos, *J. Geophys. Res.* **90**, 11,355–11,366, 1985; © copyright by the American Geophysical Union.)

be located near the bottom of the fault plane; this is apparently related to strength conditions in the crust.

The hypocenters of earthquakes outside subduction zones rarely occur deeper than 15–20 km; the deepest earthquakes define the base of the *seismogenic zone*. Within the seismogenic zone the crust deforms either by stable or unstable sliding on faults or as a brittle material that fails when subjected to stresses greater than the strength of the material. The strength depends on both temperature and pressure. For most materials that could realistically make up a significant fraction of the crust and for realistic temperature profiles, the strength increases to a depth of about 15 km and then decreases rapidly. The maximum in the strength curve is near or below the base of the seismogenic zone where unstable frictional sliding occurs. The existence of the seismogenic zone has important consequences for earthquake size. The implication is that a *maximum* fault width is available for rupture, below which stable

sliding or ductile deformation takes place. Earthquakes that rupture this entire zone can be classified as *large*, and those that only rupture part of the seismogenic zone can be classified as *small*. When we discuss scaling in the next section, we will see that the distinction of size is important.

Thus far we have described fault slip, D, in terms of *average* slip on the fault. The slip on the fault plane varies considerably, and in fact the slip often appears to be concentrated on patches. These patches, called *asperities* (by analogy to the microstructural protrusions in Amonton's early work on friction), represent zones of relatively high stress drop. As mentioned earlier, *average* stress drop appears remarkably independent of earthquake size, but parts of the fault may have stress drops an order of magnitude higher than the average. Aftershocks often appear to concentrate around the edges of asperities. Figure 9.24 shows the slip history inferred for the 1986 North Palm Springs, California, earthquake. This slip history was de-

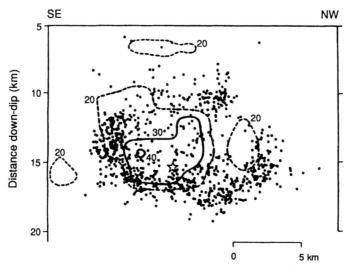

FIGURE 9.24 Aftershocks on the fault plane of the 1986 North Palm Springs earthquake. The displacement that occurred during the earthquake is indicated by the contours, showing centimeters of slip, as inferred from analyzing seismic records. The high-slip zones define several asperities. Note that the aftershocks in general outline these regions. (From Hartzell, *J. Geophys. Res.* **94**, 7515–7534, 1989; © copyright by the American Geophysical Union.)

rived from modeling near-field seismic recordings of the ground velocity. Note how the aftershocks outline the zone of high slip. The variation in frictional properties or slip history on the fault responsible for the nonuniform slip is not well understood and continues to be an active area of research.

9.7 Scaling and Earthquake Self-Similarity

The faulting parameters of most earthquakes appear to be related in a systematic and predictable manner. For example, as moment increases, the rupture duration increases in a generally predictable fashion, as shown in Figure 9.16. The interdependence of the various faulting parameters provides *scaling relations*. It is

possible to develop a theoretical basis for most scaling relations, which provides insight into the physics of rupture.

The data in Figure 9.25 are consistent with a linear relation between the logarithm of fault area and seismic moment; the slope of the line is equal to two thirds. If we assumed that these events occurred on a circular fault, then from Table 9.1

$$M_0 = \mu\, A\overline{D} = \frac{16}{7}\Delta\sigma a^3 = \left(\frac{16}{7\pi^{3/2}}\Delta\sigma\right) S^{3/2}$$

$$(9.48)$$

or, collecting terms,

$$\log M_0 = \frac{3}{2}\log S + \log\left(\frac{16}{7\pi^{3/2}}\Delta\sigma\right).$$

$$(9.49)$$

Thus, the theoretical slope is equal to the

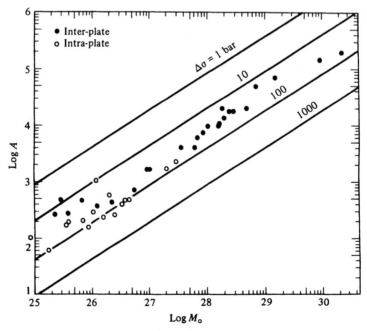

FIGURE 9.25 Area versus moment for inter- and intraplate earthquakes. Note that the interplate earthquakes show little scatter about a stress drop of 30 bars. The intraplate earthquakes have stress drops of ~ 100 bars. (Modified from Kanamori and Anderson, 1975.)

observed slope, which implies that $\Delta\sigma$ must be fairly constant for a large range of earthquakes. The value of $\Delta\sigma$ will only shift the linear fit up or down (lines of constant $\Delta\sigma$ are shown in Figure 9.25). Most earthquakes apparently have stress drops between 10 and 100 bars, although order-of-magnitude variations up or down are observed. The relative constancy of stress drop implies a similarly uniform strain drop, $\overline{D}/\overline{L}$. This fact was pointed out in Chapter 1 in the discussion of elastic rebound, when we showed that nearly all earthquakes occur at strain levels of $\sim 10^{-5}$.

In detail, the scatter in the stress drop has some systematic behavior. First of all, there is much more error in determining fault area than moment, which causes a factor of 2 to 10 uncertainty in stress drop. If the earthquake is large enough to have a well-recorded aftershock zone, area can be determined directly. Otherwise the area must be determined by measuring the source time function duration and assuming a rupture velocity or using directivity. Second, two populations of earthquakes occur in Figure 9.25: *interplate* and *intraplate* earthquakes. The interplate earthquakes are those events that occur along, or parallel to, major plate boundaries with large moment-release rates (Chapter 11 will discuss the tectonic setting of these earthquakes). Thrust earthquakes along subduction zones and strike–slip events along transform boundaries fall in this group. Earthquakes that occur away from plate boundaries, within the plate interior, are intraplate earthquakes. Interplate earthquakes have average stress drops of about 30 bars. On the other hand, intraplate earthquakes have systematically larger stress drops, about 100 bars. The higher stress drop implies larger moment release per fault length. This can be seen in Figure 9.26.

The linearity between $\log L$ and $\log M_0$ implies $\overline{D} = \alpha' L$ (where L is the fault length), which, of course, implies constant stress drop. Both interplate and intraplate earthquakes follow a linear trend, but α' is about five times larger for intraplate events. Several possible mechanisms could cause the intraplate faults to be much stronger: (1) Intraplate faults have slow slip rates and *gouge zones* composed of fracture debris on the fault surface are only 1–10 m thick, whereas plate boundary faults may have gouge zones 100–1000 m thick. This could lead to higher fault friction in intraplate environments. (2) Intraplate events have long recurrence intervals, possibly allowing the faults to "heal" through chemical processes. (3) Intraplate faults are much shorter and more discontinuous than interplate faults.

Within the class of interplate or intraplate earthquakes, a constant-stress-drop assumption leads to what are known as the conditions of *static similarity*:

$$\frac{W}{L} = k_1 \quad \text{(constant fault aspect ratio)}$$

$$(9.50)$$

$$\frac{\overline{D}}{L} = k_2 \quad \text{(constant strain)}. \quad (9.51)$$

We can combine these such that $M_0 = \mu \overline{D} W L = \mu k_1 k_2 L^3$; thus $M_0 \approx L^3$. This similarity relation implies that if you double the length of a fault, you will increase the moment by a factor of 8!

There are also conditions of *dynamic similarity*:

$$\tau_r = k_3 \tau_c = k_3 \frac{L}{v_r} \quad (9.52)$$

This implies that the rise time, and hence the total displacement, $\overline{D}$, is proportional to the length of the fault. This, of course, is exactly the relationship of constant strain. We can use this to rewrite Eq.

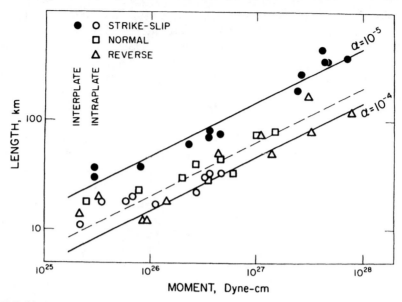

FIGURE 9.26 Fault length versus moment, showing a clear separation of intra- and inter-plate earthquakes. (From Scholz *et al.*, 1986.)

(9.23) as $\hat{u}(\omega) = M_0 F(\omega)$, where

$$F(\omega) = \left| \frac{\sin\left(\dfrac{\omega \tau_r}{2}\right)}{\dfrac{\omega \tau_r}{2}} \right| \left| \frac{\sin\left(\dfrac{\omega L}{2v_r}\right)}{\dfrac{\omega L}{2v_r}} \right|. \quad (9.53)$$

Equation (9.53) then has four regions of approximate behavior depending on the period of observation ($T = 2\pi/\omega$):

$$F(\omega) = \begin{cases} \dfrac{v_r T}{L \pi} \approx \dfrac{1}{L} & \text{if } \tau_r < \dfrac{T}{\pi}, \dfrac{L}{v_r} > \dfrac{T}{\pi} \\[2ex] \dfrac{T}{\pi \tau_r} \approx \dfrac{1}{L} & \tau_r > \dfrac{T}{\pi}, \dfrac{L}{v_r} < \dfrac{T}{\pi} \\[2ex] \dfrac{T^2}{\pi^2 \tau_r} \dfrac{v_r}{L} \approx \dfrac{1}{L^2} & \tau_r > \dfrac{T}{\pi}, \dfrac{L}{v_r} > \dfrac{T}{\pi} \\[2ex] 1 \approx L^0 & \tau_r < \dfrac{T}{\pi}, \dfrac{L}{v_r} < \dfrac{T}{\pi}. \end{cases}$$

$$(9.54)$$

Thus, the amplitude of a seismic phase will scale with fault dimension in several different ways. If the period of observation is very long compared to the source time function duration, then the amplitude is independent of fault length. This is the case when we have magnitude saturation. On the other hand, if the period of observation is short compared to the source time function, then the amplitude scales as $1/L^2$. If we make our observations of surface-wave amplitude at $T = 20$ s, we can use the relation to develop scaling for L with M_s. Figure 9.27 shows the four quadrants for $M_s \approx F(L)$. For most events, $\tau < 20/\pi$; if $L > 20$ km (which is true for events greater than $M_s = 6.5$), then $L/v_r > 20/\pi$, and $M_s \approx \log L^2$.

Static and dynamic similarity are conditions of *self-similarity*. One of the remarkable things about earthquake dynamics is that we can use self-similarity of the rupture process to accurately predict the behavior of source parameters. Self-similarity does break down when we attempt to compare *small* and *large* earthquakes as defined in the previous section. All small earthquakes have source dimensions that

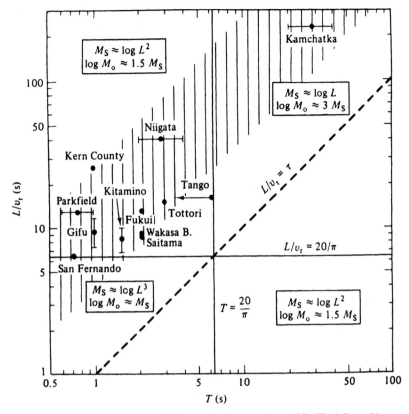

FIGURE 9.27 Scaling for τ_c, τ_r, and the seismic moment. (Modified from Kanamori and Anderson, 1975.)

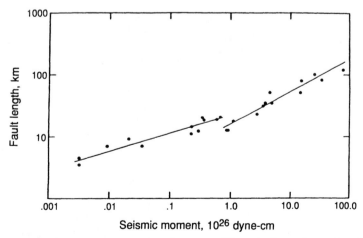

FIGURE 9.28 Difference in the scaling for "small" and "large" earthquakes. (From Shimazaki, 1986.)

are less than the width of the seismogenic zone. On the other hand, large earthquakes rupture completely through the seismogenic zone; so fault width becomes independent of earthquake size. Figure 9.28 shows that $M_0 \approx L^3$ until a moment of 7.5×10^{25} N m ($M_w = 6.6$), after which $M_0 \approx L^2$.

If we combine self-similar scaling with the basic spectral shape of the ω^2 model, we obtain the source spectra shown in Figure 9.29 for far-field displacement, velocity, and acceleration. In this case the corner frequency is inversely proportional to the fault length ($\propto \sqrt[3]{M_0}$). Note that the

velocity spectra are peaked and the acceleration spectra fall off at low frequency. These spectra characterize the variations in signals expected over the body-wave frequency band for small to great earthquakes, all as a consequence of self-similar scaling of the sources.

9.8 Earthquake Statistics

A final important scaling relationship is the relation between earthquake size and frequency of occurrence. Richter and

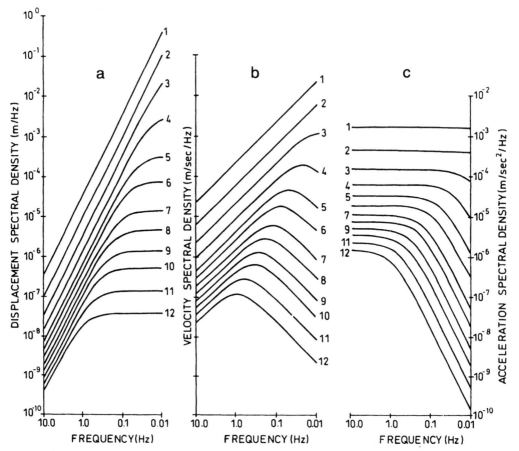

FIGURE 9.29 Spectral densities at a distance of 1 km from a double-couple point source for the ω^2 model. For a stress drop of p bars, one must multiply the ordinate by p. (From Duda, 1989.)

Gutenberg first proposed that in a given region and for a given period of time, the frequency of occurrence can be represented by

$$\log N = A - bM_s \qquad (9.55)$$

where N is the number of earthquakes with magnitudes in a fixed range around magnitude M_S, and A and b are constants. The constant b in (9.55) is called the b value. We can rearrange terms in (9.55) and substitute M_0 from (9.44) for M_S to obtain

$$N(M_0) = A'M_0^{-(b/1.5)}. \qquad (9.56)$$

This type of power-law size distribution arises from the self-similarity of earthquakes. Figure 9.30 shows the number of earthquakes that have occurred worldwide since 1977 with magnitudes larger than 5.5. The b value is 1.0. In general, b values are between $\frac{2}{3}$ and 1 and do not show much regional variability. If we consider Eq. (9.55) determined *per year*, then the A value gives the maximum expected earthquake (assuming $b = 1$). Globally, we expect one earthquake per year to be larger than $M_w = 8.0$. This implies that about 10 events of magnitude 7.0 (see Figure 1.10) and 100 or more events with magnitude 6.0 should occur per year. With this in mind, you can understand the seismologist's frustration with the question, "Is this the big one?" which is heard any time an earthquake does damage in California. Earthquakes of magnitude 6.0 are hardly unusual phenomena, at least globally.

It is interesting to use Eq. (9.55) to calculate the yearly energy release from earthquakes. The largest event in a given year usually accounts for approximately 50% of the total seismic energy release, and the events with magnitudes greater than 7.0 account for more than 75% of the total (see Figure 1.15). The total energy release from the Earth averages 1.0–2.0×10^{24} erg/yr. This number is relatively small compared to the energy release of other geophysical processes. (The 1991 eruption of Mt. Pinatubo in the Philippines released 8×10^{26} ergs!)

The one case in which the b value departs significantly from ~ 1.0 is *earthquake swarms*. Swarms are sequences of earthquakes that are clustered in space and time and are not associated with an identifiable mainshock. The b values for swarms can be as large as 2.5, which im-

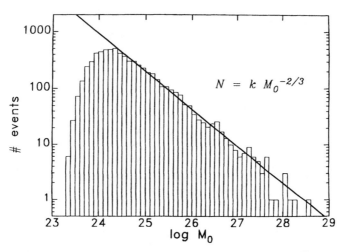

FIGURE 9.30 Number of earthquakes as a function of seismic moment. This is a global data set for shallow events since 1977. (Courtesy of Göran Ekström.)

plies that no large earthquakes accompany the occurrence of small-magnitude events. Swarms most commonly occur in volcanic regions, and the generally accepted explanation is that faults simply are not large or continuous in this environment and that stress is substantially heterogeneous. Thus, the maximum moment expected for a given earthquake is small, and many smaller events must accommodate the strain accumulation. (Recall that the energy associated with a magnitude of 5.0 is 1000 times smaller than that for a magnitude of 7.0, so we mean *many* smaller events!)

Equation (9.55) is often used in seismic hazard analysis to determine the *maximum credible earthquake* during a time window. If we assume that A and b values determined for a given period of time are self-similar in both size and frequency of occurrence, the relation can be extrapolated to larger time windows. The recurrence time of an earthquake of magnitude M is proportional to N^{-1}. For example, if we monitored a region for one year and found that $b = 1$ and $A = 3.5$, we would expect a magnitude 4.5 earthquake in the next 10 years and a magnitude 5.5 earthquake in the next century. Obviously, this type of analysis is loaded with assumptions; a maximum magnitude will eventually be reached, and a 1-yr window may hardly represent the earthquake frequency of occurrence for all magnitudes.

Box 9.4 What is the Stress on a Fault?

In our discussions of fault dynamics, we found that most processes scale with stress drop, which averages less than 100 bars. We have not talked about the absolute level of stress of a fault, which is one of the fundamental controversies in earthquake mechanics. Two basic data types provide conflicting expectations for fault stress: (1) laboratory measurements of rock friction and measurements of static strength suggest that faults are inherently strong, while (2) dynamic measurements of seismic energy and frictional heat production suggest faults are inherently weak. The dynamics arguments are based on the energy budget of the faulting event; the work done in sliding the sides of the fault past each other must equal the energy released by generation of seismic waves or converted into heat by overcoming frictional resistance to fault motion. The energy per unit fault area can be written as

$$\frac{E}{A} = \frac{E_s}{A} + \frac{E_h}{A} = \bar{\sigma}\eta D + \sigma_r D,$$ (9.4.1)

where $\bar{\sigma}$ is the average of the stress on the fault before and after the earthquake, σ_r is the frictional resistance stress, η is the seismic efficiency, and D is the final displacement on the fault. Seismic efficiency is defined as the fraction of the total energy release that is partitioned into seismic radiation:

$$\eta = \frac{E_s}{E} = \frac{\sigma_s}{\bar{\sigma}},$$ (9.4.2)

continues

where σ_s is the stress that caused the seismic radiation (presumably stress drop). σ_s is usually found by

$$\sigma_s = \mu \frac{E_s}{M_0}, \tag{9.4.3}$$

where $\log E_s = 11.8 + 1.5 M_w$. For any reasonable value of μ, σ_s is less than 5 MPa. This energy source is tiny and implies that $\bar{\sigma}$ must be small unless the seismic efficiency is less than 0.1. σ_r can be determined by measurements of the heat flow around active faults. The San Andreas fault in California has been extensively sampled for heat flow, and no significant heat-flow signature is associated with the fault, suggesting that the frictional resistance is less than 10 MPa.

Byerlee's law [Eq. (9.3)] predicts that most faults will have internal coefficients of friction on the order of 0.6. The *normal* stress on the fault at a depth of 8–10 km (the middle of the fault plane) will be related to the lithostatic load, $\rho g h$. If the crust behaves hydrostatically, the normal stress will be equal to the lithostatic stress; if the crust behaves like a *Poisson solid*, the normal stress will be ~ 0.3 times the lithostatic load. In either case, the average shear stress on the fault should exceed 50 MPa. This value is five times that allowed by the heat-flow measurements. Herein lies the controversy. We can reconcile these observations in several possible ways: (1) significant earthquake energy may go into neglected processes such as chemical or phase changes along the fault surface, (2) the heat-flow anomaly is "washed" away with ground-water transport, or (3) stick–slip sliding mechanisms need to be modified for nonlinear phenomena such as *separation phases* that reduce the normal stress on the fault, thereby lowering the frictional heat generation. At this time there is little consensus in the geophysical community on the state of stress on faults.

References

Aki, K. (1968). Seismic displacements near a fault, *J. Geophys. Res.* **73**, 5359–5375.

Brace, W. F., and Byerlee, J. D. (1966). Stick slip as a mechanism for earthquakes. *Science* **153**, 990–992.

Brune, J. N. (1970). Tectonic stress and the spectra of seismic shear waves from earthquakes. *J. Geophys. Res.* **75**, 4997–5009.

Byerlee, J. D. (1978). Friction of rock. *Pure Appl. Geophys.* **116**, 615–626.

Cagnetti, V., and Pasquale, V. (1979). The earthquake sequence in Friuli, Italy, 1976. *Bull. Seismol. Soc. Am.* **69**, 1797–1818.

Cipar, J. (1979). Source process of the Haicheng, China earthquake from observations of *P* and *S* waves. *Bull. Seismol. Soc. Am.* **69**, 1903–1916.

Duda, S .J. (1989). Earthquakes: Magnitude, energy, and intensity. *In* "Encyclopedia of Solid Earth Geophysics" (D. James, ed.), pp. 272–288. Van Nostrand–Reinhold, New York.

Geller, R. J. (1976). Scaling relations for earthquake source parameters and magnitudes. *Bull. Seismol. Soc. Am.* **66**, 1501–1523.

Hartzell, S. (1989). Comparison of seismic waveform inversion results for the rupture history of a finite fault: Application to the 1986 North Palm Springs, California, earthquake. *J. Geophys. Res.* **94**, 7515–7534.

Haskell, N. A. (1964). Total energy and energy spectra density of elastic waves from propagating faults. *Bull. Seismol. Soc. Am.* **54**, 1811–1841.

Heaton, T. (1990). Evidence for and implications of self-healing pulses of slip in earthquake rupture.

Phys. Earth Planet. Inter. **64**, 1–20.

Kanamori, H. (1977). The energy release in great earthquakes. *J. Geophys. Res.* **82**, 2981–2987.

Kanamori, H., and Anderson, D. L. (1975). Theoretical basis of some empirical relations in seismology. *Bull. Seismol. Soc. Am.* **65**, 1073–1095.

Kasahara, K. (1981). "Earthquake Mechanics." Cambridge Univ. Press, Cambridge, UK.

Kostrov, B. V. (1966). Unsteady propagation of longitudinal shear cracks. *J. Appl. Mech.* **30**, 1241–1248.

Kostrov, B. (1974). Seismic moment and energy of earthquakes, and seismic flow of rock. *Izv., Acad. Sci., USSR, Phys. Solid Earth (Engl. Transl.)* **1**, 23–40.

Richter, C. F. (1958). "Elementary Seismology." Freeman, San Francisco.

Scholz, C. H., Aviles, C. A., and Wesnousky, S. G. (1986). Scaling differences between large interplate and intraplate earthquakes. *Bull. Seismol. Soc. Am.* **76**, 65–70.

Shimazaki, K. (1986). Small and large earthquakes: The effects of the thickness of the seimogenic layer and free surface. *Maurice Ewing Ser.* **6**, 209–216.

Stein, R. S., and Barrientos, S. E. (1985). Planar high-angle faulting in the Basin and Range: Geodetic analysis of the 1983 Borah Peak, Idaho, earthquake. *J. Geophys. Res.* **90**, 11,355–11,366.

Utsu, T., and Seki, A. (1954). A relation between the area of aftershock region and the energy of main shock. *J. Seismol. Soc. Jpn.* **7**, 233–240.

Veith, K. F., and Clawson, G. E. (1972). Magnitude from short-period *P* wave data. *Bull. Seismol. Soc. Am.* **62**, 435–452.

Yamashita, T. (1976). On the dynamical process of fault motion in the presence of friction and inhomogeneous initial stress. Part I. Rupture propagation. *J. Phys. Earth* **24**, 417–444.

Additional Reading

Aki, K. (1972). Scaling laws of earthquake source time function. *Geophys. J. R. Astron. Soc.* **34**, 3–25.

Aki, K. (1984). Asperities, barriers and characteristics of earthquakes. *J. Geophys. Res.* **89**, 5867–5872.

Anderson, E. M. (1951). "The Dynamics of Faulting." W. S. Cowell Ltd., Edinburgh.

Brune, J. N. (1970). Tectonic stress and the spectra of seismic shear waves from earthquakes. *J. Geophys. Res.* **75**, 4997–5009.

Byerlee, J. D., and Brace, W. F. (1968). Stick–slip, stable sliding, and earthquakes—effect of rock type, pressure, strain rate and stiffness. *J. Geophys. Res.* **73**, 6031–6037.

Kanamori, H., and Anderson, D. (1975). Theoretical basis of some empirical relations in seismology. *Bull. Seismol. Soc. Am.* **65**, 1073–1095.

Kasahara, K. (1981). "Earthquake Mechanics." Cambridge Univ. Press, Cambridge, UK.

Kostrov, B., and Das, S. (1988). "Principles of Earthquake Source Mechanics." Cambridge Univ. Press, Cambridge, UK.

Lachenbruch, A., and Sass, J. (1980). Heat flow and energetics of the San Andreas fault zone. *J. Geophys. Res.* **85**, 6185–6222.

Mendoza, C., and Hartzell, S. H. (1988). Aftershock patterns and main shock faulting. *Bull. Seismol. Soc. Am.* **78**, 1438–1449.

Scholz, C. H. (1990). "The Mechanics of Earthquakes and Faulting." Cambridge Univ. Press, Cambridge, UK.

Sibson, R. H. (1982). Fault zone models, heat flow, and the depth distribution of earthquakes in the continental crust of the United States. *Bull. Seismol. Soc. Am.* **72**, 151–163.

10

SEISMIC WAVEFORM MODELING

Early in this text we stated that one of the goals of a seismologist is to understand every wiggle on the seismogram. The preceding chapters have dealt with phenomena that influence the structure of seismogram: propagation effects, source effects, and characteristics of the seismometer itself. It is possible to model each of these effects mathematically and, therefore, to develop a procedure to *predict* the character of a seismogram in a realistic model of the Earth. Such a mathematical construction is known as a *synthetic seismogram*. The formalism of comparing synthetic and observed seismograms is known as *waveform modeling*. Waveform modeling has become one of the most powerful tools available to seismologists for refining Earth structure models and understanding fault rupture processes. In general, waveform modeling is an iterative process in which differences between the observed and synthetic seismograms are minimized by adjusting the Earth structure or source representation.

The underlying mathematical theory for constructing synthetic seismograms is called *linear filter theory*. The seismogram is treated as the output of a sequence of linear filters. Each filter accounts for some aspect of the seismic source or propaga-

tion. Figure 10.1 shows an example of a trapezoid-shaped *P*-wave ground displacement, along with recordings on short- and long-period seismometers. The trapezoid shape can be considered to be the output of filters that account for the effects of rupture on a finite fault plane as well as any propagation effects (Chapter 9). This ground motion is then distorted by the recording characteristics of the seismometer, a linear filter that is usually well known, and the output is a seismogram.

It is possible to characterize the elements of a linear filter system by considering the response of the filter to an impulse, or delta, function. In a physical sense, this corresponds to a single, instantaneous pulse of energy at the source for which the complex resulting seismogram determines the propagation filter. If the impulse response of a particular filter is $f(t)$, its corresponding Fourier transform is $F(\omega)$. If $f(t)$ is known, the response, $y(t)$, of an arbitrary input, $x(t)$, can be calculated with the *convolution* operator (see Box 9.2). If $X(\omega)$ is the Fourier transform of $x(t)$, then the transform of the output signal, $Y(\omega)$, is given by

$$Y(\omega) = F(\omega)X(\omega). \qquad (10.1)$$

If a signal goes through a succession of

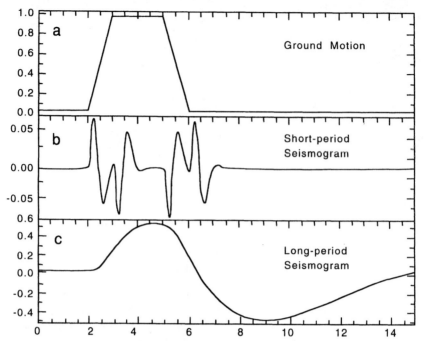

FIGURE 10.1 (Top) A trapezoid time function, corresponding to a hypothetical ground motion. (Middle) A seismogram produced by the trapezoid motion convolved with a short-period instrument response. (Bottom) A seismogram produced by the trapezoid convolved with a long-period instrument response.

filters, $f_1, f_2, \ldots f_n(t)$, the Fourier transform of the output signal is given by

$$Y(\omega) = F_1(\omega) F_2(\omega) \cdots F_n(\omega) X(\omega). \tag{10.2}$$

In other words, the output signal is given by the multiple product of the spectra of each filter and the input signal.

In seismic waveform modeling, there are three basic filters:

$$u(t) = s(t) * g(t) * i(t), \tag{10.3}$$

where $u(t)$ is the seismogram, $s(t)$ is the signal from the seismic source, $g(t)$ is the propagation filter, and $i(t)$ is the seismometer response. In actuality, $s(t)$ and $g(t)$ can be divided into several filters to account for specific effects. For example, $g(t)$ can be divided into a filter that accounts for the multiplicity of arrivals due

to reflections and refractions at material boundaries within the Earth along with a filter that accounts for the seismic-wave attenuation. Similarly, $s(t)$ can be divided into filters accounting for source radiation conditions and fault rupture characteristics.

Linear filter theory provides a very elegant methodology for waveform modeling. It is possible to isolate the effects of one specific process on the character of the seismogram. For example, the effects of $g(t)$ for teleseismic body waves are easily accounted for, so often only the character of $s(t)$ need be adjusted or timed such that a synthetic seismogram adequately predicts an observation. Most of what is known about seismic source processes has been learned by applying such a procedure. In this chapter we will explore waveform modeling and provide some examples.

10.1 Body Waveform Modeling: The Finite Fault

We can readily construct the filters on the right-hand side of Equation (10.3) for a simple *point source*. From Figure 9.3 we know that the far-field source time history of a single particle on a fault is approximately boxcar shaped. The length of the boxcar is τ_r (the rise time), and the height of the boxcar is M_0/τ_r, where M_0 is the seismic moment. We call a single-particle fault a *point source*; the body waves from a point-source dislocation would be a simple boxcar pulse if no other filters were in operation. A more realistic source would include temporal and spatial *fault finiteness*, and the source-time function is more clearly approximated by a trapezoid (see Chapter 9). The source rise time and source finiteness can be thought of as two filters, with the output being the source-time function. Figure 10.2 shows a graphical representation of the various filters that

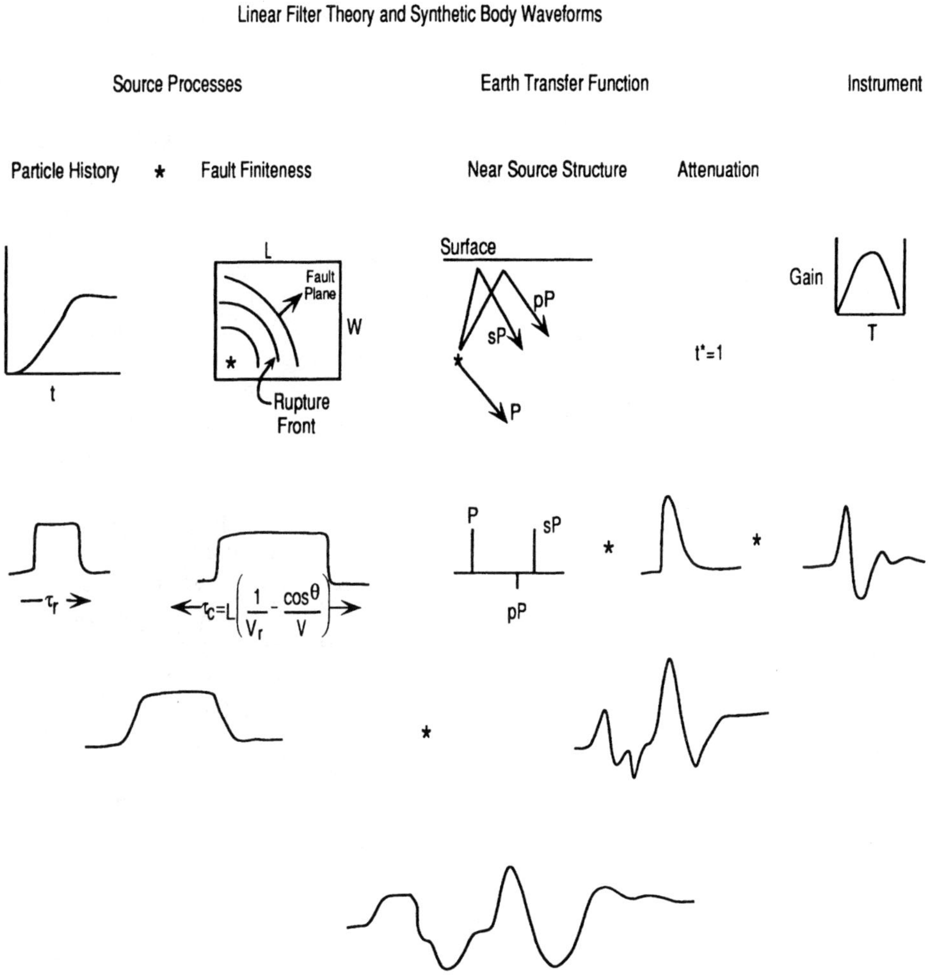

FIGURE 10.2 Schematic representation of various processes and their equivalent filter representations, which combine to give the total seismogram seen at the bottom.

produce a teleseismic body-wave seismogram, the first two of which produce the source time function.

The most complex filter in Eq. (10.3) is $g(t)$, sometimes called the *Earth transfer function*. This filter accounts for all propagation effects such as reflections, triplications, diffractions, scattering, attenuation, mode conversions, as well as geometric spreading. The usual procedure is to divide $g(t)$ into a filter that accounts for elastic phenomena, $R(t)$, and a filter that accounts for attenuation, $A(t)$. At teleseismic distances, $R(t)$ is a time series with a sequence of impulses temporally distributed to account for the variability in arrival times. At teleseismic distances, the most important P-wave arrivals are P, pP, and sP, so $R(t)$ is a "spike train" with three pulses spaced to account for the differences in arrival times. The amplitude of a given spike depends on the angle of incidence at the surface and the seismic radiation pattern. In Chapters 3 and 4, mathematical expressions were developed to calculate the amplitudes of impulse P waves. In Chapter 8, we developed the equations for a far-field P wave:

$$u_P(r,t) = \frac{1}{4\pi\rho r\alpha^3} R^P \dot{M}\left(t - \frac{r}{\alpha}\right), \quad (10.4)$$

where R^P gives the radiation pattern in terms of fault geometry and takeoff angle. We can rewrite (10.4) using the fact that any double couple can be represented as a weighted sum of three elementary faults (Section 8.5) to give

$$u_p(r,t) = \frac{1}{4\pi\rho r\alpha^3} \sum_{i=1}^{3} A_i(\phi,\lambda,\delta)c_i$$

$$* \dot{M}\left(t - \frac{r}{\alpha}\right), \quad (10.5)$$

where A_i is called the *horizontal radiation pattern* and c_i is called the *vertical radia-*

tion pattern, which are given by

$$A_1 = \sin 2\phi \cos \lambda \sin \delta$$

$$+ \tfrac{1}{2} \cos 2\phi \sin \lambda \sin 2\delta$$

$$A_2 = \cos \phi \cos \lambda \cos \delta - \sin \phi \sin \lambda \cos 2\delta$$

$$A_3 = \tfrac{1}{2} \sin \lambda \sin 2\delta, \quad (10.6)$$

where $\phi = \phi_s - \phi_f$, and

$$c_1 = -p^2$$

$$c_2 = 2\varepsilon p\eta_\alpha$$

$$\varepsilon = \begin{cases} +1 & \text{if ray is upgoing} \\ -1 & \text{if ray is downgoing.} \end{cases}$$

$$c_3 = p^2 - 2\eta_\alpha^2 \quad (10.7)$$

The three fundamental faults are (1) a vertical strike–slip fault, (2) a vertical dip–slip fault, and (3) a 45° dipping thrust fault ($\lambda = 90°$) evaluated at an azimuth of 45°. (By plugging in the appropriate strike, dip, and rake, you can see that A_2 and A_3 vanish for the first fundamental fault, A_1 and A_3 vanish for the second fundamental fault, and so on.) Equation (10.5) is extremely useful because it isolates $R(t)$ and provides a simple methodology for its calculation given an arbitrary fault orientation. If we calculate the spike train for each of the three fundamental faults, we just require a linear sum to account for the effects of any fault. Equation (10.5), as written, is only accurate for a half-space. If the P wave interacts with structure, it will undergo reflection and transmission, which depend on the angle of incidence. The c_i coefficient contains all the information about the angle, so we can rewrite (10.5) as

$$u_P(r,t) = \left(\frac{1}{4\pi\rho r\alpha^3} \sum_{k=1}^{N} \sum_{i=1}^{3} A_i c_i R_{MO_k} \Pi_k\right)$$

$$* \dot{M}\left(t - \frac{r}{\alpha}\right), \quad (10.8)$$

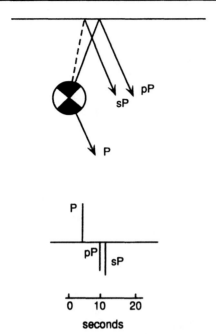

FIGURE 10.3 Primary raypaths corresponding to direct P and surface reflections pP and sP that arrive at a teleseismic station. For a shallow source these arrivals are close together in time, and together they comprise the P "arrival." The relative amplitudes of the arrivals are influenced by the source radiation pattern and the free-surface reflection coefficients. Small-amplitude differences due to extra attenuation or geometric spreading for the upgoing phases also can be accounted for if the source is deep.

where N is the number of arrivals, or rays, represented by the Earth filter; $R_{M_{O_k}}$ is the receiver function, with M being the mode type (P or S wave) of the kth ray; and O is the recording component (radial or vertical). Finally, Π_k is the product of all the transmission and reflection coefficients that the kth ray experiences on its journey from the source to receiver. The parenthetic term on the right side of Eq. (10.8) is just the $R(t)$ we need to calculate the Earth transfer function.

Although (10.8) looks complicated, it is actually straightforward to determine $R(t)$ at teleseismic distances. Figure 10.3 shows $R(t)$ for a dip–slip fault in a half-space.

The amplitudes of the depth phases are affected by both the surface reflection coefficient and the radiation pattern from the source. In the example, P and pP both leave the source with a compressional motion. Upon reflection at the free surface, pP is inverted. The combined effects of the SV radiation pattern and free-surface reflection also invert the polarity of the sP arrival relative to P. The relative arrival times of the various phases depend on the depth of the earthquake and the distance between the source and receiver (which controls the ray parameter or take-off angle). The surface-reflection delay times are given by

$$\Delta t \approx d\eta_U + d\eta_D, \qquad (10.9)$$

where η_U and η_D are the vertical slowness of the upward and downward paths of a given depth phase and d is the hypocentral depth.

The relative amplitudes of the spikes in $R(t)$ vary greatly depending on source orientation. This variability produces waveforms that are diagnostic for different fault orientations. Waveform modeling is much more powerful for constraining fault orientation than first-motion focal mechanisms because it provides more complete coverage of the focal sphere and uses relative-amplitude information. A realistic $R(t)$ actually contains many more than just three wave arrivals. For a layered Earth structure, multiple reflections and conversions occur both near the source and beneath the receiver. In general, these multiples are much less important than the primary three rays at teleseismic distances unless the earthquake occurred beneath the ocean floor. In this case *water reverberations*, rays bouncing between the surface and ocean floor, can produce significant additional spikes.

The attenuation filter, $A(t)$, is usually represented by a t^* operator (see Chapter 3). At teleseismic distances t^* is nearly constant over much of the body-wave frequency band and is thus easy to parameterize as a filter. Figure 3.38 shows an impulse convolved with short- and long-period instruments for several values of t^*. As t^* increases, the high frequencies are more effectively removed. Note that the *amplitude* of the short-period signal is affected by changes in t^* to a much greater degree than the long-period signal.

Figure 10.4 shows a suite of P synthetic waveforms for the relevant fundamental faults using all of the filter elements we have discussed. The corresponding Earth transfer function, which includes the radiation pattern, is given in the left-hand column, and three different time functions are used (all the sources have the same seismic moment, so the shortest-duration source has the "highest" stress drop). P and SH waveforms for different fault orientations differ enough to be diagnostic of the source type, although there are trade-offs between the various filters. Of course, much additional information is contained in the azimuthal pattern of motions that would be observed for each fundamental

fault. The source depth, time function, fault orientation, and seismic moment are known as the *seismic source parameters*. The goal of waveform modeling is to recover the source parameters by "fitting" the observed waveforms with synthetics. The strongest trade-off is between source depth and source time function duration. Figure 10.5 demonstrates this trade-off. Basically, a deeper source with a shorter-duration source function may be similar to a shallower source with a longer source function. Broadband data can overcome much of this trade-off for simple sources. However, the convolutional nature of linear filter theory implies that direct trade-offs must exist. Differences in source depth exactly trade off with complex source functions for a single station, although using multiple stations can again reduce, but not eliminate, the trade-offs.

From the mid-1970s through the early 1980s, many studies of earthquake source parameters were done using teleseismic waveform modeling, mainly of long-period WWSSN data. The methodology involved

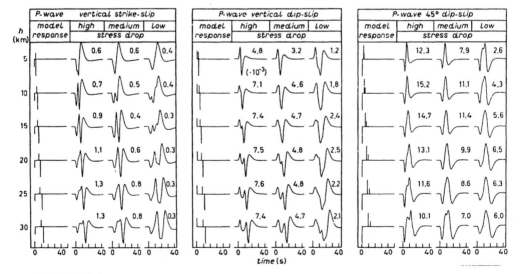

FIGURE 10.4 P-wave synthetic seismograms for the three potential terms with varying depth and time functions. The numbers in the upper right are actual potential amplitudes without the $M_0/4\pi\rho_0$, $1/R$ decay, and receiver functions included. The source time parameters, $d\tau$, are high stress drop (0.5, 1.0, 0.5), medium stress drop (1.0, 3.0, 1.0), and low stress drop (2.0, 6.0, 2.0). (From Langston and Helmberger, 1975. Reprinted with permission from the Royal Astronomical Society.)

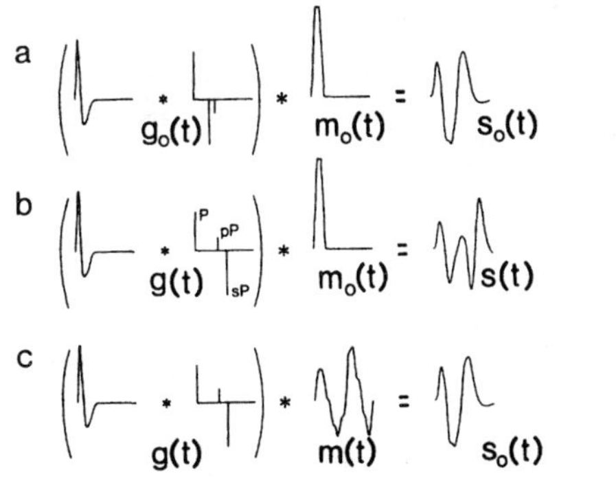

FIGURE 10.5 Illustration of the trade-off between source depth and source time function duration for teleseismic *P* waves. The synthetics have a long-period WWSSN response, convolved with a impulse response Green's function and a source time function. Note that identical waveforms can be produced for different combinations of Green's function and source time function (rows a and c). Both depth and mechanism were changed in this case, but simply changing depth can give the same result. The trade-offs can be overcome by using multiple stations, to some extent. (From Christensen and Ruff, 1985).

fitting long-period *P* and *SH* waves that were well distributed in azimuth about the source. The waveform information constrains the focal mechanism, depth, and source time function. A comparison of the predicted and observed *amplitudes* of the waveforms yields the seismic moment. In general, about a factor of 2 scatter is typically observed in moment estimates from station to station. This scatter reflects uncertainty in the filters, particularly $g(t)$. (Although some uncertainty was associated with the WWSSN instrument response, modern broadband digital data exhibit less amplitude scatter.) Once the time function is known, it is possible to infer the source dimensions if we assume a rupture velocity. Given an estimate of fault area, the average displacement $(\overline{D})$ on the fault and the stress drop can be calculated. Aftershock distribution or observed surface faulting is often used to estimate fault dimensions.

A fundamental concept underlying waveform modeling is separation of the source and propagation effects. For a double couple, (10.8) explicity achieves this. Now let us consider a full moment tensor source where all moment tensor terms have an identical source time history, $s(t)$. Using (8.84) we can rewrite Eq. (10.3) as

$$u_n(x,t) = s(t) * i(t) * \sum_{i=1}^{5} (m_i \cdot G_{in}(t))$$

(10.10)

$$m_1 = M_{11}, \qquad m_2 = M_{22}, \qquad m_3 = M_{12},$$

$$m_4 = M_{13}, \qquad m_5 = M_{23},$$

where u_n is the vertical, radial, or tangential displacement, and the Earth transfer function has been replaced by the summation operator. The summation is the product of the seismic moment tensor (here represented by m_i, the five elements left when assuming no isotropic component, i.e., $m_{33} = -(m_{11} + m_{22})$, and $G_{in}(t)$, the corresponding *Green's functions*. The moment tensor terms are simply constants to be determined. The Green's functions are impulse displacement responses for a seismic source with orientation given by each corresponding moment tensor element. Note that each moment tensor Green's function i will give *three* components (n) of displacement. Any arbitrary fault orientation can be represented by a specific linear combination of moment tensor elements (see Section 8.5), so the summation in Eq. (10.10) implies that any *Earth transfer function* can also be constructed as a linear combination of Green's functions. This is an extremely powerful representation of the seismic waveform because it requires the calculation of only *five* (or with some recombination of terms, *four*) fundamental Green's functions to produce a synthetic waveform for an arbitrary moment tensor at a given distance.

Equation (10.10) is the basis for inversion procedures to recover the seismic source parameters. It includes the purely double-couple representation in (10.8) as a special case. In the simplest case, let us assume that the source time function and source depth are known. Then $s(t)$ and $i(t)$ can be directly convolved with the Green's functions, yielding a system of linear equations:

$$u_n(x, t) = \sum_{i=1}^{5} m_i \cdot H_{in}(t), \quad (10.11)$$

where $H_{in}(t)$ are the new Green's functions (impulse response passed through an attenuation and instrument filter). We can write (10.11) in simple matrix form

$$\mathbf{u} = \mathbf{Gm}. \quad (10.12)$$

In order to match the observed seismogram in a least-squares sense, we can draw on the methods introduced in Chapter 6 to invert (10.12) for an estimate of $\mathbf{m}$

$$\mathbf{m} = \mathbf{G}^{-1}\mathbf{u}, \quad (10.13)$$

where G^{-1} is a generalized inverse.

This holds for each time step in the observed seismogram, $u_n(x, t)$. Basically, all one is doing is find the five weighting terms (moment tensor values) of functions that add up to give the seismogram. A single horizontal record that is not naturally rotated can be used to recover the full moment tensor, because each time sample helps to constrain $\mathbf{m}$. More stable estimates of the moment tensor are provided by inverting all three components at a single station. The most stable procedure is to simultaneously fit many seismograms from stations with distinctive Green's functions. For a given time t with multiple stations (10.12) can be written in vector form as

$$\begin{bmatrix} u_1 \\ u_2 \\ \vdots \\ u_k \end{bmatrix} = \begin{bmatrix} G_{11} & G_{12} & \cdots & G_{15} \\ G_{21} & G_{22} & \cdots & G_{25} \\ \vdots & \vdots & & \vdots \\ G_{k1} & G_{k2} & \cdots & G_{k5} \end{bmatrix} \begin{bmatrix} m_1 \\ m_2 \\ \vdots \\ m_5 \end{bmatrix}$$
$$k \times 1 \qquad\qquad k \times 5 \qquad\qquad 5 \times 1,$$

$$(10.14)$$

where k is the number of waveforms of interest; when $k > 5$, the system is overdetermined, and it should be possible to resolve the moment tensor. In practice, the system must be *very* overdetermined to resolve $\mathbf{m}$, which is easily achieved using multiple time samples.

Of course, we usually do not know the source time function or source depth *a priori*, so we can recast the problem as an *iterative* inversion. In this case we discretize the source time function and invert for the time series. The two most common parameterizations of the time function are a series of boxcars, or overlapping triangles (Figure 10.6). Consider the case in which the boxcar parameterization is chosen. Then we can write $s(t)$ as

$$s(t) = \sum_{j=1}^{M} B_j b(t - \tau_j), \quad (10.15)$$

where $b(t - \tau_j)$ is equal to a boxcar of width $\Delta\tau$ that begins at time τ_j and ends at $\tau_j + \Delta\tau$, B_j is the height of the boxcar, and $M\Delta\tau$ is the total length of the time function. Equation (10.15) can be used to rewrite (10.10) as

$$u_n = i(t) * \sum_{j=1}^{M} \sum_{i=1}^{5} B_j m_i \big[b(t - \tau_j) * G_{in}(t) \big].$$

$$(10.16)$$

Now this equation has two sets of unknowns: the heights of the boxcars, B_j, and the elements of the moment tensor, m_i. Since Eq. (10.16) is a nonlinear function of the unknowns, an iterative, linearized least-squares inverse can be used. We assume an initial model, construct synthetics for it, and then match the data in a least-squares sense by minimizing the difference, $\text{obs}(t) - \text{syn}(t) = \Delta d(t)$. We then solve

$$\Delta\mathbf{d} = \mathbf{A}\,\Delta\mathbf{P}, \quad (10.17)$$

where $\mathbf{A}$ is a matrix of partial derivatives ($A_{ij} = \partial u_i / \partial P_j$) of the synthetic waveform

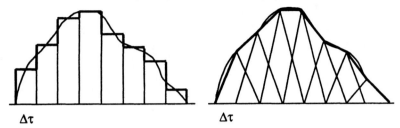

FIGURE 10.6 Two alternative parameterizations of an arbitrarily shaped source function.

(u_i) with respect to a given parameter P_j. ΔP is the model vector to be solved for, which contains the *changes* in the parameters, P_j, required to diminish the difference between the observed and synthetic seismograms. This type of linearization is valid only for small ΔP; thus it requires a good starting model, and a criterion is added to the inversion to minimize ΔP.

Equation (10.17) can be solved with the generalized inverse techniques described in Section 6.4. In general, simultaneous

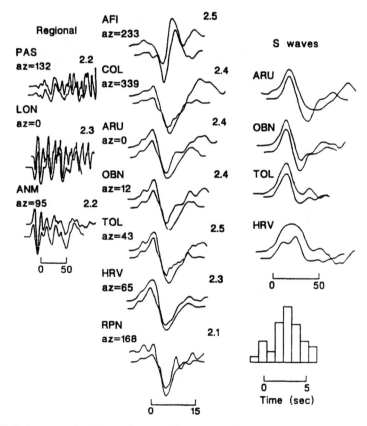

FIGURE 10.7 An example of waveform modeling for the 1989 Loma Prieta earthquake. Ground displacements are for the P_{nl} and teleseismic P and SH waves. Top trace of each seismogram pair is the observed, and bottom trace is the synthetic. The time function used is shown at the lower right. The focal mechanism determined from this inversion is $\phi_f = 128°$ $\pm 3°$, $\delta = 66° \pm 4°$, $\lambda = 133° \pm 7°$, and the moment is 2.4×10^{19} N m. (From Wallace *et al.*, 1991.)

inversion for the moment tensor constants and time function elements results in some nonlinear parameter trade-offs that can cause some singular values to be very small, but exploring the solution space by inverting with many different starting models usually yields a robust solution. The moment tensors from waveform inversion are hardly ever "perfect" double couples. The moment tensor is usually diagonalized and decomposed into a major and minor double couple or into a major double couple and a CLVD (Section 8.5 discusses these decompositions in detail). The minor double couple is usually small and is ignored; it is usually assumed that the minor double couple is the result of noise or of mapping incomplete or inaccurate Green's functions into the source parameters. Figure 10.7 shows the results of a body-wave inversion for the 1989 Loma Prieta, California, earthquake. The source time function

was parameterized in terms of boxcars. Note that it does not look like the idealized trapezoid; we will discuss source time-function complexity in Section 10.3. The moment tensor from this inversion has only a small CLVD, suggesting that representing the source as a point source double couple, with an oblique thrust focal mechanism, adequately approximates the source for teleseismic body waves.

The power of waveform modeling for determining seismic source parameters by Eq. (10.10) depends on being able to calculate the Green's functions accurately. At teleseismic distances this is usually not a problem, since the rays P, pP, and sP have simple structural interactions and turn in the lower mantle where the seismic velocity structure is smooth. Although "ringing" can occur in a sedimentary basin, for the most part teleseismic Green's functions for isolated body-wave arrivals are

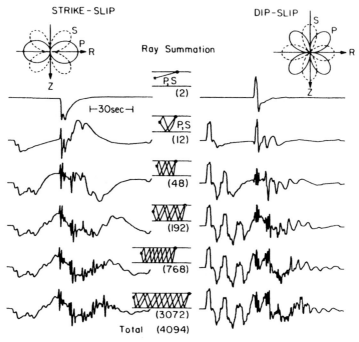

FIGURE 10.8 Vertical-component displacement seismograms for a station 1000 km from a shallow (8 km) source in a simple layer over a half-space model. No instrument response is included. (From Helmberger, 1983.)

Box 10.1 Slow Earthquakes

Although the source duration of most earthquakes scales directly with seismic moment (see Figure 9.16), there are some exceptions. In particular, *slow earthquakes* have unusually long source durations for the seismic moments associated with them. Slow earthquakes typically have an m_b that is small relative to M_s. Figure 10.B1.1 shows the effect of duration on short- and long-period body waves. The slow rise time presumably results from a very low stress drop (see Section 9.3), which controls the particle velocity. Variability in the source function occurs on all scales, from rapid events to slow creep events. Figure 10.B1.2 compares the seismic recordings of several aftershocks of the 1960 Chile earthquake. The upper two recordings are normal earthquakes, with typical fundamental mode excitation. The May 25 event has some greater complexity in the surface wave train, while the June 6 event is incredibly complex, with over an hour-long interval of surface wave excitation.

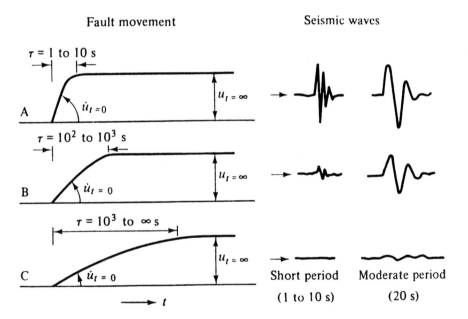

FIGURE 10.B1.1 The effect of different rise times on teleseismic signals. (From Kanamori, 1974.)

Kanamori (1972) noted that some subduction zone earthquakes produce extraordinarily large tsunamis but have moderate surface-wave amplitudes. In these cases M_S is small for the actual moment, and very slow *rupture* velocities apparently enhance the very low frequency spectrum. The physical mechanism responsible for such a slow rupture process is unknown, but in the extreme, it could produce a "silent" earthquake devoid of short-period body and surface waves. Recently, two investigators, G. Beroza and T. Jordan, suggested that several silent earthquakes

continues

occur each year that can be identified only because they produce free oscillations of the Earth. However, several sources, including large atmospheric storms and volcanic processes, can excite low-frequency oscillations, so the source of the free oscillations observed by Beroza and Jordan is somewhat uncertain, but likely to be associated with unusual earthquake dynamics.

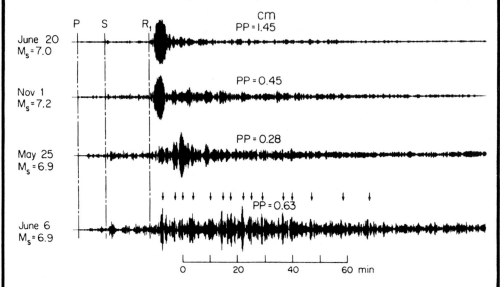

FIGURE 10.B1.2 Recordings of four aftershocks of the 1960 Chile earthquake. The upper two traces are conventional in appearance, with well-concentrated R_1 wavepackets. The lower two events have much more complex surface waves intervals, indicative of long, complex source radiation, extending over more than an hour for the June 6 event. (From Kanamori and Stewart, 1979.)

simple. This is not the case at upper-mantle and regional distances. At regional distances the crust acts as a waveguide, and hundreds of reflections between the surface and Moho are important for the waveform character. Figure 10.8 shows the vertical-component seismograms calculated for a simple layer over a half-space model for a station 1000 km from a shallow (8 km) source. Note that more than 200 rays are required before the waveform shape becomes stable. The suite of crustal reverberations following the P_n head wave comprise the P_{nl} phase. However, despite the obvious complexity in the Green's functions, the waveforms are very diagnostic of source orientation. The signature of

the seismic source on the P_{nl} waveform is robust as long as the *gross* parameters of the crustal waveguide (crustal thickness, average crustal seismic velocities, and upper-mantle P_n velocity) are well approximated. Figure 10.9 shows an example of inversion for source fault orientation from regional P_{nl} waveforms.

Regional-distance analysis is extremely important in the study of small or moderate-sized earthquakes ($m_b \leq 5.5$), which are rarely well recorded at teleseismic distances. Advances in broadband instrumentation have made it possible to determine the seismic source parameters from a *single* seismic station. The transient motion for a given double-couple orientation is

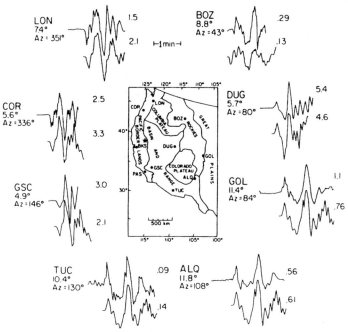

FIGURE 10.9 Comparison of regional long-period P_{nl} waveforms (upper traces) and synthetics (lower traces) obtained by waveform inversion for fault parameters of the 1966 Truckee, California, earthquake. (From Wallace and Helmberger, 1982.)

unique; thus if three components of motion are recorded, the source parameter can be determined provided that sufficiently accurate Green's functions are available (see Box 10.2).

At upper-mantle distances, triplications from the 400- and 670-km discontinuities make the body-wave Green's functions complex. Further, the mantle above the 400-km discontinuity has tremendous regional variability (Chapter 7). In general, beyond 14°, the first-arriving P wave has turned in the upper mantle, and the 400-km triplication occurs between 14° and 20°. The triplication from the 670-km discontinuity usually occurs between 16° and 23°. Figure 10.10 shows Green's functions for an upper-mantle model constructed for the western United States. The complexity and regional variability of upper-mantle-distance seismograms diminish their utility in seismic source parameter studies. Only when an earthquake occurs where the up-

per-mantle structure is very well known are the records of use for source analysis. Figure 10.11 compares observed and synthetic waveforms for the 1975 Oroville earthquake for distances from 5° to 75°, showing how well a single source model can match waveforms at regional, upper-mantle, and teleseismic distances when the structure is well known.

This text is filled with other examples of waveform modeling that have been used to illustrate various aspects of seismology. For example, Figure 9.11a shows a waveform study of the 1975 Haicheng, China, earthquake. This earthquake is well known because it was predicted by the Chinese State Seismological Bureau and the epicentral population center was evacuated, potentially saving thousands of lives. Body waves for this event show clear directivity, adding complexity to the waveforms. The pulse widths at stations to the west are much narrower than those at stations to the east,

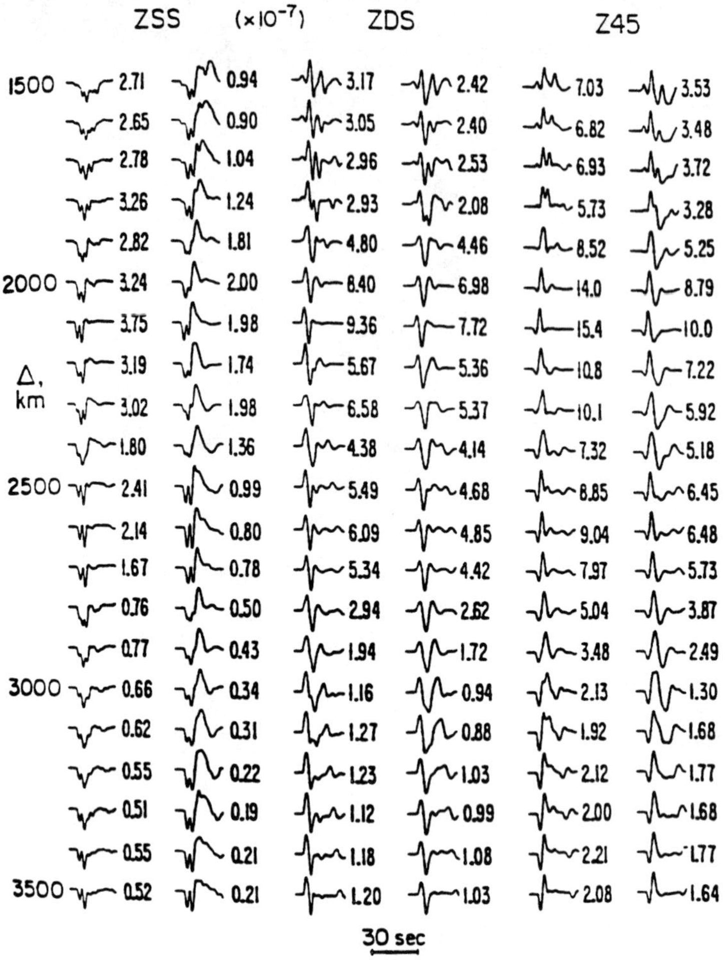

FIGURE 10.10 Upper-mantle synthetics without and with long-period WWSSN instrument for the three fundamental orientations (ZSS =vertical strike–slip; ZDS =vertical dip–slip; Z45 =45° dip–slip at an azimuth of 45°) assuming a source depth of 8 km, $t = 1$, and $\delta t_1 = \delta t_2 = \delta t_3 = 1$ for the source time history. (Modified from Helmberger, 1983.)

Box 10.2 Source Parameters from a Single Station

In Chapter 8 we showed that slip on a fault could be represented by an equivalent double-couple force system. It turns out that the displacement field from a given double couple is *unique*, which means that if we can model the entire transient displacement field at a *single* point, we should be able to recover the source orientation. In other words, a source-parameter study should require only a

continues

complete waveform inversion at a single seismic station. In practice, uncertainties in the Green's functions and source time function, limited bandwidth of recording instruments, and noise make this nearly impossible. However, at local and near-regional distances the effects of structure are easily accounted for, and the new generation of very broadband (vbb), high-dynamic-range instruments, such as the IRIS stations, makes it possible to use very sparse networks to determine accurately the source parameters of small to moderate-sized earthquakes.

Figure 10.B2.1 shows the recording of an $M_L = 4.9$ earthquake 12 km beneath a broadband station in Pasadena, California (PAS). The earthquake was well recorded on the Southern California network, and a first-motion focal mechanism was determined (see the second panel). The radial and tangential waveforms indicate that the source time function is complicated; for the synthetics, two triangles are assumed. The first-motion focal mechanism very poorly predicts the *SH* waveform and the relative sizes of the radial and tangential waveforms. A minor adjustment to the focal mechanism dramatically improves the quality of the fit of the synthetic to the observation. The main difference between the observed and synthetic waveforms is a *near-field* effect, not accounted for in the synthetic. This example shows the potential power of waveform inversion for complete seismograms.

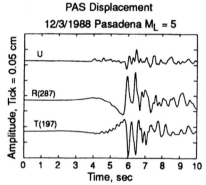

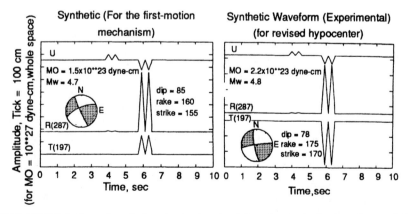

FIGURE 10.B2.1 Example of the determination of a focal mechanism by modeling the three-component data from a single station. (Modified from Kanamori *et al.*, 1990.)

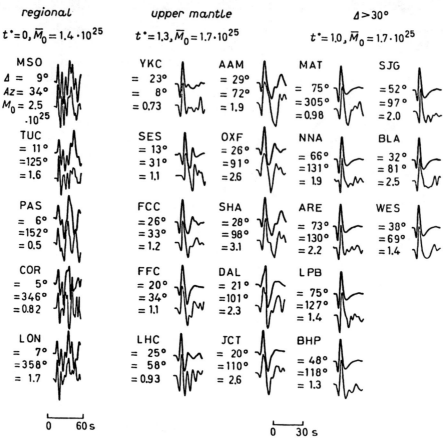

regional

$t^* = 0, \overline{M}_0 = 1.4 \cdot 10^{25}$

upper mantle

$t^* = 1.3, \overline{M}_0 = 1.7 \cdot 10^{25}$

$\Delta > 30°$

$t^* = 1.0, \overline{M}_0 = 1.7 \cdot 10^{25}$

MSO
$\Delta = \ 9°$
$Az = 34°$
$M_0 = 2.5$
$\cdot 10^{25}$

YKC
$= \ 23°$
$= \ 8°$
$= 0.73$

AAM
$= 29°$
$= 72°$
$= 1.9$

MAT
$= \ 75°$
$= 305°$
$= 0.98$

SJG
$= 52°$
$= 97°$
$= 2.0$

TUC
$= \ 11°$
$= 125°$
$= 1.6$

SES
$= \ 13°$
$= \ 31°$
$= 1.1$

OXF
$= \ 26°$
$= 91°$
$= 2.6$

NNA
$= \ 66°$
$= 131°$
$= 1.9$

BLA
$= \ 32°$
$= \ 81°$
$= 2.5$

PAS
$= \ 6°$
$= 152°$
$= 0.5$

FCC
$= 26°$
$= 33°$
$= 1.2$

SHA
$= 28°$
$= 98°$
$= 3.1$

ARE
$= \ 73°$
$= 130°$
$= 2.2$

WES
$= 38°$
$= 69°$
$= 1.4$

COR
$= \ 5°$
$= 346°$
$= 0.82$

FFC
$= \ 20°$
$= \ 34°$
$= 1.1$

DAL
$= \ 21°$
$= 101°$
$= 2.3$

LPB
$= \ 75°$
$= 127°$
$= 1.4$

LON
$= \ 7°$
$= 358°$
$= 1.7$

LHC
$= \ 25°$
$= \ 58°$
$= 0.93$

JCT
$= \ 20°$
$= 110°$
$= 2.6$

BHP
$= \ 48°$
$= 118°$
$= 1.3$

0   60 s

0   30 s

FIGURE 10.11 Comparison of synthetics with waveform data for the August 1, 1975 Oroville, California, event. The preferred model is $\phi_f = 215°$, $\lambda = -65°$, and $\delta = 48°$. Inversion results: with the 5 P_{nl} records exclusively, $\phi_f = 195°$, $\lambda = -72°$, and $\delta = 46°$; with 10 upper-mantle ranges exclusively, $\phi_f = 197°$, $\lambda = -63°$, and $\delta = 58°$; with 8 teleseismic waveforms exclusively, $\phi_f = 221°$, $\lambda = 82°$, and $\delta = 44°$. (After Yao *et al.*, 1982.)

suggesting that the fault ruptured westward along the nodal plane striking 288°. (This strike is consistent with the surface trace of the fault and the aftershock distribution.) Figure 9.11 shows the observed variability in the time function plotted as a function of azimuth, as well as a theoretical model for a fault propagating to the west for 22 km at a velocity of 3.2 km/s. The synthetic seismograms shown in Figure 9.11 were generated with directivity built into the time function.

The methodology described for inverting body waves for seismic source parameters can be applied as soon as a waveform

is "extracted" from a seismogram. Recently, the IRIS Data Management System has developed dial-up access to a significant part of the GSN (see Chapter 5). IRIS uses this remote access to implement a data-gathering system known as spyder™. When an earthquake occurs, it is located by the NEIC (National Earthquake Information Center), and an electronic message is broadcast to IRIS. The spyder system then calls GSN stations and downloads broadband seismic waveforms. These waveforms are then available via Internet to any interested seismologist. In practice, data from any earthquake greater than magnitude 6.5 are available within several hours. Thus it is possible to recover seismic source parameters within a matter of hours for large events anywhere in the world. Recent developments have made it possible to trigger spyder even more rapidly for regional networks, such as that in the western United States. It is now possible to determine focal mechanisms and seismic moments for western U.S. earthquakes with $M > 4.5$ within 1 h. This "near-real-time" analysis is used to identify the causal fault, to anticipate ensuing tsunami hazard, and to predict where strong shaking is likely to have occurred to assist in emergency response activities or shutdown of critical lifelines such as freeways and train tracks.

10.2 Surface-Wave Modeling for the Seismic Source

In Section 8.6 we discussed how fault orientation could be constrained from amplitude and phase of surface waves. It is possible to invert this information to determine the moment tensor from surface waves, but the *resolving* power for source depth and source time function is intrinsically limited. The amplitude and phase of a Rayleigh or Love wave is very dependent on the velocity structure along the travel path. This means that we must correct for the effects of velocity and attenuation heterogeneity precisely for an inversion scheme to be robust. This is equivalent to knowing the Earth transfer function in body-wave inversion procedures, but there we are not as sensitive to absolute travel time as we are for surface waves. This usually means that surface-wave inversions are best performed at very long periods (> 100 s) for which the heterogeneity is relatively well mapped. These periods are so long compared to most source durations that we can usually consider the far-field time function simply as a boxcar function

with duration τ. In this case we can write the source *spectrum* of an earthquake source as

$$V(\omega, h, \phi) = \alpha(\omega, h, \phi) + i\beta(\omega, h, \phi),$$

$$(10.18)$$

where ω is frequency, h is source depth, and ϕ is the takeoff azimuth. For Rayleigh waves the real (α) and imaginary (β) parts of the spectrum are

$$\alpha = -P_R(\omega, h) M_{12} \sin 2\phi$$

$$+ \tfrac{1}{2} P_R(\omega, h)(M_{22} - M_{11}) \cos 2\phi$$

$$- \tfrac{1}{2} S_R(\omega, h)(M_{22} + M_{11}) \quad (10.19a)$$

$$\beta = Q_R(\omega, h) M_{23} \sin \phi$$

$$+ Q_R(\omega, h) M_{13} \cos \phi \quad (10.19b)$$

and for Love waves

$$\alpha = -\tfrac{1}{2} P_L(\omega, h)(M_{22} - M_{11}) \sin 2\phi$$

$$- P_L(\omega, h) M_{12} \cos 2\phi \quad (10.20a)$$

$$\beta = -Q_L(\omega, h) M_{13} \sin \phi$$

$$+ Q_L(\omega, h) M_{23} \cos \phi. \quad (10.20b)$$

The P_R, S_R, Q_R, P_L, and Q_L terms are called *surface-wave excitation functions* (analogous to the body-wave Green's functions) and depend on the elastic properties of the source region and the source depth. Figure 10.12 shows P_L as a function of depth and period for different types of travel paths.

The spectrum, V, is calculated directly from the surface-wave seismogram if that seismogram has been corrected for instrument response and propagation effects. We can rewrite (10.18) as a matrix equation. For example, for Rayleigh waves

$$\mathbf{V} = \mathbf{BD}, \quad (10.21)$$

where

$$V = \begin{bmatrix} \alpha \\ \beta \end{bmatrix} \qquad (10.22)$$

$$B = \begin{bmatrix} -\sin 2\phi & \frac{1}{2}\cos 2\phi & -\frac{1}{2} & 0 & 0 \\ 0 & 0 & 0 & \sin\phi & \cos\phi \end{bmatrix}$$

$$(10.23)$$

$$D = \begin{bmatrix} P_R M_{12} \\ P_R(M_{22} - M_{11}) \\ S_R(M_{11} + M_{22}) \\ Q_R M_{23} \\ Q_R M_{13} \end{bmatrix}. \qquad (10.24)$$

Now **B** is a known matrix, depending only on source–receiver geometry; thus **D** contains all the unknowns. Equation (10.21) can be extended to the spectra observed at N stations. Then **B** is a $2N \times 5$ real matrix, and **V** is a real vector with dimension $2N$. This system of equations can be solved for $\mathbf{D}(\omega)$ at several frequencies. Typically the optimal choice of source duration τ is determined as that which minimizes the misfit in this inversion.

Once **D** has been determined, it is possible to decompose it into two vectors, one

containing the excitation functions and the other containing the elements of the moment tensor:

$$\Lambda = \left[D^T(\omega_1), D^T(\omega_2), \dots, D^T(\omega)_n \right]^T$$

$$\Lambda = \mathbf{EM}, \qquad (10.25)$$

where

$$E = [E_1, E_2, \dots, E_n]^T$$

$$E_i = \mathrm{diag}\big[P_R(\omega_i), P_R(\omega_i), S_R(\omega_i),$$

$$Q_R(\omega_i), Q_R(\omega_i) \big]$$

$$M = [M_{12}, M_{22} - M_{11}, M_{22}$$

$$+ M_{11}, M_{23}, M_{13}]^T. \qquad (10.26)$$

Equation (10.25) is a standard overdetermined problem that can be solved by least squares. For any real data, there will be some misfit to the spectrum, which can be measured as *error*. The excitation functions in E_i are, of course, dependent on depth, so the inversion must be repeated for several depths. A comparison of the

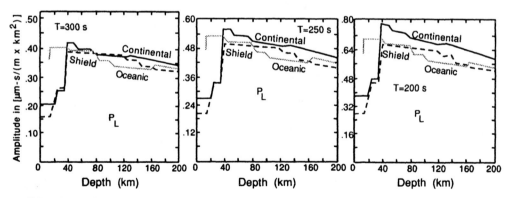

FIGURE 10.12 Dependence of the fundamental Love-mode displacement spectrum on source depth for a vertical strike–slip source. Excitation functions are shown for three different upper-mantle models, representative of shield, continental, and oceanic regions. Variations in the excitation coefficients as a function of period provide information about the source depth. (From Ben Menahem and Singh, 1981.)

errors for the different depths should result in a *minimum error*, which yields the source depth and thus the preferred moment tensor.

Let us return to the question of the source time function. We stated that the *details* of the time function do not affect the spectrum much. This is true to the extent that the source can be approximated as a point source with a boxcar source function. For large events the effective source duration will have an azimuthal pattern, as can be seen by considering the equation for source finiteness (9.20). Directivity effects are more apparent in surface waves than in body waves because their phase velocity is much slower. This source finiteness not only causes an azimuthal pattern in the phase but also reduces the amplitude of short-period waves; thus the spectrum for a large event must be corrected for source finiteness.

Figure 10.13 shows a series of moment tensor estimates from inversions of long-period surface-wave spectra (Figure 8.30) from the 1989 Loma Prieta earthquake. Several combinations of global attenuation models and source region excitation structures are considered. These inversions

---

Box 10.3 Centroid Moment Tensor Solutions

In 1981 the seismology research group at Harvard headed by Adam Dziewonski began routinely determining the seismic source parameters of all earthquakes with $M_s \geq 5.5$ using the centroid moment tensor (CMT) method. This inversion process simultaneously fits two signals: (1) the very long period ($T > 40$ s) body wave train from the *P*-wave arrival until the onset of the fundamental modes and (2) mantle waves ($T > 135$ s). These are fit for the best point-source hypocentral parameters (epicentral coordinates, depth and origin time) and the six independent moment tensor elements (not assuming a deviatoric source). The CMT solves an equation very similar to (10.10):

$$u_n(x,t) = \sum_{i=1}^{6} \psi_{in}(x, x_s, t) \cdot m_i, \qquad (10.2.1)$$

where ψ_{in} is called the *excitation kernel* and is essentially the complete seismogram Green's function for each of the moment tensor elements. The receiver is at x, and the source is at x_s (which is unknown). One initially estimates m_i, and then an iterative procedure begins that adjusts both the location and source orientation to minimize

$$u_n - u_n^0 = a_n \delta r_s + b_n \delta \theta_s + c_n \delta \phi_s + d_n \delta t_s + \sum_{i=1}^{6} \psi_{in}^0 \cdot \delta m_i, \qquad (10.2.2)$$

continues

where u_n^0 and ψ_{in}^0 are based on the initial estimate. δr_s, $\delta\theta_s$, and $\delta\phi_s$ are the changes in spatial coordinates of the hypocenter, and a_n, b_n, and c_n are the partial derivatives with respect to perturbations in the hypocentral coordinates. δt_s is the change in the origin time. The kernels are obtained by summing the normal modes of the Earth. Thus the excitations exist *a priori*, and the inversion process can be efficiently performed for many events. Figure 10.B3.1 shows the Harvard CMT catalogue for the month of July 1990. The moment tensors are not constrained to be double couples; hence many focal mechanisms are shaped more like baseballs (large CLVD components) than the expected sectioned beach balls (double couples). The largest earthquake during this month was the July 24, 1990 Philippines event (see also Figure 1.15 for more CMT solutions).

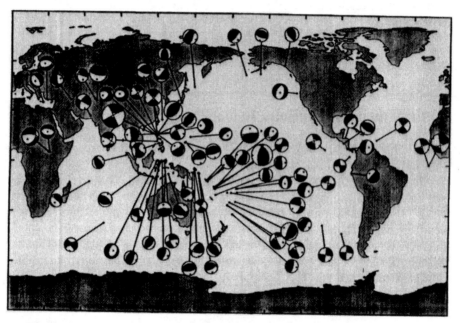

FIGURE 10.B3.1 Harvard CMT solutions for the month of July, 1990. (Based on Dziewonski *et al.*, 1991.)

provide insight into trade-offs associated with specifying source velocity and Q models. In all cases the major double couple is nearly identical to that determined from the body waveform inversion, but the minor double-couple component varies from 3% to 14% for different Earth models. This leads to a word of caution about comparing source parameters determined for different wave types. Various seismic waves are sensitive to different aspects of the rupture process, and it is *very* important to note that path corrections and the choice of attenuation will significantly affect source depths determined from surface-wave inversions. Surface waves can better constrain total seismic moment and total rupture duration than shorter-period waves can.

10.3 The Source Time Function and Fault Slip

Thus far in our discussion of faulting and radiated seismic energy, we have assumed that the rupture process is fairly smooth. This predicts a simple far-field time function approximated by a trapezoid, and slip is described by $\bar{D}$ (the average slip). In detail, the actual slip on a fault is not smoothly distributed, and source time functions deviate significantly from trapezoids. For example, consider the time function for the Loma Prieta earthquake in Figure 10.7. The irregularity of the time function is the result of temporally and spatially heterogeneous slip on the fault. Figure 10.14 shows the inferred variation in slip magnitude along the fault plane of the Loma Prieta earthquake. This slip function was derived by waveform modeling of both teleseismic P and SH waves and *strong motion* records from areas close to the fault. The slip is concentrated in two patches, with relatively small slip in the intervening regions. The regions of very high slip, known as *asperities*, are extremely important in earthquake hazard analysis because the failure of the asperities radiates most of the high-frequency seismic energy. The concentration of slip on asperities implies they are regions of high moment release, which, in turn, implies a fundamental difference in the fault behavior at the asperity compared with that of the surrounding fault. A conversion of slip to stress drop indicates that asperities are apparently regions of high strength (very large stress drop). The reason for the high relative strength could be heterogeneity in the frictional strength of the fault contact or variations in geometric orientation of the fault plane.

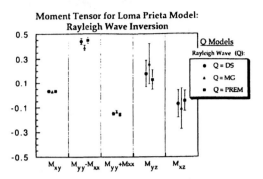

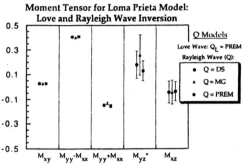

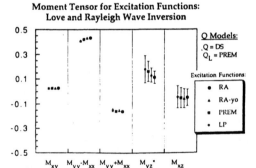

FIGURE 10.13 Moment tensor elements (Kanamori notation — see Box 8.3) for the Loma Prieta earthquake estimated from long-period Rayleigh- and Love-wave spectral inversions. Results are shown for several different attenuation models and for excitation functions from different Earth structures. (From Wallace *et al.*, 1991.)

The geometric explanation for asperities reflects the fact that faults are not perfectly planar. On all scales, faults are rough and contain jogs or steps. The orientation of the fault plane as a whole is driven by the regional stress pattern. Segments of

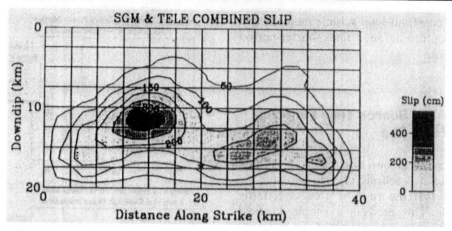

FIGURE 10.14 Slip distribution on the fault associated with the 1989 Loma Prieta earthquake (NW end on the fault on the left). There are two prominent regions of slip, known as asperities. (From Wald *et al.*, *1991*.)

Box 10.4 Tectonic Release from Underground Nuclear Explosions

    Theoretically, the seismic waves generated by an underground nuclear explosion should be very different from those generated by an earthquake. An explosive source creates an isotropic stress imbalance without the shear motion that characterizes double-couple sources. Therefore, the seismograms from an explosion should not have *SH* or Love waves, but as we saw in Figure 8.B1.1, many explosions do have *SH*-type energy. This energy is thought to be generated by a "tectonic" component, namely the release of preexisting strain by the detonation of an explosion. There are three possible mechanisms for generation of the nonisotropic seismic radiation, known as *tectonic release*: (1) triggering of slip on prestressed faults, (2) release of the tectonic strain energy stored in a volume surrounding the explosion, and (3) forced motion on joints and fractures. For all three of these mechanisms for tectonic release, the long-period teleseismic radiation pattern can be represented by an equivalent double-couple source. Depending on the orientation and size of the tectonic release, the seismic waveforms from underground explosions can be significantly modified from those we expect for an isotropic source (an explosion).

    Waveform modeling can be used to constrain the size and orientation of the tectonic release. For large explosions, it appears that tectonic release is associated with a volume of material surrounding the detonation point, and the volume is related to the size of the explosion. If an explosion is detonated within the "volume" of a previous explosion, the tectonic release is dramatically reduced. Figure 10.B4.1 shows two large underground nuclear explosions at the Nevada Test Site (NTS). BOXCAR (April 26, 1968, $m_b = 6.2$) was detonated 7 yr before COLBY (March 14, 1975, $m_b = 6.2$); the epicenters are separated by less than

continues

3 km. Although the *P* waveforms recorded at LUB are similar, there are some distinct differences. Below the BOXCAR waveform is a synthetic seismogram constructed by "adding" the waveform of a strike–slip earthquake to the waveform of COLBY. The near-perfect match between the observed and synthetic waveform for BOXCAR supports the double-couple interpretation for tectonic release.

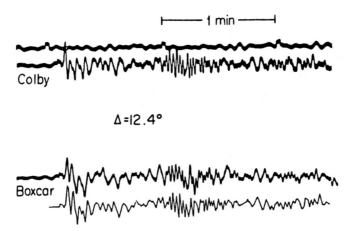

FIGURE 10.B4.1 A comparison of the *P* and P_L waveforms for BOXCAR and COLBY at LUB. Also shown is a synthetic waveform constructed by summing the COLBY waveform and a synthetic calculated for a strike–slip double couple (moment is 1.0×10^{17} N m). (From Wallace *et al.*, 1983.)

the fault that are subparallel to this orientation can have significantly higher normal stresses than surrounding regions, making them "sticking" points that resist steady, regular slip. Figure 10.15 shows a geometric irregularity that could serve as an asperity. The size and apparent strength of the asperity depend on d_s and θ_s (see Figure 10.15). At high frequencies, failure of discrete asperities may be manifested as distinct seismic arrivals. This implies that the details of source time functions may correspond to seismic radiation on particular segments of the fault. Figure 10.16 shows the source time function and inferred fault geometry for the 1978 Santa Barbara, California, earthquake ($m_b = 5.8$). The short-period *P* waves for this oblique thrust event are more complex than the long-period *P* waves. This results from the passband of the instrumentation,

which consists of WWSSN long- and short-period (1-s) seismometers, as illustrated in Figure 10.1. The long-period instrument cannot resolve the double peak apparent in the short-period signals, and the short-period records do not record the longer-period slip associated with the entire fault. The spatial distribution and orientation of the two asperities were determined from strong-ground-motion recordings. The new generation of very broadband seismometers has reduced the need for operating numerous instruments at a site to recover the details of faulting, and seismologists have begun to produce *unified source models*, which can be used to explain the entire faulting process from static offset to 10 Hz. These source models may include variation in the slip direction on the fault as well as variation in the slip magnitude. For the Loma Prieta event,

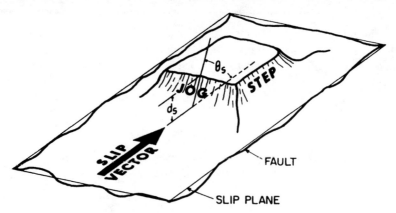

FIGURE 10.15 Geometric irregularity that could serve as an asperity. (From Scholz, 1990.)

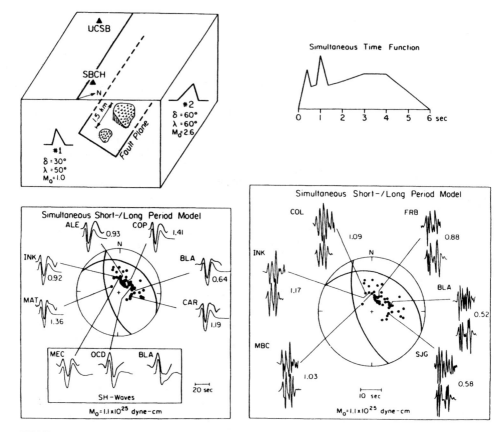

FIGURE 10.16 Source time function and inferred fault geometry for the 1978 Santa Barbara, California, earthquake (m_b =5.8). (From Wallace et al., 1981.)

Figure 1.7 shows a model of variable slip on the fault from both local and teleseismic signals.

In general, there are no near-field recordings for most earthquakes of interest, and we must infer any faulting heterogeneity from details of the far-field time function alone. As discussed in Section 10.1, the source time function is usually determined iteratively in generalized source-parameter inversion. Another approach is to recast Eq. (10.10) as a *deconvolution* problem

$$u_n(x,t) * (g(t) * i(t))^{-1} = s(t) \quad (10.27)$$

or

$$\frac{u(\omega)}{g(\omega)i(\omega)} = s(\omega). \quad (10.28)$$

This deconvolution procedure is a natural extension of linear filter theory. This is possible when the source orientation is known independently and we simply want the source time function. The major problem with this procedure is that it maps uncertainty in the Earth transfer function and source orientation into the time function. This is a problem for analysis of large earthquakes unless the Earth transfer function correctly includes the effects of fault finiteness. One way to allow for finiteness is to produce a suite of Earth transfer functions for a given geometry and write the time-domain displacement response as

$$u(x,t) = \sum_{j=1}^{M} B_j [b(t - \tau_j) * g_j], \quad (10.29)$$

where g_j is the Earth transfer function from the jth element of the fault that "turns on" at some time τ_j, which is prescribed by the rupture velocity; $b(t - \tau_j)$ is the parameterization of the time function as described in Eq. (10.15); and B_j is the variable of interest in the inversion, namely the strength of element $b(t - \tau_j)$ in the

source time function. A separate source time function is found for each element of the fault by solving for $B_j(t)$. Figure 10.17 shows an example of the forward problem for a fault that ruptures from 15 to 36 km depth. It is obvious that unless the variability in timing of the depth phases is accounted for, an inversion for the time function will be biased. Figure 10.18 shows examples of inversions for source time function based on Eq. (10.29).

The time functions in Figure 10.18 indicate very different fault behavior. The Solomon Islands earthquake had a much smoother source process than the Tokachi–Oki earthquake. The bursts of moment release during the Tokachi–Oki earthquake suggest that several asperities along the fault plane broke when the rupture front arrived. This type of time-function variability has been used to characterize segments of subduction zones. Figure 10.19 shows the source time functions from four great subduction zone earthquakes and a model for the distribution of asperities in different subduction zones. In the case of subduction zones, the variability of asperity size and distribution presumably reflects *coupling* between the subducting and overriding plates. The Aleutian subduction zone is strongly coupled along the segment that generated the 1964 Alaskan earthquake, and the Kuril region is characterized by weaker coupling and sporadic asperity distribution. The factors causing the variability in coupling are discussed in the next chapter, but the asperity model suggests that an earthquake in a strongly coupled region would be much larger than in a weakly coupled subduction zone. We will discuss coupling in much greater detail in Chapter 11.

Let us return to the heterogeneity of slip on the fault plane as shown in Figure 10.14. An important question is, What causes the rupture to stop? Along with the concept of asperities, the concept of *barriers* has been introduced for regions on the fault that have exceptional strength and

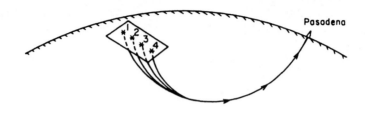

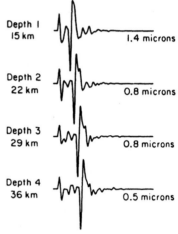

FIGURE 10.17 Earth transfer functions for a four-point source representation of a thrust-ing earthquake. The sum of the $g(t)$ convolved with time functions appropriate for each point source will give the synthetic seismogram. (From Hartzell and Heaton, 1985.)

impede or terminate rupture. Alternatively, barriers may be regions of low strength in which the rupture "dies out." This type of barrier is known as a *relaxation barrier*. The concepts of strength and relaxation barrier are generally consistent with the asperity model if adjacent segments of the fault are considered. A strength barrier that terminates rupture from an earthquake on one segment of the fault may serve as an asperity for a future earthquake. Similarly, the high-slip region of a fault during an earthquake may act as a relaxation barrier for subsequent earthquakes on adjacent segments of the fault. Aseismic creep may also produce relaxation barriers surrounding asperities that limit the rupture dimensions when the asperity fails. Unfortunately, there are also inconsistencies between the barrier and asperity models of fault behavior. In Figure 10.14 a region of moderate slip is located between the two asperities. Is this reduced slip caused by a region of previous failure, or is this a region of the fault that is primed for a future earthquake? It may be possible to resolve this question by studying the detailed spatial distribution of aftershocks. If the regions adjacent to the asperities have a concentration of aftershocks but the asperities themselves are aftershock-free, this would be inconsistent with strength barriers. There is some indication that aftershock distributions outline asperities, but there are still problems with spatial resolution that preclude strong conclusions. Aftershocks are clearly a process of relaxing stress concentrations introduced by the rupture of the mainshock, but there remains an active debate as to

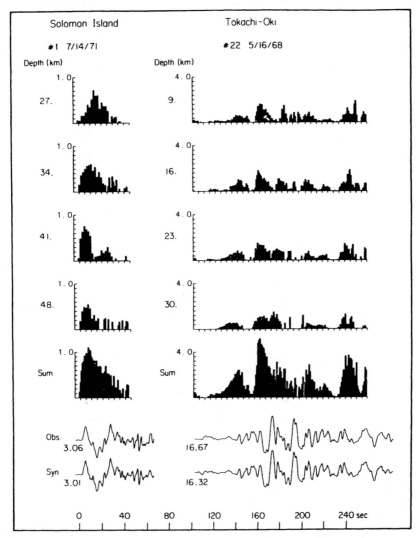

FIGURE 10.18 The source time function for two subduction zone earthquakes. Separate time functions are shown for each point source at four depths, with a sum being shown above the observed and synthetic seismograms. (From Hartzell and Heaton, 1985.)

their significance in terms of asperities and barriers. The only thing that is certain is that, averaged over long periods of time, the entire fault must slip equal amounts.

10.4 Complex Earthquakes

Fault roughness and the asperity model appear to apply to earthquakes at all scales.

When earthquakes reach a certain size, the faulting heterogeneity can be represented with the concept of *subevents*. In other words, for some large events the seismic source process can be thought of as a series of moderate-sized earthquakes. When source time functions become sufficiently complicated to suggest earthquake multiplicity, the event is known as a *complex earthquake*. Because all earthquakes

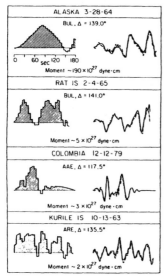

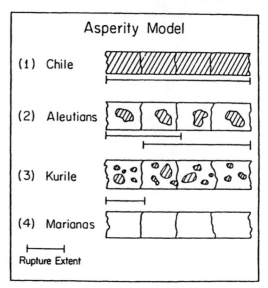

FIGURE 10.19 Source time functions from four great subduction zone earthquakes and a model for the distribution of asperities in different subduction zones. (Left is from Ruff and Kanamori, 1983; right is from Lay and Kanamori, 1981.)

Box 10.5 Modeling Tsunami Waveforms for Earthquake Source Parameters

In Chapter 4 we discussed the propagation of tsunamis, which were generated by rapid displacement of the ocean floor during the faulting process. Just as the seismic recording of a surface wave is a combination of source and propagation effects, the tidal gauge recordings of a tsunami are sensitive to the slip distribution on a fault and the ocean bathymetry along the travel path. It is possible to invert the waveform of a tsunami (ocean height as a function of time) for fault slip. The propagation effects are easily modeled because the tsunami velocity depends only on the water depth, which is usually well known. Figure 10.B5.1 shows the observed and synthetic tsunami waves from the 1968 Tokachi–Oki earthquake, which was located northeast of Honshu, Japan. Figure 10.B5.2 compares the fault slip derived from the inversion of the tidal gauge data and that determined by the analysis of surface waves. The general agreement between both models is good; slip is concentrated west and north of the epicenter (arrows on figures), while slip south of the epicenter was zero or very small.

The inversion of tsunami data is potentially very useful for pre-WWSSN data. Few high-quality seismic records exist to estimate the heterogeneous fault motion of these older events, but older tidal gauge records often exist that are as good as modern records.

continues

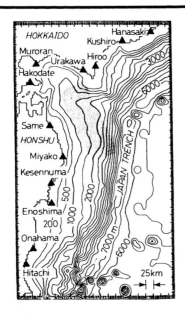

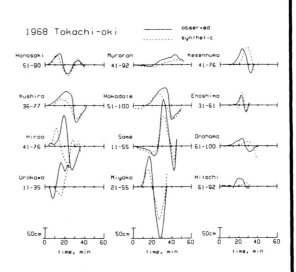

FIGURE 10.B5.1 Comparison between observed and predicted tsunamis for the 1968 Tokachi–Oki earthquake. The model fault is rectangular with heterogeneous slip. [From Satake, *J. Geophys. Res.* **94**, 5627–5636, 1989; © copyright by the American Geophysical Union.]

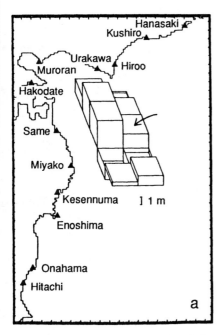

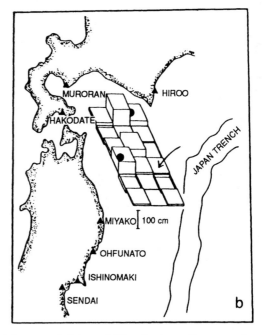

FIGURE 10.B5.2 Fault slip inferred from (a) tsunami data and (b) seismic surface waves. [From Satake, *J. Geophys. Res.* **94**, 5627–5636, 1989; © copyright by the American Geophysical Union.]

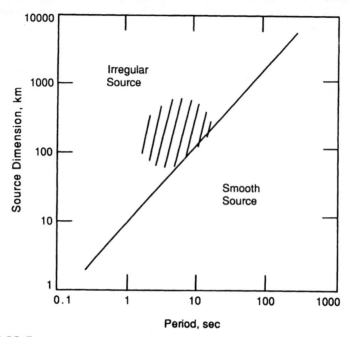

FIGURE 10.20 Empirical classification of complex earthquakes. (After H. Kanamori.)

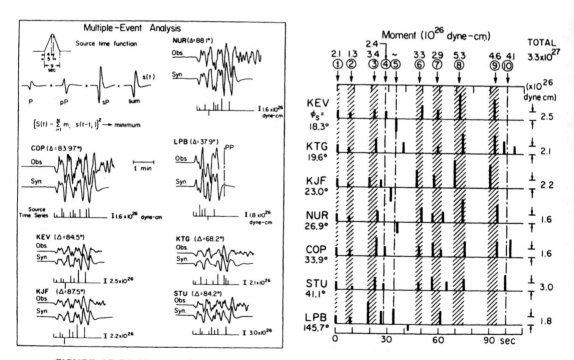

FIGURE 10.21 Multiple shock analysis for the February 4, 1976 Guatemala earthquake. (From Kanamori and Stewart, *J. Geophys. Res.* **83**, 3427–3434, 1978; © Copyright by the American Geophysical Union.)

are complex in detail, we usually reference fault complexity to the passband of observation. Figure 10.20 is an empirical classification of complex earthquakes; in the period band 5–20 s, many earthquakes with source dimensions that are greater than 100 km are complex. This is particularly true for strike–slip earthquakes. Figure 10.21 shows a multiple shock analysis for the February 4, 1976 Guatemala earthquake. A sequence of subevents is used to match each complex waveform, with consistency between the station sequences indicating the rupture complexity. The strike–slip rupture propagated bilaterally away from subevent 1, radiating pulses of energy as each fault segment failed. In this analysis it is assumed that each subevent has a specified fault orientation (the fault curves from west to east; see Figure 1.16). By matching the observed waveforms at stations azimuthally distributed around the source, we can determine the timing and moment of each subevent.

In our previous discussion of inverting for source parameters, we assumed that the rupture front progressed in a smooth and predictable manner. Clearly, in the case of the Guatemalan earthquake we have no *a priori* constraints that the rupture is smooth, nor should we expect it to be bilateral. It is possible to develop a generalized waveform inversion in which the *temporal* and *spatial* distribution of moment release can be recovered. In the simplest case, a fault can be parameterized as a series of subevents with known spatial coordinates but with unknown moment release or rupture time. Then the least-squares difference between an observed waveform and a synthetic is given by

$$\Delta = \int_0^\infty [u(t) - mw(t - t_1)]^2 \, dt, \quad (10.30)$$

where $u(t)$ is the observed seismogram, w is a synthetic seismogram calculated for a point source [w is given by (10.3), with $s(t)$, the appropriate time function for a

"unit" earthquake of moment m_0]. m is the size scaling factor, and Δ is minimized in terms of m and t_1; thus the timing and size of a subevent can be determined. We can generalize Eq. (10.30) to many subevents and multiple observations by successively "stripping away" the contribution of each subevent. In this procedure a wavelet is fit to the data, and a residual waveform is used to define a new seismogram. This residual is fit with another wavelet, stripped, and so on until the entire observed seismogram is adequately explained. This problem is usually severely underdetermined, so a "search procedure" is used to find the minima in Δ. The generalized form of (10.30) is given by

$$\Delta_k = \sum_{j=1}^M \int [x_{jk}(t) - m_k w_{jk}(t - \tau_k, f_k)]^2 \, dt,$$

$$(10.31)$$

where M is the number of stations used, x_{jk} is the residual data at the jth station after $k - 1$ iterations, m_k is the moment chosen for the kth iteration, w_{jk} is the synthetic wavelet for the jth station from the kth subevent, τ_k is the timing of the kth subevent, and f_k gives the source parameters for the kth subevent (epicenter, focal mechanism, etc.). The spatial–temporal resolution of a given subevent can be evaluated by plotting the correlation between the observed waveform and the synthetic wavelet at various allowable fault and time locations. Figure 10.22 shows the correlation for *three* iterations of such an inversion for the Guatemala earthquake. t is the time after rupture began, and l is the distance along the fault from the epicenter. For the first iteration the correlation is highest at a time of approximately 20 s and a distance of 90 km *west* of the epicenter. After this subevent is stripped away (removing the *largest* moment subevent), the process is repeated, and the largest correlation is 60 s from rupture

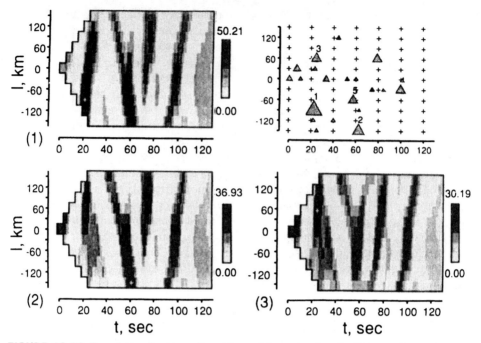

FIGURE 10.22 Correlation for three iterations of inversion for the Guatemala earthquake. The darker values indicate times and locations in which point sources can explain power in the residual seismogram from the previous iteration. The upper right shows the space–time sequence of pulses, with the size of triangles indicating the relative moment of pulses along the fault. (From Young *et al.*, 1989.)⊡

initiation, 150 km west of the epicenter. This process is repeated for a prescribed number of iterations, and the results from each iteration are combined to give the overall rupture process. It is interesting to note that the largest moment release for the Guatemala earthquake occurred near the bend in the Motagua fault, consistent with our discussion earlier in this chapter about asperities produced by irregularities.

The procedure described above has been extended to invert for source orientation of various subevents. Figure 1.16 shows corresponding results for the Guatemala earthquake with variable subevent fault orientation and moment being recovered. Such an application has a huge number of parameters and is reliable only with an extensive broadband data set.

Another waveform-modeling procedure to recover temporal changes in fault orien-

tation is to invert for a *time-dependent* moment tensor. In this case we can rewrite Eq. (10.10) as

$$u_n(x,t) = \sum_{i=1}^{5} m_i(t) * G_{in}(x,t), \quad (10.32)$$

where now the moment tensor elements are independent time series of moment release, and we incorporate the instrument response in the Green's function. Each moment tensor element now has its own time history, or time function. In the frequency domain we can write this as

$$u_n(x,\omega) = \sum_{i=1}^{5} m_i(\omega) G_{in}(x,\omega), \quad (10.33)$$

where m_i is the only unknown and it is a set of constants for each *frequency*. We can solve for m_i at a set of discrete frequency points and use the inverse Fourier transform to obtain a time-dependent moment tensor. In matrix form, Eq. (10.33) for a single frequency, f, looks like

where u_1^R and u_1^I correspond to the observed spectra at station 1 at frequency f, and u_n^R and u_n^I correspond to the spectra at station n at frequency f. The Green's

$$
\begin{bmatrix} u_1^R \\ u_1^I \\ \vdots \\ u_n^R \\ u_n^I \end{bmatrix} = \begin{bmatrix} G_{11}^R & -G_{11}^I & & G_{15}^R & -G_{15}^I \\ G_{11}^I & G_{11}^R & \cdots & G_{15}^I & G_{15}^R \\ \vdots & & \ddots & & \vdots \\ G_{n1}^R & -G_{n1}^I & & G_{n5}^R & -G_{n5}^I \\ G_{n1}^I & G_{n1}^R & \cdots & G_{n5}^I & G_{n5}^R \end{bmatrix} \begin{bmatrix} m_1^R \\ m_1^I \\ \vdots \\ m_5^R \\ m_5^I \end{bmatrix}, \qquad (10.34)
$$

Box 10.6 Empirical Green's Functions

Although Earth models have become quite sophisticated, there are many instances where our ability to compute accurate theoretical Green's functions is inadequate to allow source information to be retrieved from particular signals. This is very common for broadband recordings of secondary body waves with complex paths in the Earth (*PP*, *SSS*, etc.), as well as for short-period surface waves ($T = 5$–80 s). A strategy for exploiting these signals is to let the Earth itself calibrate the propagation effects for these signals, which are usually very complex. This is achieved by considering seismic recordings from a small earthquake located near a larger event of interest. If the source depth and focal mechanism of the two events are identical, the Earth response to each station will be the same. If the small event has a short, simple (impulse-like) source time function, its recordings approximate the Earth's Green's functions, including attenuation, propagation, instrument, and radiation pattern effects, with a corresponding seismic moment. We use these signals to model the signals for a larger event, with the differences being attributed to the greater complexity of the source time function for the larger earthquake. Often this involves deconvolving the "empirical" Green's functions from the corresponding records for the larger event. This provides an approximation of the source time function for the larger event, normalized by the seismic moment of the smaller event (Figure 10.B6.1). Isolated phases with a single ray parameter are usually deconvolved, with azimuthal and ray parameter (takeoff angle) variations in the relative source time functions providing directivity patterns that allow finiteness of the larger event to be studied. The procedure is valid for frequencies below the corner frequency of the smaller event, and in practice it is desirable to have two orders of magnitude difference in the seismic moments. Rupture processes of both tiny and great events have been studied in this way.

continues

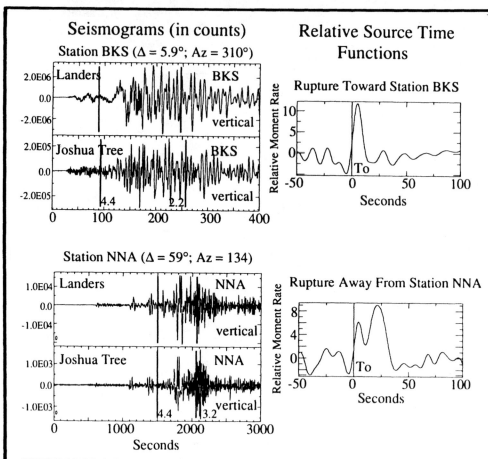

FIGURE 10.B6.1 Examples of deconvulution of recordings for a large event by recordings for a small nearby event recorded on the same station. Pairs of vertical component broadband surface wave recordings for two events are shown on the left, with the June 28, 1992 Landers ($M_w = 7.3$) event producing larger amplitudes than the nearby April 23, 1992 Joshua Tree event. Both events involved strike-slip faulting in the Mojave desert. Having similar focal mechanisms, locations, and propagation paths allows the smaller event to serve as an empirical Green's function source for the larger event. Deconvolution of the records at each station results in the simple relative source time functions shown on the right, with these giving the relatively longer source time function for the Landers event. Directivity analysis indicates that the rupture propagated toward BKS, producing a narrow pulse, and away from NNA, which has a broadened pulse. (From Velasco *et al.* 1994, with permission.)

function matrix is composed of 10 columns corresponding to the real and imaginary parts of five moment tensor elements for each station. This is required because of the complex multiplication: $(m^R + m^I)(G^R + G^I)$. The real part is $(m^R G^R - m^I G^I)$, and the imaginary part is $(m^R G^I + m^I G^R)$.

Inversion of (10.34) is typically unstable at high frequencies due to inaccuracies of the Green's functions, so only the lower frequencies are used. Figure 10.23 shows the results of a time-dependent moment-tensor inversion for the 1952 Kern County earthquake. The results show a temporal evolu-

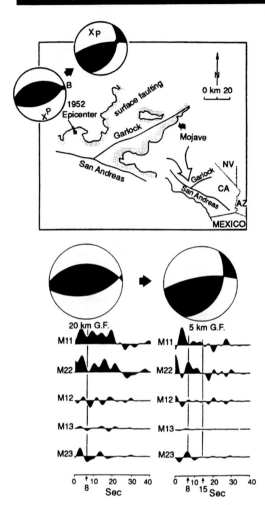

FIGURE 10.23 Results of a time-dependent moment-tensor inversion of the 1952 Kern County earthquake. Source time functions for each moment tensor element are shown for two depths. The preferred solution involves a pure thrust at 20 km depth in the first 8 s and a shallower oblique component in the next 7 s.

tion of rupture from primarily northwest–southeast thrusting to east–west oblique strike–slip motion. The geologic interpretation of the Kern County earthquake is that it started at the southwest corner of the fault at a depth of approximately 20 km. The fault ruptured to the northeast, where the fault plane became much shallower and the slip became partitioned into shortening (thrusting) and strike–slip com-

ponents. For the entire rupture, the P axes remained nearly constant, but the T axis rotated from being nearly vertical to a much more horizontal position.

10.5 Very Broadband Seismic Source Models

As the preceding discussions have indicated, seismologists use numerous methodologies to extract the details of faulting from seismic waveforms. We have tried to cast these different procedures in the context of linear filters and have concentrated on recovering the source time function. The one filter element we have largely ignored is the instrument response. This is because it is well known and can often be removed from the problem, but limited instrument bandwidth does provide an important constraint on our ability to recover source information. Given that earthquakes involve faulting with a finite spatial and temporal extent, different-frequency waves are sensitive to different characteristics of the rupture process. Further, different wave types tend to have different dominant observable frequencies as a result of interference during rupture and propagation. The net result is that wave types recorded on band-limited instruments can resolve different aspects of the fault history. Thus, inversion of the body-wave recording on WWSSN instruments may give a different picture of an earthquake than inversion of very long period surface waves recorded on a gravimeter. A truly broadband source model is required to explain rupture over a frequency range of a few hertz to static offsets. The new generation of broadband instruments help tremendously toward this end, but part of the problem is intrinsic to the physics of the seismic-wave generation. For example, the broadband waves from the 1989 Loma Prieta earthquake can be used to resolve two asperities. The fundamental-mode Rayleigh-wave analysis cannot resolve these details, but it does provide an accurate estimate of the total seismic moment. This moment is 20–30% larger than that determined by the body waves; thus the body waves are missing some of the slip process, perhaps a component of slow slip.

An ideal seismic source inversion would *simultaneously* fit the observations from different wave types over a broad frequency range. In practice, the methodology has been to perform distinct, high-resolution inversion of each wave type, thus solving for the source characteristics best resolved by a particular wave type. The distinct source characteristics are then merged to give a total model of the source. When incompatibilities in source characteristics determined by different inversions are observed, *ad hoc* procedures are used to merge the source characteristics. For

example, consider the moment discrepancy for the Loma Prieta earthquake. Figure 10.24 shows the effect of adding a long-period component of moment to the derived body-wave time function for Loma Prieta. Note that the "slow slip" component does not noticeably affect the body waves if it is spread out over more than 30 s. Although the slip model would account for the observed body and surface waves, it would require a type of fault behavior that is not observed in the laboratory. Given the uncertainty in various model assumptions, it is often difficult to judge how far to interpret these complex models from merging of results for different wave types.

A major problem with simply combining all different wave types in a single inversion is the normalization of the data. How does one weight a misfit in a P waveform as compared to a misfit of a single spectral

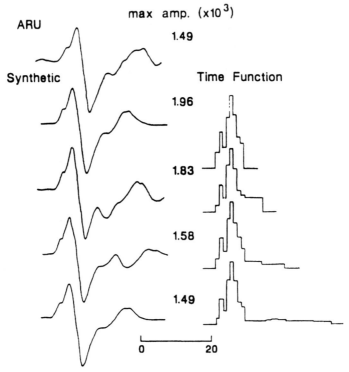

FIGURE 10.24 Effect of adding a long-period component of moment to the derived body-wave time function for the Loma Prieta earthquake. (From Wallace *et al.*, 1991.)

point for a long-period surface wave? Currently, strategies for deriving very broadband source models include iterative *feedback* inversions in which the body waves, high-resolution surface waves, and near-field strong motions are inverted independently. The results from each inversion are combined into a new starting model, which is, in turn, used in a heavily damped repeat of the independent inversions. After several iterations, all the data are combined, and the misfit is measured by a single error function, which is minimized in a final inversion. This type of procedure improves the *ad hoc* model merging and results in a model that is consistent with the sampled range of the seismic spectrum. Ultimately, it will be desirable to achieve this routinely, but this will require a better understanding of model dependence for different wave analyses.

References

Ben Menahem, A., and Singh, S. J. (1981). "Seismic Waves and Sources." 1108 pp. Springer-Verlag, New York.

Christensen, D. H., and Ruff, L. J. (1985). Analysis of the trade-off between depth and source time functions. *Bull. Seismol. Soc. Am.* **75**, 1637–1656.

Dziewonski, A. M., Ekström, G., Woodhouse, J. H., and Zwart, G. (1991). Centroid-moment tensor solutions for July–September, 1990. *Phys. Earth Planet. Inter.* **67**, 211–220.

Hartzell, S. H., and Heaton, T. H. (1985). Teleseismic time functions for large, shallow subduction zone earthquakes. *Bull. Seismol. Soc. Am.* **75**, 965–1004.

Helmberger, D. V. (1983). Theory and application of synthetic seismograms. *In* "Earthquakes: Observation, Theory and Interpretation," pp. 174–222. Soc. Ital. Fis., Bologna.

Kanamori, H. (1972). Mechanism of tsunami earthquakes. *Phys. Earth Planet. Inter.* **6**, 346–359.

Kanamori, H. (1974). A new view of earthquakes. *In* "Physics of the Earth (A Modern View of the Earth)" (in Japanese), pp. 261–282. (H. Kanamori and E. Boschi eds.), Physical Society of Japan, Maruzen, Tokyo.

Kanamori, H., and Stewart, G. S. (1978). Seismological aspects of the Guatemala earthquake of February 4, 1976. *J. Geophys. Res.* **83**, 3427–3434.

Kanamori, H., and Stewart, G. S. (1979). A slow earthquake. *Phys. Earth Planet. Inter.* **18**, 167–175.

Kanamori, H., Mori, J., and Heaton, T. H. (1990). The 3 December 1988 Pasadena earthquake ($M_L = 4.9$) recorded with the very broadband system in Pasadena. *Bull. Seismol. Soc. Am.* **80**, 483–487.

Langston, C. A., and Helmberger, D. V. (1975). A procedure for modeling shallow dislocation sources. *Geophys. J. R. Astron. Soc.* **42**, 117–130.

Lay, T., and Kanamori, H. (1981). An asperity model of large earthquake sequences. *Maurice Ewing Ser.* **4**, 579–592.

Ruff, L., and Kanamori, H. (1983). The rupture process and asperity distribution of three great earthquakes from long-period diffracted *P*-waves. *Phys. Earth Planet. Inter.* **31**, 202–230.

Satake, K. (1989). Inversion of tsunami waveforms for the estimation of heterogeneous fault motion of large submarine earthquakes: The 1968 Tokachi–Oki and 1983 Japan Sea earthquakes. *J. Geophys. Res.* **94**, 5627–5636.

Scholz, C. H. (1990). "The Mechanics of Earthquakes and Faulting." Cambridge Univ. Press, Cambridge, UK.

Velasco, A. A., Ammon, C. J., and Lay, T. (1994). Empirical Green function deconvolution of broadband surface waves: Rupture directivity of the 1992 Landers California ($m_w = 7.3$) earthquake. *Bull. Seismol. Soc. Am.* **84**, 735–750.

Wald, D. J., Helmberger, D. V., and Heaton, T. H. (1991). Rupture model of the 1989 Loma Prieta earthquake from the inversion of strong-motion and broadband teleseismic data. *Bull. Seismol. Soc. Am.* **81**, 1540–1572.

Wallace, T. C., and Helmberger, D. V. (1982). Determining source parameters of moderate-size earthquakes from regional waveform. *Phys. Earth Planet. Inter.* **30**, 185–196.

Wallace, T. C., Helmberger, D. V., and Ebel, J. E. (1981). A broadband study of the 13 August 1978 Santa Barbara earthquake. *Bull. Seismol. Soc. Am.* **71**, 1701–1718.

Wallace, T. C., Helmberger, D. V., and Engen, G. R. (1983). Evidence of tectonic release from underground nuclear explosions in long-period *P* waves. *Bull. Seismol. Soc. Am.* **73**, 593–613.

Wallace, T. C., Velasco, A., Zhang, J., and Lay, T. (1991). A broadband seismological investigation of the 1989 Loma Prieta, California, earthquake: Evidence for deep slow slip? *Bull. Seismol. Soc. Am.* **81**, 1622–1646.

Yao, Z. X., Wallace, T. C., and Helmberger, D. V. (1982). Earthquake source parameters from sparse body wave observations. *Earthquake Notes* **53**, 38.

Young, C. J., Lay, T., and Lynnes, C. S. (1989). Rupture of the 4 February 1976 Guatemalan earthquake. *Bull. Seismol. Soc. Am.* **79**, 670–689.

11

SEISMOTECTONICS

Modern geoscience is conducted in the context of *plate tectonics*, an overarching theory that describes the dynamics of the Earth's outer shell. Earthquake seismology has played a major role in developing the concept of plate tectonics, and the relationship between earthquake occurrence and tectonic processes is known as *seismotectonics*. The spatial distribution of earthquakes can be used to determine the location of plate boundaries, focal mechanisms can be used to infer the directions of relative motion between plates, and the *rate* and cumulative displacement of earthquake occurrence can be used to infer the relative velocity between plates. Because tectonic motions cause earthquakes, it is critical to understand plate tectonics when trying to reduce the societal hazard associated with faulting. In this chapter we will integrate what we have learned about the earthquake source and characteristics of wave propagation to develop constraints on the dynamics of the Earth's surface. Only a few of the many existing aspects of seismotectonics can be discussed here, and we restrict our attention to the most basic. Current journal articles are filled with many additional applications of seismology to the study of plate tectonics.

The basic concept in modern plate tectonics, introduced in the mid-1960s, is that of a mobile *lithosphere*. The lithosphere is a high-viscosity region that translates coherently on the Earth's surface. Relative motions between a mosaic of surface lithospheric plates are accommodated at plate boundaries. Figure 11.1, from a classical paper by Isacks *et al.* (1968), shows a cartoon of a mantle stratified on the basis of rheological behavior. The essence of lithospheric motions is large horizontal transport, from creation to destruction of the plate. The lithosphere behaves rigidly on geologic time scales of millions of years. This requires that the material within the lithosphere be at low enough temperatures to inhibit solid-state creep. The lower boundary of the lithosphere is sometimes defined by a temperature, or *isotherm*, usually about 1300°C, although above 600°C the material will begin to experience ductile deformation and this defines a "thermal" lithosphere. In general, the thermal lithosphere is about 100 km thick, although in some regions its thickness is essentially zero and in others it may be several hundred kilometers. Below the lithosphere is the *asthenosphere*, a region of relatively low strength that may be

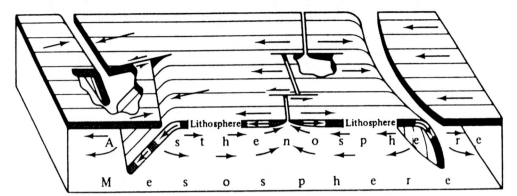

FIGURE 11.1 The basic elements of plate tectonics. The lithosphere is a region of high viscosity and strength that forms a coherently translating plate. Relative motions among plates cause them to interact along convergent, transcurrent, and divergent boundaries. The lithosphere rides on the asthenosphere, a region of low strength. (From Isacks *et al.*, *J. Geophys. Res.* **73**, 5855–5899, 1968; © copyright by the American Geophysical Union.)

nearly decoupled from the lithosphere. The asthenosphere behaves like a viscous fluid on relatively short time scales (10^4 yr). The classic example of this viscous behavior is the present-day uplift of Fennoscandia postdating the removal of the ice sheet associated with the last ice age. Beneath the asthenosphere, Isacks *et al.* (1968) defined the *mesosphere*, a region of relatively high strength that they believed to play only a passive role in tectonic processes. The concept of the mesosphere is dated, but below a depth of approximately 350 km the viscosity of the upper mantle does appear to increase. Many geophysicists also believe that the viscosity of the lower mantle is significantly higher than that of the upper mantle. In a modern context, the lithosphere is associated with the stiff upper portion of the surface thermal boundary layer of a mantle convection system.

Movements of lithospheric plates involve shearing motions at the plate boundaries. Much of this motion occurs by aseismic creep, but shear-faulting earthquakes are also produced. The release of strain energy by seismic events is restricted to regions where there is an inhomogeneous stress environment and where material is

sufficiently strong for brittle failure. For the most part, this means that earthquakes are confined to plate boundaries. Figure 11.2 shows the locations of 8800 earthquakes for which Harvard has determined a CMT solution (see Box 10.3). These correspond to the larger events, $M_w > 4.5$, in the past 15 years. This can be compared with the distribution that includes smaller events in Figure 1.11. Also shown in Figure 11.2 are the locations of most of the plates and tectonics regions discussed in this chapter. The seismicity is largely concentrated in narrow bands that represent the plate boundaries; 12 major plates are fairly obvious, but on this scale it is difficult to recognize the dozen or so additional smaller plates. Three basic types of plate boundary are characterized by different modes of plate interaction:

1. *Divergent boundaries*, where two plates are moving apart and new lithosphere is produced or old lithosphere is thinned. *Midoceanic ridges* and *continental rifts* are examples of divergent boundaries.
2. *Convergent boundaries*, where lithosphere is thickened or consumed by sinking into the mantle. *Subduction*

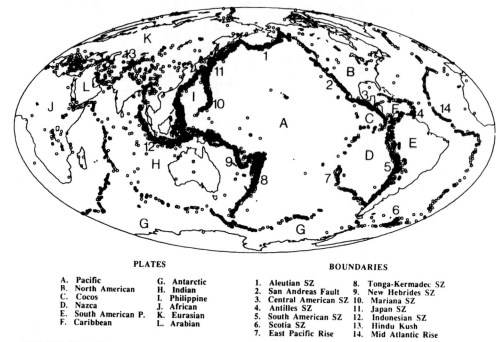

PLATES

A. Pacific	G. Antarctic
B. North American	H. Indian
C. Cocos	I. Philippine
D. Nazca	J. African
E. South American P.	K. Eurasian
F. Caribbean	L. Arabian

BOUNDARIES

1. Aleutian SZ	8. Tonga-Kermadec SZ
2. San Andreas Fault	9. New Hebrides SZ
3. Central American SZ	10. Mariana SZ
4. Antilles SZ	11. Japan SZ
5. South American SZ	12. Indonesian SZ
6. Scotia SZ	13. Hindu Kush
7. East Pacific Rise	14. Mid Atlantic Rise

FIGURE 11.2 Epicenters of 8800 earthquakes with known faulting mechanisms, which outline the important global tectonic features (SZ denotes subduction zone). There are 12 major plates (denoted by capital letters). Important plate boundaries are denoted by numbers. (Courtesy of G. Ekström.)

zones and *alpine belts* are examples of convergent plate boundaries.

3. *Transcurrent boundaries*, where plates move past one another without either convergence or divergence. *Transform faults* and other strike–slip faults are examples of transcurrent boundaries.

In general, divergent and transcurrent plate boundaries are characterized by shallow seismicity (focal depths less than 30 km). In Figure 11.2, the mid-oceanic ridge system is the best manifestation of this shallow extensional faulting. Subduction zones and regions of continental collision can have much deeper seismicity. Figure 1.11b shows only epicenters with focal depths in excess of 100 km, which isolate the subduction zones in Figure 11.2. Fully 80% of the world's seismicity is along the circum-Pacific margin and occurs mostly in subduction zones. Figure 1.11b

also shows clusters of deep Eurasian earthquakes, which are the result of the collision of India and Eurasia.

Before we develop the concept of seismotectonics further, it is best to review some of the basics of plate motion. The fundamental concept that underlies motion of rigid plates on a sphere is *Euler's theorem*, which states that the relative motion of two plates can be described by a rotation about a pole (called an *Euler pole*). Figure 11.3 shows the relative motion between two plates. If a plate boundary is *perpendicular* to a small circle about the pole of rotation, convergence or divergence must be occurring between the plates. If the boundary is *parallel* to a small circle, the relative plate motion is transcurrent. The relative velocity between two plates depends on the proximity of a plate boundary to the pole of rotation.

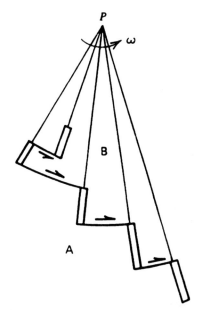

FIGURE 11.3 Motion of plate boundaries is determined by rotation about an Euler pole. Transcurrent boundaries are oriented along the arcs of small circles about the pole. Divergent and convergent boundaries are oriented along radial spokes through the pole. (Reprinted by permission from *Nature*, vol. 207, pp. 343–347; copyright © 1965 Macmillan Magazines Limited.)

The rotation is described by an angular velocity ω. As the distance to the boundary increases, the relative motion on a boundary also increases.

Several investigators have used the orientation of plate boundaries and relative velocities derived from the analysis of magnetic anomalies to construct relative plate motions for the entire world. *Absolute* plate motions are more difficult to determine because the whole surface is in motion, but it is possible if one assumes a fixed hotspot or other reference frame. Figure 11.4 shows the absolute plate velocities determined from the NUVEL-1 mode 1 (DeMets *et al.*, 1990). Note that for some midocean ridge boundaries the relative motion is as large as 20 cm/yr. If we assume that the relative motion is totally accommodated seismically, we can calcu-

late the expected moment release per year for an *average* boundary

$$M_0 = \mu \overline{D} L w, \qquad (11.1)$$

where $\overline{D}$ is 10 cm/yr. Letting L be 500 km and w be 15 km yields a moment rate of 5×10^{26} dyne cm/yr ($M_w \approx 7.0$). Not all plate motion is accommodated by seismic slip on faults, because there is not a magnitude 7 event on every 500-km-long stretch of plate boundary every year.

In this chapter we discuss the various categories of plate boundary and the associated types of seismic activity. We also develop the relationship between plate tectonic processes and earthquake recurrence. The concept that the processes driving plate tectonics are steady and long-term implies that the rate of strain accumulation and release at a plate boundary must also be relatively steady. If the mechanical response of a fault along a plate boundary has regular strike–slip behavior, a characteristic earthquake cycle may exist. This idea is of fundamental importance to earthquake hazard assessment and has given rise to the concepts of *recurrence intervals* and *seismic gaps* in earthquake prediction. One major societal goal of seismology is to predict earthquakes to minimize loss of life. Although models have been developed that can generally describe earthquake recurrence, a predictive capability is still elusive and may intrinsically be unattainable. However, earthquake damage can be reduced by modifying construction standards in regions with an established long-term seismic hazard.

11.1 Divergent Boundaries

Oceanic ridges and continental rift zones are regions of the Earth in which the relative motion across the ridge or rift is *divergent* or *extensional*. In divergent settings

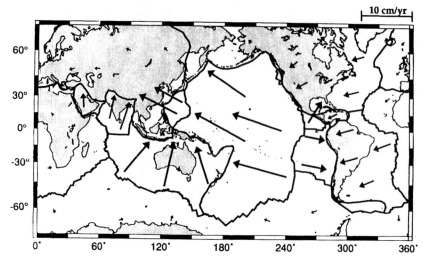

FIGURE 11.4. Absolute plate motions based on model NUVEL-1. The vectors point in the direction of plate motion relative to a fixed frame of reference (hot spots). The length of the vectors is proportional to the velocity in cm/yr. (Based on DeMets *et al.*, 1990. Figure courtesy of R. Richardson.)

the minimum principal stress, σ_3, is horizontal and directed perpendicular to the ridge or rift. The maximum principal stress, σ_1, is vertical. Most of the earthquakes in this environment involve normal faulting. Oceanic ridges are by far the most important divergent boundaries, for it is at oceanic ridges that new oceanic lithosphere is actually created and partitioned to the diverging plates on either side of the ridge. For this reason, ridges are also known as *spreading centers*.

The creation and evolution of oceanic lithosphere strongly influence the seismicity at oceanic ridges. Figure 11.5 shows a schematic cross section through a ridge. At the plate boundary the lithospheric thickness is essentially *zero*. As the material added to the plates at the spreading center cools below a critical isotherm, it acquires the mechanical property, high viscosity, of "lithosphere." As the lithospheric plate moves away from the plate boundary, it continues to cool, which has two principal consequences: (1) the lithosphere beneath a point on the plate thickens, and (2) the depth to the ocean floor increases. The

lithospheric thickness can be calculated by assuming that an isotherm for a cooling boundary layer is the lithospheric boundary:

$$th = 2c\sqrt{kt}, \qquad (11.2)$$

where c and k are heat diffusion properties of the material. This implies that th is proportional to $\sqrt{t}$, the square root of the age of the plate at a given position. If the plate moves away from the spreading center at a constant rate, called the *spreading rate*, th is proportional to $\sqrt{d}$, where d is the distance from the ridge. The map of the age of ocean plate (Figure 7.33) exhibits the systematic age increase away from ridges.

The second thermal effect, the increase of oceanic depth with age, requires the concept of isostasy. In the simplest case, isostasy requires two columns of rock at equilibrium on the Earth to have the same total mass. If one column is less dense, the column will have a higher surface elevation. The density of most Earth materials decreases as temperature increases. In

Box 11.1 Earthquake Swarms

In Section 9.8 we discussed b values and the scaling of earthquake size with frequency of occurrence. For nearly all regions of the Earth, $b \leq 1.0$; the most notable exception is for *earthquake swarms*, for which b values typically exceed 1.0 and are often as large as 2.5. The high b values imply that no well-defined mainshock occurs but, rather, many events of approximately the same size. The most frequent explanation for high b values is a weak crust, incapable of sustaining high strain levels and a very heterogeneous stress system. Most earthquake swarms occur in volcanic regions, and their occurrence has often been related to the movement of magma. Figure 11.B1.1a shows the hypocenters from a 14-yr period plotted on an east–west cross section through Kilauea caldera on the island of Hawaii. The earthquakes delineate a magma system extending from 60 km depth to shallow magma reservoirs. The narrow "pipe" of hypocenters between 40 and 12 km is thought to correspond to the major magma conduit from a deep reservoir (at approximately 60 km depth) to shallower magma chambers. Earthquakes occur at a relatively constant rate between 40 and 12 km, but at shallow depths, earthquakes occur very episodically as bursts or swarms of events. This appears to be directly related to magma chamber inflation preceding eruption. Figure 11.B1.1b shows a plot of the cumulative earthquake occurrence along the southwestern rift region of Kilauea.

These earthquake swarms are thought to represent pervasive rock cracking and block rotation. Typically, the focal mechanisms of swarms are normal slip or, occasionally, strike–slip. An exception to this is an intense 2-yr swarm in the Gilbert Islands, located in the Southwest Pacific, far from plate boundaries. Figure 11.B1.2 shows the temporal behavior of the swarm and focal mechanisms for the largest events, which are predominantly thrusting. Prior to December 1981 no

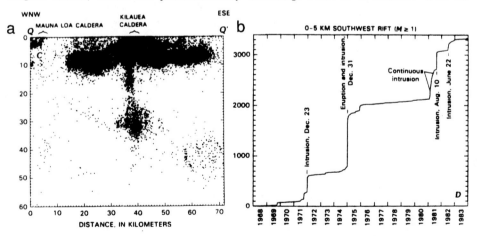

FIGURE 11.B1.1 (a) Hawaiian seismicity for a 15-yr period. Cross section runs along the southwest rift zone of Kilauea. Deep seismicity indicates that the magma source is in the mantle. (b) Time history of the shallow seismicity shown in (a). Extensive swarms of earthquakes occur just prior to eruption. (From Klein *et al.*, 1987.)

continues

seismicity had been detected in this region. Note that the magnitudes of the events do not indicate a mainshock–aftershock decay. The *b* value is approximately 1.35.

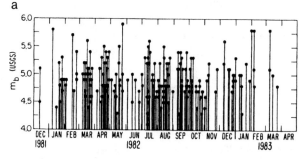

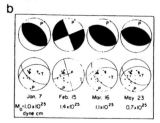

FIGURE 11.B1.2 (a) Event magnitude as a function of time for the Gilbert Islands swarm. Note that there are no clear "mainshocks." (b) Mechanisms of the Gilbert Island swarm. (From Lay and Okal, 1983.)

other words, as temperature rises, most materials expand. Conversely, as temperature decreases, density increases, and isostasy requires a *smaller* column. At oceanic ridges the temperature of the shallow mantle is very high, which is manifested as high elevation, or shallow ocean depths. As lithosphere moves away from the ridge and cools, it becomes more dense and "sinks"; thus the ocean depth increases. Figure 11.6 shows the ocean depth as a function of age for two plates and a $\sqrt{t}$ model. The correlation of *th* with ocean depth for this model is excellent, except for oceanic regions older than 100 million years.

The thermal structure of young oceanic lithosphere has a profound influence on the observed seismicity. Oceanic ridges that are spreading rapidly have a broad, smooth topographic signature. The lithospheric thickness is very small near the ridge, which means that the available "width" of a fault plane is also small. This condition results in shallow, moderate- to small-sized earthquakes. Figure 11.7 shows the maximum depth of faulting for near-oceanic ridge events plotted as a function of the half-spreading rate. The *half-spreading rate* is defined as one-half the

relative motion between plates on either side of a ridge. Half-spreading rates greater than 6 cm/yr are considered high (in Figure 11.7 the half-spreading rate of 22 cm/yr is for the Nazca plate—rapid spreading indeed!). The *centroid* is defined as the optimal point-source location for the moment release for an earthquake. This is not to be confused with the hypocenter but, rather, corresponds to the center of the slip distribution on the fault. A common assumption is that the centroid depth for large shallow events that rupture to the surface is approximately one-half the depth extent of faulting. Figure 11.7 shows that as the spreading rate increases, the maximum centroid depth shallows. Since earthquake magnitude correlates with fault dimensions, one can infer that as the spreading rate increases, the *maximum* expected magnitude of an earthquake near the ridge *decreases*. Fast spreading centers have few earthquakes, all of small magnitude, whereas slow spreading centers are capable of generating much larger earthquakes (although midoceanic ridge earthquakes rarely exceed $M_w = 7.0$).

Strictly speaking, most normal-faulting earthquakes associated with oceanic ridges

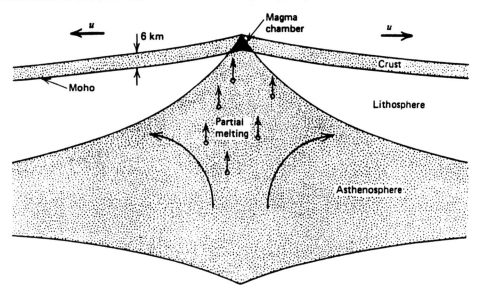

FIGURE 11.5 Schematic cross section through a spreading center. The lithosphere is defined by an isotherm, so as the plate cools and moves away from the ridge, its thickness increases. This associates the oceanic lithosphere with a thermal boundary layer. (From Turcotte and Schubert, 1982.)

do not occur exactly at the plate boundary but are on faults associated with a feature known as the *axial valley*. To illustrate this, Figure 11.8 shows the topography of a section of the Gorda Rise. A series of normal faults occurs on either side of the depressed axial valley, probably produced in response to isostatic adjustment. Figure 11.9 shows typical normal-fault focal

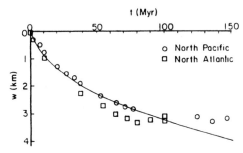

FIGURE 11.6 Ocean depth as a function of age of the seafloor. If we assume that spreading is constant at the ridge, the age of the ocean floor is proportional to the distance from the ridge. (From Turcotte and Schubert, 1982.)

mechanisms along a midocean ridge (the Carlsberg Ridge in the Indian Ocean). Occasionally, earthquake swarms (see Box 11.1) occur along oceanic ridges. Most ridge swarms are associated with fast spreading centers and are thought to be related to magmatic intrusions.

Continental rifts are also divergent plate boundaries, although we usually do not recognize the opposite sides of a rift as distinct, separate plates until rifting produces new oceanic plate. The faulting in continental rifts is usually much more complicated than that at oceanic ridges, probably reflecting the nature of continental crust and the difference between active and passive rifting. In general, continental rifts are much wider, and the seismicity is more diffuse than at oceanic ridges. The lithosphere may be thinned at a rift, but it does not go to zero until continental breakup occurs. Table 11.1 lists the major continental rifts; the "spreading" rates for these rifts are much smaller than for most oceanic ridges.

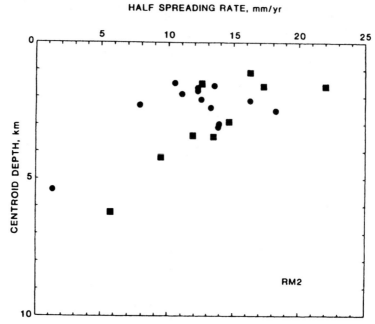

FIGURE 11.7 Maximum centroid depth of near-ocean ridge earthquakes as a function of the half-spreading rate. As the spreading rate decreases, the lithosphere near the ridge thickens, and the seismogenic zone also grows. (From Huang and Solomon, *J. Geophys. Res.* **92**, 1361–1382, 1987; © copyright by the American Geophysical Union.)

The most active continental rift is the East African Rift (EAR), which is part of a spreading system that includes the Red Sea and the Gulf of Aden, which are floored by oceanic material. This rift system broke apart the African and Arabian plates, and it is further breaking apart the African plate. Figure 11.10 shows the complex East African Rift zone. Rifting within East Africa began at the Afar triple junction approximately 25–40 million years ago and has since propagated southward at an

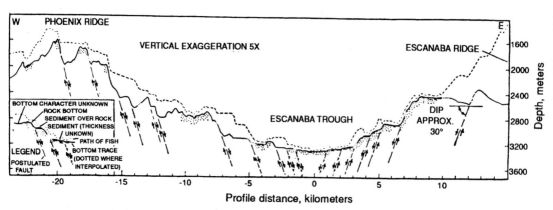

FIGURE 11.8 Topography across the Gorda Rise. The axial valley, which is approximately 30 km across, is bounded by a series of normal faults. (After Atwater and Mudie. Copyright © 1968 by the AAAS.)

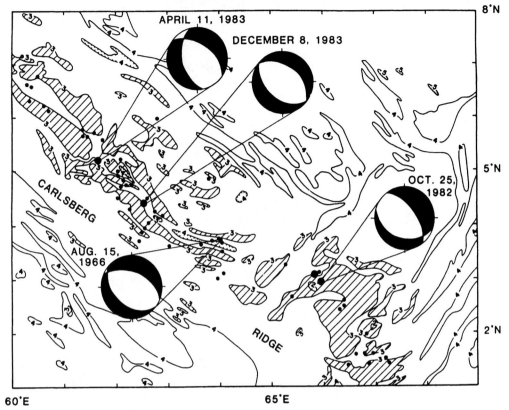

FIGURE 11.9 Focal mechanism along a midoceanic ridge in the Indian Ocean. The *T* axis is oriented parallel to the direction of spreading. (From Huang and Solomon, *J. Geophys. Res.* **92**, 1361–1382, 1987; © copyright by the American Geophysical Union.)

estimated rate of 2–5 cm/yr (Oxburgh and Turcotte, 1974). The rift splits into a western and an eastern branch just north of the equator and reconnects at about 8° south. Figure 11.11 shows the focal mechanisms of events along the EAR system, which for the most part are normal-slip earthquakes. The largest events along the

TABLE 11.1 Continental Rifts

	Location	Length (km)	Maximum earthquake size
Baikal Rift	54°N, 110°E	2000	7.5
East African Rift	15°N, 38°E	3000	7.5
Rheingraben	49°N, 8°E	300	5.0–5.5
Rio Grande Rift	35°N, 107°W	700	6.5

EAR are $M_s > 7.5$; this can be generalized to state that the maximum magnitude expected in a continental divergent setting will be significantly larger than that along an oceanic ridge. This reflects the thermomechanical properties of the continental crust. Although the heat flow in the EAR is much higher than in surrounding regions, the seismogenic zone still extends to at least 10 km.

11.2 Transcurrent Boundaries

Transcurrent boundaries, between horizontally shearing plates, are of two types: (1) *transform faults*, which offset ridge segments, and (2) strike–slip faults that connect various combinations of divergent and

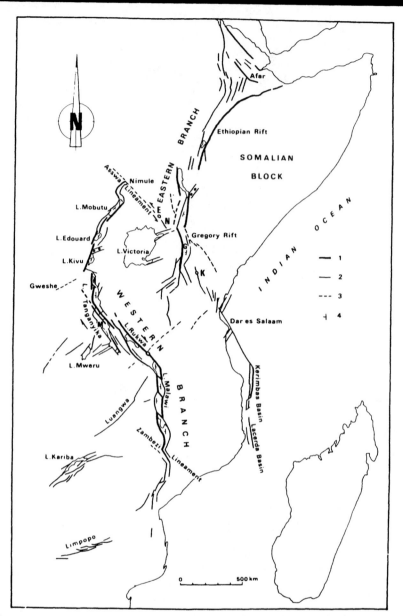

FIGURE 11.10 Major tectonic features of the East African Rift. The rift has two branches, which merge in Ethiopia; the East African Rift joins the Red Sea Rift at Afar. The Red Sea was formerly continental, but now it is completely floored by oceanic crust. (Reprinted from Chorowicz, J., 1989, pp. 203–214, with kind permission from Elsevier Science Ltd., The Boulevard, Langford Lane, Kidlington OX5 1GB, UK.)

convergent plate boundaries. One of the interesting features of midoceanic ridges is that they are *offset* by lineaments known as *fracture zones* (see Figure 11.12). The portion of the fracture zone between the ridge crests is actively slipping as plate A shears past plate B. On the basis of the apparent left-lateral offset of the ridges, one might expect the ridge segments to be separated by a left-lateral strike–slip fault. However, this is not the case, and the relative position of the ridges does not

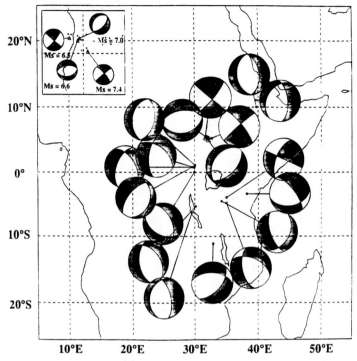

FIGURE 11.11 Focal mechanisms in central Africa. The East African Rift has two arms, as indicated by the trends of the focal mechanisms. (From Gao *et al.*, 1994.)

change with time. Wilson (1965) proposed the concept of *transform faults*, shear faults that accommodate ridge spreading, which was subsequently confirmed primarily with seismic focal mechanisms. The ridge segments initiated in their offset position largely as a result of preexisting zones of weakness in continental crust that has rifted apart. For the transform fault in Figure 11.12, the mechanisms are thus right-lateral strike–slip.

The fracture zone beyond the ridge–ridge segment is a pronounced topographic scarp, but it is wholly contained within a single plate, so no horizontal relative motion and little or no seismicity occur on the "scar." Figure 11.13 shows a schematic cross section along the transform fault pictured in Figure 11.12. As discussed in the previous section, cooling effects control the ocean floor bathymetry. In general, the fracture zone separates

material of different ages and thus of different bathymetry and lithospheric thickness. This scarp is most pronounced at a ridge crest, where the fracture zone juxtaposes material of zero age with material that has been transported from the adjoining ridge crest.

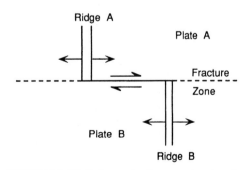

FIGURE 11.12 Schematic diagram showing the offset of two ridge segments by a transform fault.

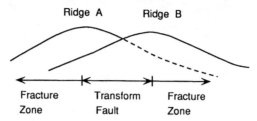

FIGURE 11.13 Schematic cross section showing topographic variations along the transform fault shown in Figure 11.12.

The size of earthquakes on a transform fault largely depends on two factors: (1) the length of offset between ridges and (2) the rate of spreading. The earthquake size is limited by the lithospheric thickness available for brittle failure. If a transform fault is long, connecting slowly spreading ridges, a substantial seismogenic source region is available. In most cases the largest expected magnitude for transform events is 7.0–7.5. Figure 11.14 shows seismic moments for earthquakes on several transform faults, and as expected, the maximum moments are smallest for fast spreading systems. Figure 11.15 shows seismicity and focal mechanisms along the Mid-Atlantic Ridge near the equator. In all cases the east–west plane is believed to be the fault plane, and vertical, right-lateral, strike–slip motion occurs on the faults.

Strictly speaking, transform faults can also offset convergent boundaries, but for simplicity, we reserve the name transform fault for ridge–ridge offsets. Two examples of other transcurrent plate boundaries are shown in Figure 11.16. In the first case, the Scotia and Sandwich plates are bounded on the north and south by strike–slip faults. The Caribbean plate is similarly situated between two convergent boundaries that are connected via a transcurrent boundary called the Cayman Trough. This fault extends through Central America, where the 1976 Guatemalan earthquake occurred (discussed in Chapter 10). It is difficult to characterize or generalize about transcurrent fault earthquakes

because they exhibit diverse behavior, but they possess several basic tendencies. While the length of the fault can increase to over 1000 km, the fault width quickly reaches a maximum given by the seismogenic zone. Increased earthquake size is associated not with increased fault width but with greater fault length and increasing rupture complexity. It is quite common for large transcurrent earthquakes to involve multiple subevents.

Large strike–slip faults that juxtapose continental material are usually complicated, and their formation is somewhat problematic—clearly they cannot form as a simple ridge–ridge offset. There are many theories on the formation of these features, such as plate boundary "jumps," subduction of ridges, highly oblique subduction, or continent–continent collision. The most famous continental strike–slip fault is the San Andreas fault in California, which separates the North American and Pacific plates. Atwater (1970) describes a plausible scheme for the development of the San Andreas fault. Thirty million years ago the Farallon plate was subducting beneath western North America. The midocean ridge separating the Pacific plate and the Farallon plate was west of the North American coastline (Figure 11.17). The relative motion between the North American and Farallon plates was convergent. The convergence rate was greater than the spreading rate between the Farallon and Pacific plates, and eventually the ridge was subducted. When this

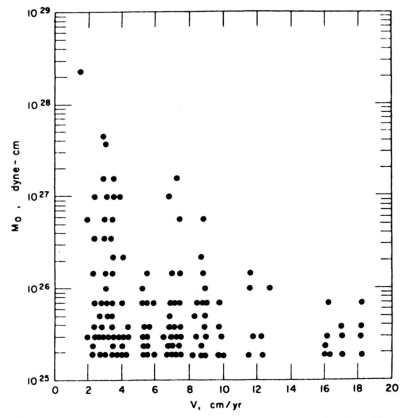

FIGURE 11.14 Plot of M_0 versus spreading rate for transform earthquakes. The moment decreases as spreading rate increases. (From Burr and Solomon, *J. Geophys. Res.* **83**, 1193–1205, 1978; © copyright by the American Geophysical Union.)

happened, the Farallon plate ceased to exist, and the faulting along the Pacific–North American boundary was subsequently dictated by the Pacific–North America relative motion, which produced a transcurrent boundary.

A consequence of this model for the evolution of the San Andreas is that the fault is *growing*, with its northern end near Cape Mendocino (Figure 11.18) propagating northward. The offset across the San Andreas is 350–400 km, and if we assume that the present rate of motion between the Pacific and North America plates (~ 5 cm/yr) is applicable to the past several million years, the San Andreas must be more than 10 million years old. A puzzling aspect of the San

Andreas is that it is more than 100 km *inboard* of the coastline in central California. The basic Atwater model predicts that the transcurrent boundary should juxtapose continental and oceanic material, but the actual situation is more complex. Figure 11.18a shows a fault map of present-day California. Although the San Andreas is the dominant fault in the state, the Pacific–North American boundary is very complex. It is probably best to think of the boundary as a *zone* of faults that accommodates the plate motion and plate boundary evolution. The present-day San Andreas fault may be positioned along a lineament of preexisting weakness within what used to be the upper plate on a convergent margin. The San Andreas actu-

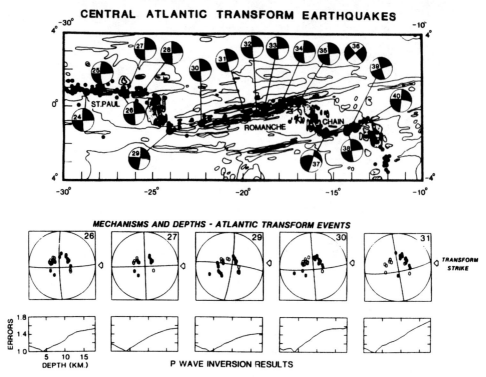

FIGURE 11.15 Seismicity and focal mechanisms along mid-Atlantic transform faults near the equator. Some of the focal mechanisms with *P*-wave first motions are shown. The waveform inversion depths are also shown (most depths are approximately 4 km). (Modified from Engeln *et al.*, *J. Geophys. Res.* **91**, 548–577, 1986; © copyright by the American Geophysical Union.)

ally accommodates only a fraction of the relative Pacific–North American plate motion, with distributed deformation accounting for the rest.

Note that the San Andreas cannot lie completely on a small circle about any simple pole of rotation. It has numerous bends and appears to have numerous *strands*, especially at the southern end. This complicates predicting the seismic behavior of the San Andreas. Some sections of the fault slip only as great earthquakes; others slip continuously through *creep*. Figure 11.18b shows the historic seismic record for California. This record contains two great San Andreas fault earthquakes: (1) the 1906 San Francisco earthquake, which ruptured the northern third of the

San Andreas, and (2) the 1857 Fort Tejon earthquake, which ruptured the south central San Andreas fault. Both events were about $M_s = 7.9$–8.0. In the decade of the 1980s, 12 significant damaging earthquakes occurred in California, and only one, the 1989 Loma Prieta earthquake, appears to be directly related to the San Andreas. Even the Loma Prieta event is problematic in that it did not rupture to the surface, and the fault-plane solution indicates oblique thrust motion, inconsistent with the San Andreas orientation. The Loma Prieta event occurred in a region with a small bend in the San Andreas, which produces compression across the fault, which is accommodated on several splay faults.

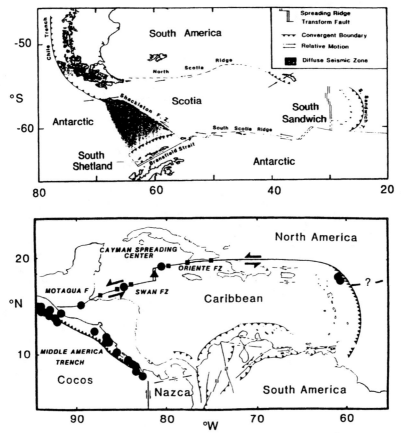

FIGURE 11.16 Two transcurrent plate boundaries (non–transform fault boundaries). (a) The Scotia plate in the South Atlantic is separated from the South American plate by a long strike–slip fault. (From Pelayo and Wiens, 1989.) (b) The Caribbean plate is separated from the North American plate by the Cayman Trough, a boundary that is *primarily* strike–slip. (From Stein *et al.* 1988.)

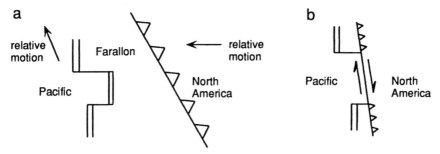

FIGURE 11.17 Plausible tectonic evolution of the San Andreas fault, which separates the North American plate from the Pacific plate. (After Atwater, 1970.)

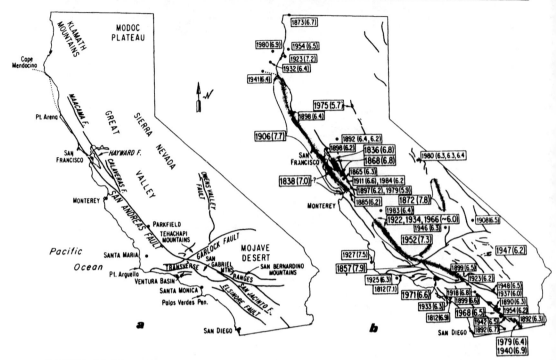

FIGURE 11.18 (a) Major Quaternary faults in California. (b) Historical earthquakes of California. (From Wesnousky, *J. Geophys. Res.* **91**, 12,587–12,631, 1986; © copyright by the American Geophysical Union.)

Transcurrent faults in continental settings are extremely important from a seismic hazard point of view. These faults typically traverse regions of high population density, and in this century earthquakes on these faults have been responsible for the largest number of fatalities. Other major plate-boundary continental transcurrent faults are the Anatolian fault in Turkey (see Figure 11.51), the Alpine fault in New Zealand, and the Sigang fault in Burma. Some continental transcurrent faults occur far from plate boundaries, and we will discuss these further in the section on intraplate earthquakes.

11.3 Convergent Boundaries

11.3.1 Subduction Zones

There are two types of convergent boundaries, *subduction zones* and *conti-*

nent–continent collision zones. Continental collision zones are rare (although extremely important for producing major mountain ranges and plateaus), and currently more than 90% of the world's convergent boundaries are subduction zones, which form at zones of oceanic–oceanic or oceanic–continental convergence. Convergence at these boundaries is accommodated by underthrusting of one lithospheric plate beneath another. The underthrust plate descends through the asthenosphere and represents "consumed" lithosphere, which balances the surface area of new lithosphere created at mid-oceanic ridges. Subduction of the lithosphere is one of the most important phenomena in global tectonics; no oceanic crust older than Jurassic (~ 200 million years) exists, yet we find continental crust 20 times as old. Summing the area of ocean floor that has opened since Jurassic time, we find that 20 billion km^3 of material has been subducted. At the present rate of subduction, an area equal to the entire surface of the Earth will be cycled into the interior in 160 million years.

The initiation of subduction is a topic of considerable controversy. Although no generally accepted model exists to explain how subduction initiates, the thermal buoyancy of the plate explains why it *continues*. If a slab of lithosphere is displaced into the asthenosphere, it will be "colder" and hence more dense than the surrounding asthenosphere. This results in a negative buoyancy force that causes the lithosphere to sink. The sinking of a subducting plate through the asthenosphere is a complex function of the age of the lithosphere, the rate of convergence, and the age of the subduction zone. As the name *thermal buoyancy* implies, the effect is a manifestation of the temperature differences between the slab and surrounding mantle. This temperature difference also has a signature on the seismic velocities. In Chap-

ter 7 we showed how tomography can be used to image the subducting slab.

As also discussed in Chapter 7, the ultimate fate of the subducted plate is very controversial; models range from continuous subduction to great depths—perhaps to the core–mantle boundary—before thermal equilibration and assimilation take place, to no subduction below 660 km depth. The seismic activity along the slab can be used to address these hypotheses.

Three categories of seismic activity occur in subduction zones. The first is the interaction between the two converging lithospheric plates. Typically, an oceanic plate is bent at a 10° to 30° angle and dives below the overriding plate. This results in a large contact zone between two plates on which frictional sliding must take place, producing interplate seismicity. The largest earthquakes occur along these contacts and involve thrust faulting. The 1960 Chilean earthquake had a fault length of more than 1000 km, a fault width of 200 km, and an *average* displacement of 24 m! The M_w

---

Box 11.2 Creeping Plate Boundaries

In Chapter 9 we discussed stick–slip as a physical mechanism for faulting and noted that under certain conditions, stable sliding could also continuously accommodate the stress field. When a fault slips continuously, we refer to this as *aseismic slip* or *creep*. The best known example of a creeping fault is a 100-km-long segment of the San Andreas between San Juan Bautista and Parkfield, California. Figure 11.B2.1 shows the rate of slip along this creeping section of the San Andreas. The fault creep was discovered when the Cienega Winery was built on the San Andreas in 1948, and the walls of the winery were progressively offset at a steady rate of approximately 11 mm/yr.

Faults that creep require special frictional behavior; either the fault has unusually low normal stresses, or the fault is lined with very ductile or weak material. The creeping segment of the San Andreas is characterized by a wide zone (more than 2 km) that has a pronounced low seismic velocity. This low-velocity zone is absent in the adjacent locked sections of the San Andreas. This suggests that the fault zone is filled with "fault gouge," a poorly consolidated, clay-rich material. For whatever reason, the creeping segment of the fault has acquired this fault lubricant; thus the majority of the fault's slip is accommodated aseismically.

continues

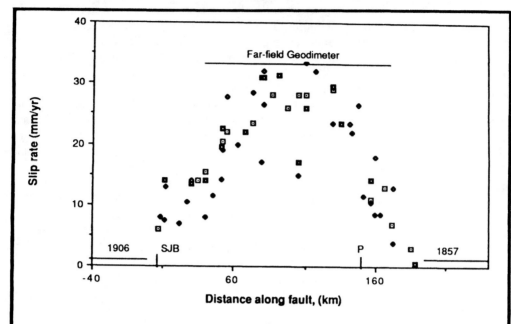

FIGURE 11.B2.1 The slip rate (mm/yr) along the creeping segment of the San Andreas fault. The different symbols refer to different types of instrumentation. SJB is San Juan Bautista, which is at the northern end of the creep segment. P is Parkfield. The ends of the 1906 and 1857 ruptures are shown with lines. (From Scholz, 1990 with permission.)

of the Chilean event was 9.5. The second category of seismicity is related to the internal deformation of the overriding plate commonly associated with back-arc extension or upper-plate compression. The final category of seismic activity has to do with internal deformation within the subducting plate that results from the slab's interaction with surrounding mantle. It is this *intraplate* seismicity that allows the seismologist to map the subducting plate and to infer the mechanical state of the mantle.

In the early 1930s, K. Wadati first observed deep zones of seismicity beneath Japan. With the advent of plate tectonics, these zones were recognized as an expression of the Pacific plate subducting beneath the Eurasian plate. H. Benioff detailed the occurrence of deep seismicity zones in many regions of the world in the 1940s, and we now refer to these deep seismic belts as Wadati–Benioff zones. Figure 11.19 shows a depth distribution of earthquakes south of the island of Fiji, along the Tonga–Kermadec subduction zone. The cross sections show a narrow, planar trend of earthquake foci—this is the Wadati–Benioff zone. This zone represents the brittle core of the slowly warm-

FIGURE 11.19 (a) Tectonic, structural and seismicity distribution along the Pacific/Indo-Australian plate boundary. The Tonga–Kermadec subduction zone stretches from 15 to 30° S. Contours of seismicity indicate the dipping slabs. (From Hamburger and Isacks, *J. Geophys. Res.* **92**, 13,841–13,854, 1987; © copyright by the American Geophysical Union.) (b) Cross sections through the Tonga–Kermadec subduction zones. The cross sections are along arcs of latitude. The seismicity is thought to outline the shape of the subducting plate. (From Giardini and Woodhouse. Reprinted with permission from *Nature*, vol. **307**, 505–509. Copyright © 1984 Macmillan Magazines Limited.)

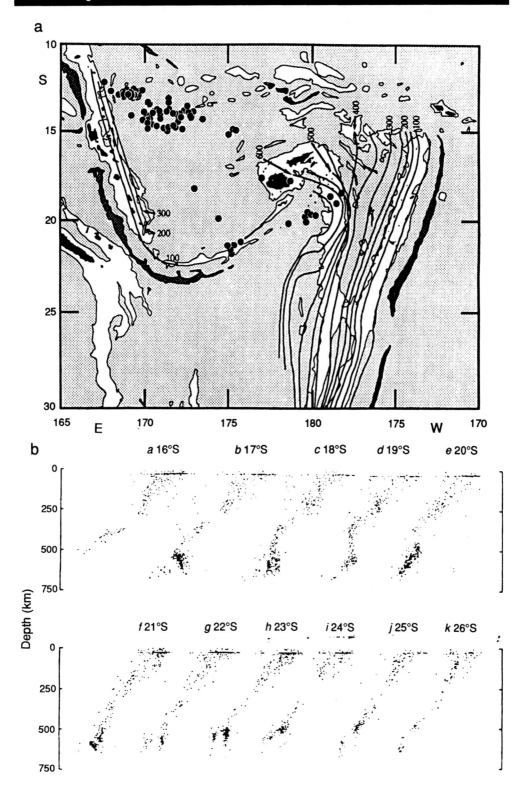

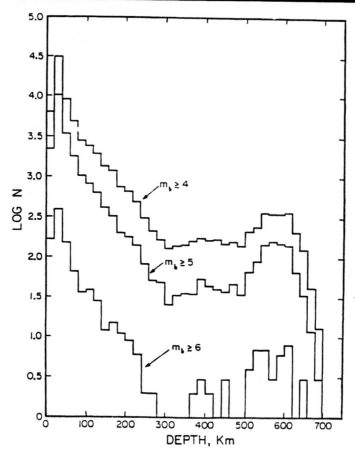

FIGURE 11.20 Logarithm of the total number of earthquakes observed globally versus depth. The three curves are for different cutoff magnitudes. Note the minimum in energy release between 250 and 500 km depth. (From Vassiliou *et al.*, 1984.)

ing lithosphere, about 30 km thick, in the subducting plate.

Internal Deformation within the Slab. The state of stress in a subducting plate depends on the balance of two forces: (1) the negative buoyancy of the descending slab and (2) the resistance force of mantle that is being displaced by the subducting plate. These forces depend strongly on the viscosity structure, phase transformations in the slab, the rate of subduction, the age of the subduction zone, and the depth of slab penetration. Figure 11.20 shows a tabulation of hypocentral depths from circum-

Pacific subduction zones. The largest number of earthquakes and largest energy release occur in the upper 200 km. This is the region of interplate interaction and slab bending. Normal frictional sliding processes dominate at this depth. Below about 50 km all events are within the plate rather than on the plate interface. The earthquake activity is at a minimum between 200 and 400 km depth, where the subducting lithosphere is interacting with *weak* asthenosphere. Frictional sliding may occur at these depths only if hydrous phases destabilize and release water or other fluids to allow high pore pressures to exist. Below 400 km, the number of earthquakes increases with depth and some slabs strongly distort. Increasing resistance to slab penetration is often inferred, but frictional sliding mechanisms are generally not expected at these depths (unless further hydrous phases exist at these depths and release fluids as they destabilize), so other mechanisms such as phase changes may be operating (Box 11.3). All earthquake activity ceases by a depth of 700 km. Some of the largest deep events are found near the maximum depth of seismicity in different slabs, so there is not a simple tapering off of activity. This maximum depth is conspicuously consistent with the velocity discontinuity near 660 km depth (see Chapter 7). A phase change in the slab may occur that suppresses earthquake failure. The termination of seismicity is a first-order observation of the fate of the subducted slab, but its implication is controversial, because an aseismic slab extension may exist in the lower mantle.

The focal mechanisms of the earthquakes along the Wadati–Benioff zone can be used to map the stress orientation in the slab. The stress orientation is controlled by the slab geometry and the balance of thermal, resistive, and negative buoyancy forces. If negative buoyancy dominates, the slab will be in downdip *extension*, and the T axis as determined from the earthquakes will be parallel to the dip of the slab. As the resistive force becomes more important, the slab experiences downdip *compression*, and the P axis will be parallel to the trend of the Wadati–Benioff zone. Figure 11.21 shows the expected focal mechanisms for downdip compression and extension, respectively. Note in both cases that one of the nodal planes is nearly vertical (assuming the slab dips at an angle of $\sim 60°$, which is true for several slabs beneath 200 km). Figure 11.22 shows that focal mechanisms for deep earthquakes in the Kuril subduction zone tend to have compression axes aligned along the slab dip direction. In the upper 300 km of most slabs, focal mechanisms are consistent with downdip extension, whereas the bottom part of the slab is in downdip compression. To understand this phenomenon, consider a long rod held against the force of gravity. If the rod is supported at the top, the rod will be in extension. If the rod is supported at the bottom, its own weight will cause the rod to be in compression. If the rod is partially supported at the top and bottom, the state

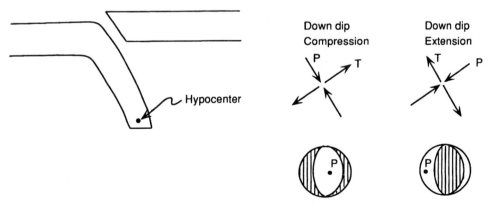

FIGURE 11.21 Focal mechanisms expected for *downdip compression* and *downdip extension*. The focal mechanism can appear to involve normal or thrust motion, depending on the slab dip and the orientation of the principal stresses. The relevant characterization of the stress regime is the direction of the *P* and *T* axes with respect to the dipping slab.

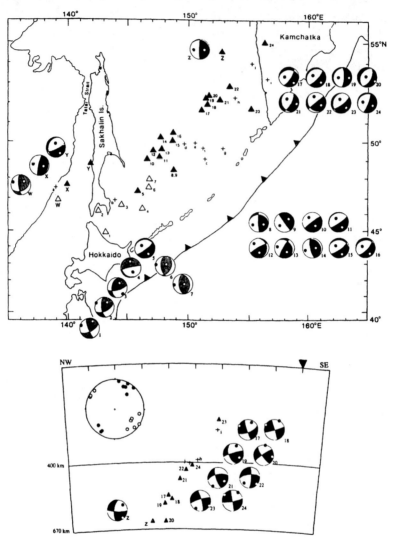

FIGURE 11.22 (Top) Map view showing the deep seismicity in the Kuril Islands. Earthquakes below 300 km are plotted, with their focal mechanism being shown. Solid quadrants are compressional *P*-wave motions, with a white dot indicating the tension axis; white quadrants are tensional *P*-wave motions, with a black dot indicating the compressional axis. (Bottom) Cross section striking approximately NW–SE for the northern Kuril slab. (From Glennon and Chen, *J. Geophys. Res.* **98**, 735–769, 1993; © copyright by the American Geophysical Union.)

of stress is transitional. Isacks and Molnar (1971) recognized a corresponding range of behavior in subducting slabs by examining numerous subduction zones. Figure 11.23 shows the stress patterns for various subduction zones and Isacks and Molnar's model for accounting for the stress condi-tions. *Strength* is a time-dependent con-cept, and most geophysicists would replace strength with viscosity in Isacks and Molnar's model. In addition, given the un-certain mechanism associated with deep earthquakes, it is not clear what stress variations are actually required.

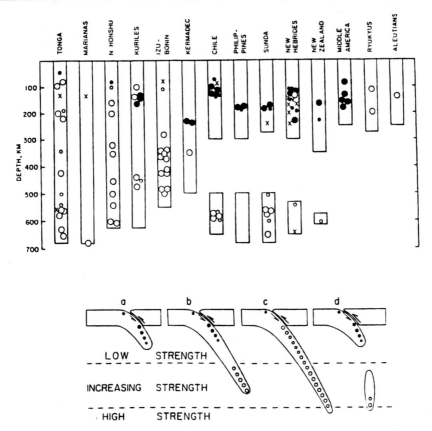

FIGURE 11.23 Distribution of downdip extension (solid hypocenters) and downdip compression (open hypocenters) in various subduction zones. The mechanical interpretation is shown below. (From Isacks and Molnar, 1971.)

Although the general model for the state of stress in a slab outlined in the previous paragraphs explains the gross features of Wadati–Benioff zone seismicity, some subduction zones show a very interesting variation at depths of 50 to 200 km. In the Japan, Kuril, and Tonga trenches, the Wadati–Benioff zone is made up of two distinct planes. Each plane is defined by a thin, well-defined cluster of epicenters; the planes are separated by 30–40 km. Figure 11.24 shows an example of the double Wadati–Benioff zones in northern Japan. The two planes have earthquakes with different focal mechanisms. The upper plane is predominantly in downdip compression, while the lower plane is in downdip extension.

There are several proposed models to explain double Wadati–Benioff zones that involve the unbending/bending of the plate and thermal stresses. In one model, bending a thin plate causes extension in the outer arc of the bend, while the underside of the plate is in compression. The extensional zone is separated from the compressional zone by a neutral surface. Shallow tensional events and deep compressional events are, in fact, observed in the outer rise of subduction zones. If the plate is suddenly released from the torquing force, it will "unbend" and experience forces opposite to those imposed during bending. This could explain the double Benioff zone stresses at intermediate depth (Figure 11.25a). A second type

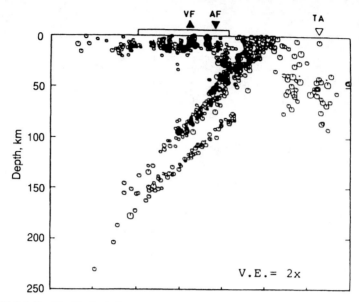

FIGURE 11.24 Double Wadati–Benioff zone in northeast Japan. (After Hasegawa *et al.*, 1978.)

of model for the double seismicity is based on a "sagging" force. Figure 11.25b shows that if a plate sinks into the mantle at a shallow angle and if the leading edge of this subduction zone is supported, the middle may sag under its own weight.

The earthquakes in double Wadati–Benioff zones are small, rarely exceeding magnitude 5.5. Double Wadati–Benioff

zones are not observed everywhere and are probably related to a balance between thermal buoyancy and rate of subduction. The Earth's present subduction zones are consuming oceanic lithosphere of all ages, but the thermal buoyancy force in old, cold lithosphere is much larger than in young, warm lithosphere. Once subduction has started in a given zone, *ridge push*, the

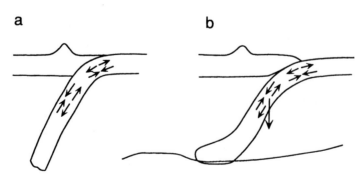

FIGURE 11.25 (a) Unbending model of double Wadati–Benioff zone. (b) Sagging model of double Wadati–Benioff zone. (After Sleep, *J. Geophys. Res.* **84**, 4565–4571, 1979; © copyright by the American Geophysical Union.)

Box 11.3 Mechanisms for Deep Earthquakes

Deep earthquakes have long posed a problem for seismologists. Laboratory experiments indicate that the pressures at a few hundred kilometers depth should prohibit brittle fracture and frictional sliding processes. Yet earthquakes as large as $M_w = 8.2$ have occurred at 650 km. The deep seismicity has many characteristics that are similar to those of shallow earthquakes. Most important, the deep earthquakes have radiation patterns consistent with double couples, which implies shear faulting. In 1945 P. Bridgman proposed that phase transformations, the reordering of solid phases to higher-density structures, could produce nonhydrostatic stresses that may cause the deep earthquakes. Recent work has shown that phase transformations and fault growth may be a self-feedback system. The growth of microfractures causes a hydrostatic stress that, in turn, promotes phase transformations. The localization of the transformation and shear strain that develops rapidly after fault growth can, in turn, cause the fault to grow.

This mechanism for deep earthquakes depends strongly on the thermal structure of the subducting plate. As we saw in Chapter 7, two major phase changes are postulated to occur in the mantle, one at 400 km depth and one at ~ 660 km depth. Since the slab is *colder* than the surrounding mantle, the 400-km phase transition (olivine $\rightarrow \beta$-spinel) should occur at shallower depths. However, experiments show that this equilibrium reaction may be kinetically suppressed, meaning that olivine can actually persist to greater depth. The olivine $\rightarrow \beta$-spinel transformation can occur rapidly on planes of maximum shear stress, giving rise to double-couple–like energy release. The net effect is a thin metastable wedge of olivine-like material in the slab; it may be this wedge that provides the source region for the deep earthquakes (see Figure 11.B3.1).

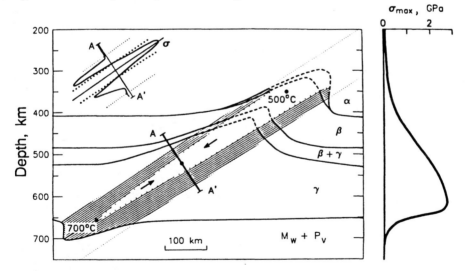

FIGURE 11.B3.1 Stress and temperature profile in the subducting slab. The central core of the slab may consist of olivine that is metastable below 400 km depth. Transformational faulting may occur along the edges of the tongue of olivine, causing deep earthquakes. (From Kirby *et al.*, Science **252**, 216–225; copyright 1991 by the AAAS.)

integrated body force due to the cooling and sinking of lithosphere away from spreading centers, will help push the subducted plate into the asthenosphere. If the subducting plate is old lithosphere and the rate of convergence at the subduction zone is low, the negative buoyancy force is the dominant force acting on the plate, and the plate will sink under its own weight. On the other hand, if the subducting plate is old but the rate of convergence is *very high*, the subducting plate is being "pushed" into the mantle faster than it would sink on its own. This is known as *forced subduction*. In the delicate balance between forced subduction and slab pull, the slab is neither strongly in compression nor in tension. This allows the residual stress field, which is much smaller than typical subduction stresses, to control the

seismicity. In this case, it may be possible to observe the plate "unbending" or "sagging." Figure 11.26 shows the correlation of age of subducted slab, rate of convergence, and state of stress (downdip compression or extension) in the slab. Subduction zones with double Wadati–Benioff zones occur in the middle girdle (between the dashed lines).

As we discussed in Chapter 7, the ultimate fate of subducted slabs penetrating as deep as 660 km remains controversial (Figure 11.27). The observation that seismicity ceases near this depth indicates that something significant happens associated with the 660-km boundary. Two basic theories explain the maximum depth of seismicity. In the first, the 660-km discontinuity is viewed as impenetrable to subduction. When the slab encounters the bound-

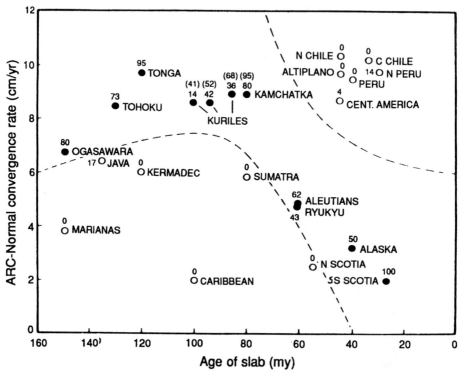

FIGURE 11.26 Correlation between the arc-normal convergence rate, the age of the subduction lithosphere, and the percentage of in-plate compressive events. (From Fujita and Kanamori, 1981. Reprinted with permission from the Royal Astronomical Society.)

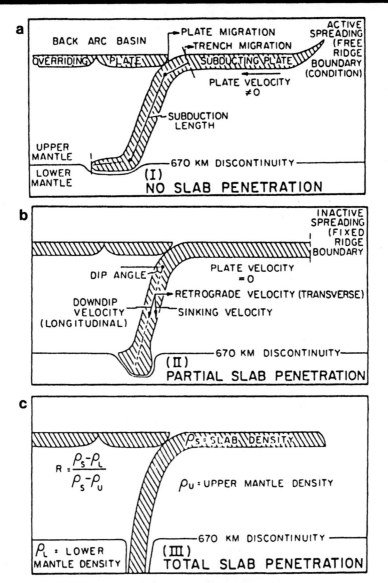

FIGURE 11.27 Schematic models for the possible fate of deep slabs. (a) A compositional contrast between the upper and lower mantle prevents slab penetration. (b) A compositional contrast exists but is insufficient to prevent slab penetration. This is called *penetrative convection*. Once the slab heats up, it will be buoyant and will emerge from the lower mantle. (c) The slab readily penetrates into the lower mantle. In this case the 660-km discontinuity is a phase change, and the slab simply transforms as it passes the boundary. (From Kincaid and Olson, *J. Geophys. Res.* **92**, 13,832–13,840, 1987; © copyright by the American Geophysical Union.)

ary, it must flatten out, although it may depress the discontinuity. What could cause such a "strong" boundary? There are two choices: (1) the 660-km discontinuity may be a boundary between chemically distinct lower and upper mantles, or (2) the viscosity across the boundary may increase by more than several orders of magnitude, enough to prevent penetration. In the case of the chemical boundary, the lower mantle must have a composition with a high enough density to exceed the thermally induced density anomaly and inertial effects of the slab. The existence of strong viscous or compositional stratification would cause mantle convection to be separated into upper- and lower-mantle convective regimes. The second theory is that the 660-km discontinuity is a phase change, with conversions to high-pressure perovskite being expected at this depth. Although viscosity may increase at this transition, the slab can usually penetrate the phase boundary, with seismicity terminating as the phase transformation occurs. Most geophysicists agree that the seismic boundary is such a phase change, but controversy still exists over whether any compositional change also occurs.

The most direct way for seismologists to determine the ultimate fate of the slab is to search for a seismic signature. Since all seismicity stops at ~ 660 km depth, only the seismic velocity difference between the slab and surrounding mantle can provide this signature. Unfortunately, accurately measuring this velocity contrast is difficult, as discussed in Chapter 7. The most recent high-resolution tomographic results suggest that penetration may depend on the individual slab. Some slabs appear to deflect and broaden at the 660-km discontinuity, while others appear to penetrate at least a few hundred kilometers. Resolving this issue will contribute to solving the problem of whole-mantle versus layered-mantle convection and will elucidate the evolution of the Earth.

Shallow Subduction Zone Seismicity. The shallow seismicity in subduction zones marks the interaction of the subducting and overriding lithospheres and accounts for 70% of the annual global seismic energy release. Nearly all *great* earthquakes (magnitude > 8.0) occur in this region. Figure 11.28 shows the general framework for great thrust earthquakes in shallow subduction zones: the high-strength lithosphere of each plate is juxtaposed over an area that may be from 50 to 200 km wide. Focal mechanisms of earthquakes from this region are typically shallow-dipping (15°–30°) thrust events. The great Alaska earthquake of 1964 occurred along the convergent boundary between the Pacific and North American plates. The moment magnitude M_W was 9.2 (seismic moment of 9×10^{22} N m), and the event had a 14-m average slip on a fault that was 700 km long and 180 km wide. Figure 11.29 shows the aftershock area of the Alaska earthquake and some of the coseismic elevation changes. Montague Island was uplifted more than 12 m on a small splay fault off the main thrust.

The variations between subduction zones in terms of maximum earthquake size and rupture length is remarkable. In Chapter 9 we introduced the concept of asperities and patches on the fault that have locally high stress drops. We can extend this concept to subduction zones and introduce the concept of *coupling*, which is a measure of the seismogenic mechanical interaction between the subducting and overriding lithospheric plates. In some subduction zones, such as the Marianas, the largest thrusting earthquakes are relatively small, and a significant component of aseismic slip must occur. In other regions, such as southern Chile, the subduction slip occurs primarily in very large earthquakes. The concept of coupling is illustrated in Figure 11.30, which shows fault segments that fail in large ruptures and the size of asperities

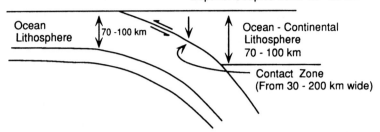

FIGURE 11.28 Schematic cross section through the shallow portion of a subduction zone.

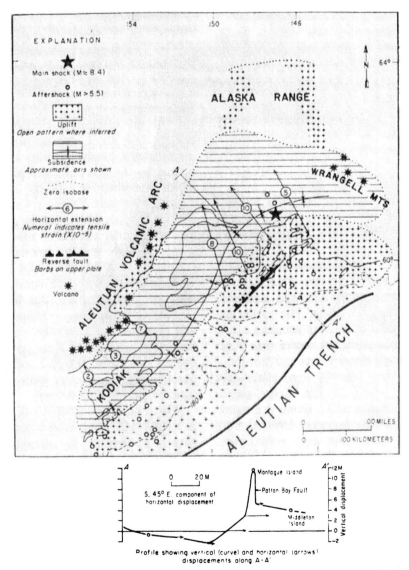

FIGURE 11.29 Tectonic displacements and seismicity associated with the 1964 Alaska earthquake relative to the Aleutian trench and volcanic arc. Tectonic data after Plafker (1965) and unpublished USCGS data; epicenters from Page (1967). (From Plafker, 1972.)

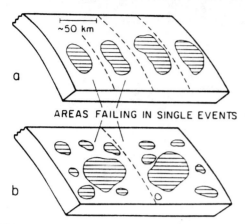

FIGURE 11.30 Schematic diagram of the asperity model, for stress heterogeneity on the fault plane. The hachured regions have a high stress drop and relatively large displacements during major earthquakes. Some regions have more heterogeneous stress conditions, which give rise to more complex earthquake ruptures. (From Lay and Kanamori, 1981.)

within those segments. (Note that this definition of asperity need not correspond to protrusions on the fault, as associated with asperities on the small scale discussed in Chapter 9.) A strongly coupled subduction zone will have a greater portion of its interface covered by asperities compared with less coupled zones. The motivation for this model of asperity coupling comes from the complexity of source time functions for large thrust events and the size of the fault zone inferred from aftershocks and surface-wave models of the source finiteness. Figure 11.31 shows the difference in the time functions inferred from two great subduction zone earthquakes. The 1964 Alaska earthquake is characterized by a smooth rupture, with a single large asperity at one end of the fault plane. The Kuril Islands earthquake was a multiple-event sequence. These observed differences in rupture have been used to characterize subduction zones.

Figure 10.19 showed four categories of subduction zones proposed by Lay and

Kanamori (1981) based on the relative size of the asperities. Category 1 is exemplified by southern Chile, where great earthquakes tend to occur regularly in time over approximately the same rupture zone. Category 2 includes subduction zones with smaller rupture dimensions, as typified by the central portion of the Aleutian subduction zone. Other regions that fall into this category are Colombia, northern Kamchatka, and southern Japan. Earthquakes in category 2 subduction zones demonstrate temporal variation in rupture mode, with occasional very large ruptures spanning segments of the trench that fail individually at other times. Category 3 is typified by the Kuril Islands subduction zone, which produces large earthquakes from repeated ruptures of the same portion of the subduction zone but which seldom fail simultaneously to generate larger events. Finally, category 4 is characterized by the Marianas subduction zone, which is distinguished by the absence of great earthquakes. Category 4 zones have a large component of aseismic slip. Figure 11.32 shows the aftershock zones and regions of high seismic moment release (asperities in this context) for 21 major earthquakes, 16 of which occurred along subduction zones. From this figure it is obvious that coupling is highly variable, and even the four categories are a simplification. In general, only category 1 earthquakes produce $M > 9.0$ earthquakes. Seismologists continue to study large ruptures to add to this data base.

What causes the variability in coupling? A number of factors, such as age of lithosphere, contribute to coupling. Figure 11.33 shows the relation between the maximum magnitude (M_w) event in a given zone, the convergence rate, and the age of the subducted lithosphere. It is clear that the maximum observed earthquake size increases with increasing convergence rate and decreasing lithospheric age. Ruff and Kanamori (1980) performed a linear regression on these data and obtained the

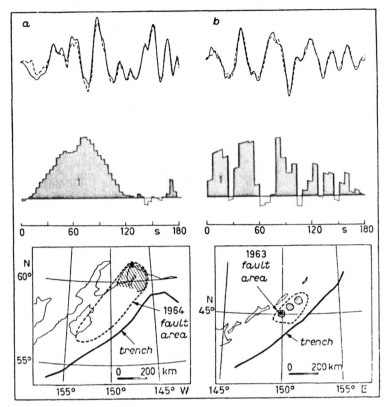

FIGURE 11.31 The source time functions for the 1964 Alaska and 1963 Kuril Islands earthquakes. Note the variation in temporal moment release, which is interpreted in terms of different fault zone stress heterogeneity, or asperities. (From Ruff, 1983.)

following relationship:

$$M_w = -0.00889T + 0.134V + 7.96,$$

$$(11.3)$$

where T is the age of the subducted plate in millions of years and V is the convergence rate in cm/yr. Figure 11.34 shows M_w predicted by (11.3) versus the actual value of M_w for 22 subduction zones. The solid dots are subduction zones with *back arcs*, which have spreading centers formed behind the subduction zone. Most of these regions are weakly coupled—it appears that the presence of a back arc leads to or results from reduced seismic coupling. Three of the four earthquakes that produced earthquakes of $M_w \geq 9.0$ have Wadati–Benioff zones that extend only to

shallow depths (≤ 200 km). Two other factors that are common to the subduction zones with large events are (1) shallow dip of the uppermost part of the slab (between 10° and 2° versus averages of about 30°), and (2) the subducted plate has smooth topography ("rough" topography on plates can be due to fracture zones, ridges, or seamounts).

Equation (11.3) can also be used to predict the size of an earthquake expected in a subduction zone. The *Cascadia* subduction zone along the coast of the Pacific Northwest (Oregon and Washington) is historically quiescent. However, the convergence rate between North America and the Juan de Fuca plate, along with the young age of the Juan de Fuca plate, suggests that a magnitude 9 + event is credi-

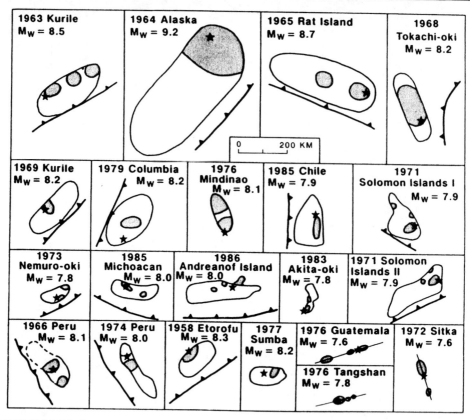

FIGURE 11.32 Schematic map-view summaries of aftershock zones (solid lines) and regions of high seismic moment release (stippled regions) inferred from seismic-wave analysis for 21 major earthquakes. Subduction zone events are shown with the nearby trench axis. (From Thatcher, *J. Geophys. Res.* **95**, 2609–2623, 1990; © copyright by the American Geophysical Union.)

ble. Geological evidence for past great earthquakes in this region is now accumulating as well.

The nature of coupling on the shallow thrust plane in subduction zones appears to be influenced by the *history* of the subduction zone. Figure 11.35 shows the evolutionary subduction model proposed by Kanamori (1977). Shallow-dipping, broad, strongly coupled zones (e.g., Alaska) produce extensive ruptures. The thrust zone may be weakened and partially decoupled by repeated fracturing, yielding smaller rupture lengths and asperities. Large normal-faulting events that fracture the descending lithosphere represent a transition to tensional stress in the slab and complete decoupling of the plate interface, which may result in the development of a back-arc basin by trench retreat.

These ideas about seismic coupling are very qualitative, and the actual frictional properties on the thrust faults are undoubtedly further complicated by the history of prior slip, hydrological variations, and thermal regime of the plate interface. The notion of asperities is useful primarily as a qualitative characterization of the stress heterogeneity, not as a model for dynamic slip processes.

Seismicity Away from the Subducting Slab. The stress system associated with the subduction process produces deformation well

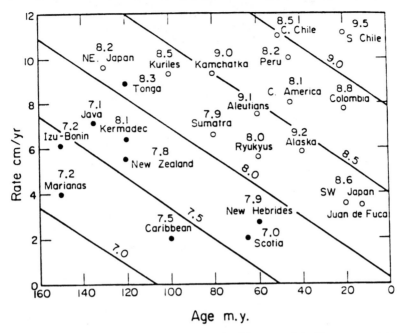

FIGURE 11.33 Maximum earthquake magnitude in different subduction zones plotted as a function of convergence rate and age of subducting lithosphere. (Modified from Ruff and Kanamori, 1980.)

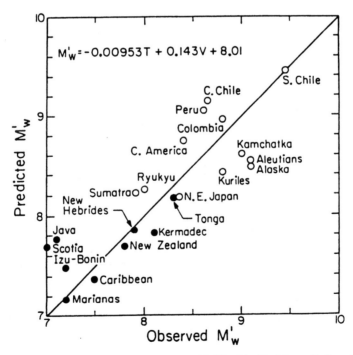

FIGURE 11.34 Linear fit to the data in Figure 11.33. (Modified from Ruff and Kanamori, 1980.)

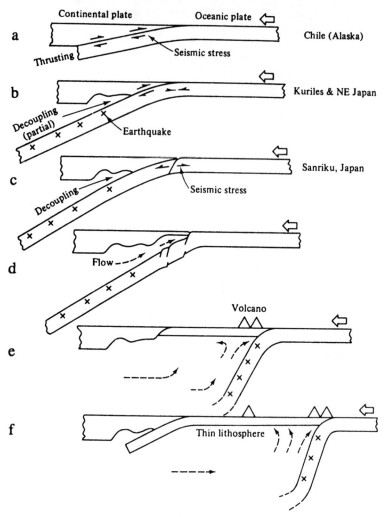

FIGURE 11.35 Schematic model of interplate coupling and decoupling for evolving subduction zones. (From Kanamori, 1977.)

beyond the trench. In the extreme, the deformation in the overriding plate can be as extensive as that within the Andes of South America. The Altiplano (central Andes) in southern Peru, Bolivia, and northern Chile is a large plateau with an average elevation of 4 km. This plateau has been formed, at least in part, by compressional crustal shortening. Much of the seismicity reflects this ongoing compression and has thrust or strike–slip mechanisms. Figure 11.36 summarizes the focal

mechanisms in Peru, which are dominated by P axes (compressive stress direction) oriented along the direction of convergence between South America and the Nazca plate. In the highest elevations of the Altiplano, normal faults also occur, which are probably related to gravitational body forces acting on topography. As with all subduction zone seismicity, it is difficult to make sweeping characterizations about upper-plate activity. In Japan the internal deformation within the overriding plate

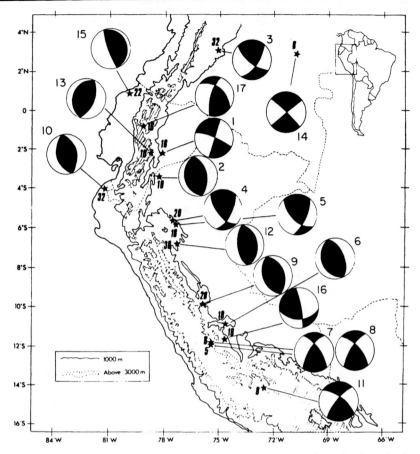

FIGURE 11.36 Map of the Peruvian Andes and a summary of focal mechanisms in the upper plate, indicating compressional deformation. (From Suárez *et al.*, *J. Geophys. Res.* **88**, 10,403–10,428, 1983; © copyright by the American Geophysical Union.)

appears to be strike–slip, while clearly in Peru it is thrusting. In general, recurrence intervals on these types of faults are much longer than for the trench thrust events.

Within the subducting plate relatively few earthquakes occur that are associated with the bending of the plate, oceanward of the trench, in what is called the outer rise. For trenches in which the subducting and overriding plates are weakly coupled, these intraplate events can be very large. Tensional stress from the negative buoyancy of the slab concentrates in the outer rise. This tensional stress results in large normal-faulting earthquakes, which may actually "break" the subducting lithosphere. The 1933 Sanriku and 1977

Sumbawa earthquakes are examples of this type of earthquake. Figure 11.37 shows the tectonic setting for these normal-faulting events.

FIGURE 11.37 Schematic figure of the sense of motion on large normal-faulting events in subduction zones. In regions with weak interplate coupling, the normal faulting can involve great earthquakes, effectively detaching the deep plate.

11.3.2 Continental Collisions

A second type of convergent boundary is a continent–continent collision. Unlike oceanic lithosphere, continental lithosphere is too buoyant to subduct. When two continental masses come into contact along a convergent boundary, the relative motion between the plates is taken up in lithospheric shortening and thickening or, in extreme cases, lateral expulsion of lithosphere away from the collision along strike–slip faults. Although continent–continent collisions account for only a small fraction of the present convergent boundaries, they profoundly affect topography. The continent–continent collision between India and Eurasia produced the Himalayas and the Tibetan Plateau, which are the most conspicuous topographic features in the world. The two major drainages from the Himalayas (the Ganges and Indus) carry more than 50% of the world's river-delivered sediment load. Former continent–continent collisions are recognized as suture zones. The Appalachian and Ural mountains are examples of ancient collisions.

The seismotectonics of a continent–continent collision are complicated. Usually, faulting is dominated by thrusting, which is a manifestation of the lithospheric shortening. In many of the collisions, well-developed, low-angle thrusts occur beneath the suture zone. These low-angle faults can generate very large earthquakes ($M_W > 8.5$). The effects of collision can also be propagated large distances inboard of the suture. Strike–slip faulting in China and normal faulting at Lake Baikal are the result of the Indian–Eurasian collision 2000 km away.

Let us consider the Indian–Eurasian collision in more detail. Fifty million years ago oceanic lithosphere north of the Indian shield was subducting beneath Eurasia. By 40 million years ago the oceanic material had been completely consumed, and the Indian shield began to collide with the Eurasian continental mass. Because the continental lithosphere of India is too buoyant to subduct, the collision has been accommodated by uplift of the Tibetan Plateau (with a doubling of the crustal thickness) and "ejection" of Southeast Asia away from the collision along a series of strike–slip faults. India continues to move northward relative to Eurasia at a rate of 5 cm/yr, making the India–Eurasian collision zone one of the most seismically active regions in the world.

Figure 11.38 shows a tectonic map of the India–Eurasia collision zone. The boundary between India and Eurasia used to be the Indus–Tsangpo suture. Now the boundary is marked by an arcuate system of thrust faults over 2000 km long known as the Main Boundary Thrust (MBT) and the Main Central Thrust (MCT). At either end of the collisional arc are sharp "kinks" known as syntaxes. In the northeast the Indian plate is terminated by the Indo–Burma syntaxis. In the west it is terminated by the Pamir and Chaman faults in Pakistan. The Tibetan Plateau is a region of extreme elevation (in excess of 5 km) and thick crust (in places, in excess of 70 km). Although the exact mechanism of the uplift of Tibet is still a source of controversy, it represents significant lithospheric shortening. A series of strike–slip faults occur north and east of Tibet that "move" Eurasia away from the collision zone and elevate Tibet. These are some of the most active strike–slip faults in the world, and the long written history in China allows us to investigate the earthquake cycle.

The area affected by the India–Eurasia collision is extremely active seismically. The collision belt accounts for approximately 15% of yearly global seismic energy release. The majority of the energy is released along the MBT and MCT during major earthquakes. The Indian shield is being thrust beneath the Himalaya along a

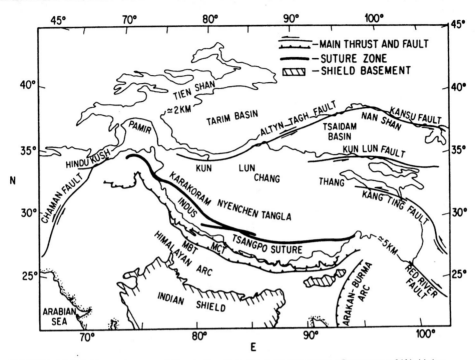

FIGURE 11.38 Tectonic map of the India–Eurasia collision zone. (Courtesy of W. Holt.)

plane dipping 10°–20°N. Figure 11.39 shows focal mechanisms of moderate earthquakes along the thrust zone. There have been four great earthquakes along the MBT in the past 100 yr (1897, 1905, 1934, 1950). The four great earthquakes have rupture zones up to 300 km long, and this area has a very serious seismic hazard. The 1897 event, located near the Indo–Burma syntaxis, is one of the largest earthquakes in recorded history. The east–west extent of the rupture zone is at least 300 km and may be as large as 500 km. The width of the fault is probably about 250 km. If we assume that these great thrust faults relieve nearly all the strain accumulation due to the movement of the Indian plate, events on each segment should repeat every 200–500 years.

The strike–slip faulting within Asia is not well understood. Two competing hypotheses for the existence of this faulting are as follows: (1) These faults are a major feature and take up most of the motion of the collision. They have been active nearly as long as the collision has been in existence. (2) They are minor, relatively recent features, which are a manifestation of the extremely overthickened crust of Tibet. The crust of Tibet is so thick that it now exerts a significant horizontal force, and tensional events occur within the plateau. At least a dozen magnitude 8.0 earthquakes have occurred along the strike–slip system of faults since 1500 AD. The Xianshuihe fault (300 km long) has had at least 11 magnitude 6.75+ events since 1725. Figure 11.40 shows the location of historic epicenters on the Xianshuihe fault. The earthquakes along the Altyu–Tagh–Nan Shan fault in northern China are frequently associated with very large surface displacements. The 1957 Gobi–Altai earthquake had displacements as large as 10 m.

Other continent–continent collisions include the Zagros thrust system in Iran and the Alpine arc in eastern Europe.

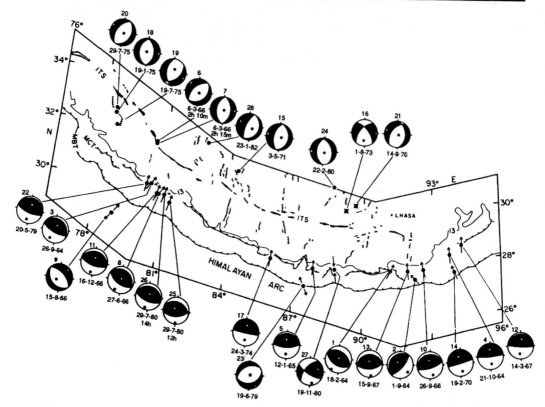

FIGURE 11.39 Focal mechanism and aftershock areas for events along the India–Eurasia collision zone. Note the east–west extension in the Tibetan Plateau, in contrast to the underthrusting mechanisms of the Himalayn front. (From Ni and Barazangi, *J. Geophys. Res.* **89**, 1147–1163, 1984; © copyright by the American Geophysical Union.)

11.4 Intraplate Earthquakes

Although the vast majority of seismic moment release occurs at plate boundaries, some regions of seismicity are far removed from plate boundaries. This seismicity is referred to as *intraplate*, and it represents internal deformation of a plate. The nature of intraplate seismicity is often quite complicated, and the tectonic driving mechanisms are poorly understood. Intraplate and interplate earthquakes differ in two important ways. First, the recurrence interval of intraplate events is generally much longer than that of interplate events, and second, intraplate events typically have much higher stress drops. These two observations may be coupled; since intraplate faults fail infrequently, they appear to be "stronger" than interplate faults. There are several possible explanations for this. Faults that move frequently with a high slip rate will produce a well-defined gouge zone, which weakens the fault. Further, *interseismic healing*, involving chemical processes that progressively weld a fault, will be more effective for faults with long recurrence times.

Intraplate seismicity is difficult to characterize seismically because the recurrence times are long and we have few examples of repeat events. Often, earthquakes occur on old zones of crustal weakness such as sutures or rifts that are *reactivated* by the present stress field (Sykes, 1978). The origin of the stress field that causes intraplate

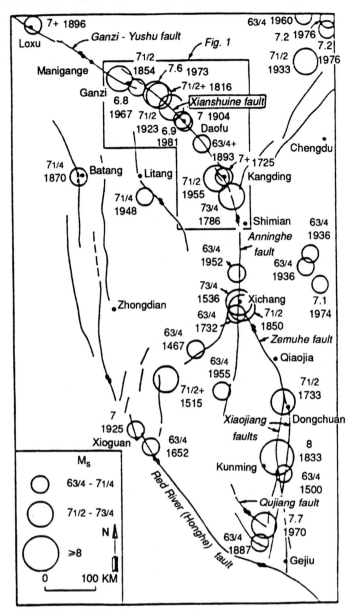

FIGURE 11.40 Earthquake history of the Xianshuihe fault in western China (Yunan). (From Allen *et al.*, 1991.)

earthquakes is somewhat problematic. It appears to stem mainly from the driving forces of plate tectonics, such as slab pull or ridge push, far from where faulting takes place. The stress is probably "localized" or concentrated by weak structures. The majority of intraplate events have P and T axes that are consistent with regional stress provinces. A few localized driving forces, such as removal of a glacial load, stresses induced by surface topography, loading of reservoirs, ascent of mantle plumes, or delamination of the deep crust, may also cause intraplate earthquakes.

The faults that produce intraplate earthquakes are not easily recognized at the surface. This is because the faulting is usually several kilometers deep, and little cumulative offset occurs because of the long recurrence intervals. The eastern and midcontinent regions of the United States are usually thought of as "stable" and seismically inactive, but in fact, some very large earthquakes have occurred in these regions. In 1886 an earthquake devastated the Charleston, South Carolina, area. Figure 11.41a shows the intensity map (intensity is a measure of perceptible shaking or damage from an earthquake) for the Charleston earthquake; it was felt as far away as Chicago. Considerable controversy surrounds the mechanism of the Charleston event, but it has been suggested that it is the reactivation of a detachment (horizontal fault) beneath the southern Appalachians. This detachment may be a feature similar to the MBT in the Himalaya.

A sequence of earthquakes near New Madrid, Missouri, during the winter of 1811–1812 represents the largest seismic energy release episode in the historical record of the eastern United States. Figure 11.41b shows the isoseismals (contours of equal shaking damage) for the first of the New Madrid events (three major events with magnitudes greater than ~ 8.0 occurred on December 16, 1811, January 23, 1812, and February 7, 1812). The earthquakes were felt as far away as New York City. These events caused a major change in the course of the Mississippi River, and the land along the banks of the Mississippi sank up to 5 m in several places. Reelfoot Lake in western Tennessee was formed by the subsidence of swamp land (the lake is 30 km long, 10 km wide, and 2 to 4 m deep). Figure 11.42 shows the recent (1974–1990) seismicity in the New Madrid seismic zone. There are two prominent trends in seismicity: (1) a 120-km-long lineament trending from Arkansas to the tip of Kentucky and (2) a nearly orthogonal segment in southeastern Missouri. The first

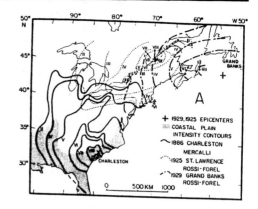

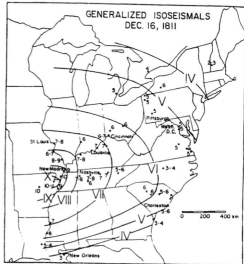

FIGURE 11.41 (a) Isoseismals for the 1886 Charleston, South Carolina, earthquake. (From Seeber and Armbruster, 1981.) (b) Isoseismals for the December 16, 1811 earthquake in New Madrid, Missouri. (From Nuttli, 1973.)

feature parallels the Mississippi Embayment, a trough that increases in depth to the south. The Mississippi Embayment has been postulated to be a failed arm of a triple junction, which would indicate that it is a rift-like structure. Focal mechanisms of microearthquakes within the New Madrid seismic zone indicate that the fault slip is a mixture of thrusting and right-lateral strike–slip. The usual explanation for the earthquakes is reactivation of an ancient zone of weakness under the action

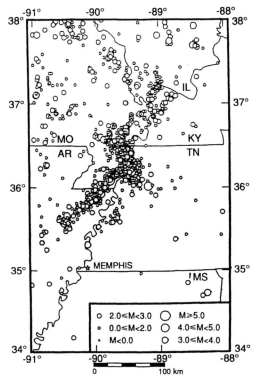

FIGURE 11.42 Seismicity in the New Madrid seismic zone. (Courtesy of R. Herrmann.)

Some investigators have suggested that it is not a true intraplate setting but a complex "back arc" to a convergent boundary. On the other hand, the earthquake recurrence rate on most of the faults is longer than 1000 years, and the stress drops of most recent events are consistent with those of intraplate earthquakes. Table 11.2 gives the largest earthquakes to occur in the Basin and Range since 1872. The earthquakes appear to be clustered around the edges of the province. Figure 11.43 shows the location of the largest events in the western United States that have occurred since 1769 and their surface fault scarps. Although the San Andreas is the most active feature, the Basin and Range events account for nearly the same total energy release. Of course, a magnitude 9.5 thrust event along the Oregon–Washington coast, which is not implausible, would dwarf all of these events.

The apparent clustering of Basin and Range seismicity around the margins of the province may be an artifact of a short period of observation, but extension of the Basin and Range probably takes place at its edges. There are two prominent "seismic belts," the Nevada Seismic Belt in west central Nevada, and the Wasatch fault zone in central Utah (no historic events shown in Figure 11.43, but many fault scarps are found in the region, even in Salt Lake City). Although individual faults in the Basin and Range have long recurrence intervals (> 1000 years), it appears that large seismic events may cluster temporally within a restricted geographic region. For example, note the "burst" of activity along the Nevada Seismic Belt in the last century (1887, 1915, 1932, 1934, and three events in 1954). This region may have been quiet for the previous millennia. Temporal clustering has profound implications for seismic hazard evaluation. During a temporal burst a region will have an elevated hazard; once the burst ceases, the hazard is greatly reduced. There is no good explanation for the apparent cyclic seismic

of the current regional stress system, which is completely different from the one that formed the Mississippi Embayment.

The large size of the 1811–1812 earthquakes makes the New Madrid seismic zone an important seismic hazard. It is very difficult to determine the recurrence interval for the 1811–1812 sequence, since there is no surface expression of the faulting. Using the *b* values determined from the instrumental record in the last 30 years, the recurrence is thought to be on the order of 1000 years.

The western United States is one of the most seismically active intraplate environments. The seismicity is located primarily in the Basin and Range Province, and nearly all the earthquakes are associated with normal faulting or a mixture of normal and strike–slip faulting. The Basin and Range is a greatly extended terrain.

TABLE 11.2 Large Basin and Range Earthquakes

Name	Date	Latitude (°N)	Longitude (°W)	Magnitude (M_s)
Owens Valley	March 26, 1872	36.5	118.3	8.3
Pitaycachi	May 3, 1887	31.0	109.2	7.2
Pleasant Valley	October 10, 1915	40.5	117.5	7.5
Clarkson	June 27, 1925	46.0	111.2	6.7
Cedar Mountain	December 21, 1932	38.8	118.0	7.2
Excelsior Mountain	January 30, 1934	38.3	118.5	6.3
Harsel Valley	March 13, 1934	41.8	112.9	6.6
Virginia City	October 18, 1935	46.6	112.0	6.2
Fallon	July 6, 1954	39.3	118.4	6.6
Fallon	August 24, 1954	39.4	118.4	6.8
Fairview Peak	December 16, 1954	39.2	118.0	7.0
Dixie Valley	December 16, 1954	39.7	117.9	6.8
Dixie Valley	March 23, 1959	39.4	118.0	6.1
Hebgen Lake	August 17, 1959	44.8	111.1	7.1
Pocatello Valley	March 28, 1975	42.2	112.5	6.0
Mammoth Lakes	May 25, 1980	37.6	118.6	6.4, 6.3
Mammoth Lakes	May 27, 1980	37.6	118.6	6.3
Borah Peak	October 28, 1983	44.0	113.9	7.0

energy release, but apparently the strain release must be related to regional deformation in the lowermost crust or the upper mantle.

Some intraplate earthquakes are *induced*; this means that human activity triggers the earthquakes. This phenomenon was first observed with the filling of Lake Mead behind Hoover Dam beginning in 1935. Before the reservoir was filled, the background seismicity along the Nevada–Arizona border was low; beginning in 1936 the seismicity began to rise, culminating in a magnitude 5.0 earthquake in 1940. After this largest event, the seismicity has declined. The epicenters of the earthquakes near Lake Mead are all shallow (most less than 6 km depth) and appear to cluster on steep faults on the east side of the lake. Since the Lake Mead experience, more than 30 examples of reservoir-induced seismicity have been documented. Figure 11.44 shows the correlation between reservoir level and local seismicity for the Koyna dam in the Deccan traps of western India. Impounding of water began at Koyna in 1962, in an area that appears to have been

aseismic for at least the previous hundred years. In early 1964, earthquakes began to occur under the lake at shallow depths. In 1967, earthquakes with $M = 5.5$ and 6.5 occurred, the largest event causing 117 casualties. Since that time, the seismicity level has decreased, although a peak in seismicity appears to occur 1 to 2 months after the rainy season when the reservoir is at its highest level.

What causes reservoirs to generate earthquakes? It is unlikely that the cause is simply the weight of the water, which would only add a tiny fraction of the total stress 2 or 3 km below the surface. A more likely explanation is that the *pore pressure* increases because of the hydrostatic head of the reservoir. In Figure 9.1 we showed a Mohr circle and stated that faulting occurs when the circle intersects the failure-criteria envelope. The effect of increasing pore pressure will move the Mohr circle toward the origin (the radius of the circle remains the same), thus closer to failure. The fact that for some dams, such as Lake Mead, the seismicity returns to prereservoir levels after a large event implies that

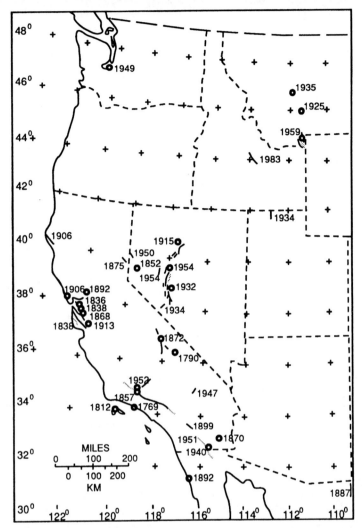

FIGURE 11.43 Fault scarps of the historic western United States earthquakes. Note that most are *not* on the plate boundary, which is defined by the San Andreas fault in central and Southern California and the Cascadian subduction zone along Northern California, Oregon and Washington.

once the regional strain accumulation is relieved (triggered), the radius of the Mohr circle decreases.

11.5 The Earthquake Cycle

In our discussions of the earthquake process up to this point we have dealt with earthquakes as isolated, unique episodes.

But the fact is, nearly all shallow earthquakes with magnitudes greater than 6.0 occur on *preexisting* faults. This implies that the vast majority of earthquakes represent a repeat rupture of an established fault. This was first recognized by H. F. Reid, who was commissioned by the Carnegie Institute to study the 1906 San Francisco earthquake. Reid came up with the *theory of elastic rebound*, which is illus-

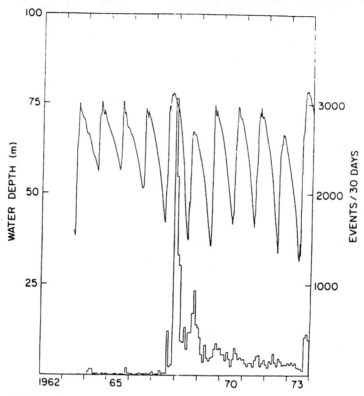

FIGURE 11.44 Correlation between the reservoir level at Koyna dam and the seismicity in the area. (From Simpson *et al.*, 1988.)

trated in Figure 1.4. Strain accumulates along the fault because of relative motions on either side of the fault. The fault is "locked" due to friction and does not move. Once the strain reaches a critical value, on the order of 10^{-5}, the fault overcomes the frictional resistance and the strain is relaxed. Once the fault has finished slipping, the process of strain accumulation begins anew. The idea that faults fail repeatedly, with some periodic behavior, is extremely important for earthquake prediction and mitigation of earthquake hazards. The key aspect connecting elastic rebound and earthquake periodicity is the steady application of loading motions and accumulation of strain. In our discussions of plate tectonics in the last few sections, we discussed relative plate velocities. If

the tectonic processes that cause these relative motions are steady, the strain accumulation should also be steady. Thus, if we assume fault behavior is time independent, earthquakes should occur along faults at regular intervals known as *recurrence intervals*.

Figure 11.45a shows the predicted fault behavior and strain release for such a system. τ_1 represents the strength of the fault; once this value is reached, the fault fails, and the strain is reduced to some minimum value τ_2, which depends on the fault friction. The fault offset, or displacement across the fault plane, would be the same for each event, and the recurrence interval is constant. This similarity of each earthquake gives rise to the concept of a *characteristic earthquake*. The characteristic

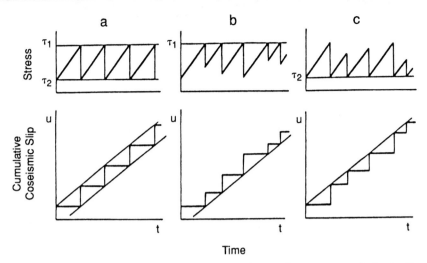

FIGURE 11.45 Various scenarios for buildup and release of stress on a fault. (a) Regular stick–slip faulting. (b) Time-predictable model. (c) Slip-predictable model. (From Shimazaki and Nakata, *Geophys. Res. Lett.* **7**, 279–282, 1980; © copyright by the American Geophysical Union.)

earthquake model suggests that faults are *segmented*, and individual segments behave in a predictable fashion.

Unfortunately, although the characteristic earthquake model is important for framing the problem of earthquake recurrence, it is rarely observed in nature. The reasons are many: apparently fault mechanical behavior is not constant, the strain accumulation on a fault is not purely elastic, and plate tectonic processes may not be steady in the short term. In Figure 11.45 the two stresses, τ_1 and τ_2, control the behavior of the fault. Given the nonlinear nature of frictional sliding and occurrence of adjacent events and nonuniform slip, it would be remarkable for τ_1 and τ_2 to be constants. Figures 11.45b and 11.45c show two more general types of behavior in which either τ_1 or τ_2 is allowed to vary. When τ_1 is constant, the earthquake behavior is said to be *time predictable*. The stress drop may vary from event to event; thus the time to the next earthquake will vary. This time can be predicted from the time–displacement line for steady fault motion. In the second model, earthquakes occur over a range of

τ_1, but the fault always relaxes to a constant τ_2. This is referred to as *slip predictable*.

If the time-predictable model is correct, then the amount of displacement in an event will specify the time interval to the next event. On the other hand, if the slip-predictable model is correct, the lapse time since the last event specifies the potential fault displacement at any given time. In both cases an estimate of the long-term relative motion is required. Numerous faults have been investigated to determine whether either of these models is applicable. It is clear that neither model is perfect, a situation some seismologists jokingly refer to as the "unpredictable" model. More faults have a weak tendency to exhibit time-predictable behavior. Figure 11.46 shows the cumulative slip and moment release along the Calaveras fault near San Francisco Bay, California. This fault appears to have time-predictable behavior, but only a short time window is available to characterize the statistics.

One of the biggest difficulties in determining the characteristics of fault behavior is that we rarely have more than one or

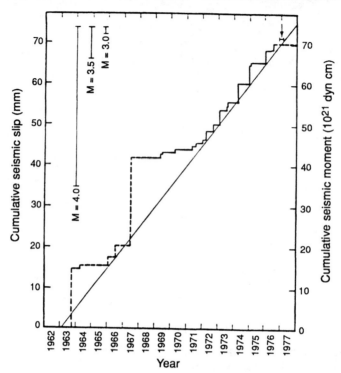

FIGURE 11.46 Cumulative moment and seismic slip in a zone of the Calaveras fault (1962–1977) in central California. (From Bufe *et al.*, 1977.)

two cycles in the historical record. Box 11.4 describes a field known as *paleoseismology*, in which geologic disturbances associated with fault movement (creation of fault scarps, offset of streams, landslides, soil liquefaction) can be assessed to increase the earthquake record of a fault. Figure 11.47 summarizes paleoseismic work on a section of the San Andreas fault near Pallet Creek, California. The mean recurrence time from 10 dated earthquakes is 131 years. The last event occurred in 1857, so this mean value is of much concern. It is clear that there are significant deviations from this recurrence interval; further, the earthquakes may form four clusters (events 1, 2, and 3; 4, 5, and 6; 7 and 8; 9 and 10). In these clusters the earthquakes occur over a short interval, then a long interval of quiescence takes place. How does this relate to the time-predictable model? It would seem that these models for earthquake recurrence are too simplistic. In general, the *mean*

recurrence time may be well defined, but significant stochastic fluctuations occur. Recent research on earthquake recurrence has begun to parameterize the dynamics of the stress loading cycle in terms of nonlinear dynamics. The solutions to nonlinear systems often exhibit *chaotic* behavior, which is very difficult to predict. This has important repercussions for those who desire to predict earthquakes with exacting precision of time, place, and size.

Despite the variability and difficulties of estimating earthquake recurrence, we can make some basic generalizations. First, total slip along a fault must balance the plate motions. In other words, if a fault is segmented, the total slip on the segments over many cycles must be equal and consistent with expected slip. Second, plate tectonic driving processes are approximately steady, and average recurrence times can be estimated. Figure 11.48 shows the earthquake history between 1918 and 1973 in the Kuril–Japan subduction zone.

The entire length of the subduction zone broke in a series of nonoverlapping earthquakes, and presumably the process will repeat itself. Consider what Figure 11.48 would have looked like in 1970; the entire fault would have ruptured except for a region between the source areas of the 1952 and 1969 earthquakes. This region would be referred to as a *seismic gap*. This term is used for regions that are known to have had previous earthquakes (so it is known that they are not aseismic) but have not had an event for an interval that is a large fraction of the average regional recurrence time. Figure 11.49 shows a longer time sequence of ruptures along the Nankai trough in southern Japan. A seis-

mic gap is present near Tokai. If one can identify seismic gaps, those regions can be assigned a relatively high earthquake potential. Nishenko produced a map of seismic gaps in the circum-Pacific region, which is reproduced in Figure 11.50. The procedure has fairly successfully identified areas that have subsequently had large earthquakes, but numerous surprises have also occurred, where events reruptured a zone that was not considered to be a "mature" gap.

Another interesting observation of some earthquake sequences is an occasional apparent "migration" of activity. Figure 11.51 shows the north Anatolian fault in Turkey. This fault is a major transcurrent bound-

Box 11.4 Paleoseismology

Accurate assessment of seismic hazard in any region requires careful estimation of seismic energy release. Unfortunately, for most regions on the Earth, "significant" seismicity in the instrumental record (post-1960) is insufficient to characterize adequately the nature of the moment release. In these areas we must resort to *paleoseismic* studies, which are investigations of the geologic signature of earthquakes that occurred in the Holocene. When a large earthquake occurs, it may produce a fault scarp, trigger landslides, cause soil liquefaction, offset streams, or produce broad regions of uplift or downdrop. All of these features can be preserved in the geologic record and can be recognized hundreds, and sometimes thousands, of years later. Identifying these features allows seismologists to construct a much longer seismic time line.

Fault scarps are an excellent example of paleoseismic indicators. Consider a vertical dip–slip earthquake; when the fault ruptures the surface, it produces an abrupt offset. The scarp is steep and well defined immediately following the earthquake. As time passes, the scarp erodes and changes shape. Figure 11.B4.1 shows a hypothetical time history of such a scarp. It turns out that the change in shape due to erosion is governed by the diffusion equation, and the degrading scarp can be modeled with an *error function*. If we know the initial conditions of the scarp (dip angle) and prefaulted surface (slope and existence of previous scarps), and if we have control on the "erodibility" of the scarp, we can *date* when the scarp formed. Figure 11.B4.2 shows such an analysis for a scarp on the Lost River fault in southern Idaho. The best fit to the profiles gives an age of 6000 to 3000 years before the present for the scarp formation.

continues

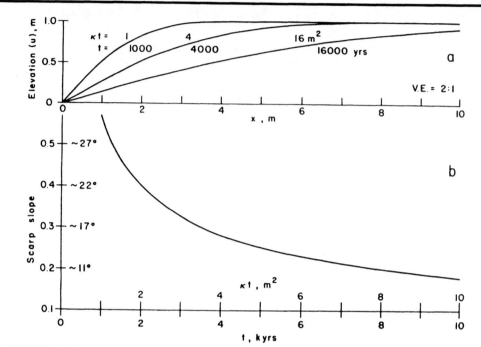

FIGURE 11.B4.1 The evolution of a hypothetical fault scarp with time. The change in scarp shape is controlled by the diffusion equation. (From Hanks *et al.*, 1984) *J. Geophys. Res.* **89**, 5771–5790; copyright by the American Geophysical Union.)

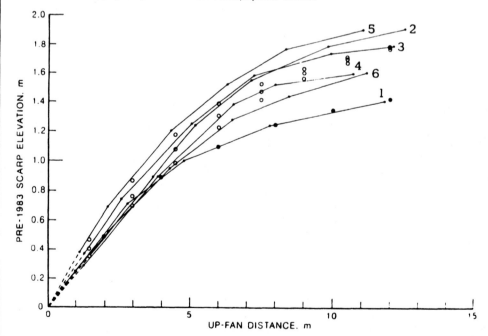

FIGURE 11.B4.2 Profiles across the Borah Peak fault scarp before the 1983, $M_S = 7.1$ earthquake. The "average" profile suggests the last event before 1983 occurred approximately 5000 years ago. (From Hanks and Schwartz, 1987).

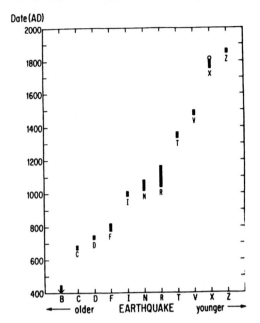

Date (AD)

FIGURE 11.47 Dates of the inferred last 10 earthquakes on a section of the southern San Andreas fault at Pallet Creek. The dates were determined from organic materials trapped along disturbed strata. (From Sieh *et al.*, *J. Geophys. Res.* **94**, 603–624, 1989; © copyright by the American Geophysical Union.)

ary, which is often compared to the San Andreas fault. This century the earthquake activity has proceeded roughly east to west. Although this example may be a coincidence, enough other examples give migration some credibility. Identifying seismic gaps and migration is problematic in intraplate environments. Some intraplate regions appear to be subjected to episodic strain rates, as described earlier for the western Basin and Range in Nevada. In the Middle East, earthquakes appear to migrate from one intraplate block to another, concentrating earthquake activity on that block for several centuries before entering a very long quiescent period. Slow strain pulses migrating in the lithosphere may account for this behavior, but no observations of such deformation "waves" have yet been made.

11.6 Earthquake Prediction

One of the most important societal goals of seismologists is the prediction of earthquakes. Accurate predictions could save not only lives but also billions of dollars worth of equipment and buildings. This goal has yet to be fulfilled, but there has been some progress in understanding effects that might be expected before an earthquake. However, earthquake prediction is also a research field that has had a checkered reputation. Examples are too numerous in which the public has been made to fear an impending earthquake on the basis of a "prediction" that was either observationally or theoretically flawed.

It is best to start a discussion of earthquake prediction with a basic definition. A *prediction* is *successful* when it provides an accurate assessment of the *time*, *place*, and *size* of an earthquake. Of course, predictions are usually categorized on the basis of bounds on these three parameters. The basic categories are (1) long-term (made up to years in advance), (2) intermediate-term (made weeks in advance), and (3) short-term (made hours or days in advance). The accuracy of the time, place, and size of the predicted earthquake differs with the type of prediction. Society responds to these different types of predictions in very different manners: long-term predictions can affect urban planning, resulting in programs to mitigate the effects of an earthquake, such as reinforcing buildings. Intermediate-term predictions should promote emergency preparedness and planning. Finally, short-term predictions can lead to evacuation orders and shutting down industries that would be critically damaged during an earthquake. Predictions are inherently a *social* exercise, and it is important not to couch predictions as purely a scientific endeavor. Consider the social consequences of any prediction: it may lead to reduction in property value, business losses, and gen-

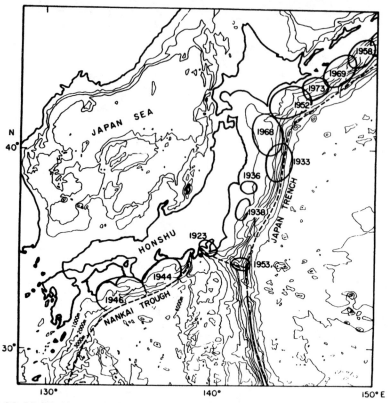

FIGURE 11.48 Slip history of the Kuril–Japan subduction zone. Note how the ruptures this century have filled in the thrust zone. Events in the previous century tended to occur in the same regions. (From Mogi, 1985.)

eral economic depression. This puts extraordinary pressure on the seismologist to be correct in a field that is intrinsically imprecise, but it also focuses attention on the social importance of seismology.

Long-term predictions are largely based on identification of fault characteristics such as segmentation, recurrence interval, and the time of the last earthquake. The identification of seismic gaps is an example of a long-term prediction. Intermediate- and short-term predictions, on the other hand, are based on *precursory phenomena*. By this, we mean observable changes caused by strain accumulation leading up to the earthquake. It has been observed that the volumetric properties of rock change as a function of strain. Models to

explain these changes are known as *dilatancy models*. Bridgman (1949) first noted in the laboratory that rocks that are subject to uniaxial loads experience changing stress–strain behavior. Initial loading produces more volumetric compaction than would be expected from simple solid elasticity. The closing of microfractures is thought to cause this. After these cracks are closed, the stress–strain relation is linear, as expected in a solid. At stresses of approximately one half the fracture stress, rocks typically dilate or expand. A simple example of this behavior is walking on wet sand at the beach. Stepping on the sand disturbs the sand grain compaction, and for a short time, the sand is *undersaturated* due to a *local* porosity increase. In rocks,

Box 11.5 Earthquake Probability

The 1971 San Fernando, California earthquake caused more than $1.2 billion of damage and killed 64 people. It was painfully evident that the United States was not addressing the tremendous potential economic disaster associated with earthquakes. Congress created the National Earthquake Hazards Reduction Program (NEHRP) to address scientific, engineering, and social issues associated with earthquake occurrence. One of the principal goals of NEHRP was the *prediction* of earthquakes. Although much progress has been made toward understanding the earthquake process, a foolproof methodology for accurately forecasting events remains elusive. However, many regions in the United States are now assigned "earthquake probabilities" by the U.S. Geological Survey. The probabilities of rupture on a given fault are based on knowing the recurrence interval and the time of the last major earthquake.

Figure 11.B5.1 shows recurrence information for a number of plate boundaries where three or more earthquakes have occurred on the same fault segment (Nishenko and Buland, 1987). Shown is the ratio of the recurrence interval of a given earthquake to the mean recurrence interval of the region. If earthquakes behaved in a purely periodic fashion, then T/T_{ave} would always be 1. Although the mean recurrence is well defined, individual recurrence rates vary significantly. The probability that an earthquake will occur at the *mean* recurrence interval depends on the shape of this diagram. Seismologists have attempted to fit Gaussian, lognormal, and Weibull functions to this distribution, with reasonable success.

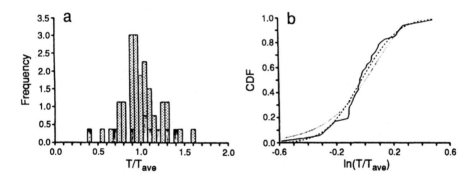

FIGURE 11.B5.1 (a) The distribution of recurrence intervals normalized by the average recurrence. (b) The probability density function for the recurrence distribution shown in (a). (From Nishenko and Buland, 1987.)

Once the type of function is chosen, the probability that an earthquake will occur at some time t in the interval $T < t < T + \Delta T$ is given by

$$P(T \leq t \leq T + \Delta T) = \int_{T}^{T+\Delta T} f(t)\, dt, \qquad (11.5.1)$$

continues

where $f(t)$ is defined from the functional form of the recurrence distribution (Figure 11.B5.1). The probability in Eq. (11.5.1) can be modified to calculate a conditional probability if the time of the last event is known:

$$P(T \leq t \leq T + \Delta T | t > T) = \int_{T}^{T+\Delta T} f(t)\, dt \Big/ \int_{T}^{t} f(t)\, dt. \qquad (11.5.2)$$

The conditional probability needs to be updated as T increases. For example, a section of the San Andreas fault may have a probability of an earthquake occurring in the next decade of 0.3; if the earthquake does not occur within 5 yr, the process can be repeated, and the probability will *increase* for the next 10-yr interval. In 1988 the Working Group on California Earthquake Probabilities produced a conditional probability map for the San Andreas fault for the ensuing three decades (shown in Figure 11.B5.2). A key assumption in this was the segmentation of the fault into independent rupture zones. The area near Parkfield currently has a high probability of a $M_W \approx 6.0$ event occurring. Fault rupture in this region has a recurrence interval of about 22 yr, with the last event occurring in 1966. Several other segments have a 20–30% probability of magnitude 7 to 8 earthquakes over the next 30 yr.

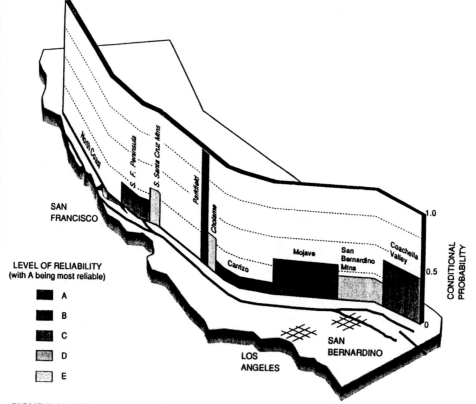

FIGURE 11.B5.2 The conditional probability of major earthquakes along different segments of the San Andreas fault. The probability illustrated is for the time interval 1988–2018. (From Agnew *et al.*, 1988).

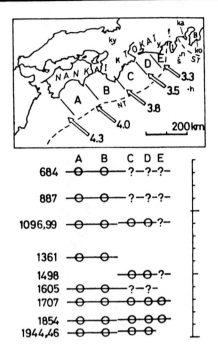

FIGURE 11.49 History of large earthquakes along southeast Japan. Segments *AB* always rupture together. Segments *CD* rupture together, sometimes with *E*. In 1707 all segments ruptured in a single event. Segment *E* is a seismic gap called the Tokai gap. Most of the Japanese earthquake prediction effort is concentrated on this gap because it is close to Tokyo. (From Ishibashi, 1981.)

the most plausible explanation for this nonelastic volume increase is the development of microcracking throughout the rock, increasing the void space. Finally, these microcracks coalesce, leading to the formation of a fault.

The dilatancy model provides a physical framework to investigate expected precursory behavior. Figure 11.52 shows changes of a rock under compressive loading. Note the volume increase ($\Delta V/V_0$), which occurs dramatically at about 50% of the failure stress. This represents the development of cracks, which in turn causes variations in many of the physical properties of the medium. The behavior of rocks during dilatancy depends strongly on whether the cracks are dry or wet. In a wet model, the

diffusional water plays an important role. Figure 11.53 shows the expected physical parameter changes for wet and dry dilatancy models. There are several predicted changes that the seismologist should be able to measure. For example, both models predict changes in seismic velocity. In the beginning stages of dilatancy, the development of cracks causes this because they reduce the elastic modulus. From laboratory measurements, these cracks appear to affect α much more than β, which results in a decrease of α/β by 10–20%. During this stage, land uplift and ground tilt are also expected. In late stages, the velocity ratio returns to normal due to either water saturation of the cracks or closing of the porosity.

Another important predicted change is the level of emission of the radioactive gas radon. Radon is only one example of a class of phenomena known as *geochemical precursors*. Most geochemical precursors are associated with ground water and involve dissolved ions or gases. Groundwater anomalies were among the earliest and most frequently reported earthquake precursors. Changes in taste and temperature in wells and springs prior to large earthquakes were documented in Japan more than 300 years ago. Most seismologists believe that microfracturing prior to major earthquakes in late stages of dilatancy causes increases in ion and gas concentrations in ground water. Laboratory studies of dilatant rocks show that porosity increases from 20% to as much as 40% prior to rupture. Dissolution or alteration of fresh rock surfaces could significantly increase the ion concentrations in the ground water. In the case of radon, which is produced by the radioactive decay of radium, the fresh fractures allow more of the gas to escape. Most precursory radon anomalies appear a few weeks to days before the impending earthquake. Radon is commonly used because it is easily measured. Other gas anomalies include hydrogen and helium. Figure 11.54 shows the

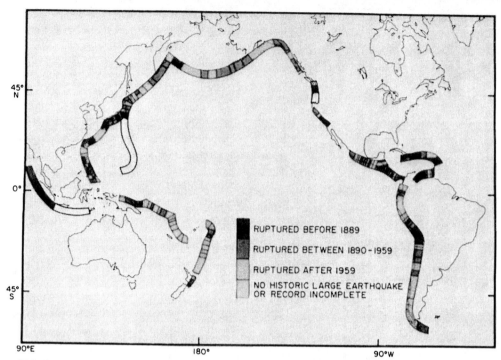

FIGURE 11.50 Seismic gaps and seismic potential for the circum-Pacific region. (From Nishenko, 1991.)

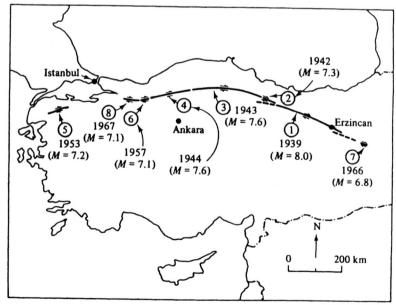

FIGURE 11.51 Fault displacements along the North Anatolian fault associated with major earthquakes since 1939. (From Allen, 1969.)

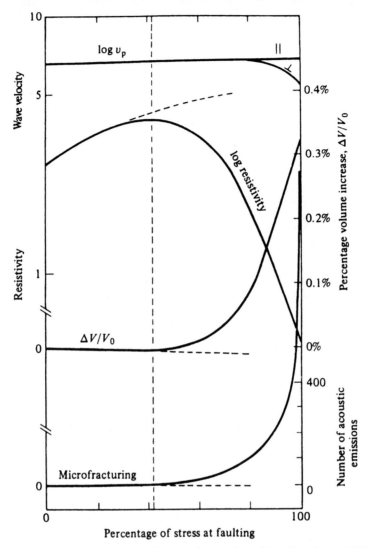

FIGURE 11.52 The behavior of rock under loading. (From Kasahara, 1981.)

precursory changes of radon, water temperature, water level, and strain before the 1978 Izu–Oshima–Kinkai earthquake (M_s = 7.0). The precursor is clear and well defined. Unfortunately, there are numerous examples of *no* geochemical precursors associated with other earthquakes, and comparable fluctuations in all of these measures occur that are not associated with an earthquake.

The easiest precursory indicator to measure is the seismicity pattern. Within a

source region of a large earthquake, numerous small faults or heterogeneities on the main fault probably exist that can produce earthquakes in response to the loading cycle. It has been found that after the aftershock sequence of a major earthquake is complete, the seismicity level drops to a low background level. Prior to the next event, the background activity sometimes drops almost to zero for an interval of months or years. This phenomenon is known as *seismic quiescence*

Box 11.6 The Crustal Deformation Cycle

It is difficult to evaluate the cyclic nature of earthquakes without considering the entire process of crustal deformation. The *crustal deformation cycle* is often divided into four phases: (1) interseismic, (2) preseismic, (3) coseismic, and (4) postseismic. If Reid's model of elastic rebound were completely correct, only two phases of deformation would occur: coseismic and interseismic. The deformation cycle is more complex because the Earth does not respond uniformly to strain. A more complete representation of the deformation process is given by an elastic shell, which contains the fault, over a viscoelastic asthenosphere. The elastic shell, which is the lithosphere, is driven by a remote tectonic process such as ridge spreading. The fault in the lithosphere is divided into two parts: (1) the seismogenic zone, where strain is released only in the sudden slip of an earthquake, and (2) a deep region where the fault creeps continuously. As the seismogenic zone cycles to failure, a strain is imparted to the rest of the system. Immediately after failure, this strain is concentrated close to the fault. As time increases, the stress diffuses outward in the viscous material. Figure 11.B6.1 shows the vertical land movement determined by leveling in the region near the 1947 Nankai ($M = 8.1$) earthquake. Note that for 3 yr after the earthquake, deformation occurred at a much higher rate than that expected from plate motion. This is a manifestation of the viscous relaxation of the stress.

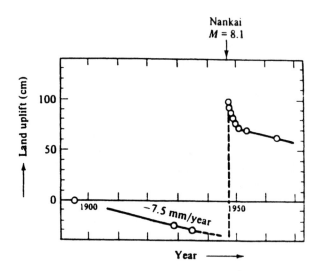

FIGURE 11.B6.1 Vertical ground motion preceding and following the 1947 Nankai earthquake. (From Okada and Nagata, 1953.)

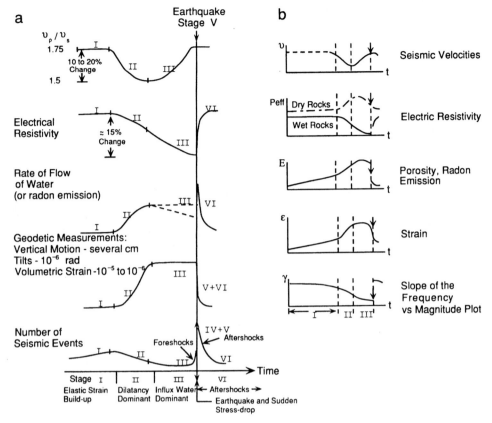

FIGURE 11.53 Expected physical changes for dilatancy. (a) Wet model. (b) Dry model. (From Kasahara, 1981.)

and is fairly common. The quiescence is often broken by a buildup of activity, with swarms of activity before the mainshock, which is known as a preshock sequence. K. Mogi noted that the earthquake activity before the mainshock tends to concentrate on the periphery of the impending rupture zone. This "outlining" of a rupture zone often looks like a doughnut, and this pattern is known as a *Mogi doughnut*. Figure 11.55 shows a space–time plot of seismicity for the Kuril subduction zone. The largest event along this trench, the 1963 $M_w = 8.5$ earthquake, was preceded by a distinct period of quiescence. The length of the 1963 rupture zone can be inferred from the lineament of aftershocks (nearly 400 km long). In 1961, the seismic activity

became very high along a substantial length of the Kuril arc, but the year before the event the entire region was quiet.

Seismic quiescence is usually ended by *foreshock* activity, earthquakes that occur just before the mainshock. Of earthquakes with $M > 7.0$, almost 70% are preceded by foreshocks even if no quiescence occurs (Jones and Molnar, 1979). The nature of foreshock activity varies greatly, but it typically begins 5–10 days before the mainshock. This is consistent with the dilatancy models that include rapid microcracking just before failure.

In summary, many phenomena appear to be precursory to earthquakes, so various levels of earthquake prediction should be possible. Unfortunately, each earthquake

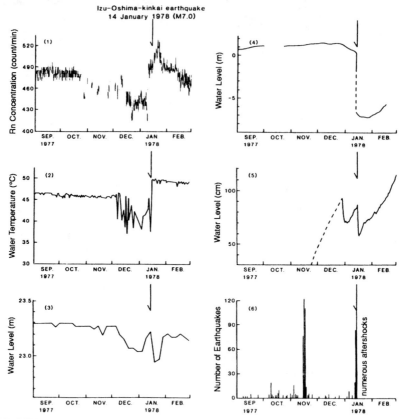

FIGURE 1.54 Precursory changes in radon concentration, water temperature, water levels, and number of minor earthquakes prior to the Izu–Oshima earthquake of 1978. (From Wakita, 1981.)

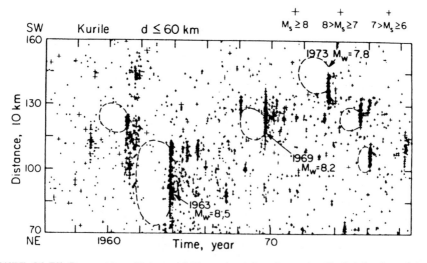

FIGURE 11.55 Space–time history of the seismicity along the Kuril Islands subduction zone. Periods of quiescence preceding large earthquakes are circled. (From Kanamori, 1981.)

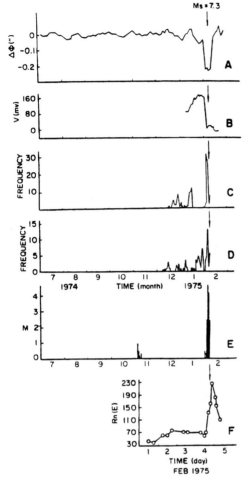

FIGURE 11.56 Changes in short term anomalies before the 1976 Haicheng earthquake. (From Zhang and Fu, 1981.) A) Ground tilt at Shenyang station (Δ = 150 km); B) Electrical potential difference at Yang Kou station (Δ = 15 km); C) Number of unusual animal behavior reports in the Dandong area (Δ = 150 km); D) Number of unusual groundwater phenomena in Dandong area (Δ = 150 km); E) Foreshocks; F) Radon content at the Liuhetang Hot Spring (Δ = 72 km).

animal behavior (Figure 11.56). A prediction, based on the foreshocks, was issued by the Provincial Seismological Bureau on the day of the earthquake, and buildings and communes were evacuated. The earthquake was very destructive, but almost no one died. This success caused great excitement in the earthquake prediction community. Unfortunately, a year later a magnitude 7.7 earthquake destroyed Tangshan, a city only 200 km southwest of Haicheng, and killed more than a quarter of a million people. In that case no precursors were detected, and no prediction was issued. Seismologists clearly have their work cut out for them.

References

Agnew, D., Allen, C. R., Cluff, L. S., Dieterich, J. H., Ellsworth, W. L., Keeney, R. L., Lindh, A. G., Nishenko, S. P., Schwartz, D. P., Sieh, K. E., Thatcher, W., and Wesson, R. L. (1988). Probabilities of large earthquakes occurring in California on the San Andreas fault. *Geol. Surv. Open-File Rep. (U.S.)* **88–398**.

Allen, C. R. (1969). "Active Faulting in Northern Turkey," Contrib. No. 1577. Div. Geol. Sci., Calif. Inst. Technol., Pasadena.

Allen, C. R., Luo, Z., Qian, H., Wen, X., Zhou, H., and Huang, W. (1991). Field study of a highly active fault zone: The Xianshuihe fault of southwestern China. *Geol. Soc. Am. Bull.* **103**, 1178–1199.

Atwater, T. (1970). Implications of plate tectonics for the Cenozoic tectonic evolution of western North America. *Bull. Geol. Soc. Am.* **81**, 3513–3536.

Atwater, T., and Mudie, J. D. (1968). Block faulting on the Gorda rise. *Science* **159**, 729–731.

Bridgman, P. W. (1949). Polymorphic transitions and geologic phenomena. *Am. J. Sci.* **243A**, 90–97.

Bufe, C. G., Harsh, P. W., and Burford, R. O. (1977). Steady state seismic slip—A precise recurrence model. *Geophys. Res. Lett.* **4**, 91–94.

Burr, N. C., and Solomon, S. C. (1978). The relationship of source parameters of oceanic transform earthquakes to plate velocity and transform length. *J. Geophys. Res.* **83**, 1193–1205.

Chorowicz, J. (1989). Transfer and transform fault zones in continental rifts: Examples in the Afro–Arabian Rift system. Implication of crust breaking. *J. Af. Earth Sci.* **8**, 203–214.

shows extraordinary variability in its precursory behavior, including total absence of any known precursors. The 1975 Haicheng earthquake (M_s = 7.3) in northeast China was the first major earthquake to be predicted. It was preceded by many different kinds of precursors: ground-water changes, tilting, foreshocks, and strange

DeMets, C., Gordon, R. G., Argus, D. F., and Stein, S. (1990). Current plate motions. *Geophys. J. Int.* **101**, 425–478.

Engeln, J. F., Wiens, D. A., and Stein, S. (1986). Mechanisms and depths of Atlantic transform earthquakes. *J. Geophys. Res.* **91**, 548–577.

Fujita, I., and Kanamori, H. (1981). Double seismic zones and stresses of intermediate depth earthquakes *Geophys. J. R. Astron. Soc.* **66**, 131–156.

Gao, L., Zhang, J., Wallace, T., and Lay, T. (1994). The 1990 southern Sudan earthquake sequence. *J. Geophys. Res.* (in press).

Giardini, D., and Woodhouse, J. H. (1984). Deep seismicity and modes of deformation in Tonga subduction zone. *Nature (London)* **307**, 505–509.

Glennon, M. A., and Chen, W.-P. (1993). Systematics of deep-focus earthquakes along the Kurile–Kamchatka arc and their implications on mantle dynamics. *J. Geophys. Res.* **98**, 735–769.

Hamburger, M. W., and Isacks, B. L. (1987). Deep earthquakes in the Southwest Pacific: A tectonic interpretation. *J. Geophys. Res.* **92**, 13,841–13,854.

Hanks, T. C., and Schwartz, D. P. (1987). Morphologic dating of the pre-1983 fault scarp on the Lost River fault at Doublespring Pass Road, Custer County, Idaho. *Bull. Seismol. Soc. Am.* **77**, 837–846.

Hanks, T. C., Bucknam, R. C., Lajoie, K. R., and Wallace, R. E. (1984). Modification of wave-cut and faulting controlled landforms. *J. Geophys. Res.* **89**, 5771–5790.

Hasegawa, A., Umino, N., and Takagi, A. (1978). Double-planed structure of the deep seismic zone in the northeastern Japan arc. *Tectonophysics* **47**, 43–58.

Huang, P., and Solomon, S. (1987). Centroid depths and mechanisms of mid-oceanic ridge earthquakes in the Indian Ocean, Gulf of Aden and Red Sea. *J. Geophys. Res.* **92**, 1361–1382.

Isacks, B. L., and Molnar, P. (1971). Distribution of stresses in the descending lithosphere from a global survey of focal mechanism solutions of mantle earthquakes. *Rev. Geophys. Space Phys.* **9**, 103–174.

Isacks, B., Oliver, J., and Sykes, L. R. (1968). Seismology and the new global tectonics. *J. Geophys. Res.* **73**, 5855–5899.

Ishibashi, K. (1981). Specification of a soon-to-occur seismic faulting in the Tokai district, central Japan, based upon seismotectonics. *Maurice Ewing Ser.* **4**, 297–332.

Jones, L. M., and Molnar, P. (1979). Source characteristics of foreshocks and their possible relationship to earthquake prediction and premonitory slip on faults. *J. Geophys. Res.* **84**, 3596–3608.

Kanamori, H. (1977). Seismic and aseismic slip along subduction zones and their tectonic implications. *Maurice Ewing Ser.* **1**, 162–174.

Kanamori, H. (1981). The nature of seismicity patterns before large earthquakes. *Maurice Ewing Ser.* **4**, 1–19.

Kasahara, K. (1981). "Earthquake Mechanics." Cambridge Univ. Press, Cambridge, UK.

Kincaid, C., and Olson, P. (1987). An experimental study of subduction and slab migration. *J. Geophys. Res.* **92**, 13,832–13,840.

Kirby, S. H., Durham, W. B., and Stern, L. A. (1991). Mantle phase changes and deep earthquake faulting in subducting lithosphere. *Science* **252**, 216–225.

Klein, F. W., Koyanagi, R. Y., Nakata, J. S., and Tanigawa, W. R. (1987). The seismicity of Kilauea's magma system. *Geol. Surv. Prof. Pap. (U.S.)* **1350**, 1019–1185.

Lay, T., and Kanamori, H. (1981). Earthquake doublets in the Solomon Islands. *Phys. Earth Planet. Inter.* **21**, 283–304.

Lay, T., and Okal, E. (1983). The Gilbert Islands (Republic of Kiribati) earthquake swarm of 1981–1983. *Phys. Earth Planet. Inter.* **33**, 284–303.

Mogi, K. (1985). "Earthquake Prediction." Academic Press, New York.

Ni, J., and Barazangi, M. (1984). Seismotectonics of the Himalayan collision zone: Geometry of the underthrusting Indian plate beneath the Himalaya. *J. Geophys. Res.* **89**, 1147–1163.

Nishenko, S. P. (1991). Circum-Pacific seismic potential: 1989–1999. *Pure Appl. Geophys.* **135**, 169–259.

Nishenko, S. P., and Buland, R. (1987). A generic recurrence interval distribution for earthquake forecasting. *Bull. Seism. Soc. Am.* **77**, 1382–1399.

Nuttli, O. W. (1973). The Mississippi Valley earthquakes of 1811 and 1812: Intensities ground motion and magnitudes. *Bull. Seismol. Soc. Am.* **63**, 227–248.

Okada, A., and Nagata, T. (1953). Land deformation of the neighborhood of Muroto Point after the Nankaido great earthquake in 1946. *Bull. Earthq. Res. Inst., Tokyo University*, **31**, 169–177.

Oxburgh, E. R., and Turcotte, D. L. (1974). Membrane tectonics and the East African rift. *Earth Planet. Sci. Lett.* **22**, 133–140.

Page, R. A. (1967). Aftershocks and microaftershocks of the great Alaska earthquake of 1964. Ph.D. Thesis, Columbia University, Palisades, NY.

Pelayo, A. M., and Wiens, D. A. (1989). Seismotectonics and relative plate motions in the Scotia Sea region. *J. Geophys. Res.* **94**, 7293–7320.

Plafker, G. (1965). Tectonic deformation associated with the 1964 Alaska earthquake. *Science* **148**, 1675–168.

Plafker, G. (1972). Alaska earthquake of 1964 and Chilean earthquake of 1960: Implications for the tectonics. *J. Geophys. Res.* **77**, 901–925.

Ruff, L. J. (1983). Fault asperities inferred from seismic waves. *In* "Earthquakes: Observation, Theory and Interpretation," (H. Kanamori and E. Boschi, eds.), pp. 251–276. Soc. Ital. di Fis., Bologna.

Ruff, L. J., and Kanamori, H. (1980). Seismicity and the subduction process. *Phys. Earth Planet. Inter.* **23**, 240–252.

Scholz, C. H. (1990). "The Mechanics of Earthquakes and Faulting." Cambridge Univ. Press, Cambridge, UK.

Seeber, L., and Armbruster, J. G. (1981). The 1886 Charleston, South Carolina earthquake and the Appalachian detachment. *J. Geophys. Res.* **86**, 7874–7894.

Shimazaki, K., and Nakata, T. (1980). Time-predictable recurrence model for large earthquakes. *Geophys. Res. Lett.* **7**, 279–282.

Sieh, K., Stuirer, M., and Brillinger, D. (1989). A more precise chronology of earthquakes produced by the San Andreas fault in southern California. *J. Geophys. Res.* **94**, 603–624.

Simpson, D. W., Leith, W. S., and Scholz, C. H. (1988). Two types of reservoir induced seismicity. *Bull. Seismol. Soc. Am.* **78**, 2025–2040.

Sleep, N. Y. (1979). The double seismic zone in downgoing slabs and the viscosity of the mesosphere. *J. Geophys. Res.* **84**, 4565–4571.

Stein, S., DeMets., C., Gordon, R. G., Brodholt, J., Argus, D., Engeln, J. F., Lundgren, P., Stein, C., Wiens, D. A., and Woods, D. F. (1988). A test of alternate Caribbean Plate relative motion models. *J. Geophys. Res.* **93**, 3041–3050.

Suárez, G., Molnar, P., and Burchfiel, B. C. (1983). Seismicity, fault plane solutions, depth of faulting, and active tectonics of the Andes of Peru, Ecuador, and southern Colombia. *J. Geophys. Res.* **88**, 10,403–10,428.

Sykes, L. R. (1978). Intraplate seismicity, reactivation of preexisting zones of weakness, alkaline magmatism, and other tectonism postdating continental fragmentation. *Rev. Geophys. Space Phys.* **16**, 621–688.

Thatcher, W. (1990). Order and diversity in the modes of circum-Pacific earthquake recurrence. *J. Geophys. Res.* **95**, 2609–2623.

Turcotte, D. L., and Schubert, G. (1982). "Geodynamics: Applications of Continuum Physics to Geological Problems." Wiley, New York.

Vassiliou, M. S., Hager, B. H., and Raefsky, A. (1984). The distribution of earthquakes with depth and stress in subducting slabs. *J. Geodyn.* **1**, 11–28.

Wakita, H. (1981). Precursory changes in groundwater prior to the 1978 Izu–Oshima–Kinkai earthquake. *Maurice Ewing Ser.* **4**, 527–532.

Wesnousky, S. G. (1986). Earthquakes, Quaternary faults, and seismic hazard in California. *J. Geophys. Res.* **91**, 12,587–12,631.

Wilson, J. T. (1965). A new class of faults and their bearing on continental drift. *Nature (London)* **207**, 343–347.

Zhang, G., and Fu, Z. (1981). Some features of medium and short term anomalies before great earthquakes, *Maurice Ewing Ser.* **4**, 497–509.

Additional Reading

Bird, J. M., ed. (1980). "Plate Tectonics." Am. Geophys. Union, Washington, DC.

Cox, A., ed. (1973). "Plate Tectonics and Geomagnetic Reversals." Freeman, San Francisco.

Gubbins, D. (1990). "Seismology and Plate Tectonics." Cambridge Univ. Press, Cambridge, UK.

Gutenberg, B., and Richter, C. F. (1949). "Seismicity of the Earth." Princeton Univ. Press, Princeton, NJ.

Isacks, B., Oliver, J., and Sykes, L. R. (1968). Seismology and the new global tectonics. *J. Geophys. Res.* **73**, 5855–5899.

Kanamori, H. (1986). Rupture process of subduction-zone earthquakes. *Annu. Res. Earth Planet. Sci.* **14**, 293–322.

Le Pichon, X., Franchteau, J., and Bournin, J. (1973). "Plate Tectonics." Elsevier, Amsterdam.

Lomnitz, C. (1994). "Fundamentals of Earthquake Prediction," John Wiley & Sons, Inc., New York.

Ma, Zongjin, Fu, Zhengxiang, Zhang, Yingzhen, Wang, Chengmin, Zhang, Guomln, and Liu, Defu (eds.) (1990). "Earthquake Prediction," 332 pp. Seismological Press/ Springer-Verlag, Beijing/ Heidelberg.

Rikitake, T. (1976). "Earthquake Prediction." Developments in Solid Earth Geophysics 9. Elsevier, Amsterdam.

Shimazaki, K., and Stuart, W. (1985). "Earthquake Prediction,." Birkhäuser, Basel.

Simpson, D. W., and Richards, P. G., eds. (1981). "Earthquake Prediction—An International Review," Maurice Ewing Ser., Vol. 4. Am. Geophys. Union, Washington, DC.

INDEX

International Geophysics Series

EDITED BY

RENATA DMOWSKA
Division of Applied Sciences
Harvard University
Cambridge, Massachusetts

JAMES R. HOLTON
Department of Atmospheric Sciences
University of Washington
Seattle, Washington

*Out of print.